Sequence Stratigraphy, Paleoclimate, and Tectonics of Coal-Bearing Strata

Edited by
Jack C. Pashin
and
Robert A. Gastaldo

AAPG Studies in Geology No. 51

Published by

The American Association of Petroleum Geologists
Tulsa, Oklahoma, U.S.A.

and

The AAPG Energy Minerals Division

Printed in the U.S.A.

ISBN: 0-89181-058-7

AAPG Editor: John C. Lorenz
Geoscience Director: J. B. "Jack" Thomas

This publication is available from:

The AAPG Bookstore
P.O. Box 979
Tulsa, OK U.S.A. 74101-0979
Phone: 1-918-584-2555 or 1-800-364-AAPG (U.S.A. only)
Fax: 1-918-560-2652 or 1-800-898-2274 (U.S.A. only)
E-mail: bookstore@aapg.org
www.aapg.org

Table of Contents

About the Editors

Jack C. Pashin is head of the Energy and Minerals Group at the Geological Survey of Alabama. He received a B.S. degree in geology from Bradley University in 1982, and M.S. and Ph.D. degrees in geology from the University of Kentucky in 1985 and 1990, respectively. Over the past 16 years, Jack has published numerous papers on the geology of conventional and unconventional hydrocarbon reservoirs, and has won a variety of awards for his research on the stratigraphy and structure of coalbed methane reservoirs in the Black Warrior basin and conventional reservoirs in the Gulf of Mexico basin. His current research focuses on the stratigraphic, structural, and hydrologic characterization of coalbed methane reservoirs and the potential for sequestration of carbon dioxide in coal. Jack is active in several geological societies and committees. He has served as Vice President of the Energy Minerals Division of AAPG, as Chairman of the Antoinette Lierman Medlin Scholarship Committe of the Geological Society of America's Coal Geology Division, and as a Technical Editor of the Journal of Paleontology. He is currently Chairman of the AAPG Energy Minerals Division's Coal Committee, an Associate Editor of the AAPG Bulletin, President of the Alabama Geological Society, and a member of the Executive Committee of the International Coalbed Methane Symposium.

Robert A. Gastaldo became the Whipple-Coddington Professor of Geology and Chair of the Department at Colby College, Waterville, Maine, in 1999 after serving as Alumni Professor of Geology at Auburn University, Alabama. Gastaldo received his Ph.D. from Southern Illinois University at Carbondalc, and has been a Fulbright Research Fellow (Netherlands) and awarded a Forshungspreis by the Alexander von Humboldt Foundation, Germany. He served as an Associate Editor for the Geological Society of America Bulletin, Associate Editor and then Co-Editor of PALAIOS (1996–2002), and is on the editorial boards of the Review of Palaeobotany and Palynology and the Journal of Taphonomy. His research spans the Devonian to Recent and is focused on the utilization of plant fossil assemblages to solve sedimentologic, paleoecologic, paleoclimatic, and tectonic problems. His current research focuses on plant taphonomy and Devonian ecosystems associated with the Acadian orogeny, the ecological stability of mid-Carboniferous terrestrial systems, the paleontology and sedimentology of plant-bearing continental rocks across the Permian-Triassic boundary in South Africa, and peat-accumulating systems in Holocene carbonate settings (among other things).

Acknowledgments

This book is the product of a theme session sponsored by the AAPG Energy Minerals Division (EMD) called "Sequence Stratigraphy of Coal-Bearing Strata," which the editors chaired at the 2000 AAPG Annual Convention in New Orleans, Louisiana, on April 17, 2000. We especially are grateful to the EMD leadership and committee chairs, without whose support this volume would not have been possible. EMD leaders deserving special mention include Chacko John, Scott Peters, Margaret Anne Rogers, and Andrew R. Scott. We also would like to thank AAPG Elected Editor John Lorenz and the many fine individuals in the Publications Department at AAPG headquarters whose encouragement and assistance were indispensable. The chapters in this book were reviewed rigorously by multiple experts, and we appreciate their candid comments and constructive suggestions, all of which improved the quality of this publication. The reviewers included Allen Archer, Walter Ayers, Jack Beuthin, Mitch Blake, Kevin Bohacs, John Calder, Richard Carroll, Tim Demko, Ted Dyman, Cortland Eble, Frank Ettensohn, Phil Heckel, George Klein, John Nelson, David Pocknall, Jim Staub, and two anonymous reviewers. Some of these individuals were kind enough to review more than one manuscript. Finally, we would like to thank the authors of the chapters, whose expertise and diligence made this volume possible.

Preface

The professional association of the editorial team began during the Southeastern Section meeting of the Geological Society of America in 1990, which was held in Tuscaloosa, Alabama. Jack Pashin led a pre-meeting field trip to examine the Mary Lee coal zone of the Pottsville Formation (lower Pennsylvanian) in the frontal structures of the Appalachian thrust belt, which form the southeast margin of the coal-bed methane fairway of the Black Warrior Basin. Bob Gastaldo, conversely, led a postmeeting field trip to the northwestern outcrops of the Mary Lee coal zone, which were in the heart of surface-mining country. This was an exciting time: coal-bed methane development was burgeoning, and the productivity of Alabama's coal mines was at record levels. This also was a time when sequence-stratigraphic principles were beginning to be applied to coal-bearing strata, and we were applying these principles from very different perspectives. Indeed, Jack's views were influenced by the terrestrial character of Mary Lee strata along the Appalachian front, as well as his interest in tectonics and coal-bed methane development, whereas Bob's views were influenced by the marginal marine character of Mary Lee strata in the surface-mining area, as well as his experience in plant taphonomy and paleobotany. Although our views came from opposite ends of the basin, we shared a lot of common ground because we were both using the language of sequence statigraphy to address the basic issues of sedimentologic, paleoclimatic, and tectonic history.

Moving ahead nearly a decade, Jack was attending the fall leadership meeting of the AAPG Energy Minerals Division (EMD) in Tulsa, where activities were being planned for the upcoming 2000 AAPG Annual Convention in New Orleans. The Energy Minerals Division is the AAPG division devoted to energy resources outside of conventional oil and gas. Accordingly, EMD addresses diverse topics, including coal, coal-bed methane, geothermal energy, oil shale, tar sand, gas hydrates, remote sensing, and energy economics. An EMD-sponsored session called "Sequence Stratigraphy of Coal-Bearing Strata" seemed timely, considering the emerging interest in coal-bed meth-ane and the CO_2 sequestration potential of coal. It also seemed like a potential topic for an AAPG book as there was a dearth of publications applying these principles to coal-bearing sequences. Shortly thereafter, Jack collared Bob to co-chair the session, speakers were invited, and the next thing we knew, we were all in New Orleans at the convention (yes, there was time to enjoy the Big Easy, too). As the session unfolded, it was increasingly apparent that all of the presenters were using the paradigm of sequence stratigraphy to solve the full range of sedimentologic, paleoclimatic, and tectonic problems that coal geologists have been pondering for more than a century. Clearly, sequence stratigraphy had matured to the point where researchers with backgrounds in a variety of geologic subdisciplines were comfortable using the paradigm as a tool to solve broad geologic problems. Because of this, we felt the session was highly successful and confirmed that the principles originally articulated by Peter Vail and his coworkers are fulfilling their promise.

Because stratigraphy, paleoclimate, and tectonics are so inextricably linked, we felt this book should be titled accordingly. Several publications focus on the definition and demonstration of the sequence-stratigraphic paradigm in coal-bearing strata, including a book by Claus Diessel (1992), short course notes by Peter McCabe (1993), and some journal articles. For this reason, we chose to orient the volume toward collected papers that use sequence-stratigraphic principles to address paleoclimatic and tectonic problems in coal-bearing successions. This volume features 10 chapters on coal-bearing strata of Pennsylvanian through Tertiary age in North America, South America, and New Zealand. Many of the original presenters provided manuscripts for this volume, and additional contributions were solicited that help round out the field. Most chapters highlight the Pennsylvanian strata of eastern North America, which reflects the content of the AAPG session, as well as the primary background of the editors. A pair of chapters on the Conemaugh Group of the Appalachian basin and a pair of chapters on the Pottsville Formation of the Black Warrior Basin serve as bookends for the volume and underscore the contrasting approaches and interpretations of the authors. Indeed, each chapter presents a unique view of the sequence stratigraphy, paleoclimate, and tectonics of coal-bearing strata. In a sense, we are all solving the same types of geologic problems from opposite ends of the basin.

Jack C. Pashin
Robert A. Gastaldo

REFERENCES CITED

Diessel, C. F. K., 1992, Coal-bearing depositional systems: Berlin, Springer-Verlag, 721 p.

McCabe, P. J., 1993, Sequence stratigraphy of coal-bearing strata: Short Course Notes, AAPG Annual Meeting, New Orleans, Louisiana, 81 p.

Martino, R. L., 2004, Sequence stratigraphy of the Glenshaw Formation (middle–late Pennsylvanian) in the central Appalachian basin, *in* J. C. Pashin and R. A. Gastaldo, eds., Sequence stratigraphy, paleoclimate, and tectonics of coal-bearing strata: AAPG Studies in Geology 51, p. 1–28.

Sequence Stratigraphy of the Glenshaw Formation (Middle–Late Pennsylvanian) in the Central Appalachian Basin

Ronald L. Martino
Department of Geology, Marshall University, Huntington, West Virginia, U.S.A.

ABSTRACT

The Glenshaw Formation consists predominantly of sandstones and mudrocks with thin limestones and coals, which are thought to have accumulated in alluvial, deltaic, and shallow-marine environments. Analysis of 87 Glenshaw outcrops from southern Ohio, eastern Kentucky, and southern West Virginia has revealed widespread, well-developed paleosols. These paleosols are used, along with marine units and erosional disconformities, to develop a high-resolution sequence-statigraphic framework. The tops of the paleosols constitute boundaries for nine allocycles, which are interpreted as fifth-order depositional sequences. Allocycles in this framework correlate with similar allocycles described from the northern Appalachian basin.

A sequence-stratigraphic model is proposed that provides a framework for interpreting facies architecture in terms of base-level dynamics linked to relative sea level changes. Lowered base level caused valley incision along drainage lines and sediment bypassing of interfluves, which led to development of well-drained paleosols. Rising base level produced valley filling by fluvioestuarine systems (lowstand systems tract/transgressive systems tract), whereas pedogenesis continued on interfluves. As drainage systems aggraded, the coastal plain water table rose, and interfluvial paleosols were onlapped by paludal and lacustrine deposits. Histosols succeeded and partially overprinted paleosols with vertic and calcic features. Highstand systems tract (HST) facies in the coastal plain consist of widely separated, high-sinuosity fluvial channel and estuarine channel sandstones encased in overbank mudstones, whereas within marine units, HST facies with coarsening-upward regressive deltaic and interdeltaic facies are developed.

The sequence-stratigraphic framework provides the basis for a better understanding of the depositional systems, base-level dynamics, and climatic changes that influenced the infilling of the central Appalachian basin. The paleoenvironmental and sequence-stratigraphic context of channel and valley fills may benefit future petroleum exploration in the Appalachian basin and other analogous settings.

INTRODUCTION

Glenshaw Stratigraphy

The Glenshaw Formation constitutes the lower 66–80 m (217–262 ft) of the Conemaugh Group in the central Appalachian basin (Martino et al., 1996). The current stratigraphic framework is based on key beds, including laterally persistent coal seams and marine units (Figures 1, 2). In West Virginia, four widespread marine units have been distinguished, including, in ascending order, the lower Brush Creek, upper Brush Creek (Pine Creek), Cambridge, and Ames (Arkle et al., 1979; Merrill, 1986; Martino et al., 1996).

Paleogeographic Setting

The Dunkard basin (Figure 3) is a synclinorium containing strata from the cratonward portion of the central Appalachian foreland basin. These features were formed by warping and thrust loading and, during continental collision, associated with the Alleghanian orogeny (Quinlan and Beaumont, 1984). During the late Pennsylvanian (Stephanian), the center of what is now the Dunkard basin was positioned approximately 0–6° south of the equator (Opdyke and Divenere, 1994). The beginning of the Stephanian was marked by a long-term climate change toward increased aridity in the Appalachian basin attributable to orogenesis and associated rainshadow effects. Short- and intermediate-term (tens of thousands to hundreds of thousands of years) climate cycles may have produced the alternation of geochemically distinct stratigraphic intervals, including paleosols, coal beds, and nonmarine limestones (Cecil, 1990; Cecil et al., 1994).

Depositional Environments

The Glenshaw Formation contains channel-fill sandstones, red and olive mudrocks, and coal beds that accumulated on an alluvial coastal plain with a northwest paleoslope (Arkle, 1974; Donaldson, 1979). Nearshore and deltaic facies have been interpreted for sparsely fossiliferous olive-gray mudrocks and limestone and burrowed sandstones in the middle and uppermost portions of the Glenshaw (Donaldson, 1979; Martino et al., 1985; Merrill, 1986, 1988; Martino et al., 1996).

Shallow seas advanced at least eight times from the southwest into the Appalachian basin during Glenshaw deposition (Busch, 1984), whereas regressions were accompanied by the development of deltas that advanced northwestward across the basin (Donaldson, 1979).

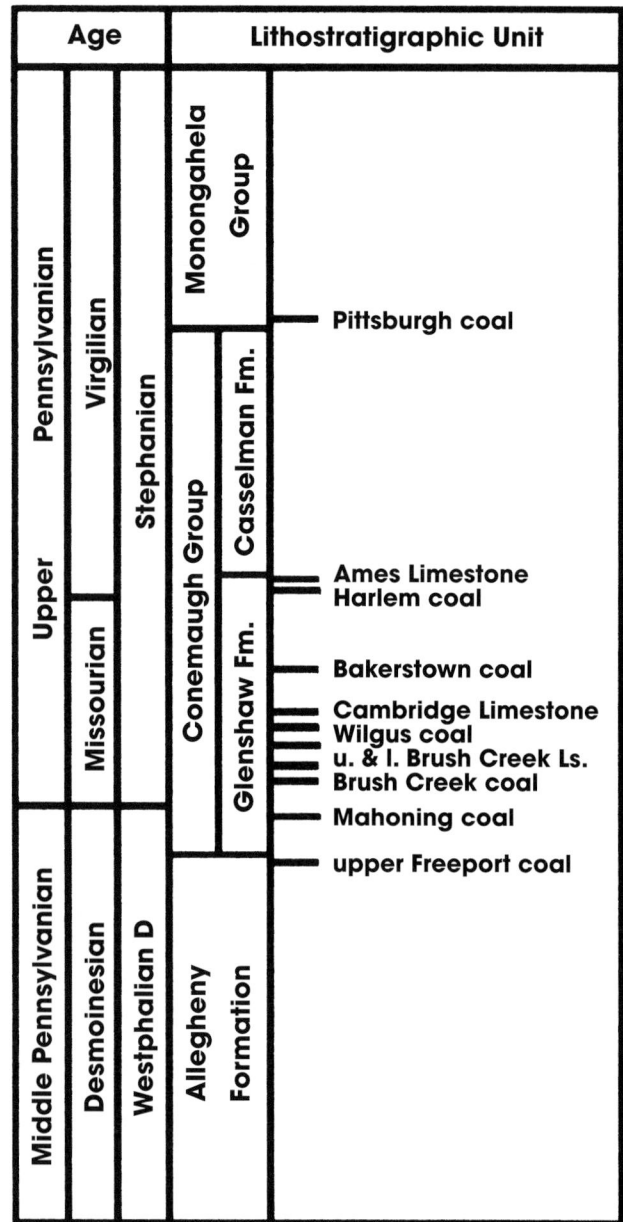

Figure 1. Stratigraphic framework for the Glenshaw Formation in the southern portion of the Dunkard basin (from Martino et al., 1996).

Economic Significance

The Glenshaw Formation contains several high-volatile bituminous coals that are generally too thin to mine, except along the northeastern edge of the Dunkard basin. These coals tend to be high in total sulfur (1–3%; Repine et al., 1993).

Two oil-producing sandstones occur in the Glenshaw Formation, including (1) the First Cow Run and (2) the Big Dunkard sands (driller's terms). These correspond to the (1) Saltsburg and Buffalo Sandstones and (2) the Mahoning Sandstone (Cardwell and Avery,

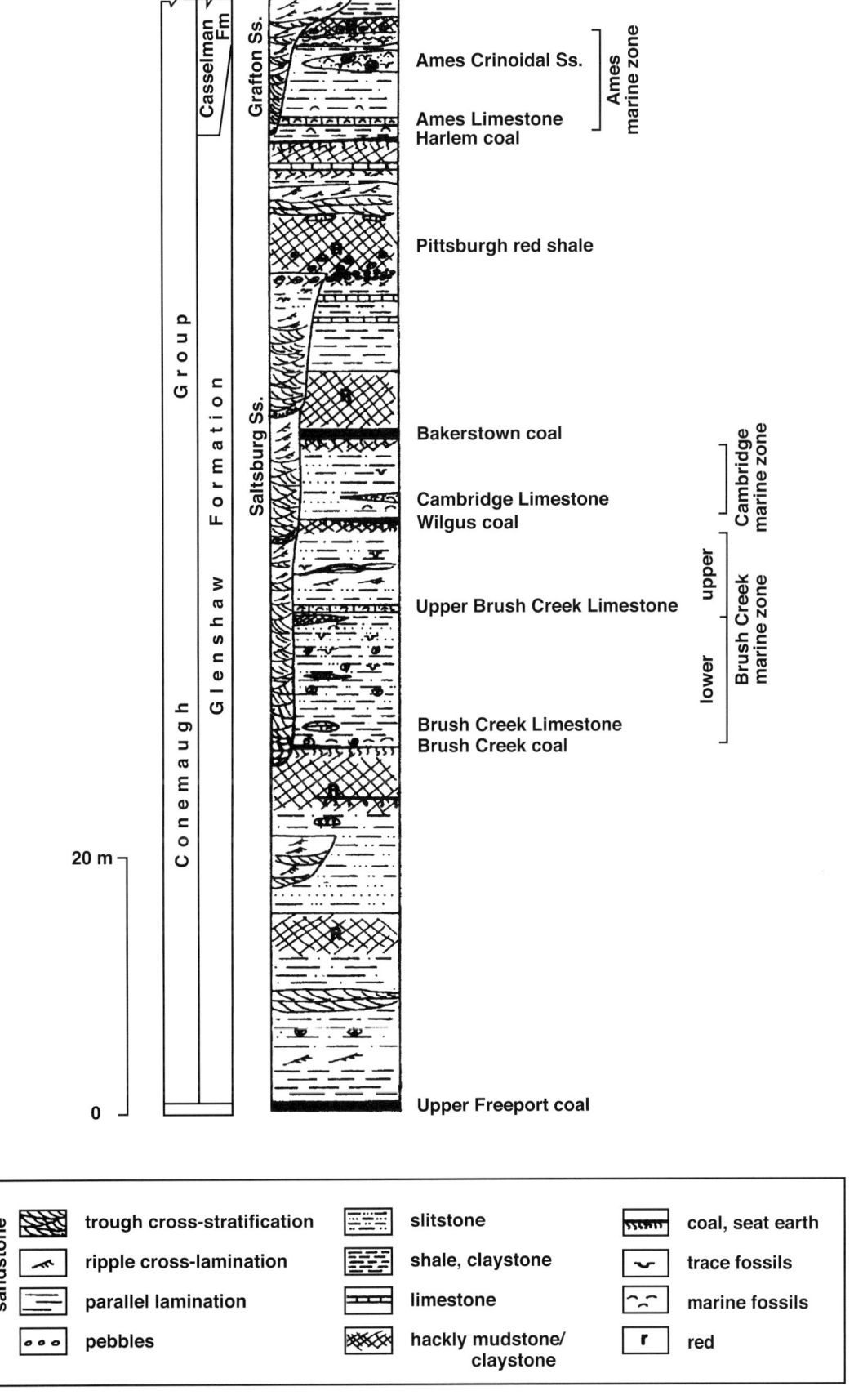

Figure 2. Composite stratigraphic section for the Glenshaw Formation in the vicinity of Huntington, West Virginia (modified from Martino et al., 1996).

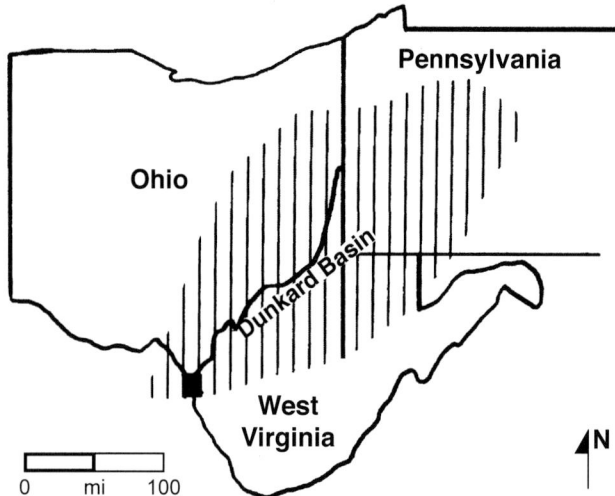

Figure 3. Aerial extent of the Dunkard basin, with Huntington, West Virginia area, at southwest end represented by black square. Black rectangle at southern end of the basin defines the area of the present study (see Figure 4).

1982). There is still some minor production from these reservoirs, but they are shallow targets that were of greater importance during early development in the late 1800s (Haught, 1963). Oil typically occurs in combination traps resulting from broad folds with gently dipping limbs in conjunction with a permeability barrier (Cardwell and Avary, 1982).

CYCLIC STRATIGRAPHY

Autocycles and Allocycles

Local repetitions of facies in vertical sequences are typically the product of processes that are intrinsic to a particular depositional system. Meandering rivers can produce a series of fining-upward cycles composed of alternating channel and flood-plain facies. Abandonment and reactivation of delta lobes can cause aerially restricted transgressive-regressive (T-R) cycles. Such autocycles were interpreted by many workers in the 1970s to be characteristic of Pennsylvanian strata of the central Appalachian basin (e.g., Ferm, 1970; Donaldson, 1979).

In contrast, allocycles are broader, regional or interbasinal cycles that result from factors that are extrinsic to the depositional system. External factors include tectonics, eustasy, and climate. Tectonics and eustasy control accommodation space through baselevel changes, whereas tectonics and climate control sediment supply.

Allostratigraphy

Allostratigraphy is one method of describing and evaluating basinwide or interbasinal cycles. Allostratigraphic units are mappable units bounded by discontinuities that reflect extrinsic processes that "…initiate and terminate the deposition of a sedimentologically related succession of facies" (Walker, 1992, p. 9). Commonly used bounding surfaces include disconformities and the tops of paleosols (North American Commission on Stratigraphic Nomenclature [NACSN], 1983). Allostratigraphy is distinguished from sequence stratigraphy in that it is descriptive. There are no implicit assumptions regarding which specific processes caused the stratigraphic features.

Sequence Stratigraphy

Currently, there are two conceptual frameworks for sequence stratigraphy. One developed by Exxon researchers uses unconformities as the boundaries of sequences (Vail et al., 1977; Van Wagoner et al., 1988), whereas the other uses maximum flooding surfaces as genetic sequence boundaries (Galloway, 1989). Both types of sequences include the deposits of one complete relative sea level cycle.

Sequence-stratigraphic concepts provide a powerful tool for unraveling the evolution of sedimentary basins by dividing the basin fill into genetic packages (Van Wagoner et al., 1988). Recognition of individual sequences requires identification of subaerial unconformities of regional or interregional extent and their correlative marine unconformities and conformities. The deposits of depositional systems become progressively stacked to form sequences as a result of the interplay among subsidence, sediment input, and sea level change. Sequence stratigraphy provides a chronostratigraphic framework for the correlation and mapping of sedimentary facies. It also acts as a tool for stratigraphic prediction (Emery and Myers, 1996). A potential problem in employing sequence-stratigraphic analysis is the assumption of eustatic control, particularly where other potentially overriding allocyclic factors may be involved (Walker, 1992).

Type 1 sequence boundaries are the most common type of boundary in siliciclastic basins (Van Wagoner et al., 1990). They are laterally continuous, basinwide or interbasinal surfaces that are distinguished in shelf settings by the presence of erosional truncation, subaerial exposure, and a basinward shift in facies. Incised drainage lines formed during lowstands of sea level produce erosional truncation. The erosional surface associated with incised valleys passes laterally into a correlative paleosol formed during subaerial exposure (Van Wagoner et al., 1990). These paleosols represent "condensed sections" formed in terrestrial settings caused by low rate of deposition or nondeposition. Avulsion of channel systems out of an area may cause local paleosol development, but widespread, strongly developed paleosols are more likely to form from extrinsic controls on sediment input. Rejuvenation and incision of rivers would reduce or sharply eliminate sediment to interfluvial uplands and would be expected to accompany falling base level.

Valley fills are elongate cut-and-fill bodies that are larger than a single channel and may be formed by relative sea level changes, inland tectonic uplift, and climate change (Dalrymple et al., 1994). Paleovalley fills range from 8 to 100 m (26 to 328 ft) thick and from 0.5 to 64 km (0.3 to 40 mi) in maximum width (Schumm and Ethridge, 1994). In valley fills formed by relative sea level change, the valley wall and floor represent type 1 sequence boundaries, and the fills typically include some evidence for marine influence. Simple valley fills are produced by a single relative sea level cycle. In compound valley fills, more than one sea level cycle may take place during valley filling. This can produce a complex mixture of fluvial facies of varying styles, estuarine facies, and possibly shallow-marine facies in the valley fill.

Channel fills may also develop during regression as coastal plain drainage systems override shoreline and shallow-marine deposits. In these cases, channel sands will represent coarse members in coarsening-upward sequences deposited during sea level highstands. This pattern of sedimentation is commonly terminated by a fall in base level and fluvial incision (Miall, 1997).

Avulsion events could introduce channel systems into flood basins producing incision. In this instance, the depth of incision probably would not exceed the channel depth if this took place in a highstand systems tract (HST). Channel deposits of this type would differ from those produced during lowstand incision and infilling during transgressive systems tract (TST), in that greater relief in the latter would be expected along the erosional sequence boundary.

Previous Work on Pennsylvanian Depositional Cycles of the Central and Northern Appalachian Basin

Depositional cycles have been recognized in the Pennsylvanian strata of the Appalachian basin since the 1930s (Weller, 1930; Stout, 1931; Wanless and Weller, 1932). Eight cycles (cyclothems) were distinguished in the Glenshaw portion of the Conemaugh Group in Ohio (Sturgeon and Hoare, 1968). Wanless and Shephard (1936) attributed these types of cycles to global sea level changes caused by fluctuations in ice volume of Gondwanan ice sheets.

Through the 1960s and 1970s, an increasing awareness of inherent behavior of alluvial and deltaic depo-systems led many workers to reinterpret cyclothems previously distinguished in the Appalachian basin as autocyclic in origin (e.g., Ferm, 1970; Donaldson, 1979). Attempts to correlate mid-continent cycles with those in the Appalachian basin were inhibited by lack of a detailed biostratigraphic framework for the Appalachian basin, and by the prevailing view that eustasy would be masked or overshadowed by tectonic and/or autocyclic processes that were not prevalent in other cratonic basins.

During the 1980s, conflicting views emerged regarding the presence and origin of Glenshaw cycles. Busch (1984) and Busch and Rollins (1984) described 11 fifth-order, allocyclic, T-R units associated with the Glenshaw Formation. Their proposed cycles closely corresponded to cyclothems that were described earlier (Flint, 1951; Sturgeon and Hoare, 1968). The T-R cycles of Busch and Rollins are 5–30 m (16–98 ft) thick and were interpreted to have resulted from glacioeustatic sea level changes. Their study was based mainly on data from Ohio and Pennsylvania in the northern portion of the Appalachian basin. Busch and West (1987) included both T-R units and climate-change surfaces in their fifth-order allocycles. Climate-change surfaces were defined as contacts between continental strata formed under arid subaerial conditions (e.g., aridosols and vertisols) and overlying coal and lacustrine limestone formed under more humid conditions. Busch and West (1987) correlated the Glenshaw allocycles with those known from the mid-continent, maintaining that they were the product of glacioeustasy.

Heckel (1995) also correlated Appalachian basin allocycles with mid-continent cycles. He used conodonts and palynomorphs and attributed both T-R cycles and climate-change cycles to a glacioeustasy. Veevers and Powell (1987) indicated that the maximum extent of glacial ice on Gondwanaland occurred during the late Pennsylvanian. Heckel (1995) maintained that moisture needed to feed the wet portions of cycles in totally nonmarine sequences was provided by the proximity of greatly expanded cratonic seas produced during glacioeustatic highstands. Heckel's work relied largely on the stratigraphic foundation of Busch (1984) and Busch and Rollins (1984), which was developed in the northern Appalachian basin.

Cyclic variations in paleoclimate are thought by some to be reflected in late Pennsylvanian, 100,000–400,000-yr sedimentary cycles recorded in the Appalachian basin (Cecil, 1990; Cecil et al., 1994; Cecil and Dulong, 1998). The wetter portions of the cycles were interpreted as sea level lowstand phases and are represented by laterally extensive coal beds deposited in topographic lows and contemporaneous upland ultisol-like paleosols. Drier portions of the climate cycles were thought to be recorded by lacustrine limestone that grades laterally into highly calcareous paleovertisols formed during sea level highstands. This perspective sharply contrasts with those expressed by Busch and Rollins (1984), Busch and West (1987), and Heckel (1995), who maintained that wetter climate phases corresponded to highstands, whereas increased dryness was associated with lowstands.

Merrill (1986) described the lithostratigraphy of Conemaugh outcrops along the West Virginia–Kentucky border. He concluded that the cyclothem concept could not be applied to Conemaugh strata in this area because of the limited lateral extent of individual lithosomes and packages of lithosomes. He favored Ferm's (1970) view that differential rates of deltaic growth and abandonment generally caused the development of aerially limited marine units. Martino et al. (1996) argued that whereas individual beds (such as a particular marine limestone) were commonly laterally restricted, genetic packages of strata did appear to be widely developed in the southern Dunkard basin.

Donaldson and Eble (1991) interpreted the Conemaugh Group to contain (1) an intermediate allocycle 90 m (295 ft) thick and 3.5 m.y. long resulting from tectonic processes (uplift/thrust loading) and (2) minor allocycles with an average duration of 0.4 m.y., which were probably caused by glacioeustatic sea level fluctuations. These minor allocycles correspond to the fifth-order cycles of Busch and Rollins (1984). Donaldson and Eble (1991) suggested that autocycles formed by river avulsion could occur embedded in minor allocycles.

STATEMENT OF THE PROBLEM

The preceding literature review indicates not only considerable interest but also varied viewpoints concerning the presence, character, and origin of Glenshaw depositional cycles. Many workers (Wanless and Shepard, 1936; Busch and Rollins, 1984; Busch and West, 1987; Heckel, 1995) maintain the predominance of eustatic sea level changes (over tectonic and climatic factors) in causing allocycles. If correct, then Exxon's sequence-stratigraphic approach would be appropriate. However, a significant portion of the Appalachian basin (West Virginia, Kentucky, southernmost Ohio) remains understudied at the scale needed to verify the basinwide extent of these cycles and the dominance of eustasy in their origin.

Problems addressed by this study include the following:

1) Are there widely developed depositional cycles in the central Appalachian basin?
2) If present, can they be correlated with those distinguished in the northern part of the basin (Busch and Rollins, 1984)? If not, what factors differed in the central portion of the basin that could account for this? If depositional sequences are distinguishable, can they be correlated with those distinguished in other basins, strengthening the case for eustasy as the dominant controlling factor?
3) What expression (if any) do these cycles have in entirely nonmarine sections of the Conemaugh in the interior regions of West Virginia? Can sequence-stratigraphic elements be distinguished and correlated toward the southeast into entirely terrestrial facies?

RESULTS AND DISCUSSION

Genetic Facies Assemblages

Glenshaw outcrops used in this study (Figure 4) consist mainly of road cuts in the central Appalachian basin. Glenshaw sedimentary facies can be grouped into the following broad divisions: fluvioestuarine channel, coastal plain flood basin, and shallow to marginal marine. A summary of their characteristic features is given in Table 1.

Large-scale Channel-fills

Fluvial and deltaic channels consist of channel-form bodies with unimodal paleoflow as indicated by cross-bedding (Figure 5). The mean flow direction is toward the north-northwest, which is away from the Alleghanian orogen and consistent with fluviodeltaic reconstructions of earlier workers (Arkle, 1974; Donaldson, 1979; Donaldson et al. 1985). Local flow directions range from west to northeast; such variability is likely in high-sinuosity channel systems and

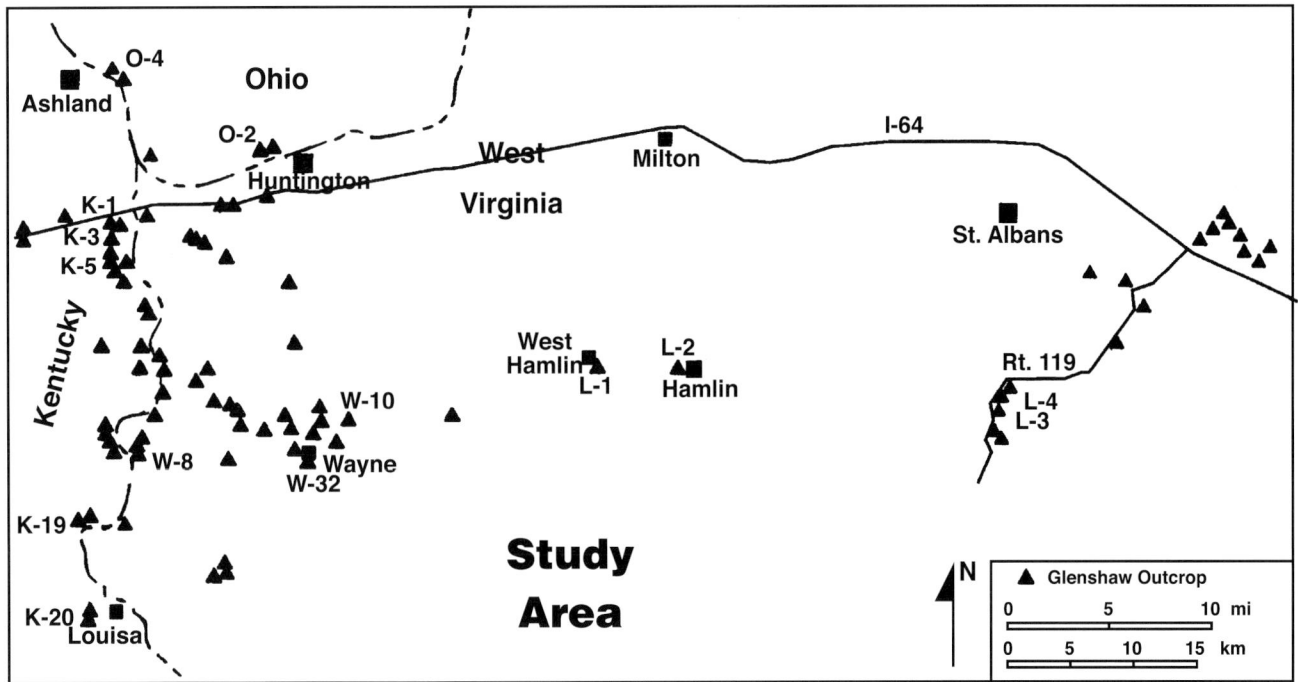

Figure 4. Outcrop locations from Kentucky, Ohio, and West Virginia that were used in this study. Stratigraphic sections were measured and described at each location. See Figure 3 for regional perspective. Labeled locations are used in cross sections or photos.

radiating deltaic distributaries. Channel sandstones commonly exhibit large-scale bar accretion surfaces that commonly dip at a high angle (e.g., 70–90°) with respect to the paleoflow directions indicated by internal cross-strata, a feature that is common in meandering channel systems. Mud-filled channels resulted from avulsion and channel deactivation. In meandering systems, this produces oxbow lakes. Both single-story and multistory channel fills are common (Figures 6, 7).

A deltaic origin is indicated where the channel system directly influenced the underlying or laterally equivalent marine facies. This would be evidenced by subaqueous splays or levees that thicken laterally toward the channel facies, or by mouth bar deposits that are localized in the vicinity of the channel system. A good example of this occurs in the lower Brush Creek marine zone near Wayne, West Virginia (Figure 4, W-10) (Martino et al., 1996).

Estuarine channel fills of the Glenshaw closely resemble those of fluvial channels in most respects. They are distinguished by the presence of cross-bedding with clay-draped foresets, cross-bedding with southeastward flow (up the paleoslope; Figure 5), and by parallel-laminated silt-clay couplets that exhibit thick and thin bundling (Figure 8A). These features

develop in tidal settings as the result of periodic slackwater periods and varying competence associated with daily, fortnightly, monthly, and seasonal tidal cycles. Similar estuarine channel deposits have been recognized in the middle Pennsylvanian Kanawha Formation of southern West Virginia (Martino, 1996).

Flood-Basin Facies

Low-lying areas between active drainage lines consist of a mosaic of depositional environments, including lakes, splay channels and levees, and clastic and peat swamps. These facies have been modeled previously in detail for Pennsylvanian strata in the Appalachian basin in the context of fluvially dominated deltaic deposystems (e.g., Horne et al., 1978; Donaldson, 1979).

Splay and levee sandstone wedge out away from active fluvial and distributary channels. Internal textures and structures indicate waning flow and deposition of single or stacked flood deposits.

Dark gray shales that overlie and laterally intergrade with coals commonly contain plant fossils and lack root traces, and are interpreted as clastic lacustrine deposits. Nonmarine limestones, such as the Twomile Limestone of I. C. White (1885), have been

Table 1. Summary of Glenshaw facies attributes.

Facies	Lithology	Sedimentary Structures	Geometry	Fossils
Fluvial and deltaic channel	very fine to very coarse sandstone, uncommonly conglomeratic; shale/mudstone plugs	cross-stratification: trough + tabular sets; compound (epsilon) cross-beds; laminated/hackly	channel-form ribbon; 6–10 m (20–33 ft) thick	uncommon burrows; plants
Estuarine channel	medium to very fine sandstone	cross-stratification with clay-draped foresets; southeast flow parallel lamination with thick to thin bundles	6–10 m (20–33 ft)	uncommon burrows
Flood Basin				
(1) Crevasse splay/levee	fine to very fine sandstone-shale	scour-fill trough sets, ripple cross-lamination; parallel lamination	tabular-wedge shaped; channels 1–4 m (3.3–13 ft) thick	plant fragments *Lockeia, Limnopus Sinusites, Diplichnites*
(2) Lakes	shale, siltstone, micritic limestone	laminated to massive	pods, sheets	ostracodes, *Spirorbis*, Conchostracans plants, coprolites, stromatolites, vertebrate fragments
(3) Mires	coal, carbonaceous shale	laminated	pods, sheets	
Offshore marine	shale, limestone, siltstone	thin-bedded, graded bedding	sheets	brachiopods, bivalves, gastropods, echinoderms, bryozoans, cephalopods, conodonts, fusilinids
Shoreface	shale, siltstone, very fine sandstone	thin-bedded, ripple bedding, parallel lamination, hummocky cross-stratification	elongate/lobate	*Wilkingea, Paleophycus, Aulichnites, Teichichnus, Rhizocorallium*
Mouth bar	shale, siltstone, very fine sandstone, siderite nodules	thin-bedded, parallel lamination, graded bedding, parting lineations, hummocky cross-stratification	lobate	*Curvolithus*, plant fossils
Sand bar/shoal	very fine sandstone	highly bioturbated; uncommon trough cross-stratification	elongate	crinoids, brachiopods, gastropods +

recognized in the study area (Henry and Gordon, 1979; Merrill, 1986). The occurrence of micritic lacustrine limestone has been attributed by previous workers to drier climates (Cecil et al., 1985; Cecil 1990). However, lacustrine carbonates can also form under more humid climatic settings (Talbot and Allen, 1996). Glenshaw facies sequences suggest both carbonate and clastic lake development took place during transgressive stages and associated rising water tables. The water table in tropical settings has been correlated with Milankovitch climate cycles (Kutzbach and Street-Perrott, 1985); lowered water table and dry conditions occur during high-latitude glaciation, whereas rising water tables and formation or deepening of lakes occurs during interglacial stages.

Marine Facies

Marginal-marine facies generally lack body fossils except for occasional burrowing bivalves like *Wilkingea* that could cope with pulses of rapid deposition. Trace fossils, such as *Paleophycus, Aulichnites,*

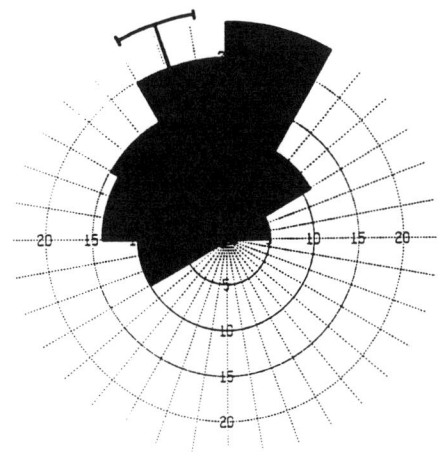

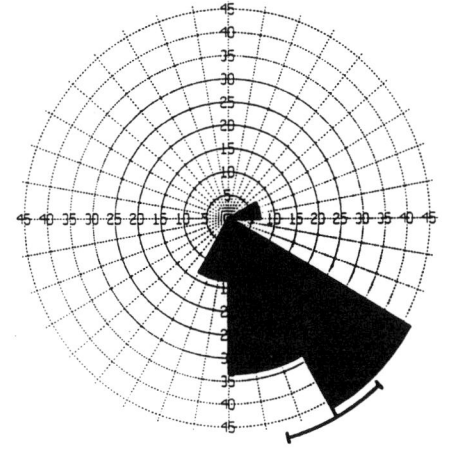

Figure 5. Paleocurrent roses and statistical data for fluvio-estuarine channels. Unimodal flow occurs in individual channel fills, and most exhibit fluvial or ebb tide-dominated flow toward the northwest (left rose). Local occurrence of flood-dominated (southeastward) flow occurs in tidal channel and sand-flat facies in the Mahoning Sandstone (right rose).

Calculation method	Frequency
Class interval	30°
Filtering	deactivated
Data type	unidirectional
Rotation amount	0°
Population	101
Maximum percentage	23.8%
Mean percentage	11.1%
Standard deviation	7.92%
Vector mean	341.12°
Confidence interval	10.3°
R-mag	0.66

Calculation method	Frequency
Class interval	30°
Filtering	deactivated
Data type	unidirectional
Rotation amount	0°
Population	15
Maximum percentage	46.7%
Mean percentage	25%
Standard deviation	18.36%
Vector mean	150.92°
Confidence interval	13.94°
R-mag	0.89

Teichichnus, and *Curvolithus*, are useful indicators of marine influence. Hummocky cross-stratification and symmetric ripple bedding from shoreface deposits reflect wave-generated currents, but their rarity and the generally thin nature of sandy nearshore facies probably reflect limited wave energy that would be expected for a narrow seaway located on the equator in the doldrums. Distributary mouth bar deposits, such as those in the lower Brush Creek marine zone, produce local thickening and coarsening in the marine unit that is laterally limited to 1–2 km (0.62–1.24 mi) along depositional strike (Figure 8C). Offshore marine shales and limestones contain stenohaline marine invertebrates (Table 1).

Various depth-related biofacies have been recognized in Pennsylvanian cyclothems (Boardman et al. 1984) that can be used to help delineate the maximum flooding surface. In the lower Brush Creek T-R cycle, a vertical succession of biofacies dominated by (1) *Lingula*, (2) nearshore mollusks, and (3) open-marine stenotopic organisms defines progressively deeper water in the upper portion of the TST, whereas the HST is generally barren of macrofossils because of increased rate of deposition, turbidity, and/or dilution. Locally, the base of the lower Brush Creek HST contains a return to the nearshore molluskan association.

Paleosols

Recognition of paleosols is based on the presence of soil horizons, soil structure, and/or root traces (Retallack, 1988). Glenshaw paleosols are very distinctive in outcrop because of their easily weathered, hackly, variegated appearance and horizonization (Martino, 1992) (Tables 3–5). The paleosol type and degree of development are important in assessing its paleoclimatic and sequence-stratigraphic significance.

The developmental stage that a paleosol has reached is a significant indicator of exposure time, although parent material composition and soil type also need to be considered. Glenshaw paleosols developed on mixed clayey and sandy alluvium and less commonly on sand-clay mixtures of marine facies. Thin paleosols that lack horizons, lack well-developed soil structure, and contain preserved primary stratification disturbed by root traces that represent very weak development and relatively brief periods of slow or no deposition. In weakly to moderately developed soils, peds and cutans are found, but primary stratification persists. Strong soil development results in obliteration of bedding, whereas in very strong development, the clayey (Bt) horizon is significantly thicker than 1 m (3.3 ft) and is commonly associated with major geological unconformities (Retallack, 1990) (Figure 9; Table 2).

Figure 6. Roadcut near Chesapeake, Ohio (location O-2) exposing about 30 m (98 ft) of Glenshaw strata from the Cambridge Limestone through the Ames Limestone and shale at the top of the cut. Note the large-scale accretion surfaces in the Saltsburg Sandstone, a fluvial-estuarine channel fill about 10 m (33 ft) thick.

Entisols (very weakly developed) and inceptisols (weakly developed; Retallack, 1988) (Figure 9) are laterally discontinuous in Glenshaw flood-basin facies. Good examples occur between the upper Freeport and Brush Creek coals. These soil types typically form over brief periods of time (tens to hundreds of years; Retallack, 1990). Histosols are distinguished by thick surface organic (O) horizons and are represented by coals and carbonaceous shales where the precompaction thickness was at least 40 cm (15.7 in.). Histosols develop in low-lying, poorly drained areas where organic production exceeds decomposition. The local, pod-like geometry of many Glenshaw coals (e.g., Mahoning and Wilgus coals) suggests a rolling or undulatory topography.

Most of the strongly developed paleosols in the Glenshaw exhibit features associated with vertisols and aridosols (Tables 3–5). An excellent example has

been described from above the Saltsburg Sandstone along Route 23 at Savage Branch, Kentucky (Martino, 1992) (Figure 8D; Table 3). The paleosol is 5.17 m (17 ft) thick at this location and is a widespread unit that correlates with the Pittsburgh shale of West Virginia and western Pennsylvania and the Round Knob shale of Ohio. It contains hummock-and-swale structure (mukkara), prominent slickensides and clastic dikes, and evidence of abundant swelling clay; these features are characteristic of vertisols (Retallack, 1988). The profile also contains features that develop in aridosols, including high Munsell values and extensive pedogenic carbonate (caliche). Vertisols are uniform, thick (>50 cm [>20 in.]), clay-rich paleosols with deep wide cracks for part of the year. Hummock-and-swale topography (i.e., gilgai microrelief) results from swelling and upward-buckling of the soil along hummocks with deep fissures (Retallack, 1990). The swales

Figure 7. Outcrop along Kentucky State Route 23 about 8 mi (13 km) south of Ashland (K-3). Multistory fluvial-estuarine channel complex represents Saltsburg and possibly Buffalo Sandstones. A mud-filled, abandoned channel is exposed in the first story (right). Lateral accretion surfaces are evident in the upper story. The three stacked channels are 26 m (85 ft) thick and are capped by thick paleosol of the Pittsburgh shale at the top of the cut.

may receive sediment eroded from adjacent hummocks, as well as chemical precipitates from ephemeral lakes. The concave-upward lenses of carbonates at the top of the paleosol appear to represent the latter. Seasonal deposition of carbonate-filled fractures as alkaline waters filled deep open fissures (Figure 8D). The carbonate lenses are spaced laterally at regular intervals of about 6–7 m (20–23 ft) along the outcrop. Conjugate systems of slickensides also developed as a result of clay heave.

Vertisols typically are associated with low-relief terrain and subhumid to semiarid climates (18–152 cm [7–60 in.] rainfall/yr) with a pronounced dry season. Aridosols develop in semiarid to arid regions and commonly have shallow calcareous horizons (Retallack, 1990). Vertisols may develop in as little as a few hundred years on smectitic claystones and shales, whereas strongly developed aridosols require tens of thousands of years to develop (Birkland, 1984; Retallack, 1990).

Compound paleosol zones occur where individual paleosol units are vertically stacked. Two kinds of paleosol zones occur in the Glenshaw Formation: (1) stacked paleosols of the same type and (2) stacked

profiles representing two different types of soils. In both cases, pedogenesis was interrupted by influx of sediment. The first type is evident in multiple-bedded coal seams with rooted shale or sandstone splits. Stacking of well-drained paleosols (aridosols/vertisols) also occurs. The second type of paleosol zone is commonly represented by aridosols or vertisols that are capped by histosols. These zones represent two phases of soil development and reflect a rise in water table, which, in many cases, was associated with rising sea level (indicated by shallow-marine to estuarine roof rock). Previous workers have recognized these types of compound paleosols elsewhere in late Pennsylvanian strata of the Appalachian basin (Cecil, 1990; Fedorko, 1998).

Glenshaw Sequence-stratigraphic Model

The development of a Glenshaw stratigraphic sequence can be illustrated using the model shown in Figure 10:

1) During lowered base level associated with sea level lowstands, coastal plain rivers downcut 20–35 m (66–115 ft), which led to valley incision. Sediment

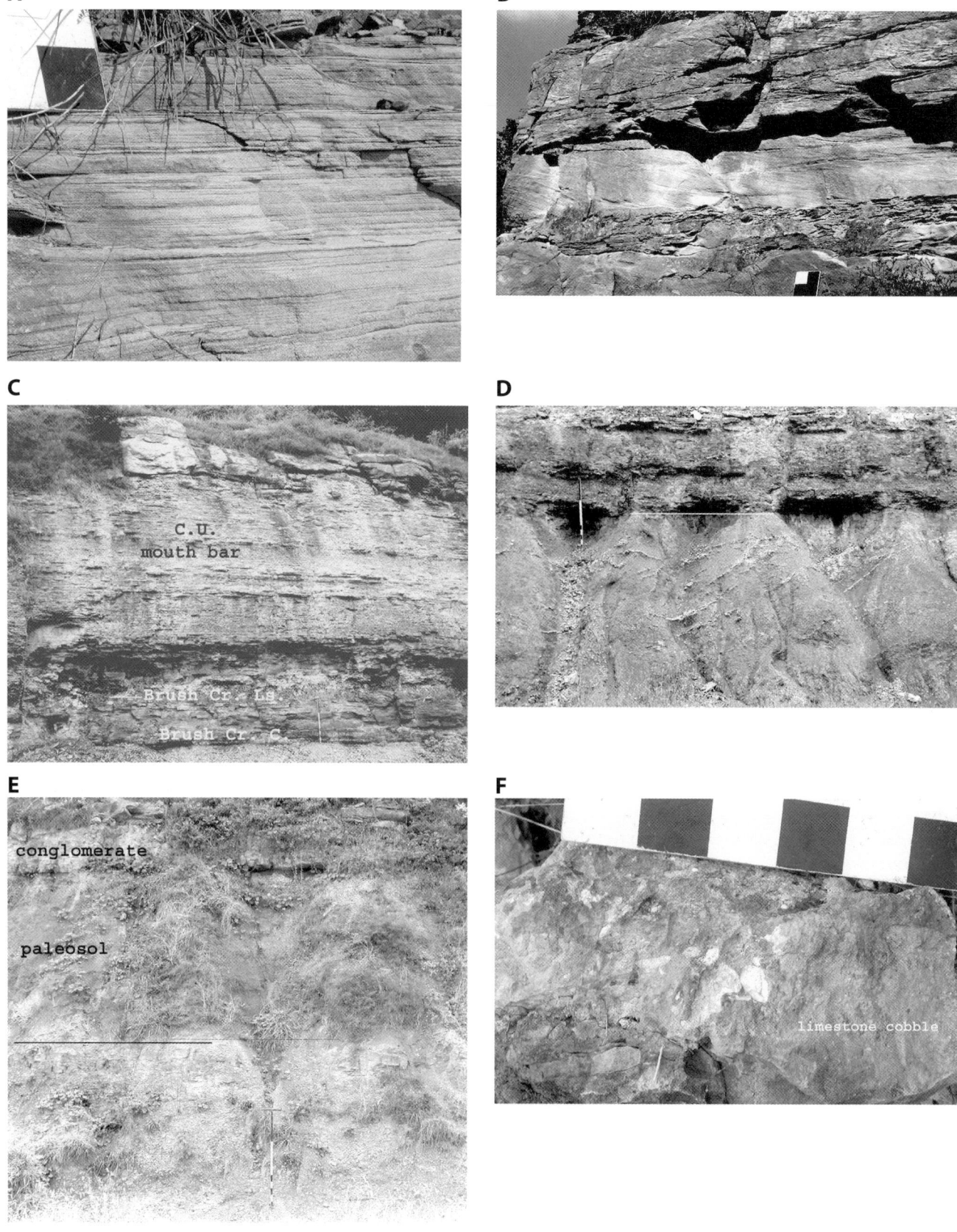

A

B

C

C.U.
mouth bar

Brush Cr. Ls.

Brush Cr. C.

D

E

conglomerate

paleosol

F

limestone cobble

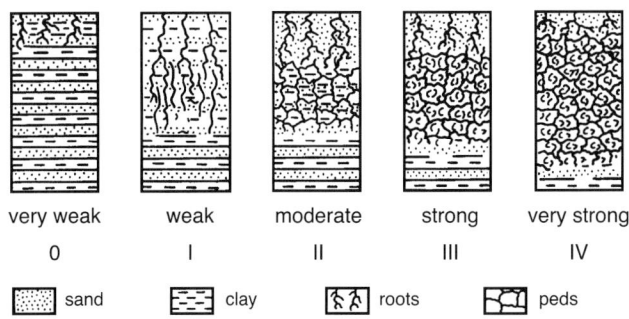

very weak   weak   moderate   strong   very strong
0   I   II   III   IV

sand   clay   roots   peds

Figure 9. Stages in the formation of clayey subsurface soil horizons (Bt) in mixed clayey and sandy alluvium (from Retallack, 1990).

bypassing of interfluves led to pedogenesis. The lower water table (following falling river and sea level) caused well-drained conditions for soil development. A type 1 sequence boundary formed, which is marked by an erosional disconformity along paleovalley and by a nondepositional disconformity on the interfluves (top of a well-drained, mature paleosol).

2) Rising sea level and base level initiated aggradation of fluvial system in paleovalley (lowstand systems tract [LST]); pedogenesis continued on interfluves, which allowed the maturation of vertisols and aridosols.

3) Continued rise in sea level or base level led to a rising water table; standing shallow water resulted in peat accumulation where clastic influx remained low; late-stage valley-fill associated with high-sinuosity streams commonly preserves evidence of tidal influence (TST). Completion of valley filling allowed alluvium to spread out over interfluves that had been sediment starved up to that point. Clastic and carbonate lakes developed in the coastal plain where water depth became too great to support standing vegetation, whereas in downdip locations, marginal to shallow-marine environments onlapped interfluvial paleosols and valley fills (TST).

4) During sea level highstand, rapid aggradation of the coastal plain occurred in association with high

accommodation space. This produced isolated, high-sinuosity fluvial channel deposits encased in overbank fines (Shanley and McCabe, 1993) (Figure 11). Regression occurred in marine units during highstand once estuarine sediment sinks became filled. Deltaic channel fills and mouth bars formed locally during late highstand.

5) Incision of fluvial drainage lines into HST coastal plain and sea-fill deposits occurred in response to falling sea level or base level. Between rivers, withdrawal of the sea led to erosion or exposure and pedogenesis of shallow-marine or flood-basin facies, which produced the next sequence boundary.

Rising base level that allowed infilling of valleys also ultimately led to peat accumulation. Coal beds that formed in this way would develop across the region, but the same seam would overlie well-drained paleosols on former interfluves interfluvial sequence boundary (ISB) and hydromorphic paleosols above the valley fills (Figure 10).

A characteristic feature of Glenshaw paleosols that mark sequence boundaries is that they exhibit evidence of well-drained conditions followed by "drowning." This is illustrated by the facies sequence (1) vertisol or aridosol, (2) coal/histosol, (3a) lacustrine shale and/or limestone, or (3b) marine shale and/or limestone. In some cases, facies sequence 2 is missing, and lacustrine or marine facies directly overlie a vertisol or aridosol. This indicates an initially low water table that subsequently rose to inundate the topography. Busch (1984) described similar Glenshaw paleosols representing allocycle boundaries in the northern Dunkard basin. Rising sea level may have led to a wetter climate, which also helped to raise the water table in the coastal plain (Busch, 1984; Busch and West, 1987; Heckel, 1995).

In cases in the Glenshaw Formation where relief produced by channel incision is less than 10 m (33 ft), HST avulsion channel fill would be difficult to distinguish from LST/TST incised-valley fills (IVFs). If facies reflected rising sea level were found (braided fluvial to meandering fluvial to estuarine), a LST/TST origin

Figure 8. (A) tidal rhythmites in Mahoning Sandstone along West Virginia State Route 52 south of Prichard (W-8). Scale divisions are 5 cm (2 in.). (B) Tabular planar cross-beds from estuarine sand-flat facies, Mahoning Sandstone. (C) Coarsening-upward distributary mouth bar sequence in lower Brush Creek cyclothem along Kentucky Route 23, about 8 mi (13 km) south of Ashland (K-4). Jacobs staff = 1.5 m (4.9 ft). (D) Paleovertisol 5.17 m (17 ft) thick (=Pittsburgh shale) along Kentucky Route 23, about 9 mi (14.5 km) south of Ashland Conjugate shears are filled with micrite (K-7). (E) Stacked paleosol zone along Kentucky State Route 3 about 6 mi (6.7 km) south of Ashland (K-1). Paleosol and overlying Brush Creek coal truncated by a 1.5-m (5-ft)-thick cobble conglomerate. (F) Conglomerate in (E). Note limestone cobble at left, probably from upper Brush Creek Limestone. Scale divisions are 5 cm (2 in.).

Table 2. Stages of paleosol development.*

Stage	Features
Very weakly developed	little evidence of soil development except for root traces; abundant sedimentary textures remaining from parent material
Weakly developed	with a surface-rooted zone (A horizon) as well as incipient subsurface clayey, calcareous, sequioxidic or humic, or surface organic horizons, but not developed to the extent necessary for qualification as a U.S. Department of Agriculture (USDA) argillic, spodic, or calcic horizon or histic epipedon
Moderately developed	with surface-rooted zone and obvious subsurface clayey, sequioxidic, humic or calcareous, or surface organic horizons: qualifying as USDA argillic, spodic or calcic horizon or histic epipedon and developed to an extent at least equivalent to stage II carbonate accumulation (few to common carbonate nodules and veinlets with powdery and filamentous carbonate in places between nodules)
Strongly developed	with especially thick, red, clayey, or humic subsurface (B) horizons or surface organic horizons (coals or lignites) or especially well-developed soil structure or calcic horizons at accumulation stages III–V (III: carbonate forming continuous layer comprised of coalescing nodules with isolated nodules and powdery carbonate outside main horizon; IV: upper part of carbonate layer with weakly developed platy or lamellar structure capping less pervasively calcareous parts of the soil profile; and V: platy or lamellar cap to carbonate layer strongly expressed, brecciated in places, and with pisolites of carbonate)
Very strongly developed	uncommonly thick subsurface (B) horizons or surface organic horizons (coals or lignites) or calcic horizons of accumulation stage VI (brecciation and recementation, as well as pisoliths common in association with the lamellar upper layer): such a degree of soil development is mostly found in major geological unconformities

Modified from Retallack, 1990.

Table 3. Paleosol description for Pittsburgh shale along Route 23, milepost 7.9.*

Depth of Paleosol [in Centimeters]	(Horizon)	Composition/features
0–1.3	(A)	mudstone, dark greenish gray (N4 5GY 4/1), noncalcareous, sharp erosional upper contact
1.3–14	(A)	mudstone, dark greenish gray, weathers greenish gray, noncalcareous; laterally equivalent to light gray micritic limestone lenses which are concave-upward, and laterally spaced at 4.8–5.6-m (16–18-ft) intervals
14–144	(Bk)	claystone, dark greenish gray (N4 5GY 4/1), moderately to strongly calcareous; micritic carbonate present as infillings of steeply inclined fractures 3.8–5.1 cm (1.5–2 in.) in width, and as greenish gray (N6 5 GY 6/1) to light gray (N7 5Y 7/1) nodules
144–296	(Bk)	mudstone, variegated, dusky red (7.5 YR N 4/2) and dark greenish gray (N4 4GY 4/1); weakly to moderately calcareous; slickensides common to abundant; fine to medium angular blocky peds; micritic carbonate nodules (light gray to greenish gray)
296–357	(K)	micritic limestone, gray (N6 5Y 6/1) with angular fragments of greenish gray (N5 5G 5/1), calcareous mudstone; comprised of coalescing nodules
357–517	(C)	sandstone, fine grained, calcareous, sideritic; clastic dikes(?); micritic and sideritic nodules; calcite-filled fractures at top

All Munsell colors are from fresh, unweathered surfaces (from Martino, 1992).

Table 4. Description of paleosol directly beneath Ames Member, Route 23, Savage Branch.*

Depth of Paleosol [in Centimeters]	(Horizon)	Composition/features
0–2	(O)	coal, bright, mostly vitrain/clarain
2–17	(E)	underclay, gray (N6) with yellow-brown mottles; platy soil structure, uncommon slickensides
17–112	(Bt)	claystone, dark gray (N4) in upper part to dark greenish gray (N4 5GY 4/1) in lower part; black metallic staining [manganese(?)]; small ironstone nodules (2–3 mm [0.07–0.12 in.]); siderite nodules in lower part; generally noncalcareous except for micritic nodules up to 5 cm (2 in.) in diameter in lower 15 cm (6 in.)
112–127	(K)	micrite, dark greenish gray (N4 5GY 4/1), light greenish gray, and weak red (10R 4/3), angular fragments
127–187	(Bk)	mudstone, dark greenish gray (N4 5GY 4/1), moderately calcareous, sandy, with blocky angular peds
187–197	(K?)	micritic limestone, brecciated

*Milepost 8.1; Martino, 1992.

would be favored. Where relief along the erosional surface is greater than 10 m (33 ft), a LST/TST IVF also would be more likely.

Sequence-stratigraphic Analysis of Glenshaw Formation

A generalized illustration of Glenshaw stratigraphy is given in Figure 12. Nine fifth-order stratigraphic sequences are identified. The tops of well-drained paleosols are sequence boundaries and correspond to allocycles previously reported by Martino (1998) (Figure 13). Sequence boundaries and systems tracts are numbered in vertical succession (SB1, LST1, TST1, HST1; SB2, LST2, etc.).

The first depositional sequence identified in this study begins in the upper Allegheny Formation and is well exposed in Ohio Route 52 roadcuts near Ashland, Kentucky (Figures 4, O-4; 14; 15). Multistory

Table 5. Paleosol directly above Ames Member, Route 23 milepost 8.4.*

Depth of Paleosol [in Centimeters]	(Horizon)	Compostion/features
0–0.5	(O)	carbonaceous shale, claystone, dark gray (N4), sharp upper contact with overlying crudely bedded, greenish gray mudstone
0.5–38	(E)	claystone, greenish gray (N5 5G 5/1), silty, with dark greenish gray peds and brownish gray cutans; slickensides common; peds 1–2 cm (0.4–0.8 in.) in diameter and subangular
38–118	(F.B)	claystone, greenish gray (N5 5G 5/1) to dark greenish gray (N5 5GY 5/1), with subordinate weak and dusky red mottles (7.5 YR 4/2 and 7.5 YR 3/2); angular blocky peds, transitional top and base; weakly calcareous with 1–2 mm (0.04–0.08 in.) carbonate nodules
118–228	(Bt)	claystone, dusky red (10R 3/3) to olive (5Y 4/3), weakly to moderately calcareous; well-developed soil structure, including angular blocky peds 1–4 cm (0.4–1.6 in.) in diameter with red clay cutans; abundant slickensides
228–238	(B)	clay, light olive green, plastic
238–298	(B)	mudstone, dark greenish gray, with angular peds
298–318	(B)	clay, light olive green, weathers rust orange

Top of Ames Member, very fine sandstone.

*Martino, 1992.

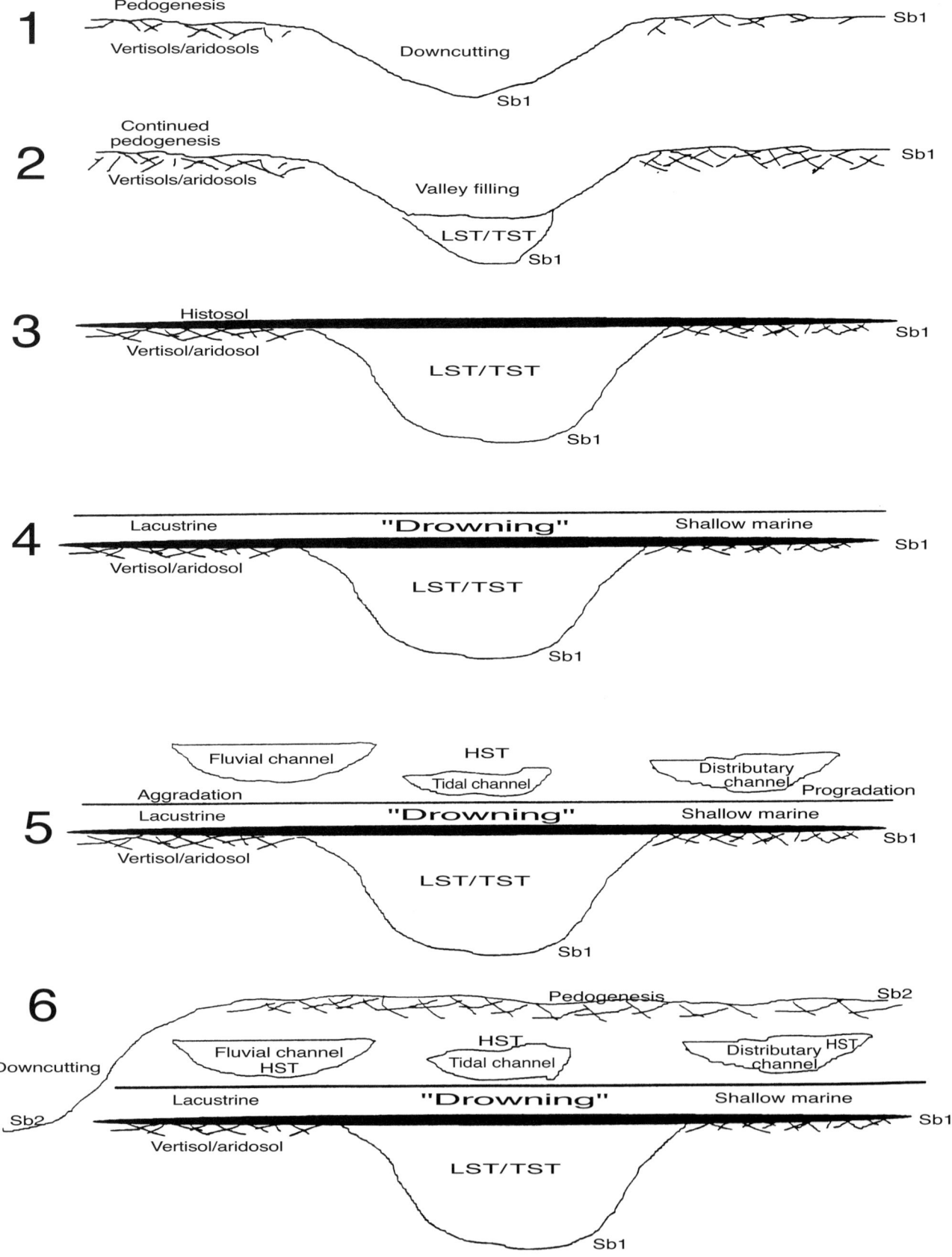

Figure 10. Sequence-stratigraphic model for Glenshaw Formation. Sequence boundaries (Sb) developed at the top of well-drained paleosols formed on interfluves during lowstands and then passed laterally into erosional disconfomities of incised valleys. LST, TST, and HST are lowstand, transgressive, and highstand systems tracts, respectively.

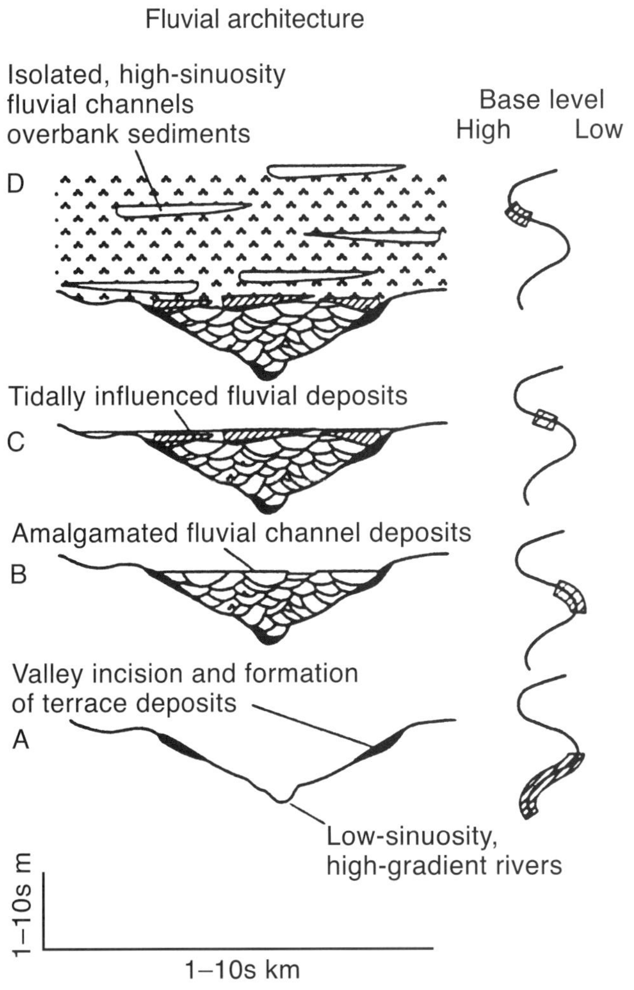

Fluvial architecture

Isolated, high-sinuosity
fluvial channels
overbank sediments

Base level
High   Low

D

Tidally influenced fluvial deposits

C

Amalgamated fluvial channel deposits

B

Valley incision and formation
of terrace deposits

A

Low-sinuosity,
high-gradient rivers

1–10s m

1–10s km

Figure 11. Model for development of an incised-valley fill (Shanley and McCabe, 1993). Tidal influence occurs as estuaries succeed fluvial systems within valley. Once aggradation expands onto interfluves, a much lower channel/overbank ratio occurs.

fluvial-estuarine channels beneath the upper Freeport coal comprise an IVF (LST/TST1) with a maximum thickness of about 25 m (82 ft), the base of which corresponds to SB1 in Figure 12. The upper Freeport coal is part of the TST1. McCabe (1993) noted that during rising base level, peats can form that onlap sequence boundaries. Therefore, coal beds may cap well-drained, interfluvial paleosols that represent sequence boundaries. The roof shale of the upper Freeport contains the nonmarine bivalve *Anthaconaia provosti* (E. Belt, 2000, personal communication), indicating the drowning of the swamp and the formation of lacustrine conditions during late TST. The common occurrence of isolated meandering river and tidal channel facies encased in flood-plain fines between the upper Freeport and Mahoning coal horizons is indicative of high rates of sedimentation and aggradation that accompanied increased sea and base level

and high clastic sediment supply. These facies are interpreted as HST1. This fluvioestuarine package is truncated by stacked channel fills interpretable as incised valleys because of falling base level. Thus, the lower Mahoning Sandstone (i.e., sandstone between the upper Freeport and Mahoning coal horizons) includes both channel fills that are part of HST1, whereas others in the same interval are deposits of LST2/TST2 (Figure 12).

Other examples of IVFs include portions of the Buffalo Sandstone, Saltsburg Sandstone (Figure 8), and Grafton Sandstone (Figure 16). The maximum thickness of the IVFs is from 20 to 35 m (66 to 115 ft), although greater apparent thicknesses occur where IVFs incise into one another. This is illustrated by the Saltsburg-Buffalo IVF, a compound valley fill that extends from the base of the Pittsburgh shale to at least several meters below the Brush Creek coal (Figures 2, 12). One consequence of valley cutting was mass wasting of oversteepened valley walls. A spectacular example of a huge slump block of this type is well exposed near Prichard, West Virginia (Figure 17).

Intrabasinal Correlation

The nine paleosol-bounded allocycles of the Huntington area can be traced northward to the Ashland area and southward to Louisa (Figure 18). Correlations are facilitated by the presence of four marine units. These marine units pinch out eastward between Wayne and West Hamlin (Figures 4, 19). The disappearance of the lower Brush Creek, upper Brush Creek, and Cambridge marine units appears to coincide with the appearance of "probable nonmarine limestones" such as the Twomile Limestone described by Henry and Gordon (1979) in the Charleston area. These limestones commonly contain *Spirorbis*, which occur only in brackish to marine waters today (Tasch, 1980). Perhaps these were coastal lakes with intermittent connection to the sea.

Eight of the eleven allocycles interpreted for the northern Appalachian basin can be recognized in the central Appalachian basin (Busch and West, 1987) (Figure 20). Busch and West (1987) interpreted an allocycle to be represented between the upper Freeport coal and upper Freeport Rider coal, but it was only found in western Pennsylvania. The local occurrence of this cycle and its association with the upper Freeport Rider coal suggest that it may be more likely to have originated from autocyclic or local tectonic processes. Busch and West (1987) described five allocycles bounded by "transgressive" surfaces between the base of the Cambridge and the base of

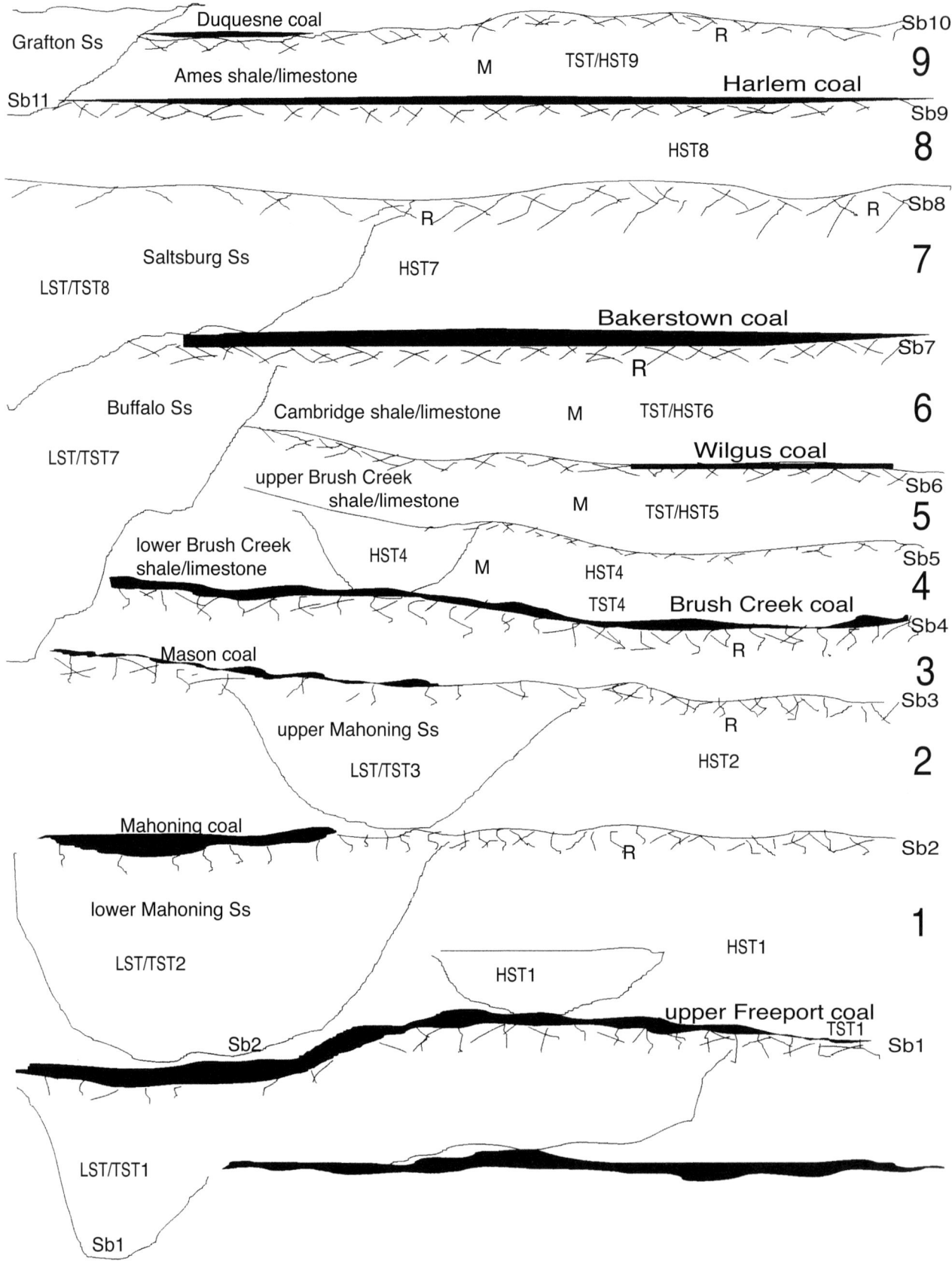

Figure 12. Sequence-stratigraphic interpretation for the Glenshaw Formation and upper Allegheny Formation at the southwestern end of the Dunkard basin. Numbers 1–9 at the right identify Glenshaw Formation sequences and correspond to paleosol-bounded allocycles reported by Martino (1998, figure 13).

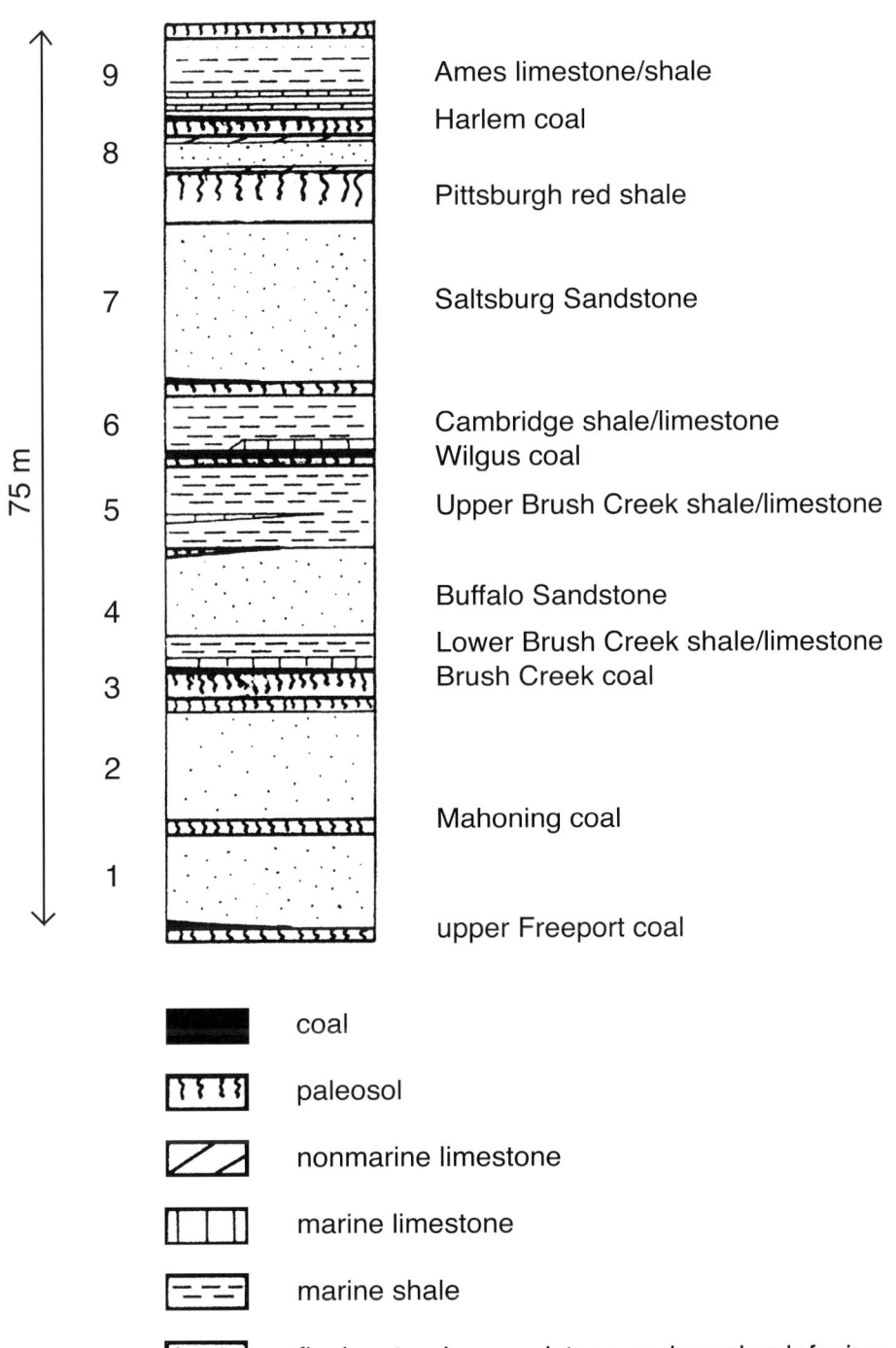

75 m

9 — Ames limestone/shale
8 — Harlem coal
— Pittsburgh red shale
7 — Saltsburg Sandstone
6 — Cambridge shale/limestone
— Wilgus coal
5 — Upper Brush Creek shale/limestone
4 — Buffalo Sandstone
— Lower Brush Creek shale/limestone
3 — Brush Creek coal
2
— Mahoning coal
1
— upper Freeport coal

Figure 13. Composite section of the Glenshaw Formation in the Huntington, West Virginia area, showing nine paleosol-bounded allocycles (Martino, 1998). Channel and valley fills are omitted for simplicity. Section is 75 m (246 ft) thick.

coal

paleosol

nonmarine limestone

marine limestone

marine shale

fluvioestuarine sandstone and overbank facies

the Ames marine units in the northern Appalachian basin, whereas only three were found in the area of this study. Decreasing accommodation space across the basin could lead to convergence of sequence boundaries, but this seems unlikely, as vertically stacked paleosols in this interval were not evident.

Interbasinal Correlation

Busch and West (1987) and Heckel (1995) proposed correlation of Glenshaw allocycles from the northern Appalachian basin with those in the Illinois basin and in Kansas (Figures 20, 21). There is a lack of agreement as to which cycles correlate. For example, more of the lower Brush Creek, upper Brush Creek, and Cambridge marine units are correlated with younger mid-continent units by Heckel (Swope, Dennis, and Dewey Limestones) than by Busch and West (Hertha, Swope, and Dennis). The Ames is correlated with the late Missourian Stanton Limestone by Busch and West, and the early Virgilian Oread Limestone by

Figure 14. Roadcut along Ohio State Route 52 about 1 mi (1.6 km) northeast of Ashland, Kentucky (O-4). The lower portion of outcrop exposes a 17-m (56-ft)-thick multistory, fluvioestuarine channel complex interpreted as an incised-valley fill (LST/TST1) which locally truncates the no. 5 (lower Kittanning) coal at this location. The upper Freeport coal on bench overlies a thick paleosol with vertical features and is overlain by flood-basin lacustrine shales and splay sands. These are truncated by moderate-sized channels below the tree line. The total stratigraphic interval from the Brush Creek coal to road level is 46 m (151 ft).

Figure 15. Outcrop along Ohio State Route 52 just north of Ashland, Kentucky. Multistory fluvioestuarine channel sandstone interpreted as IVF (LST/TST1, Figure 12) downcuts toward the left nearly to the Vanport Limestone. Upper Freeport coal (TST1) and flood-basin facies (HST1) are exposed in the upper half of the cut. Total thickness of section is about 50 m (164 ft).

Figure 16. Upper Glenshaw Formation, Kentucky State Route 23 (K-5). Grafton Sandstone (of lower Casselman Formation) truncates Ames marine unit and down to the Harlem coal at this location. The Grafton is interpreted as an IVF. Its lower contact (SB11) truncates a mature paleosol (SB10; Figure 12) described in Table 3 that caps the Ames. These relations make it apparent that the Grafton fluvial system postdates the Ames T-R cycle. Reconnaissance suggests similar relations at most other Ames locations, calling into question the idea that the Grafton delta infilled the Ames Sea (Donaldson et al., 1985).

Heckel. An early Virgilian age for the Ames is supported by conodonts (Merrill, 1986). Heckel's (1995) correlations relied on biostratigraphy using conodonts from marine units and palynomorphs from coal beds and include data not available in Heckel (1986), which Busch and West (1987) employed in their analysis.

Ross and Ross (1988) distinguished 60 unconformity-bounded sequences of Permian–Carboniferous strata in cratonic basins on a global scale. A comparison of biostratigraphically equivalent horizons indicated in Heckel (1995) (Figure 22) for the upper Freeport coal and the Ames Limestone with the mid-

continent sequence stratigraphy of Ross and Ross (1988) is shown in Figure 20. Ten unconformity-bounded sequences developed during the time interval of Glenshaw deposition in other basins. If glacio-eustatic control of fifth-order Glenshaw allocycles occurred, then these depositional sequences should be developed in the Appalachian basin. Nine Glenshaw sequences are distinguished in this study. The fewer number in the Appalachian basin could be the result of the basin's high shelf position (Heckel, 1994). Discrete lowstand exposure surfaces in the mid-continent region may merge into single-exposure surfaces as accommodation space decreases up the shelf into the

Figure 17. View of West Virginia State Route 52 roadcut 1–2 mi (1.6–3.2 km) south of Prichard as seen from Kentucky. Strata total about 100 m (328 ft) in thickness and expose the entire Glenshaw Formation and the lower 20 m (66 ft) of Casselman Formation. Note the large slump block (left) interpreted as the result of oversteepening of paleovalley wall.

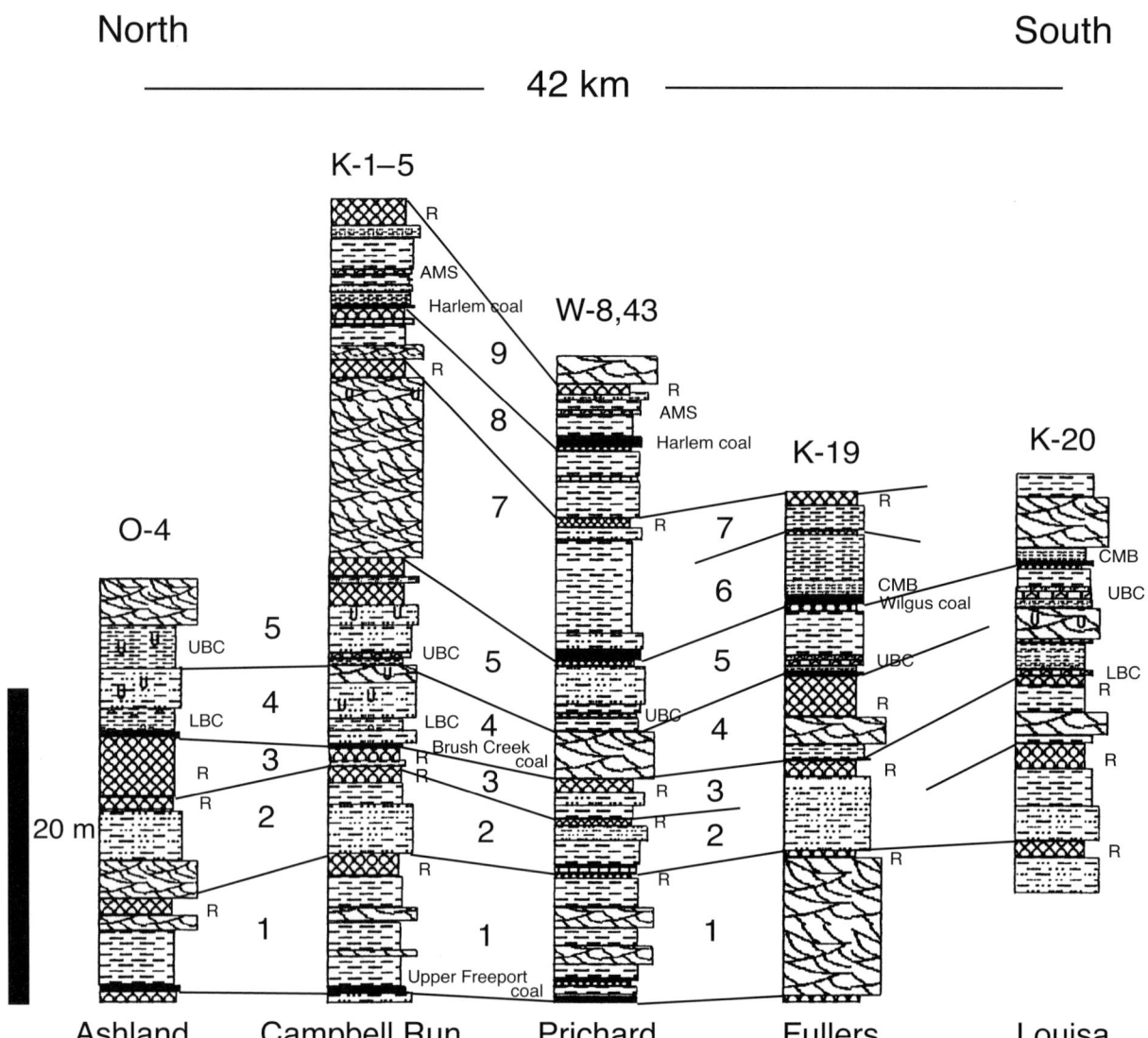

Figure 18. Correlation of Glenshaw sections from north to south through the western portion of the study area using tops of paleosols. See Figure 4 for locations and Figure 19 for key. Marine units: LBC = lower Brush Creek, UBC = upper Brush Creek, CMB = Cambridge, AMS = Ames.

Appalachian basin. The presence of a minor hiatus in the Appalachian basin has been suggested between the Mahoning and Mason coals at the Missourian–Stephanian boundary (Peppers, 1997). The first occurrence of lower Stepahanian taxa, including *Triticites* and *Thymospora obscura*, are in cyclothems that are closer to the stage boundary in the Appalachian basin than in the mid-continent. Despite the minor differences in number of Glenshaw allocycles, it appears probable that base-level changes inherent in these cycles were controlled by glacioeustaic sea level fluctuations.

The IVFs described in this study are similar in depth, character, and age to those reported from the Douglass Group of Kansas (Archer et al, 1994; Feld-

man et al., 1995). Two IVFs, including the Tonga-noxie Sandstone and the Ireland Sandstone, occur in close proximity to the Missourian–Virgilian boundary and underlie the Oread Limestone (Ames equivalent of Heckel, 1995).

CONCLUSIONS

The main contributions of this study may be summarized as follows:

1) The Glenshaw Formation in the central Appalachian basin contains widespread mature paleosols with features that indicate well-drained conditions. These paleosols are interfluvial sequence

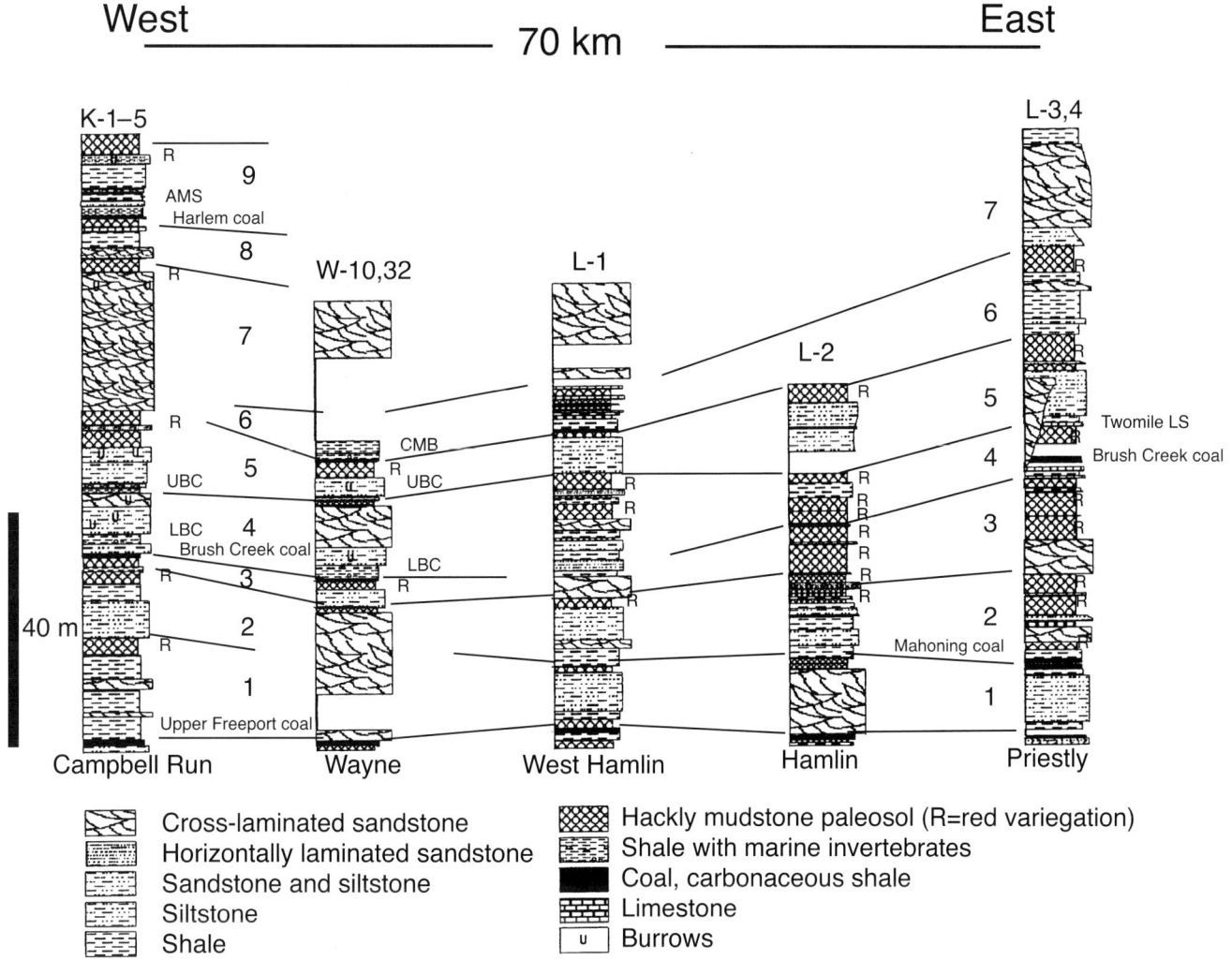

Figure 19. Correlation of Glenshaw sections from west to east through the study area, using tops of mature paleosols as sequence boundaries.

boundaries that divide the Glenshaw into nine allocycles. Four of these allocycles contain basin-wide marine units (lower Brush Creek, upper Brush Creek, Cambridge, and Ames), indicating a direct connection between sea level and base-level cycles. Tidally influenced strata occur in most of the other allocycles, suggesting sea level changes were instrumental in their development as well.

2) Incised valley-fills from 20 to 35 m (66 to 115 ft) thick occur in the Glenshaw Formation and adjacent strata; these valley fills contain multistory fluvioestuarine channel facies which are similar in depth, age, and character to IVFs reported from the Illinois basin.

3) The recognition of widespread, well-drained paleosols as sequence boundaries enables a more accurate interpretation of the origin of channel systems. Deltaic deposits are present, but not as widespread as indicated by previous workers. Only the lower Brush Creek cyclothem contains clear evidence for deltaic distributary systems in the area of this study. Deltas would be expected to develop during highstands after estuaries had filled. Estuarine facies are important constituents of coastal plain deposits formed during eustatic sea level cycles, yet they eluded detection by most previous workers.

4) The fewer number of allocycles recognized in this study compared to the work by Busch (1984) may be explained by less accommodation space in this study area, which might have caused thinning of allocycles toward the south and merging of bounding paleosols. The fewer number of marine units (four) in the Huntington area compared to the northern Appalachian basin (eight) may indicate higher rates of sediment influx from the

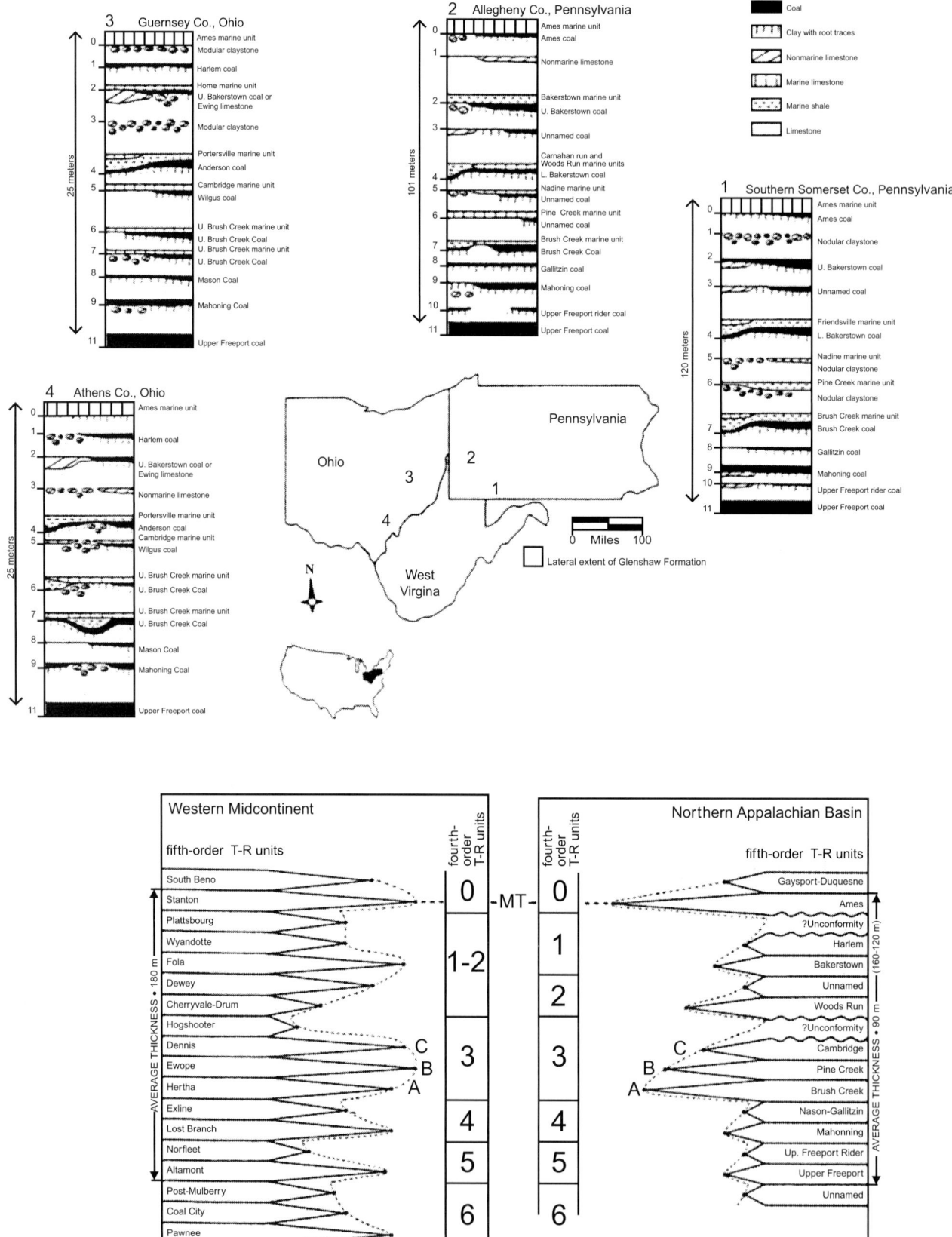

Figure 20. Eleven fifth-order T-R allocycles of Busch and West (1987) recognized in the Glenshaw Formation of Pennsylvania and Ohio.

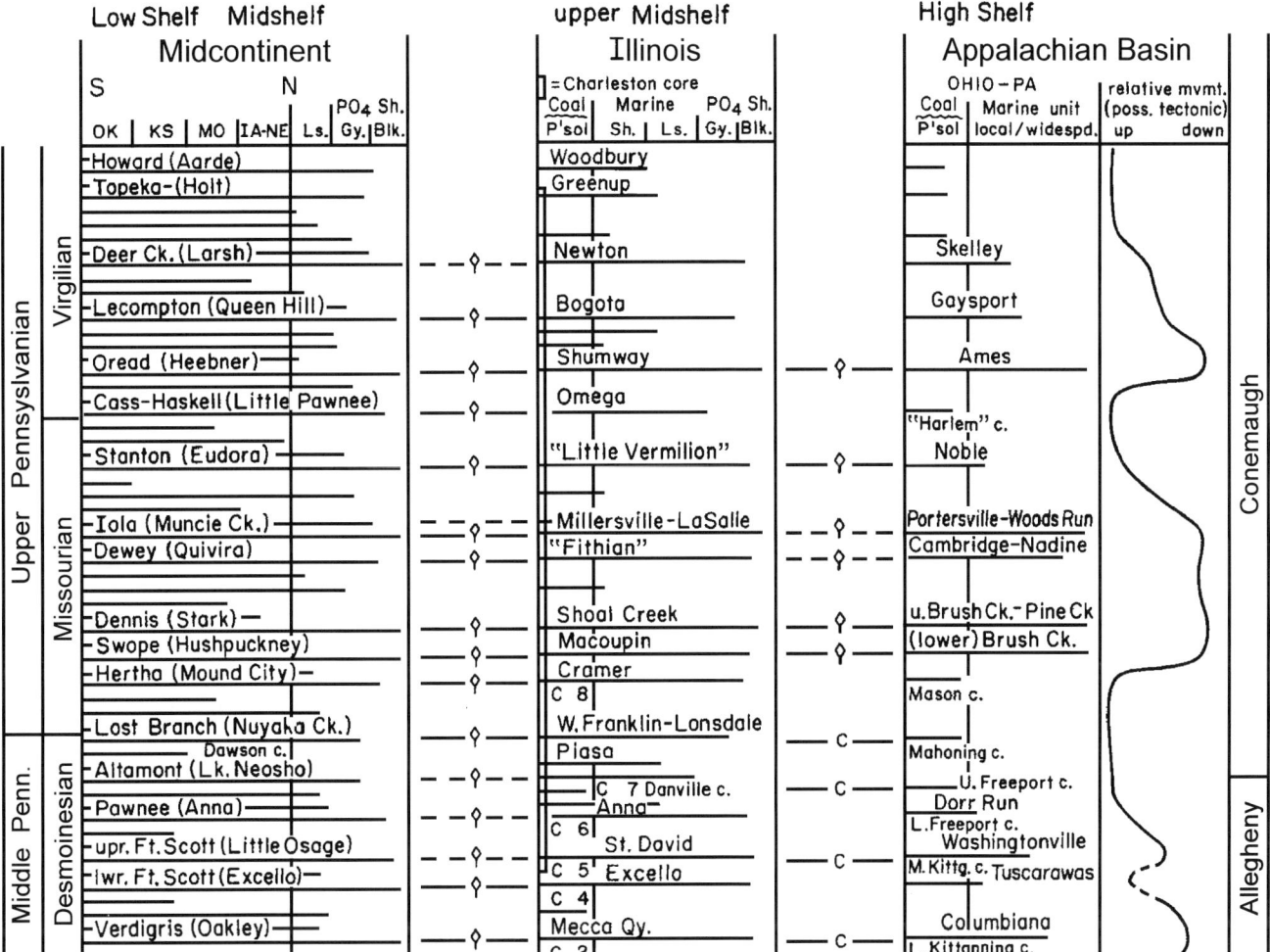

Figure 21. Interbasinal correlation of coal beds and marine units based on conodont data and palynofloras. Glenshaw Formation in Appalachian basin extends from top of upper Freeport coal to the top of the Ames Limestone. Width of horizontal lines representing marine units corresponds to the geographic extent of the transgressions. Only the more extensive transgressions in the mid-continent (low shelf) reached the Appalachian basin (high shelf; after Heckel, 1995).

southeast, which kept pace with, or outpaced rising relative sea level.

5) Changes in the Glenshaw allocycles from west to east occur across the study area. All four marine units pinch out toward the east. Lacustrine limestones, such as the Twomile Limestone, cap thick paleosols and appear to represent early highstand deposits in the coastal plain that are coeval with seas that periodically occupied the western portion of the study area.

ACKNOWLEDGMENTS

Various phases of this project have been supported by grants from the West Virginia Geological Survey, Marshall University Graduate School, and the Petroleum Research Fund administered by the American Chemical Society (grant 34516-B8). Glen Merrill and Cortland Eble provided valuable biostratigraphic information for marine limestones and coals. John Ferguson assisted in stratigraphic analysis of the Twomile Limestone. Helpful reviews by Mitch Blake and Walt Ayers are also gratefully acknowledged.

REFERENCES CITED

Archer, A. W., W. P. Lanier, and H. R. Feldman, 1994, Stratigraphy and depositional history within incised-paleovalley fills and related facies, Douglass Group (Missourian/Virgilian; Upper Carboniferous) of Kansas, U.S.A., in R. W. Dalrymple, R. Boyd, and B. A. Zaitlin, eds., Incised-valley systems: Origin and

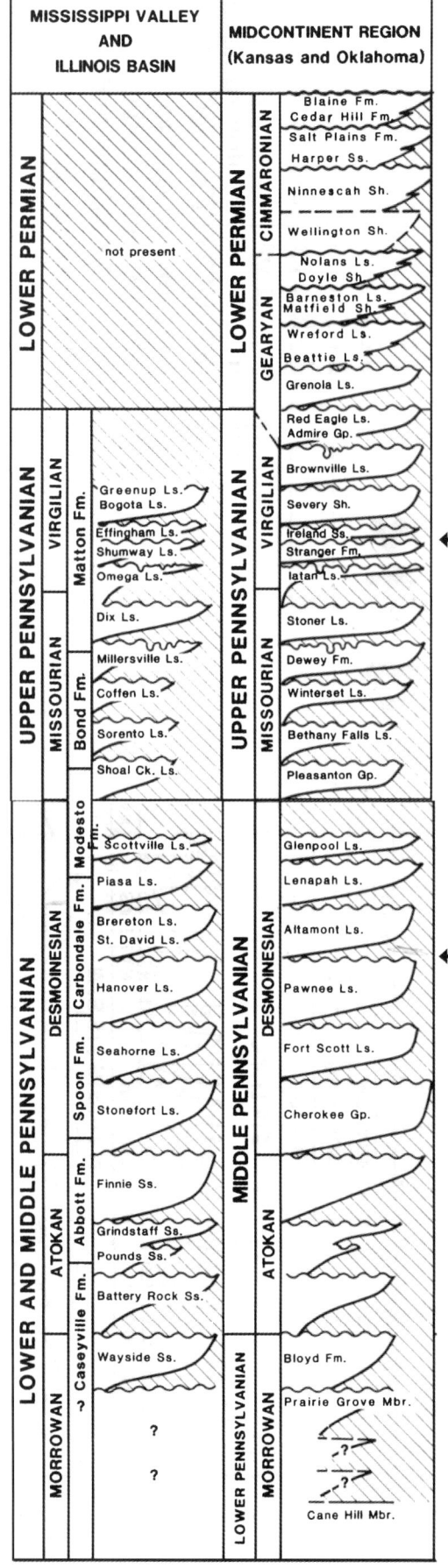

sedimentary sequences: SEPM Special Publication 51, p. 175–190.

Arkle, T. A., Jr., 1974, Stratigraphy of the Pennsylvanian and Permian systems of the central Appalachians, *in* G. Briggs, ed., Carboniferous of the southeastern United States: Geological Society of America Special Paper 148, p. 5–29.

Arkle, T., Jr., D. R. Beissel, R. E. Larese, E. B. Nufer, D. G. Patchen, R. A. Smosna, W. H. Gillespie, R. Lund, C. W. Norton, and H. W. Pfefferkorn, 1979, The Mississippian and Pennsylvanian (Carboniferous) systems in the United States— West Virginia and Maryland: U.S. Geological Survey Professional Paper 1110-D, 35 p.

Birkland, P. W., 1984, Soils and geomorphology: New York, Oxford University Press, 372 p.

Boardman, D. R., II, R. H. Mapes, T. E. Yancey, and J. M. Malikny, 1984, A new model for depth-related allogenic community succession within North American Pennsylvanian cyclothems and implications on the black shale problem, *in* N. J. Hyne, ed., Limestones of the mid-continent: Tulsa Geological Society Special Publication 2, p. 141–182.

Busch, R. M., 1984, Sea level and structural controls on paleogeography and sedimentation during deposition of the upper Pennsylvanian Glenshaw Formation of the northern Appalachian basin, *in* R. M. Busch and D. K. Brezinski, ed., Stratigraphic analysis of Carboniferous rocks in southwestern Pennsylvania using a hierarchy of transgressive-regressive units— A guidebook: AAPG Eastern Section Meeting Field Trip III, p. 56–81.

Busch, R. M., and H. B. Rollins, 1984, Correlation of Carboniferous strata using a hierarchy of transgressive-regressive units: Geology, v. 12, p. 471–474.

Busch, R. M., and R. R. West, 1987, Hierarchal genetic stratigraphy: A framework for paleoeceanography: Paleoceanography, v. 2, p. 141–164.

Cardwell, D. H., and K. L. Avary, 1982, Oil and gas fields of West Virginia: West Virginia Geological and Economic Survey, Mineral Resources Series No. MRS-7B, Part A, 45 p.

Cecil, C. B., 1990, Paleoclimate controls on stratigraphic repetition of chemical and siliciclastic rocks: Geology, v. 18, p. 533–536.

Cecil, C. B., and F. T. Dulong, 1998, Pennsylvanian paleoclimates, sediment flux, and lithostratigraphy, Appalachian basin (abs.): Geological Society of America Abstracts with Programs, v. 30, p. 9.

Cecil, C. B., R. W. Stanton, S. G. Neuzil, F. T. Dulong,

Figure 22. Unconformity-bounded depositional sequences from the Illinois basin and mid-continent for the Pennsylvanian and Permian (modified from Ross and Ross, 1988). Sequences are interpreted as correlative with worldwide eustatic sea level changes. Arrows at right show stratigraphic interval that corresponds to the Glenshaw Formation in the Dunkard basin, using the biostratigraphic correlations of Heckel (1995) for the upper Freeport coal and Ames Limestone.

L. F. Ruppert, and B. F. Pierce, 1985, Paleoclimate controls on Late Paleozoic sedimentation and peat formation in the central Appalachian basin, U.S.A.: International Journal of Coal Geology, v. 5, p. 195–230.

Cecil, C. B., F. T. Dulong, N. T. Edgar, and T. S. Albrandt, 1994, Carboniferous paleoclimates, sedimentation and stratigraphy, in C. B. Cecil and N. T. Edgar, eds., Predictive stratigraphic analysis— Concept and application: U.S. Geological Survey Bulletin, v. 2110, p. 27–28.

Dalrymple, R. W., R. Boyd, and B. A. Zaitlin, 1994, History of research, types and internal organization of incised-valley systems: Introduction to the volume, in R. Dalrymple, R. Boyd, and B. Zaitlin, eds., Incised-valley systems: Origin and sedimentary sequences: SEPM Special Publication 51, p. 353–368.

Donaldson, A. C., 1979, Depositional environments of the upper Pennsylvanian series, in K. J. Englund, H. H. Arndt, and T. W. Henry, eds., Proposed Pennsylvanian system stratotype Virginia and West Virginia: American Geologic Institute, Selected Guidebook Series 1, p. 123–132.

Donaldson, A. C., and C. Eble, 1991, Pennsylvanian coals of central and eastern United States, in H. J. Gluskoter, D. D. Rice, and R. B. Taylor, eds., Economic geology, U.S.: The geology of North America: Geological Society of America, v. P-2., p. 523–545.

Donaldson, A. C., J. J. Renton, and M. W. Presley, 1985, Pennsylvanian deposystems and paleoclimates of the Appalachians: International Journal of Coal Geology, v. 5, p. 167–193.

Emery, D., and K. J. Myers, 1996, Sequence stratigraphy: London, Blackwell Science Ltd., 297 p.

Fedorko, N., III, 1998, Investigation of a paleocatena across a late Pennsylvanian landscape comprised of organic and mineral paleosols: Ph.D. dissertation, Morgantown, West Virginia, 237 p.

Feldman, H. R., M. R. Gibling, A. W. Archer, W. G. Wightman, and W. P. Lanier, 1995, Stratigraphic architecture of the Tonganoxie paleovalley fill (lower Virgillian) in northeastern Kansas: AAPG Bulletin, v. 79, p. 1019–1043.

Ferm, J C., 1970, Allegheny deltaic deposits, in J. P. Morgan, ed., Deltaic sedimentation— Modern and ancient: SEPM Special Publication 15, p. 246–255.

Flint, N. K., 1951, Geology of Perry County: Ohio Geological Survey Bulletin, v. 48, 234 p.

Galloway, W. E., 1989, Genetic stratigraphic sequences in basin analysis: Architecture and genesis flooding-surface bounded depositional units: AAPG Bulletin, v. 73, p. 125–142.

Haught, O. L., 1963, Oil and gas fields in West Virginia: West Virginia Geological and Economic Survey, Education Series, 30 p.

Heckel, P. E., 1986, Sea level curve for Pennsylvanian eustatic marine transgressive-regressive depositional cycles along Mid-continent outcrop belt, North America: Geology 14, p. 330–334.

Heckel, P. H., 1994, Evaluation of evidence for glacio-eustatic control over marine Pennsylvanian cyclothems in North America and consideration of possible tectonic effects, in J. M. Dennison and F. R. Ettensohn, eds., Tectonic and eustatic controls on sedimentary cycles: SEPM Concepts in Sedimentology and Paleontology, v. 4, p. 65–87.

Heckel, P. H., 1995, Glacioeustatic base-level-climate model for late middle to late Pennsylvanian coal-bed formation in the Appalachian basin: Journal of Sedimentary Research, v. B65, p. 348–356.

Henry, T. W., and M. Gordon Jr., 1979, Late Devonian through Early Permian(?) invertebrate faunas in the proposed Pennsylvanian system stereotype area, in K. J. Englund, H. H. Arndt, and T. W. Henry, eds., Proposed Pennsylvanian system stratotype, Virginia and West Virginia. American Geological Institute Selected Guidebook Series 1, p. 97–104.

Horne, J. C., J. C. Ferm, F. T. Cariccio, and B. P. Baganz, 1978, Depositional models in coal exploration and mine planning: AAPG Bulletin, v. 62, p. 2379–2411.

Kutzbach, J. E., and F. A. Street-Perrott, 1985, Milankovitch forcing of fluctuations in the level of tropical lakes from 18 to 0 kyr BP: Nature, v. 317, p. 130–134.

Martino, R. L., 1992, Conemaugh Group strata in the tristate area, in C. B. Cecil and C. F. Eble, eds., Paleoclimate controls on Carboniferous sedimentation and cyclic stratigraphy in the Appalachian basin: U.S. Geological Survey Open-File Report 92-546, p. 71–76.

Martino, R. L., 1996, Stratigraphy and depositional environments of the Kanawha Formation (Middle Pennsylvanian), southern West Virginia, U.S.A.: International Journal of Coal Geology, v. 31, p. 217–248.

Martino, R. L., 1998, Facies architecture and depositional patterns in the Glenshaw Formation (late Pennsylvanian), southern Dunkard basin (abs.): Geological Society of America Abstracts with Programs, v. 30, p. 59.

Martino, R. L., M. B. Watson, K. Adkins, and G. A. Smith, 1985, Sedimentology and paleohydrology of the fluvio-deltaic Conemaugh Group (late Pennsylvanian) along the Big Sandy River, West Virginia–Kentucky: West Virginia Academy of Science Proceedings, v. 57, p. 79–90.

Martino, R. L., M. A. McCullough, and T. L. Hamrick, 1996, Stratigraphic and depositional framework of the Glenshaw Formation (late Pennsylvanian) in central Wayne County, West Virginia: Southeastern Geology, v. 36, p. 65–83.

McCabe, P. J., 1993, Sequence stratigraphy of coal-bearing strata: Short Course Notes, AAPG Annual Meeting, New Orleans, 81 p.

Merrill, G. K., 1986, Lithostratigraphy and lithogenesis of Conemaugh (Carboniferous) depositional systems near Huntington, West Virginia: Southeastern Geology, v. 26, p. 155–171.

Merrill, G. K., 1988, Marine transgression and syndepositional tectonics: Ames Member (Glenshaw Formation, Conemaugh Group, Upper Carboniferous) near Huntington, West Virginia: Southeastern Geology, v. 28, p. 153–166.

Miall, A. D., 1997, The geology of stratigraphic sequences. Springer-Verlag, New York, 433 p.

North American Commission on Stratigraphic Nomenclature (NACSN), 1983, North American stratigraphic code: AAPG Bulletin, v. 67, p. 841–875.

Opdyke, N. D., and V. J. DiVenere, 1994, Paleomagnetism and Carboniferous climate, in C. B. Cecil and N. T. Edgar, eds., Predictive stratigraphic analysis— Concept and application, U.S. Geological Survey Bulletin, v. 2110, p. 8–10.

Peppers, R. A., 1997, Palynology of the Lost Branch Formation of Kansas— New insights on the major floral transition at the middle–upper Pennsylvanian boundary: Review of Paleobotany and Palynology, v. 23, p. 223–246.

Quinlan, G. M., and C. Beaumont, 1984, Appalachian thrusting, lithospheric flexure, and the Paleozoic stratigraphy of the eastern interior of North America: Canadian Journal of Earth Sciences, v. 21, p. 973–996.

Repine, T. E., Jr., B. M. Blake, K. C. Ashton, N. Fedorko III, A. F. Keiser, E. I. Loud, C. J. Smith, S. W. McClelland, and G. H. McColloch, 1993, Regional and economic geology of Pennsylvanian age coal beds of West Virginia: International Journal of Coal Geology, v. 23, p. 75–101.

Retallack, G. J., 1988, Field recognition of paleosols, in J. Reinhardt and W. R. Sigleo, eds., Paleosols and weathering through geologic time: Geological Society of America Special Paper 216, p. 1–20.

Retallack, G. J., 1990, Soils of the past: Boston, Unwin Hyman, 520 p.

Ross, C. A., and J. R. P. Ross., 1988, Late Paleozoic transgressive-regressive deposition, in C. K. Wilgus, B. S. Hastings, C. G. St. C. Kendall, H. W. Posamentier, C. A. Ross, and J. C. Van Wagoner, eds., Sea-level changes— An integrated approach: SEPM Special Publication 42, p. 227–247.

Schumm, S. A., and F. G. Ethridge, 1994, Origin, evolution and morphology of fluvial valleys, in R. Dalrymple, R. Boyd, and B. Zaitlin, eds., Incised-valley systems: Origin and sedimentary sequences: SEPM Special Publication 51, p. 11–27.

Shanley, K. W., and P. J. McCabe, 1993, Alluvial architecture in a sequence-stratigraphic framework: A case history from the Upper Cretaceous from southern Utah, in S. S. Flint and I. D. Bryant, eds., The geologic modeling of hydrocarbon reservoirs and outcrop analogues: International Association of Sedimentologists Special Publication 15, p. 21–56.

Stout, W. 1931, Pennsylvanian Cycles in Ohio: Illinois Geological Survey Bulletin, v. 60, p. 195–216.

Sturgeon, M. T., 1968, and R. D. Hoare, 1968, Pennsylvanian brachiopods of Ohio: Ohio Geological Survey Bulletin, v. 63, p. 4–11.

Talbot, M. R., and P. A. Allen, 1996, Lakes, in H. G. Reading, ed., Sedimentary environments: Processes, facies, and stratigraphy: Cambridge, Blackwell Science, p. 83–124.

Tasch, P., 1980, Paleobiology of the invertebrates, data retrieval from the fossil record: New York, John Wiley and Sons, 975 p.

Vail, P. R., R. M. Mitchum, R. G. Todd, J. M. Widmier, S. Thompson, J. B. Sangree, J. N. Bubb, and W. G. Hatfield, 1977, Seismic stratigraphy and global changes in sea level, in C. E. Payton, ed., Seismic stratigraphy— Applications to hydrocarbon exploration: AAPG Memoir 26, p. 49–212.

Van Wagoner, J. C., H. W. Posamentier, R. M. Mitchum, Jr., P. R. Vail, J. F. Sarg, T. S. Loutit, and J. Hardenbol, 1988, An overview of the fundamentals of sequence stratigraphy and key definitions, in C. K. Wilgus, B. S. Hastings, C. G. St. C. Kendall, H. W. Posamentier, C. A. Ross, and J. C. Van Wagoner, eds., Sea-level changes: An integrated approach: SEPM Special Publication 42, p. 39–45.

Van Wagoner, J. C., R. M. Mitchum, K. M. Campion, and V. D. Rahmanian, 1990, Siliciclastic sequence stratigraphy in well logs, cores, and outcrops: Concepts for high-resolution correlation of time and facies: AAPG Methods in Exploration Series, No. 7, 55 p.

Veevers, J. J., and C. M. Powell, 1987, Late Paleozoic glacial episodes in Gondwanaland reflected in transgressive-regressive depositional sequences in Euramerica: Geological Society of America Bulletin, v. 98, p. 475–487.

Walker, R. G., 1992, Facies, facies models, and modern stratigraphic concepts, in R. G. Walker and N. P. James, eds., Facies models— Response to sea level change: Geological Association of Canada, p. 1–14.

Wanless, H. R., and F. P. Shepard, 1936, Sea level and climatic changes related to late Paleozoic cycles: Geological Society of America Bulletin, v. 47, p. 1177–1206.

Wanless, H. R., and J. M. Weller, 1932, Correlation and extent of Pennsylvanian cyclothems: Geological Society of America Bulletin, v. 43, p. 1003–1016.

Weller, J. M., 1930, Cyclic sedimentation of the Pennsylvanian period and its significance: Journal of Geology v. 38, p. 97–135.

Nadon, G. C., and R. R. Kelly, 2004, The constraints of glacial eustasy and low accommodation on sequence-stratigraphic interpretations of Pennsylvanian strata, Conemaugh Group, Appalachian basin, U.S.A., *in* J. C. Pashin and R. A. Gastaldo, eds., Sequence stratigraphy, paleoclimate, and tectonics of coal-bearing strata: AAPG Studies in Geology 51, p. 29–44.

2

The Constraints of Glacial Eustasy and Low Accommodation on Sequence-stratigraphic Interpretations of Pennsylvanian Strata, Conemaugh Group, Appalachian Basin, U.S.A.

Gregory C. Nadon
Department of Geological Sciences, Ohio University, Athens, Ohio, U.S.A.

Russell R. Kelly
Samson Resources, Tulsa, Oklahoma, U.S.A.

ABSTRACT

A major advantage in using sequence stratigraphy is the ability to predict facies changes in sequences using the systems tracts approach. Glacial eustasy, however, imposes unique constraints on the facies distribution within and between sequences that challenge the predictive utility of sequence-stratigraphic models. The asymmetric rates of ice volume fluctuation result in a sawtooth eustatic curve and the formation of rapid transgressions and slow regressions. When subsidence rates are high, the augmented rate of formation of accommodation space allows the deposition and preservation of complete sequences in which there may be a repeated vertical succession of facies. A low subsidence rate, combined with variations in the rates of eustasy and sedimentation, results in unpredictable lateral and vertical facies patterns.

Data from the upper Pennsylvanian (Westphalian D to Stephanian) Conemaugh Group in the distal foreland of the Appalachian basin illustrate this problem. Within sequences, even the limited predictive ability associated with process facies models in the highstand systems tract is compromised by the temporal variations in the eustatic component and the presence of incised channels and valleys formed during late highstand and lowstand times. Furthermore, temporal variations in the driving functions mean that the presence of a particular facies in a systems tract of one sequence (e.g., coal) cannot be used to predict its presence or absence in a similar tract of an adjacent sequence. The result is that there are stratigraphic intervals, such as the Conemaugh Group, in which the detailed lithostratigraphic data from core or outcrop cannot be confidently extended beyond a few tens of meters within a sequence and not at all between sequences.

INTRODUCTION

Sequence stratigraphy is now a widely used technique for interpreting sedimentary sequences in basins that have a marine or lacustrine component, in part because of the predictive ability of the technique. The partitioning of sediments into systems tracts has allowed an increased precision in the interpretation of the controls on facies distribution (Posamentier and Allen, 1999). The development of sequence stratigraphy from its predecessor, seismic stratigraphy, has been well documented (e.g., Van Wagoner et al., 1990; Miall, 1997); however, few discussions focus on the assumptions or boundary conditions that were inherited from the earlier technique. Seismic stratigraphy was originally developed to interpret thick sediment packages deposited on passive margins. The emphasis on the stratal geometry of the deposits provided the rationale to extend the technique to thinner, higher-frequency transgressive-regressive packages (Posamentier and Vail, 1988; Posamentier et al., 1988; Van Wagoner et al., 1990).

One of the important aspects of sequence stratigraphy is the recognition that it is the rate of formation of accommodation space, instead of simply the total accommodation space, that controls facies distribution. The rate of formation of accommodation space is interpreted to be a function of the rates of change of only three main variables: eustasy, subsidence, and sediment supply. Two of these variables, subsidence and sediment supply, commonly are assumed to be constant at the temporal scale of the sequence being considered (Posamentier and Vail, 1988). The assumption of constant tectonic subsidence is a logical consequence of the passive-margin tectonic setting and thickness of the strata in the original seismic data set. Similarly, variations in sediment supply were assumed to be minimal because of the long time interval in the lower-order sequences studied on passive margins. The constant rates of subsidence and sedimentation required eustatic fluctuations to explain relative sea level changes.

Sequence-stratigraphic models were eventually modified to take into account the facies patterns formed in ramp settings, but the primary assumptions were never modified (e.g., Van Wagoner et al., 1990; Posamentier and Allen, 1999). Does the application of high-resolution sequence stratigraphy to sediments deposited under different boundary conditions yield meaningful results? As a test of the effects of differing boundary conditions on the utility of high-resolution sequence stratigraphy, the model was applied to the late Pennsylvanian in the distal portion of the Appalachian basin.

The Pennsylvanian strata in the distal Appalachian basin show that the interpretation of relative sea level change incorporates variations in the magnitude of accommodation space and frequency of eustatic sea level change in addition to the rate of change of accommodation space. The combination of a low-accommodation setting and a high-frequency, large-magnitude eustatic driving function creates a complex pattern of erosional truncation, nondeposition, or postdepositional pedogenic alteration that severely limits the predictive ability of sequence stratigraphy. The result is that, although the facies in any one section may be interpreted in terms of relative sea level change, correlation of sequences from one section to another at length scales of more than a few meters is not possible in areas where total subsidence was limited.

THE DISTAL APPALACHIAN BASIN

The northern Appalachian foreland basin formed in response to the collision of the North American and African plates during the assembly of Pangea in the late Paleozoic (Figure 1A). The Pennsylvanian sediments are divided into six major lithostratigraphic groups that thin to the west as they onlap the Cincinnati arch (Donaldson et al., 1985; Figure 1B). The initial subdivision of the Pennsylvanian section was based largely on the presence or absence of coal. Each group contains different proportions of marine and nonmarine strata that record numerous transgressive-regressive events. The presence of fluvial deposits throughout the Pennsylvanian indicates that the Appalachian basin was overfilled throughout its history. Regional paleocurrent and paleomagnetic data show that rivers flowed to the northwest in a tropical environment (Donaldson et al., 1985; Opdyke and DiVenere, 1994). Deposition was also affected by eustatic sea level fluctuations on a range of temporal

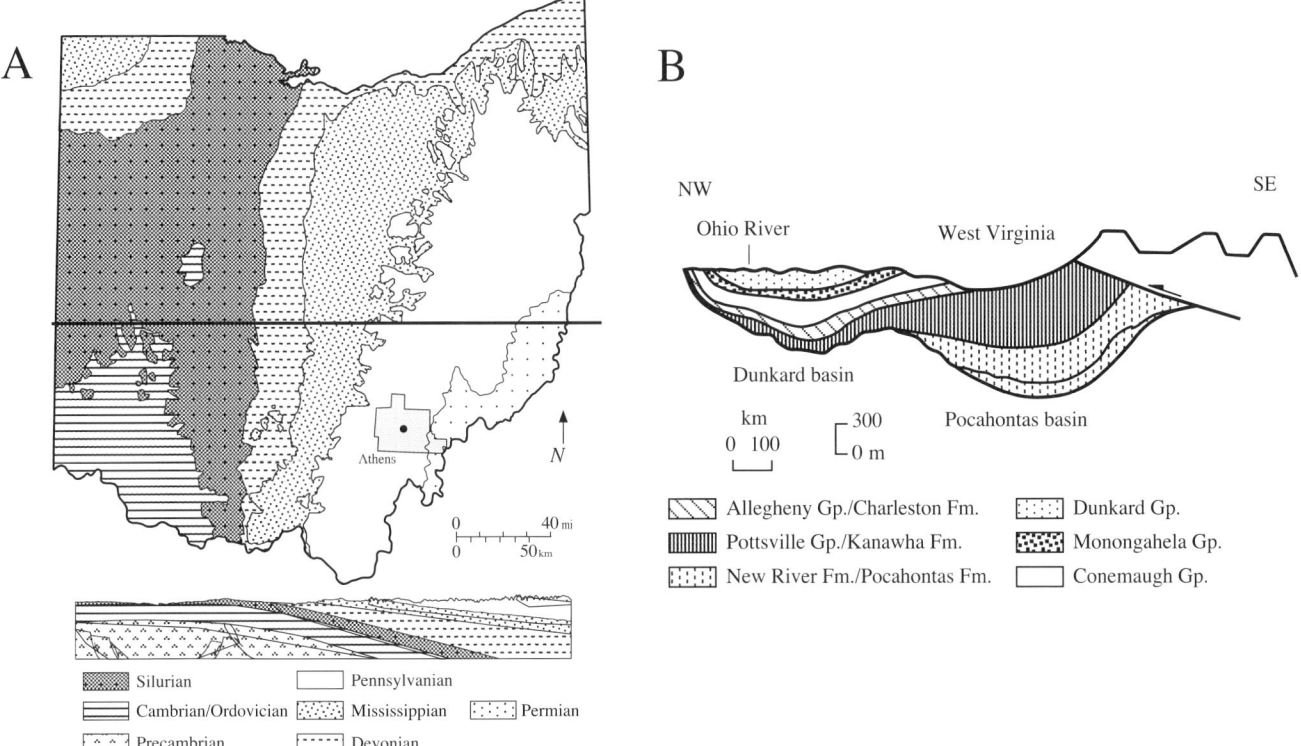

Figure 1. (A) Simplified geologic map and cross section through Ohio showing the position of Athens County on the distal margin of the Appalachian basin. Dip in southern Ohio averages approximately 0.3° to the southeast (Sturgeon, 1958). (B) Generalized cross section through the northern Appalachian foreland basin showing lithostratigraphic subdivisions (after Donaldson et al., 1985).

scales (Ross and Ross, 1987). The highest-frequency eustatic signal is attributed to the waxing and waning of continental ice sheets on Gondwana during this time interval (e.g., Heckel, 1994, 1995).

The complexity of the facies changes in each group led to the development of the cyclothem concept as an attempt to delineate mappable units. The cyclothem concept suggested that there was a recognizable succession of facies more or less related to water depth that occurred repeatedly in a vertical section (Udden, 1912; Wanless and Weller, 1932). However, it has been argued that this early facies model was flawed from its inception (Nadon, 1999), and even the type area contains no regular repetition of most of the facies (Kivet and Merrill, 1999).

The stratigraphic section in southeastern Ohio was deposited in a ramp setting of the distal Appalachian foreland basin (Figure 1B). The thickness of the Pennsylvanian section here is approximately 20% that of the equivalent strata in the proximal basin. Correlations in the distal basin have been established mainly from coal seam stratigraphy and a limited number of thin marine limestones (Condit, 1912; Stout et al., 1923; Sturgeon, 1958). Intrabasinal cor-

relations have been established primarily through conodont biostratigraphy (e.g., Merrill, 1974; Heckel 1986) and, to a lesser extent, by palynology (Peppers, 1996). Backstripping of the section in southern Ohio shows that the subsidence history of the distal portion of the basin is highly variable (Figure 2). Accumulation rates and, by inference, the rate of subsidence varied significantly through time, with the lowest rates occurring during deposition of the Conemaugh Group (Nadon et al., 1999). The consequence of these low subsidence rates, i.e., low accommodation space in a distal basin setting, is a boundary condition not present in the original sequence-stratigraphic model.

The interaction of long-term eustatic sea level changes with subsidence is evident from the distribution of marine bands in the section. There is an overall decrease in the number of marine units upsection in the Conemaugh Group (Figure 3) that corresponds to the drop in the third-order sea level curve (Figure 2). The sedimentary strata of the Conemaugh Group are divided into the lower Glenshaw and upper Casselman Formations in neighboring West Virginia (Martino et al., 1996) by a marine band named the Ames Limestone. The Ames is composed

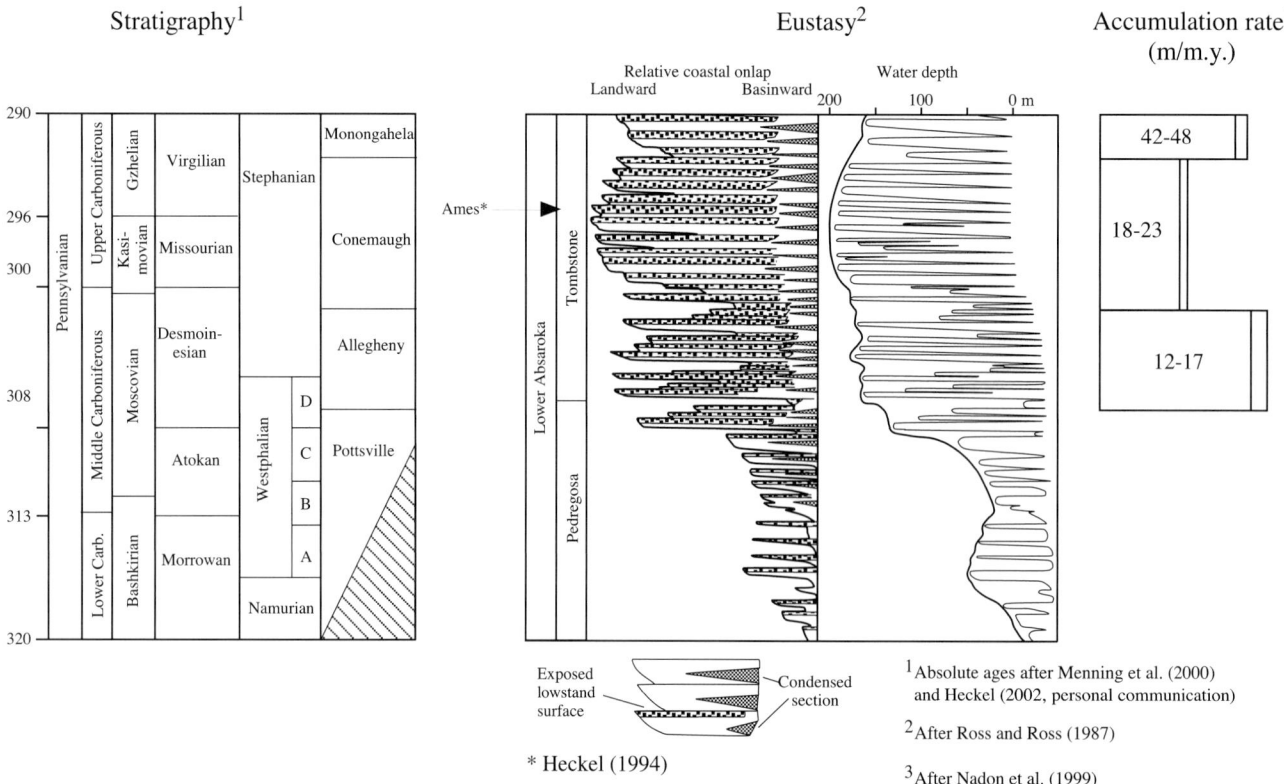

Figure 2. Variations in eustatic sea level for Pennsylvanian strata of North America and calculated accumulation rates for the Allegheny, Conemaugh, and Monongahela Groups in the study area. The eustatic curve is from data in the Midcontinent and Permian basin of Texas (after Ross and Ross, 1987). Radiometric ages after Menning et al. (2000) and P. H. Heckel (2002, personal communication). Accumulation rates were calculated from backstripping the Athens section in Athens County (Sturgeon, 1958) using the techniques of Angevine et al. (1990) (maximum values) and Allen and Allen (1990) (minimum values), with initial porosity estimates of Angevine et al. for the marine sediments and Nadon and Issler (1997) for the nonmarine strata.

of 0.3–3 m (1–10 ft) of packstone to grainstone (Merrill, 1993). This marine band (Virgilian) generally is considered to be the most extensive marker bed in the Conemaugh (Sturgeon, 1958), and therefore, the Ames interval should be the easiest to interpret in terms of high-resolution sequence stratigraphy. However, closely spaced, detailed measured sections in the vicinity of Athens, Ohio (Figure 4), show that the Ames interval is considerably more complex than previously described.

THE AMES INTERVAL

Description

In the study area (Figure 4), the Ames interval ranges in thickness from 0 to 9 m (0 to 30 ft). The base of the interval overlies red to gray blocky mudstones, which can exceed 8 m (26 ft) in thickness. The blocks, which range in size from 1 to 10 cm (0.40 to 4 in.) in

diameter, commonly exhibit slickensided surfaces. The mudstones also contain larger curved fractures, greater than 0.6 m (2 ft) in length, with well-developed slickensides. Small (<2 cm [<0.8 in.] diameter), irregularly shaped carbonate nodules occur dispersed in the mudstones.

These blocky mudstones are sharply overlain by dark gray, fissile shales that occasionally contain small mollusk shells (Figure 5). The dark gray shales are overlain by a complex series of light gray mudstones that locally contain thinner (<0.1 m [<0.3 ft] thick) beds of dark gray shales. Sharp-based sandstones of varying number and thickness abruptly overlie the mudstones. These sandstones range from very fine- to medium-grained and contain a wide variety of sedimentary structures. The sandstone in Figure 5 fines upward from a massive base to low-angle, laminated, hummocky cross-stratified, and parallel laminated sandstones; elsewhere, the sandstone is entirely wave-ripple cross-laminated.

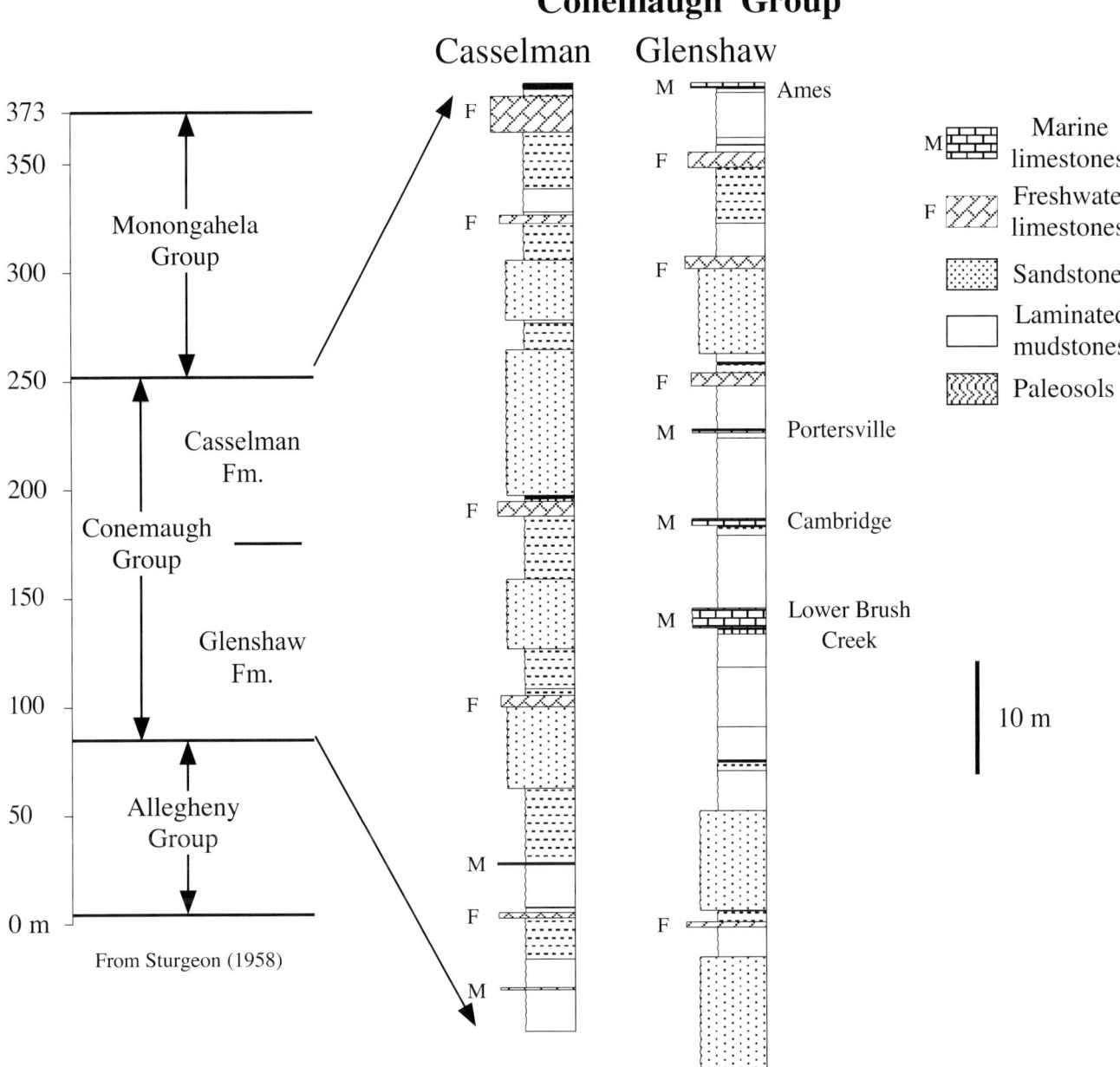

Figure 3. General section through the Conemaugh Group in the area of Athens, Ohio. Thicknesses are average values for the lithologies. The Conemaugh contains the last marine limestones in the Pennsylvanian of the Appalachian basin (after Sturgeon, 1958).

The Ames Limestone is typically a resistant bed containing brachiopods, crinoid fragments, gastropods, solitary corals, fusilinids, and conodonts scattered in a carbonate mudstone matrix (Sturgeon, 1958; Merrill, 1974). The base of the limestone commonly exhibits large branching burrows, probably *Thalassinoides*. In some locations, however, the Ames has a reddish hue at the top. Slabs that cut through the Ames at these locations reveal a mottling throughout the bed. Thin sections of the Ames where

this mottling occurs show recrystallization of the echinoid fragments, inclined patches of small siderite crystals, and thin organic layers.

Fissile shales, siltstones, and quartz-rich limestones that may contain marine mollusk shells near the base overlie the Ames Limestone. At one location, there are small (1–3 cm [0.4–1.2 in.]) clasts with black rinds scattered in thin layers within 0.3 m (1 ft) of the Ames upper contact. The laminated mudstones grade upward into another bed of blocky mudstones that

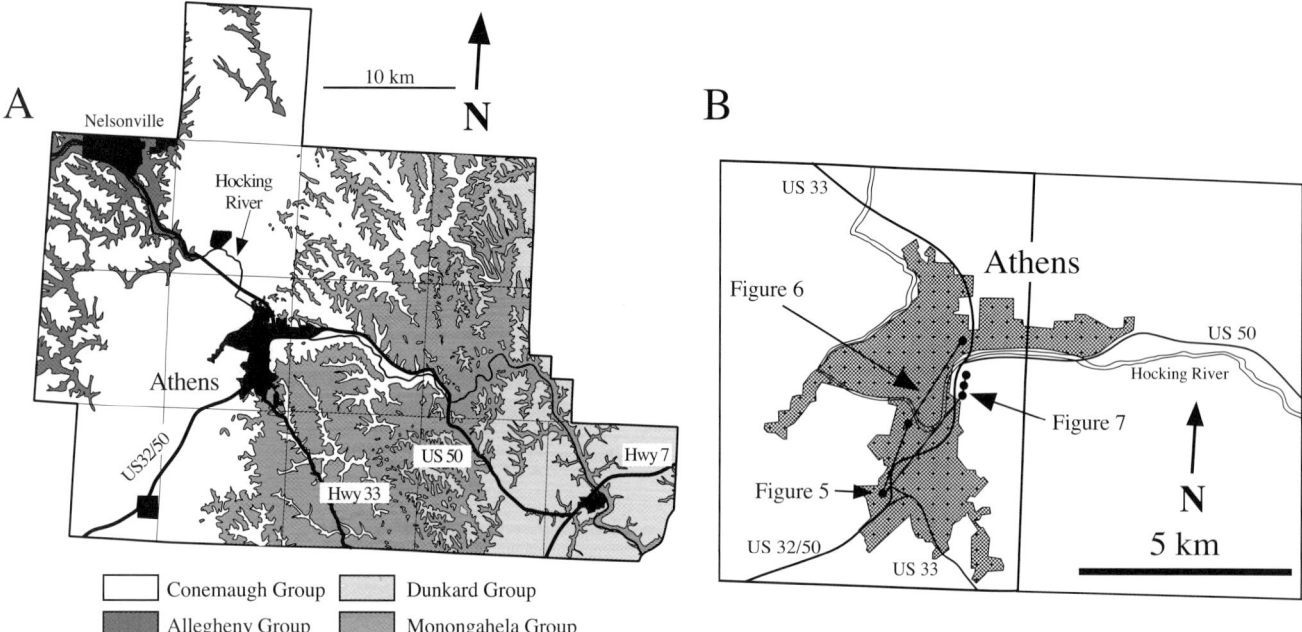

Figure 4. Location map of study area. (A) Generalized geological map of Athens County (after Sturgeon, 1958). (B) Location of measured sections described in the text.

caps the Ames interval. These upper blocky mudstones are similar in structure to those below the black shales, but are a mottled red-purple color and contain small carbonate concretions.

Correlation of these strata over a 4-km (2.5-mi) interval reveals that this modified interpretation of the Ames interval is far from complete. There are abrupt lateral changes of all facies in the interval at scales varying from a few meters to several kilometers (Figures 6, 7). The lower paleosol profile of the Ames sequence in the section labelled 682 on Figure 6 is marked by a much thinner paleosol, bright red in color, which overlies laminated sediments of a previous sequence not present in the other two sections. These lower sediments consist of laminated shales overlying an older paleosol that grade upward into a series of thickening- and coarsening-upward sandstones. The red paleosol disappears through erosion less than 70 m (230 ft) to the east prior to deposition of the basal sequence of Figure 5. Where the paleosol is absent, the laminated shales and siltstones below the Ames Limestone are uncommonly thick, but otherwise, the two sequences cannot be differentiated in the field. The Ames in the 682 locality shows no evidence of red mottling, but this coloration is obvious in the northernmost outcrop in Figure 6.

Additional complexity is evident from the sections in Figure 7. The dark gray shales and lighter gray mudstones overlie blocky gray mudstones as in Figure 5. However, the thick sandstone present in the 32/50 section (Figure 5) is replaced by thin fine-grained sandstone that may or may not be overlain by the Ames Limestone. Despite being thicker than average, the Ames Limestone in this location is present only as a discontinuous bed. The Ames contains a similar suite of marine fossils as the limestone in Figure 5, but at this location, the upper 0.2 m (0.7 ft) of the Ames is a mottled light red color, and the top 0.1 m (0.36 ft) in section 4 of Figure 7 has an unusual weathering pattern (Figure 8). Above the Ames is a series of mudstones, siltstones, and fine-grained sandstones. The lower mudstones are laminated and contain layers of concretions and pebbles with black rinds. The overlying mudstones and sandstones display a different pattern of fracturing than the underlying beds and lie on an erosion surface cut into the lower unit (Figure 8).

Elsewhere, the sediments above the thin sandstone are composed of mudstone with a coal deposited at the same stratigraphic position as the Ames Limestone. The mudstones above the sandstones have a markedly different fracture pattern than the laminated sediments below. Where the sandstone is absent, only the difference in fracture pattern marks the change in facies (Figure 9). Above the fractured facies in sections 1 to 4 are blocky mudstones, with coloration varying widely from green and gray to purple hues and with variable amounts of carbonate concretions.

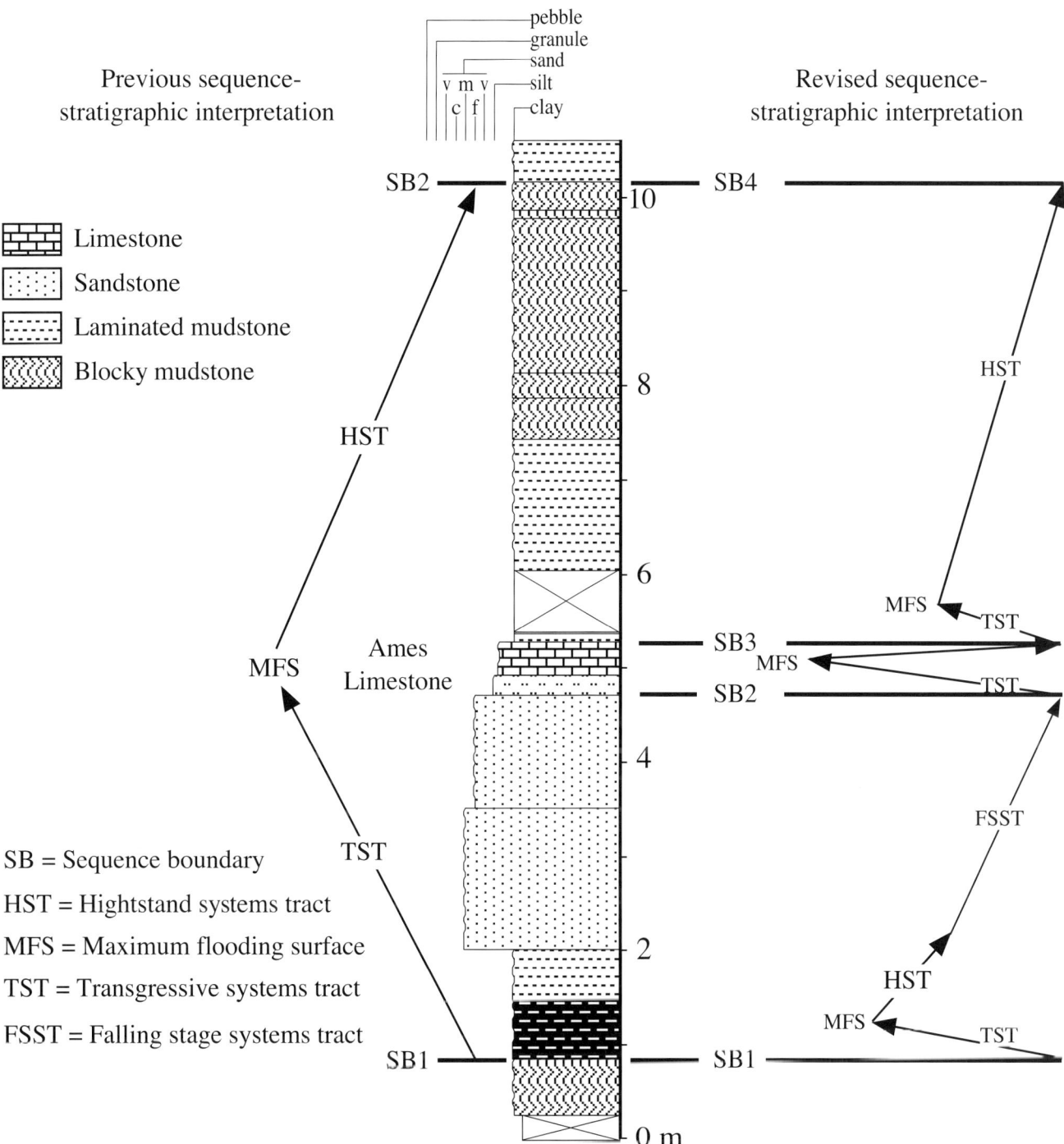

Figure 5. Detailed measured section through Ames interval along US Highway 32/50 at the junction with US Highway 33 south of Athens, Ohio. Sequence boundaries are placed at the contact of blocky mudstones (paleosols) and laminated sediments. TST intervals, which are thin in glacial-eustatic sequences, range from black fissile shale to marine limestones. The thick sandstone is interpreted as a forced regression shoreface (Kelly and Nadon, 2000).

The blocky mudstones are overlain by dark gray mudstones in sections 1, 2, and 3. South of section 4, most of the interval above a lens of conglomerate and coarse-grained sandstones replaces the Ames.

Interpretation

The blocky mudstone facies in the section are interpreted as paleosol profiles (Retallack, 1988) that were covered by marine shales following a transgression.

Figure 6. Cross section through the Ames interval showing variations in the basal paleosol. The lower sequence present in the 682 section is absent in both other sections. The thick paleosols present at both ends of the section probably contain multiple sequences. The thick sandstone in the Highway 32/50 section pinches out in the 682 section. The top of the Ames Limestone is reddened in the northern measured section.

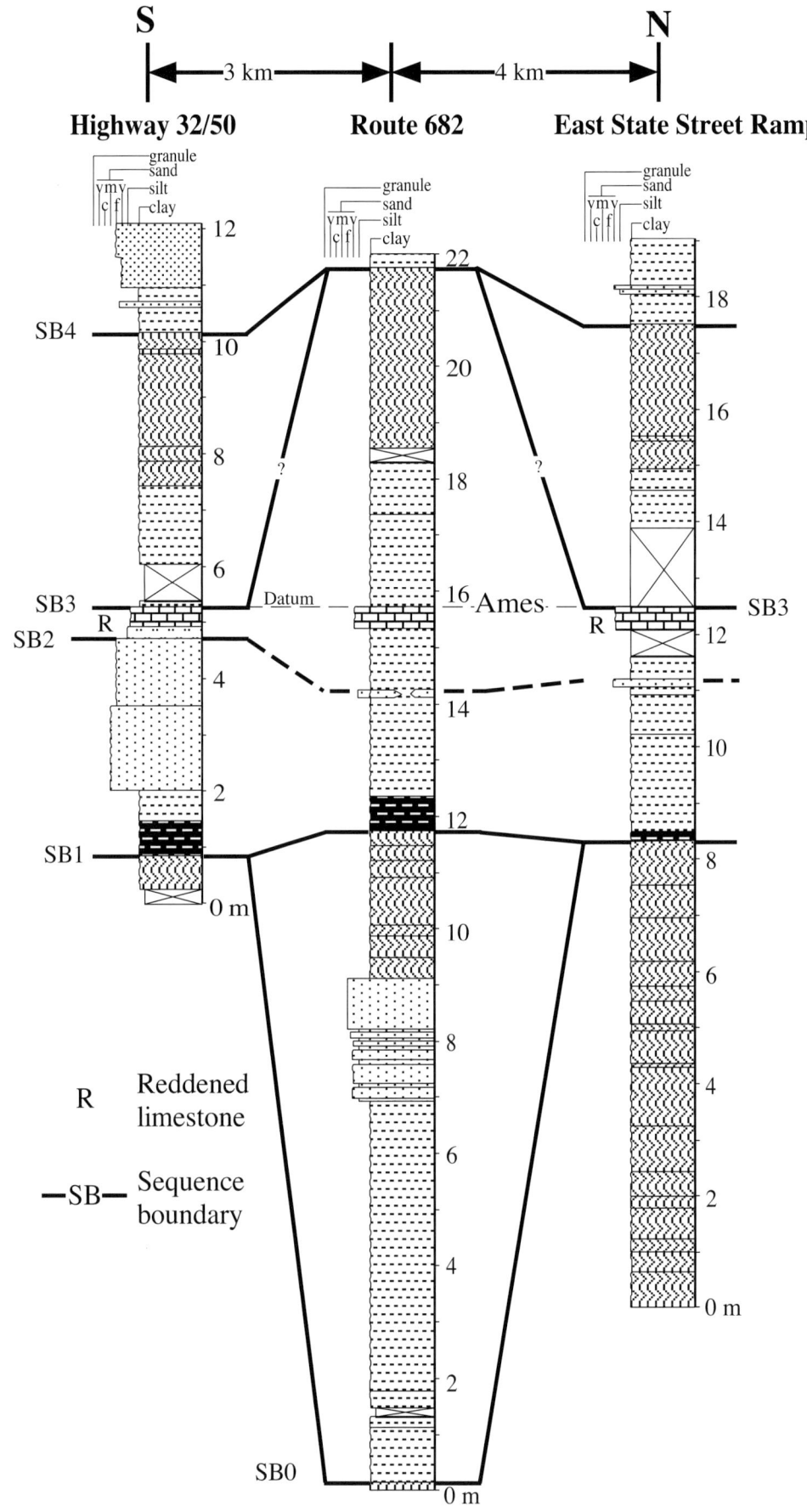

The light gray color of the basal paleosol suggests a gleyed profile, but it is not clear if the color was primary or a result of retrograde iron reduction following the transgression (Schutter and Heckel, 1985). The red coloration in the 682 section implies a strongly oxidizing environment. The carbonate concretions in the upper paleosol are interpreted as glaebules probably formed around roots. The laminated black shales represent an abrupt transgression. The lack of any material coarser than silt between the lower paleosol and the overlying laminated sediments means that a retrogradational parasequence set that is commonly associated with transgressions elsewhere (e.g., Van Wagoner et al., 1990) is absent. The increase in silt content in the overlying gray shales indicates that the black shales were deposited in relatively deep water; i.e., it contains a maximum flooding surface (MFS) interval. The presence of a thinning-upward succession of black

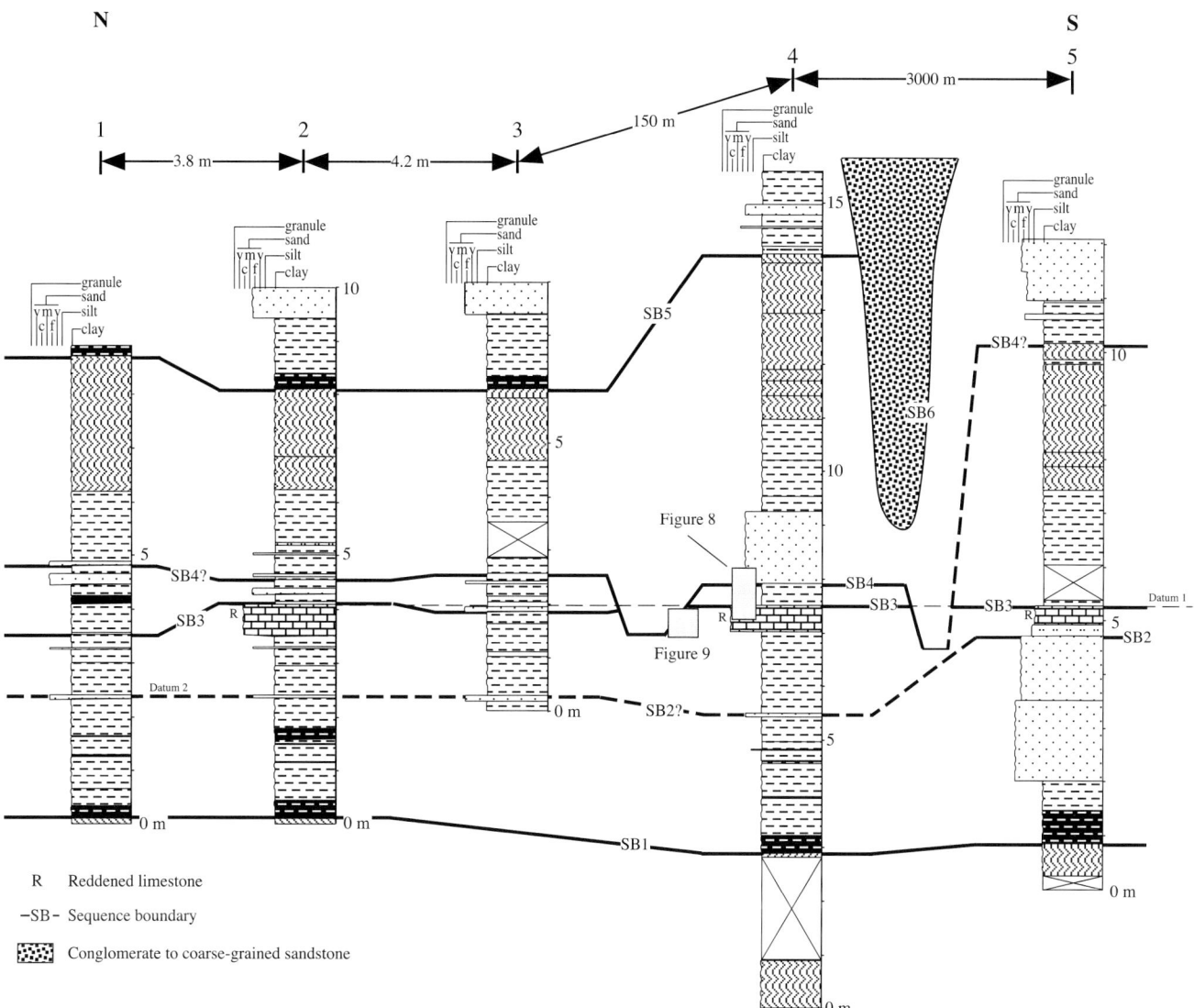

Figure 7. Cross section through the Ames interval along US Highway 32/50 showing lateral variations in facies at different length scales. Sections 1 to 3 are spaced only meters apart, but only the lower portions of the sections can be easily correlated. The Ames Limestone, which is reddened at the top, is present as a thin lens. The Ames is present in section 4 but absent 20 m (66 ft) to the north. Most of the section above the Ames is replaced by coarse-grained sandstone 70 m (230 ft) to the south. The section from Figure 4 is shown for reference. The locations of Figures 8 and 9 are shown near section 1.

shale beds is interpreted to represent a progradational parasequence set.

The variations in structures and thickness in the sandstone below the Ames in Figure 5 has been interpreted as a forced regression shoreface that records an additional drop in sea level in this interval (Kelly and Nadon, 2000; Kelly, 2002). The close proximity of the Ames Limestone above the sandstone is interpreted to represent another MFS interval. The Ames represents a more significant flooding event than the underlying black shales. The presence of the red mottling, the dissolution of the echinoid debris, and the possible root structures in the

Ames Limestone (e.g., Figures 6, 7) mean that this marine limestone has undergone subaerial exposure, forming an incipient terra rossa paleosol (e.g., Driese et al., 1998; Dickson et al., 2001) and, therefore, another sequence boundary.

The interpretation of the section between the paleosols generally has been one of increasing water depth to an MFS represented by the Ames Limestone followed by regression (left curve of Figure 5). The revised sequence-stratigraphic interpretation contains not one but three complete sequences (right curve of Figure 5). This new interpretation is consistent with a glacial-eustatic setting, where the MFS interval

Figure 8. Close-up of the top of the Ames Limestone in section 4 of Figure 7. The upper 0.2 m (0.6 ft) of the limestone is reddened and weathers into small fragments. Note the abrupt transition above the Ames from laminated to hackly fractured siltstones. This change is interpreted to represent a sequence boundary (SB3).

is expected to occur close to the basal sequence boundary (Nadon, 1999). The transgressive systems tract (TST) should be thin, and the highstand systems tract (HST) should be thick.

The lower portion of sections 1 to 4 in Figure 7 is similar to that of Figure 5, although it is unclear if the thin sandstone is a distal equivalent of the forced regression shoreface below the Ames Limestone. The red color of the Ames depicted in Figure 7 shows that subaerial exposure of the limestone was common in the study area. The transgressive event that covered the reddened Ames Limestone is truncated by an erosional surface (Figure 8) and is not present in all sections. The presence of multiple intersecting sequence boundaries and the lateral changes in facies makes correlation difficult even between closely spaced sections that contain marine strata. For example, the lateral facies variations above the Ames Limestone in Figure 7 are not the result of facies variations within sequences but of facies changes between different sequences.

A total of six different sequences are interpreted to be present based on the facies patterns within the sections in Figure 7. The presence of different facies representing the MFS interval (Figure 10) and the variations in exposure surface suggest that the magnitude of sea level change and the duration of exposure was different in each sequence. These variations do not provide any indication of sequence duration. However, exposure of the Ames Limestone following a drop in sea level suggests a long-term drawdown in sea level of sufficient magnitude that we interpret the sequence boundaries in this interval to be of the same order and produced by 100 ka glacial-eustatic sea level changes.

CONTROLS ON SEQUENCE DEVELOPMENT AND PRESERVATION

One of the major contributions of sequence stratigraphy is the ability to predict facies patterns beyond

Figure 9. The juxtaposition of laminated and hackly fractured mudstones approximately 20 m (66 ft) north of section 4 in Figure 8. The contact marks a sequence boundary that can be missed easily in the field without continuous exposure.

larger, and the preservation of facies in each sequence is consequently more limited in extent. The deposition of facies may be controlled by the rate of formation of accommodation space, but the preservation potential of the facies is a function of the total amount of subsidence relative to the frequency of relative sea level change (Figure 11B).

The problem of sequence correlation created by low facies preservation potential in the study area is compounded by both the frequency and variation in magnitude of the eustatic fluctuations. Glacial eustasy, by definition, is a result of the waxing and waning of continental ice sheets. Successive ice sheets incorporate and release different volumes of water. Therefore, although each eustatic cycle may have similar rates of transgression and regression, the magnitude of each varies (Figure 10). This variation in magnitude means that each sequence boundary in a section likely represents very different exposure times in the distal basin. The results of this difference in exposure are (1) variations in the magnitude of incision of river systems, forming either incised channels or valleys between sequences, and (2) the intensity of paleosol formation on interfluves varying between sequences.

The frequency of glacial-eustatic fluctuations poses a major problem for the correlation of sequences. Miall (1992) demonstrated that the more sequences present in a given period, the lower the probability of correctly correlating sections. His example used 40 sequences. If the periodicity of the ice ages in the late Paleozoic was similar to the Cenozoic, then the duration of Pennsylvanian sequences would have ranged from 100 to 400 ka (Imbrie et al., 1993). Based on recent estimates of stage durations of the Pennsylvanian (Menning et al., 2000; P. H. Heckel, 2002,

a simple application of Walther's Law using unconformities as the bounding surfaces. The concept that the rate of formation of accommodation space, instead of simply the amount of accommodation space, controls the deposition of facies permits the subdivision of a sequence into systems tracts. The identification of these systems tracts allows for more accurate prediction of lateral and vertical facies trends. However, the data presented above show that prediction of lateral or vertical facies trends in at least some of the Pennsylvanian strata, even from sections spaced only meters apart, is not possible. This limitation is caused by the low total accommodation space formed in this portion of the basin, combined with the magnitude and frequency of the glacial-eustatic sea level fluctuations.

The low total accommodation space in a distal foreland setting means that small variations in subsidence rate have a large effect on the preservation of facies within sequences. The original models of seismic and sequence stratigraphy were developed in regions where the total accommodation space was large. In those cases, the ratio of depth of incision occurring on a sequence boundary to the thickness of any sequence was small. Even when the conceptual model was extended to ramp settings, the depth of incised channels or valleys was viewed as considerably less than the thickness of a sequence (Figure 11A). In regions of low subsidence, this ratio is much

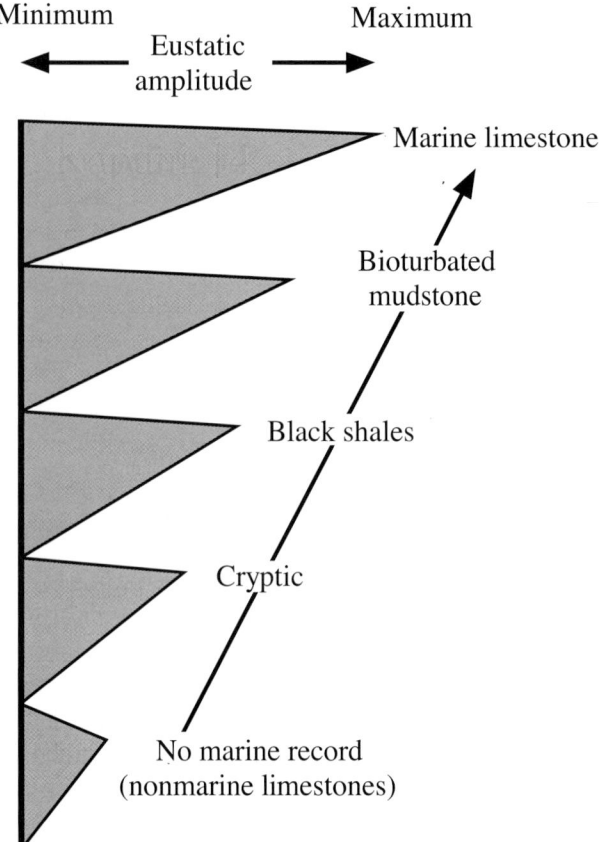

Minimum   Maximum

Eustatic amplitude

Marine limestone

Bioturbated mudstone

Black shales

Cryptic

No marine record
(nonmarine limestones)

Figure 10. Variation in the marine facies (MFS) in a vertical section at a single locality solely caused by variations in magnitude of eustatic sea level rise and fall based on the measured sections in this study. The different facies comprising the MFS interval simply can be a function of the magnitude of sea level rise (assuming equal rates of sediment supply and subsidence during each cycle) instead of being a product of different orbital periodicities.

personal communication), each stage could contain between 6 and 70 sequences (Table 1). The combination of high numbers of sequences, low total accommodation space, and high glacial-eustatic frequency makes intrabasin correlation of individual sequences extremely difficult and interbasin correlation impossible.

DISCUSSION

Although still the subject of debate, the calculated frequency of glacial eustasy in late Paleozoic in North America ranges between 100 and 400 ka in areas of reasonably complete sections (Heckel, 1980, 1986; Klein, 1990; Maynard and Leeder, 1992; Miller and Eriksson, 1999). Estimation of cycle frequency has employed techniques ranging from division of the estimated duration of an interval by the number of cycles to spectral analysis. There are problems inherent in each method, but the goal is important, because if the periodicity of Pennsylvanian sediments is 100 ka instead of 400 ka, then the sections used for the analysis are far less complete than previously thought (Table 1), and that has implications for high-resolution correlation between sections.

Heckel (1986) calculated ranges of periodicities by dividing the section into major, intermediate, and minor cycles based on the variations in lithology of the maximum flooding zones. Others have used all preserved cycles in a section (e.g. Connolly and Stanton, 1992; Miller and Eriksson, 1999). The main problems with dividing a section into a cycle hierarchy are the assumptions that all major glacial-eustatic events leave the same footprint in a basin, and that only glacial eustasy is recorded in the sediments. The a priori assumption that the glaciations that formed major cycles were of similar magnitude each time the ice sheets formed is not supported by data in the Pleistocene record (Imbrie et al., 1992, 1993) and is therefore unlikely in the late Paleozoic. In addition, repetitive patterns of sedimentation can be formed with periodicities within Milankovitch frequencies from different processes. Differentiating among causes of cyclic sedimentation requires more sophisticated mathematical analysis (Algeo and Wilkinson, 1988). Finally, the presence of paleosols shows that exposure of the Midcontinent, Illinois, and Appalachian basins occurred at numerous times, creating the possibility of missing sections. Simple division of the number of cycles by total time elapsed, therefore, will only yield the minimum possible frequency; the actual frequency may be higher. When more rigorous statistical approaches are applied, the frequency of sea level variation is found to be closer to 100 ka (Maynard and Leeder, 1992).

High-frequency variations in sea level are consistent with observations from Quaternary sections in which glacial-eustatic sea level changes have occurred, with a frequency of 100 ka over the past 600–800 ka (Imbrie et al., 1992, 1993). Despite theoretical predictions, there is no obvious record of the 400-ka eccentricity signal in late Quaternary sediments. An elegant solution to the "400-ka problem" of Imbrie and Imbrie (1980) was proposed by Riall (1999), which states that the eccentricity signal acts as a carrier wave that is detectable only as side frequencies in the 100-ka precession signal (Riall, 1999, Figures 1, 4). Riall's hypothesis could explain the higher-frequency signals in Late Carboniferous strata. The Carboniferous records probably contain some combination of

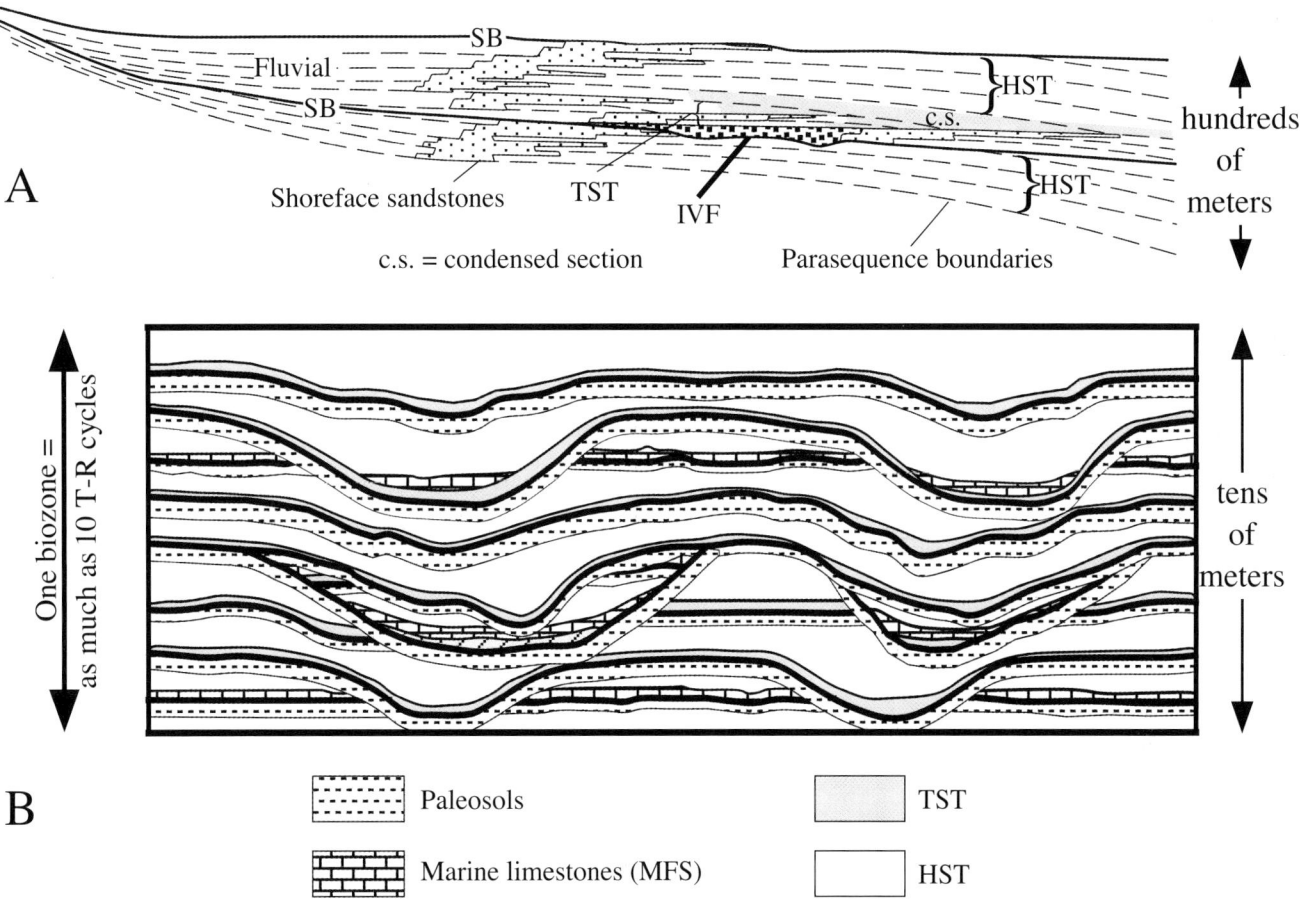

Figure 11. The role of total accommodation space in the preservation and correlation of sequences. (A) The model of ramp sedimentation in a proximal foreland basin presented in Van Wagoner et al. (1990). Note that the incisions present on the sequence boundaries are small compared to the thickness of the section. (B) Heckel et al. (1998) noted that regions of low total accommodation space allow incision during lowstand, which removes portions of underlying sequences, causing problems in lateral correlation. In addition, different duration of subaerial exposure related to variations in eustatic magnitude creates different depths of pedogenic modification of facies. The presence of as much as 10 cycles in a single conodont zone makes it difficult to impossible to differentiate between the carbonates of the MFS intervals (e.g., Miall, 1992).

intervals during which the 100- and 400-ka signals are dominant. If so, then the number of possible glacial-eustatic sequences ranges between 230 and 58 (100 and 400 ka). If all the recognized transgressive-regressive events were the result of glacial eustasy, then the sections described from the Midcontinent, which has the most complete record, ranges from 20 to 100% complete (Table 1). The implication is that a large number of sequences are missing or unrecognized in Pennsylvanian sections in North America.

The presence of paleosol horizons in all Pennsylvanian sections in North America leaves open the possibility of missing cycles. This is consistent with the location of the basins on continental crust instead of a deep-sea setting and with the variation in subsidence rates in the basins. The application of high-resolution sequence stratigraphy to the Ames interval shows that there is also the potential for a far larger number of sequences in the Appalachian basin sections than previously estimated. The only way to differentiate the sequences is through biostratigraphic evidence from MFS intervals. However, the conodont zones that provide the highest resolution still span approximately 1 m.y. (Merrill, 1974). It is possible for as much as 10 sequences to be present in each zone (Figure 10B). Because the preservation of every sequence is unlikely, it becomes impossible to correlate even closely spaced sections with a high degree of confidence. The presence of missing sections and the large number of sequences means that the problems of detailed intrabasinal correlations raised by Heckel et al. (1998), and illustrated in Figure 5,

Table 1. Comparison of the Reported Number and Duration of Pennsylvanian of Cycles from the Mid-continent and Western North America.

Basin (Author)	Interval	Duration (m.y.)	Number of Cycles Reported	Completeness of Section (%)	
				100 ka/cycle	400 ka/cycle
Midcontinent (Ross and Ross, 1987)	Atokan–Virgilian	21.5	54	25	100
Midcontinent (Ross and Ross, 1987)	Desmoinesian	7	22	31	126
Midcontinent (Connolly and Stanton, 1992)	Desmoinesian	7	29	41	161
Arizona (Connolly and Stanton, 1992)	Desmoinesian	7	41	59	228
Midcontinent (Heckel, 1986)	Missourian	2.5	17	68	272
Midcontinent (Ross and Ross, 1987)	Missourian	2.5	5	20	83

extend to interbasinal correlation of the sections as well (Miall, 1992).

CONCLUSIONS

The detailed measured sections of the interval surrounding a single marine band in the Conemaugh Group shows that (1) the number of sequences is larger than previously supposed, and (2) the accurate correlation of these sequences, even between sections spaced only a few meters apart, is problematic. The combination of a high-frequency glacial-eustatic driving function and low total accommodation space results in variable rates of duration and extent of each marine inundation of the basin, and in varying amounts of exposure during subsequent late highstands and lowstands. The dissection of each sequence in the distal basin by fluvial processes during low sea levels created a mosaic of intersecting sequence boundaries over a wide range of spatial scales. The result was partial preservation of most, if not all, the transgressive-regressive events in this region. Correlation of high-frequency sequences in intervals or areas of lower total accommodation space in the Appalachian or other basins is rendered difficult to impossible by the high-frequency glacial-eustatic sea level changes. The large number of potential sequences possible, the formation of sequences below the capabilities of present biostratigraphic control, and the lack of complete sections in all the Pennsylvanian basins prevent high-resolution interbasinal correlation.

ACKNOWLEDGMENTS

Phil Heckel, John Calder, Jack Pashin, and D. Sack provided useful critiques of an earlier version of the paper.

REFERENCES CITED

Algeo, T. J., and B. Wilkinson, 1988, Periodicity of mesoscale sedimentary cycles and the role of Milankovitch orbital modulation: Journal of Geology, v. 96, p. 313–322.

Allen, P. A., and J. R. L. Allen, 1990, Basin analysis: Principles and applications: Boston, Blackwell Scientific Publications, 451 p.

Angevine, C. L., P. L. Heller, and C. Paola, 1990, Quantitative sedimentary basin modeling: AAPG Short Course Note Series 32, 132 p.

Condit, D. D., 1912, The Conemaugh Formation in Ohio: Ohio Geological Survey Bulletin, 4th Series, v. 17, 363 p.

Connolly, W. M., and R. J. Stanton Jr., 1992, Interbasinal cyclostratigraphic correlation of Milankovitch band transgressive-regressive cycles: Correlation of Desmoinesian–Missourian strata between southeastern Arizona and the Midcontinent of North America: Geology, v. 20, p. 999–1002.

Dickson, J. A. D., A. H. Saller, and O. Hiroshi, 2001, Recognition of subaerial exposure through the style of echinoderm preservation: AAPG Annual Convention Program, v. 10, p. A50.

Donaldson, A. C., J. J. Renton, and M. W. Presley, 1985, Pennsylvanian deposystems and paleoclimates of the Appalachians: International Journal of Coal Geology, v. 5, p. 167–193.

Driese, S. G., M. R. Caudill, and K. Srinivasan, 1998, Late Mississippian to early Pennsylvanian paleokarst in east-central Tennessee: Field, petrographic, and stable isotope evidence: Southeastern Geology, v. 37, p. 189–204.

Heckel, P. H., 1980, Paleogeography of eustatic model for deposition of midcontinent Upper Pennsylvanian cyclothems, in T. D. Fouch and E. R. Magathan, eds., Paleozoic paleogeography of the west-central United States: SEPM, Rocky Mountain Section, Rocky Mountain Paleogeography Symposium 1, p. 197–215.

Heckel, P. H., 1986, Sea-level curve for Pennsylvanian eustatic marine transgressive-regressive depositional cycles along Midcontinent outcrop belt, North America: Geology, v. 14, p. 330–334.

Heckel, P. H., 1994, Evaluation of evidence for glacio-eustatic control over marine Pennsylvanian cyclothems in North America and consideration of possible tectonic effects, in J. M. Dennison and F. R. Ettensohn, eds., Tectonic and eustatic controls on sedimentary cycles: SEPM Concepts in Sedimentology and Paleontology, v. 4, p. 65–87.

Heckel, P. H., 1995, Glacial-eustatic base level-climatic model for the middle to late Pennsylvanian coal bed formation in the Appalachian basin: Journal of Sedimentary Research, v. 65B, p. 348–356.

Heckel, P. H., M. R. Gibling, and N. R. King, 1998, Stratigraphic model for glacial-eustatic Pennsylvanian cyclothems in highstand nearshore detrital regimes: Journal of Geology, v. 106, p. 373–383.

Imbrie, J., and J. Z. Imbrie, 1980, Modeling the climatic response to orbital variations: Science, v. 207, p. 943–953.

Imbrie, J. et al., 1992, On the structure and origin of major glaciation cycles: 1. Linear responses to Milankovitch forcing: Paleoceanography, v. 7, p. 701–738.

Imbrie, J. et al., 1993, On the structure and origin of major glaciation cycles: 2. The 100,000-year cycle: Paleoceanography, v. 8, p. 699–735.

Kelly, R. R., 2002, Lateral variability of the Ames interval, Conemaugh Group (Pennsylvanian), Athens and Morgan Counties, Ohio: M.S. thesis, Ohio University, 121 p.

Kelly, R., and G. C. Nadon, 2000, Complexity of the Ames sequence: Preservation of a forced-regression shoreface within the upper Pennsylvanian, Appalachian basin: Geological Society of America Abstracts with Programs, v. 32, no. 3, p. A-132.

Kivet, S. J., and G. K. Merrill, 1999, Stratal order in Pennsylvanian cyclothems: Geological Society of America Annual Meeting Abstracts with programs, v. 31, no. 7, p. A-183.

Klein, G. DeV., 1990, Pennsylvanian time scale and cycle periods: Geology, v. 18, p. 455–457.

Martino, R. L., M. A. McCullough, and T. L. Hamrick, 1996, Stratigraphic and depositional framework of the Glenshaw Formation (late Pennsylvanian) in central Wayne County, West Virginia: Southeastern Geology, v. 36, p. 65–83.

Maynard, J. R., and M. R. Leeder, 1992, On the periodicity and magnitude of Late Carboniferous glacio-eustatic sea-level changes: Journal of the Geological Society (London), v. 149, p. 303–311.

Menning, M., D. Weyer, G. Drozdzewski, H. W. J. Amerom, and I. Wendt, 2000, A Carboniferous time scale 2000: Discussion and use of geological parameters as time indicators from Central and Western Europe: Geologisches Jahrbuch, Hanover, v. A156, p. 3–44.

Merrill, G. K., 1974, Pennsylvanian conodont localities in northeastern Ohio: Ohio Department of Natural Resources, Division of Geological Survey, Guidebook, no. 3, 24 p.

Merrill, G. K., 1993, Late Carboniferous paleoecology along a tectonically active basin margin: Ames Member near Huntington, West Virginia: Southeastern Geology, v. 33, p. 111–129.

Miall, A. D., 1992, The Exxon global cycle chart: An event for every occasion?: Geology, v. 20, p. 787–790.

Miall, A. D., 1997, The geology of stratigraphic sequences: Springer-Verlag, New York, 433 p.

Miller, D. J., and K. A. Eriksson, 1999, Linked sequence development and global climate change: the upper Mississippian record in the Appalachian basin: Geology, v. 27, p. 35–38.

Nadon, G. C., 1999, Ice house sequence stratigraphy: The constraints of glacial eustasy on Pennsylvanian sedimentation: Geological Society of America Annual Meeting Abstracts with Programs, v. 31, no. 7, p. A-182.

Nadon, G. C., and D. R. Issler, 1997, The compaction floodplain sediments: What's wrong with this picture? Geoscience Canada, v. 10, p. 38–42.

Nadon, G. C., M. Cobb, and J. P. Smith, 1999, Partitioning the controls on rates of formation of accommodation space in a distal foreland basin: AAPG Annual Meeting Abstracts with Program, v. 8, p. A-98.

Opdyke, N. D., and V. J. DiVenere, 1994, Paleomagnetism and Carboniferous climate, in C. B. Cecil and N. T. Edgar, eds., Predictive stratigraphic analysis—Concept and application: United States Geological Survey Bulletin, v. 2110, p. 8 10.

Peppers, R. A., 1996, Palynological correlation of major Pennsylvanian (Middle and Upper Carboniferous) chronostratigraphic boundaries in the Illinois and other coal basins: Geological Society of America Memoir 188, 111 p.

Posamentier, H. W., and P. R. Vail, 1988, Eustatic controls on clastic deposition II—Sequence and systems tract models, in C. K. Wilgus, B. S. Hastings, C. G. St. C. Kendall, H. W. Posamentier, C. A. Ross, and J. C. Van Wagoner, eds., Sea-level changes: An integrated approach: SEPM Special Publication 42, p. 125–154.

Posamentier, H. W., and G. P. Allen, 1999, Siliciclastic sequence stratigraphy—Concepts and applications: SEPM Concepts in Sedimentology and Paleontology, v. 7, 204 p.

Posamentier, H. W., M. T. Jervey, and P. R. Vail, 1988,

Eustatic controls on clastic deposition I— Conceptual framework, *in* C. K. Wilgus, B. S. Hastings, C. G. St. C. Kendall, H. W. Posamentier, C. A. Ross, and J. C. Van Wagoner, eds., Sea-level changes: An integrated approach: SEPM Special Publication 42, p. 109–124.

Retallack, G. J., 1988, Field recognition of paleosols, *in* J. Reinhardt and W. R. Sigleo, eds., Paleosols and weathering through geologic time: Principles and applications: Geological Society of America Special Paper 216, p. 1–20.

Riall, J. A., 1999, Pacemaking the ice ages by frequency modulation of Earth's orbital eccentricity: Science, v. 285, p. 564–568.

Ross, C. A., and J. R. Ross, 1987, Late Paleozoic sea levels and depositional sequences, *in* C. A. Ross and D. Haman, eds., Timing and depositional history of eustatic sequences; Constraints on seismic stratigraphy: Cushman Foundation Special Publication 24, p. 137–149.

Schutter, S. R., and P. H. Heckel, 1985, Missourian (early late Pennsylvanian) climate in mid-continent North America: International Journal of Coal Geology, v. 5, p. 111–140.

Stout, W. B., R. T. Stull, W. J. McCaughey, and D. J. Demorest, 1923, Coal formation clays of Ohio: Ohio Geological Survey Bulletin, 4th Series, v. 26, 588 p.

Sturgeon, M. T., 1958, The geology and mineral resources of Athens County, Ohio: Department of Natural Resources, Division of Geological Survey, Bulletin, v. 57, 600 p.

Udden, J. A., 1912, Geology and mineral resources of the Peoria quadrangle: U.S. Geological Survey Bulletin, v. 506, 103 p.

Van Wagoner, J. C., R. M. Mitchum, K. M. Campion, and V. D. Rahmanian, 1990, Siliciclastic sequence stratigraphy in well logs, core and outcrops: AAPG Methods in Exploration Series 7, 55 p.

Wanless, H. R., and J. M. Weller, 1932, Correlation and extent of Pennsylvanian cyclothems: Geological Society of America Bulletin, v. 43, p. 1003–1016.

Flores, R. M., 2004, Coal buildup in tide-influenced coastal plains in the
Eocene Kapuni Group, Taranaki Basin, New Zealand, *in* J. C. Pashin and
R. A. Gastaldo, eds., Sequence stratigraphy, paleoclimate, and tectonics
of coal-bearing strata: AAPG Studies in Geology 51, p. 45–70.

3

Coal Buildup in Tide-influenced Coastal Plains in the Eocene Kapuni Group, Taranaki Basin, New Zealand

Romeo M. Flores
U.S. Geological Survey, Denver, Colorado, U.S.A.

ABSTRACT

The Eocene Kapuni Group in the Taranaki Basin, New Zealand, consists of the coal-bearing Kaimiro and Mangahewa Formations. These formations contain alternating cycles of stacked, coarsening-upward, marine-shoreface mudstone, siltstone, and sandstone, or parasequence sets laterally interfingering with fluvial-tidal coal, carbonaceous shale, mudstone, siltstone, sandstone, and conglomeratic sandstone. Regional erosion surfaces, or sequence boundaries, are found between these formations and reflect drops in relative base level with the paleoshoreline regressing several tens of kilometers. The tide-influenced coastal plains were formed during the intermittent transgressions of paleoshorelines to the south and southwest of the Taranaki Basin. Coal buildup in tide-influenced coastal marshes and alluvial belt mires was controlled by third- and fourth-order fluctuations of sea level, changes in depositional environments, basin subsidence, and accommodation space. Here, coal beds are vertically stacked and thicken upward behind landward-stepping marine-shoreface sandstones or parasequence sets.

INTRODUCTION

The coal-bearing Eocene Kapuni Group is a proven exploration play for oil, condensate, and gas in the Taranaki Basin in the North and South Islands of New Zealand (Figure 1), with production from sandstone reservoirs in 19 onshore and offshore fields (Cook and Gregg, 1997). The successful development of hydrocarbons in the Taranaki Basin has resulted in a large number of cores and geophysical logs of the Kapuni Group (Figure 2), and these have served as a basis for this study.

Several workers have studied cores and related geophysical logs of the Kapuni Group from various fields in the basin and have interpreted the depositional environments in the context of sequence stratigraphy. In general, the Kapuni Group has been interpreted as deposits of lower coastal plain and marginal marine environments (Beggs and Pocknall, 1992; Beggs et al., 1992; Pocknall et al., 1992; Flores et al., 1993; Bryant et al., 1994; King and Thrasher, 1996). A particular focus has been on the Kaimiro and Mangahewa Formations in various oil and gas fields in the basin (see Figure 1).

Figure 1. Map showing location of the wells studied in the Maui and Kapuni fields and Toru 1 well, Taranaki Basin, New Zealand. The map also shows the structural setting of the basin.

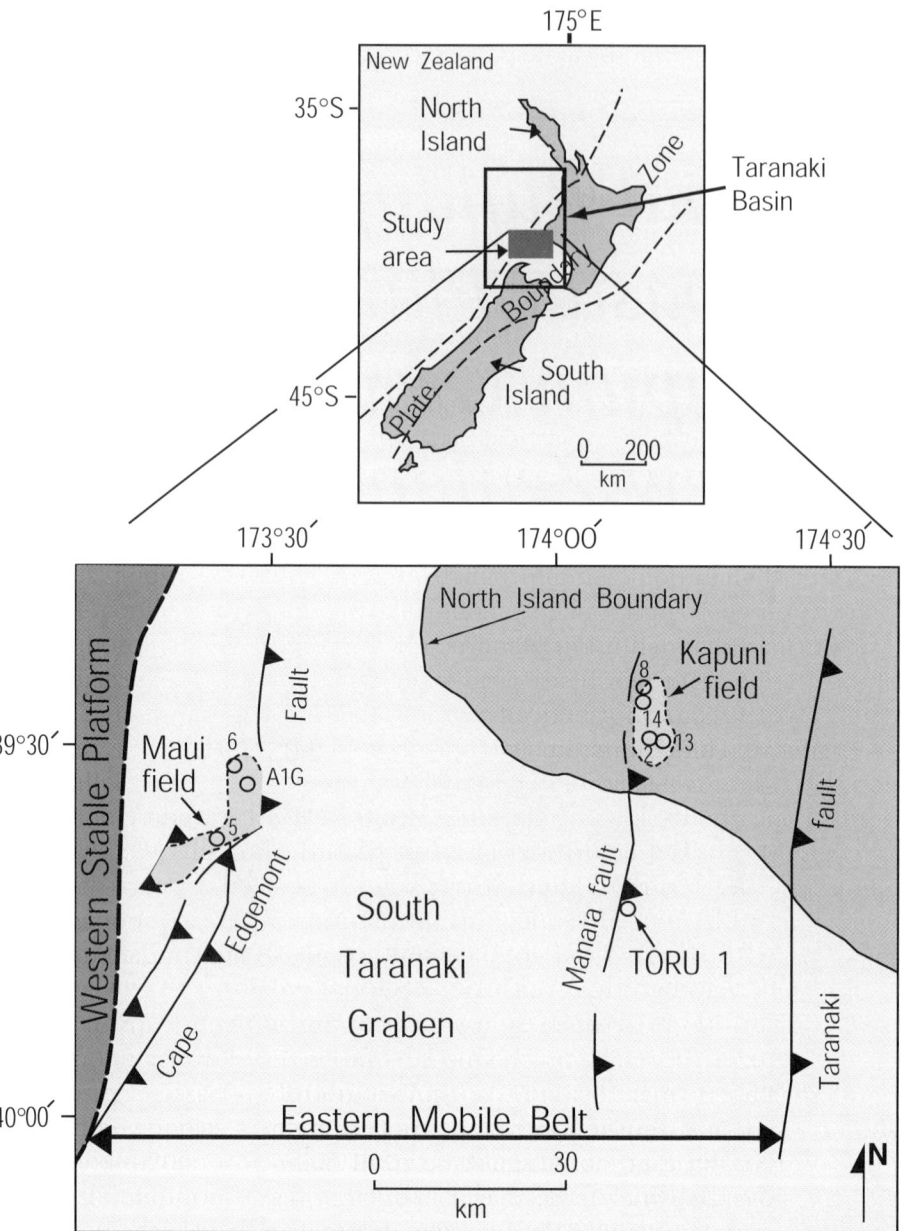

In the Maui field, New Zealand, the Kaimiro Formation (see Figures 1, 2) was interpreted by Flores et al. (1993) as consisting mainly of stacked tidal inlet and tidal creek sandstone with interbedded intertidal-subtidal sandstone, siltstone, and mudstone. These lithofacies associations were identified by Beggs and Pocknall (1992) as marine, transitional, and nonmarine. The marine lithofacies association includes shelf-shoreface mudstone and sandstone. The transitional lithofacies association includes estuarine tidal-channel and lagoonal sandstone and mudstone. The nonmarine lithofacies association consists of fluvial channel sandstone, floodplain mudstone, swamp organic deposit (coal and carbonaceous shale), and lacustrine mudstone. Beggs and Pocknall (1992) and Pocknall et al. (1992) interpreted the dominance of the estuarine tidal-channel and tidal-flat facies as resulting from backfilling incised lowstand drainage systems. These workers identified sequence boundaries in the Kaimiro Formation in the Maui field that Bryant et al. (1994) later recognized as local tidal-channel scours. However, Beggs et al. (1992) recognized a more regional sequence boundary at the base of the Kaimiro Formation, which resulted from a basinward shift of the lithofacies association, probably caused by regression because of uplift and erosion of the sediment source area to the south.

In the Maui field, the lower part of the Mangahewa Formation was interpreted by Bryant et al. (1994) as a lowstand systems tract consisting of fluvial channel lithofacies (sandstone) and a highstand systems tract consisting of lagoonal lithofacies (mudstone). Ma-

rine flooding surfaces associated with a transgressive systems tract were also recognized by Bryant et al. (1994). The upper part of the Mangahewa is interpreted as a landward succession of sandstone-dominated sequences. The coarsening-upward sandstones are shoreface lithofacies associations in a series of parasequence sets.

In the Kapuni and Toru fields, which are landward of the Maui field (see Figure 2), Flores et al. (1993) recognized a predominant fluvial-tidal lithofacies association. The estuarine tidal lithofacies (sandstone) were interpreted by Bryant et al. (1994) as incised-valley fill assigned with a lowstand system tract. King and Thrasher (1996) reinterpreted this lithofacies as resulting from a transgressive systems tract, and the

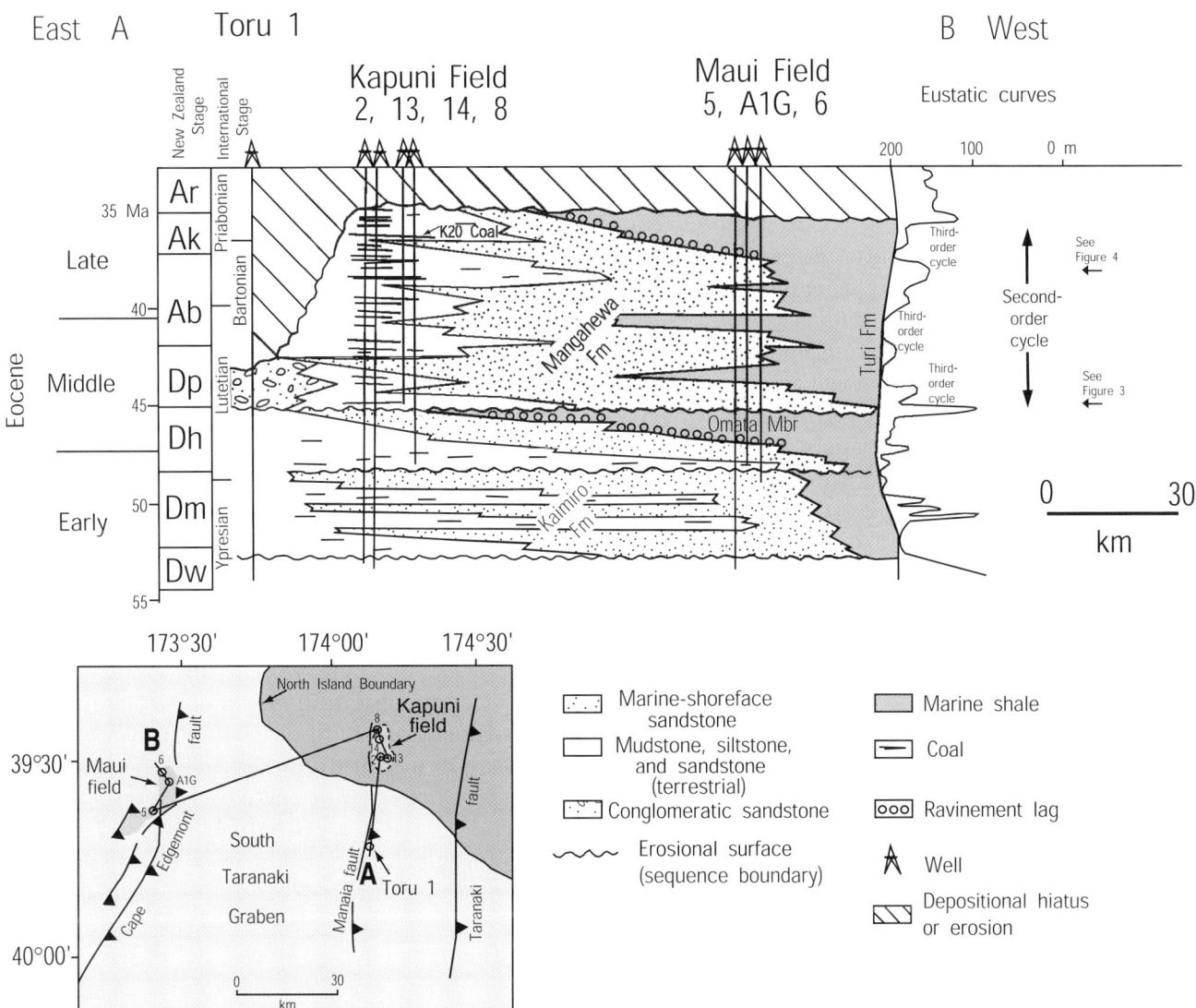

Figure 2. Stratigraphic cross section of the Eocene Kapuni Group from Toru 1 through Kapuni to Maui fields. The eustatic curves are adapted from Haq et al. (1987, 1988). The New Zealand Stage names include DM = Mangaorapan; Dh = Heretaungan; Dp = Porangan; Ab = Bortonian; Ak = Kaiatan; and Ar = Runangan. The International Stage names are adapted from Berggren et al. (1995). Fm = Formation. Mbr = Member. Diagram modified from Flores et al. (1993).

coal measures in the Mangahewa as representing highstand systems tract deposits.

Figure 2 shows sequence boundaries not only separating formational contacts but also between parasequence sets. The vertical distribution of the parasequence sets, in turn, is coincident with the rise and fall of sea level (see eustatic curves in Figure 2), as indicated by the marine shale of the Turi Formation and the fluvial sandstone and conglomerate of the Omata Member of the Turi Formation. Stacking of these parasequence sets, in turn, controlled coal buildup in tidal and alluvial mires.

The purpose of this chapter is to contribute detailed observations of the depositional lithofacies of the Kapuni and Mangahewa Formations in the Maui

and Kapuni fields, and from the Toru 1 well in the Taranaki Basin. More importantly, this investigation integrates lithofacies associations of temporal and diachronous chronostratigraphic successions and interprets their depositional settings across these fields. Lastly, the buildup or accumulation of coal deposits in these formations is interpreted in the context of relative sea level fluctuations.

METHODS

To analyze the depositional lithofacies of the Kapuni Group, cores were described from 11 wells in the Maui (Maui A1G, 5, 6, and 7) and Kapuni (Kapuni 1, 3, 8, 12, 13, and 14) fields and the Toru 1 well. Only the cores of Toru 1, Kapuni 8, Kapuni 14, Maui 6, Maui 5, and Maui A1G are discussed in this chapter, and the well locations are shown in Figure 1. The geophysical logs of Kapuni 2 and Kapuni 13 were used in this chapter, and well locations are shown in Figure 1. These cores were studied in 1992 in cooperation with the Institute of Geological and Nuclear Sciences Limited in Lower Hutt, New Zealand. In total, more than 1 km (0.62 mi) of 6.5-cm (2.6-in.)-diameter continuous core was described to include rock type, color, nature of contact, grain size and roundness, mineral composition, sorting, sedimentary structures, trace and body fossil content, and fractures. Photographs of the continuous cores and partial cores for detailed analysis were taken to complement lithologic core descriptions.

Vertical lithofacies logs of the continuous cores were constructed from rock descriptions and used to interpret depositional processes and environments. Geophysical logs, which consist of gamma-ray (GR), spontaneous potential (SP), laterolog (LLD), and deep induction (ILD) curves, were used to complement interpretations of the vertical lithofacies variations. The SP and GR log characters served as signposts for the grain-size changes in lithofacies profiles (e.g., coarsening- and fining-upward). Core descriptions by Hitchings (1986; 1987a, b) and Chatellier and Hitchings (1987) were used as guides in the analyses.

Micropaleontological and palynological assessments available from open-file reports (appendix 1.2 in King and Thrasher, 1996) were integrated and used to construct the chronostratigraphic correlations of the Kapuni Group between various study wells and fields illustrated in Figure 2. A palynofacies study by Pocknall and Beggs (1990) and the biostratigraphic framework noted by Pocknall et al. (1992) served as detailed guides to correlation between wells in the Maui field. In addition to paleontological data, this investigation used widespread erosional surfaces to establish depositional sequences.

GEOLOGICAL SETTING

The Taranaki Basin (Figure 1) is divided structurally into a western stable platform and an eastern mobile belt (King, 1991; King et al., 1991). The Maui and Kapuni fields and Toru 1 well are located in the eastern mobile belt. This eastern mobile belt consisted of depocenters localized in subsiding subbasins, which comprised the Taranaki rift (Thrasher, 1992) in association with the opening of the Tasman Sea during the Late Cretaceous. Rifting terminated

during the Paleocene, with the development of isolated, extensional-faulted subbasins (King and Thrasher, 1996), which served as depocenters of a thick sedimentary package. The Eocene was marked by continued basin subsidence and accumulation of the Kaimiro and Mangahewa Formations, relatively undisrupted by contemporaneous faulting.

In general, the Eocene sedimentation patterns reflect infilling of alluvial plains in structurally controlled subbasins in the southern Taranaki Basin, which was contemporaneous with marine incursions in the northern part of the basin (King and Thrasher, 1996) (Figures 3, 4). More specifically, these authors indicated that the Paleocene sediments were deposited in the subbasins by northward-flowing fluvial systems that drained the coastal plains in the northern part of the basin (Figures 3, 4). The headwaters of these fluvial systems originated mainly in granitic terranes (e.g., Karamea and Separation Point provenances) (Smale and Morton, 1987; Smale, 1992, 1996), which underwent deep weathering and erosion or peneplanation. During the Eocene, deepening of the marine Taranaki Basin to the north was accompanied by marine incursions to the south of the basin (Figures 3, 4). These marine incursions, represented by sediments of the Moa Group (King and Thrasher, 1996), strongly influenced the fluctuating base levels of the coastal plains (Figures 3, 4). The marine incursions or floodings were recorded by marine transgressive or ravinement surfaces defined by ravinement lags on top of the Kaimiro and Mangahewa Formations (Figure 2). The marine incursions, in turn, controlled the fluvial-tidal sedimentation during deposition of the Kapuni Group. Moreover, the marine incursions reworked older fluvial-tidal sediments, which were redeposited as marine-shoreface sediments, giving rise to the stacked parasequence sets along transgressive paleoshoreline complexes. Figure 4 shows deposition of sand by longshore drift along the paleoshoreline.

The Mangahewa Formation in the Toru 1 well reflects alluvial-plain deposits in the Taranaki Basin (Figures 2, 4). The Kaimiro and Mangahewa Formations in the Kapuni and Maui fields are located in the coastal plain and marine-shoreface environments in the northern part of the basin (Figures 2–4).

LITHOFACIES ANALYSES AND INTERPRETATIONS

In this section, representative lithofacies associations are described from the alluvial paleovalley, coastal

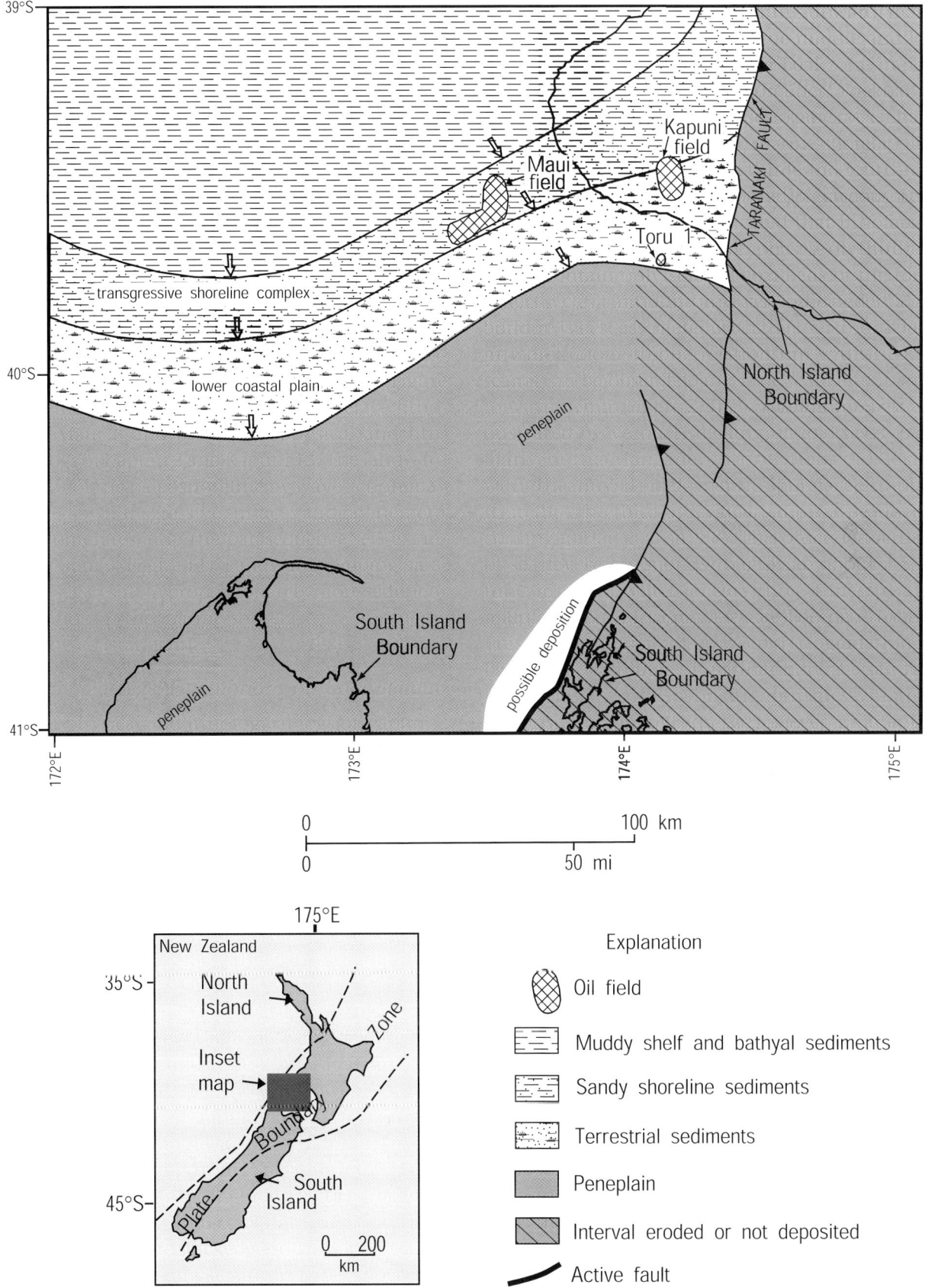

Figure 3. Paleogeographic map showing the middle Eocene depositional environments in the Taranaki Basin (adapted from King and Thrasher, 1996). The locations of the Maui and Kapuni fields and Toru 1 are shown. Arrows indicate transgressive paleoshoreline.

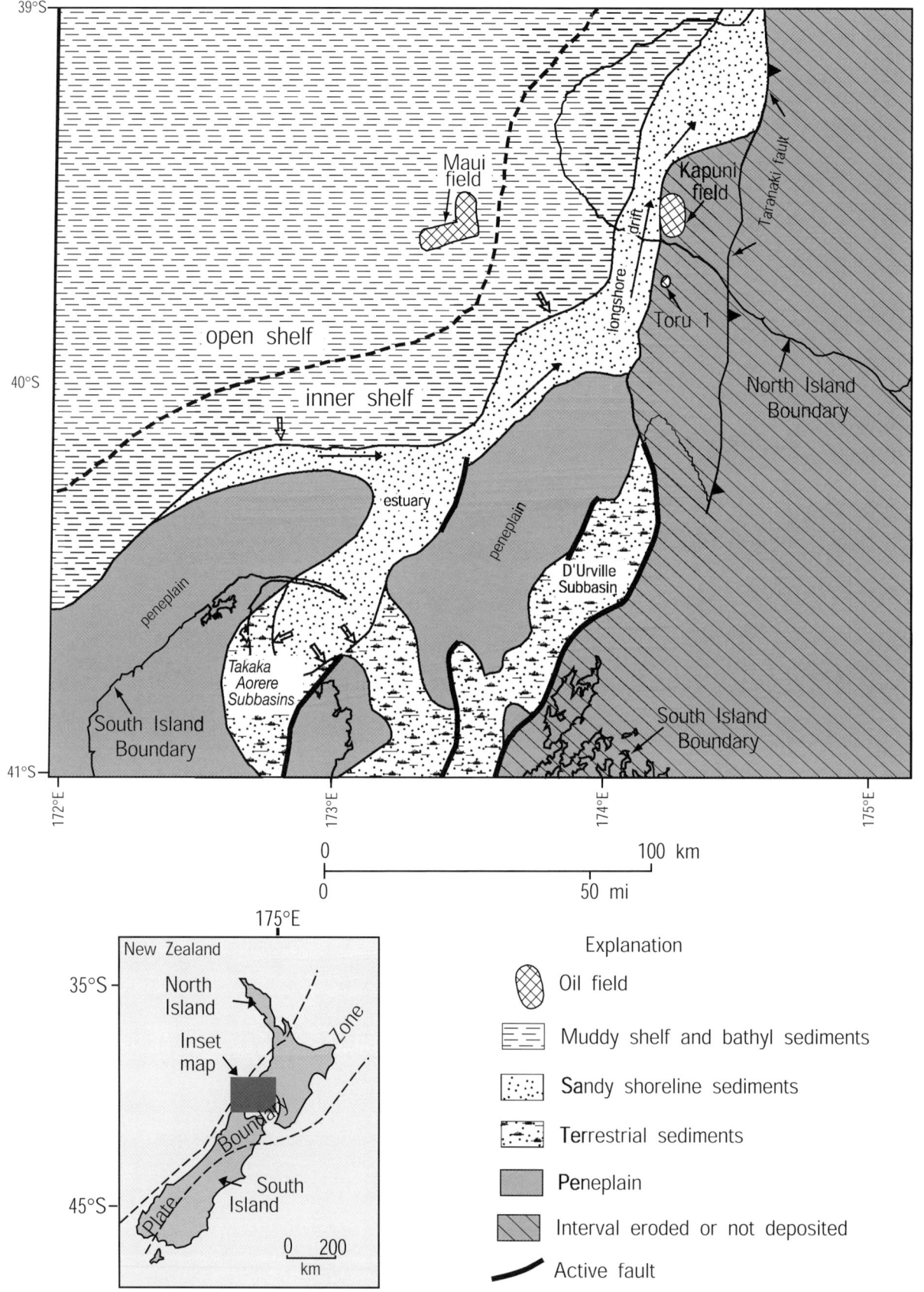

Figure 4. Paleogeographic map showing the late Eocene depositional environments in the Taranaki Basin (adapted from King and Thrasher, 1996). The locations of the Maui and Kapuni fields and Toru 1 are shown. Short arrows indicate transgressive paleoshoreline, and long arrows indicate sediment transport direction.

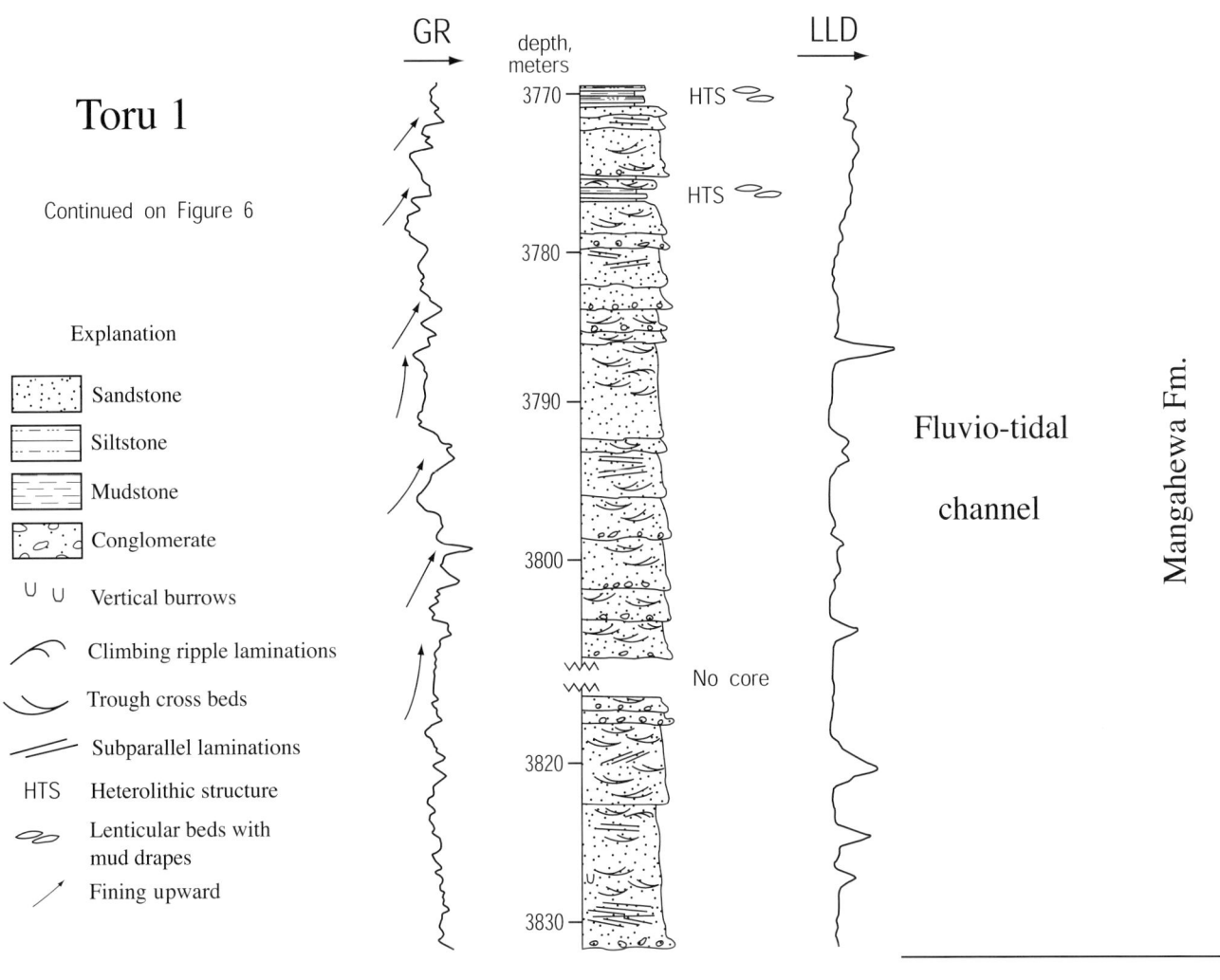

Continued on Figure 6

Figure 5. Vertical lithofacies profile and environmental interpretation of the study interval of the Mangahewa Formation in Toru 1. Related geophysical logs are also shown. Fm = Formation; GR = gamma ray; LLD = laterolog.

plain, and marine-shoreface environments. In addition, interpretations of the depositional processes and environments of these lithofacies are discussed to relate them to the regional setting for coal buildup in the Kapuni Group. The stratigraphic variation of the lithofacies associations of the Eocene Kaimiro and Mangahewa Formations in Toru 1 and the Maui field are shown in Figure 2.

Alluvial Paleovalley Lithofacies

The alluvial paleovalley lithofacies are best shown in the lower part of the Mangahewa Formation in Toru 1 (see Figures 2, 5). Toru 1 also includes the upper part of the Kaimiro Formation (Figures 2, 6).

Toru 1

The lithofacies are dominated by sandstone (Figure 5). The sandstone is buff to light gray, medium to very coarse grained, quartzose (>75% subrounded quartz), moderately sorted, and erosional-based. A lag conglomerate of coal spars, mudstone clasts, and grit-size quartz and feldspar fragments mark the erosional bases. Sedimentary structures consist of abundant trough cross-beds and massive beds, and common planar cross-beds and ripple laminations (e.g., climbing ripples, flaser beds, and lenticular beds). The trough cross-beds range from 3 to 58 cm (1.2 to 23 in.) in height. A few vertical burrows, observed at depth of 3828 m (12,559 ft), partly destroyed foresets of a trough cross-bed. The planar cross-beds range from 5 to 48 cm (2 to 19 in.) in height. The massive beds are normally graded from a lag conglomerate to a medium-grained sandstone and range from 0.02 to 1 m (0.07 to 3.3 ft) in thickness.

The thin to thick sandstone beds form fining- and thinning-upward units that exhibit a multistory vertical

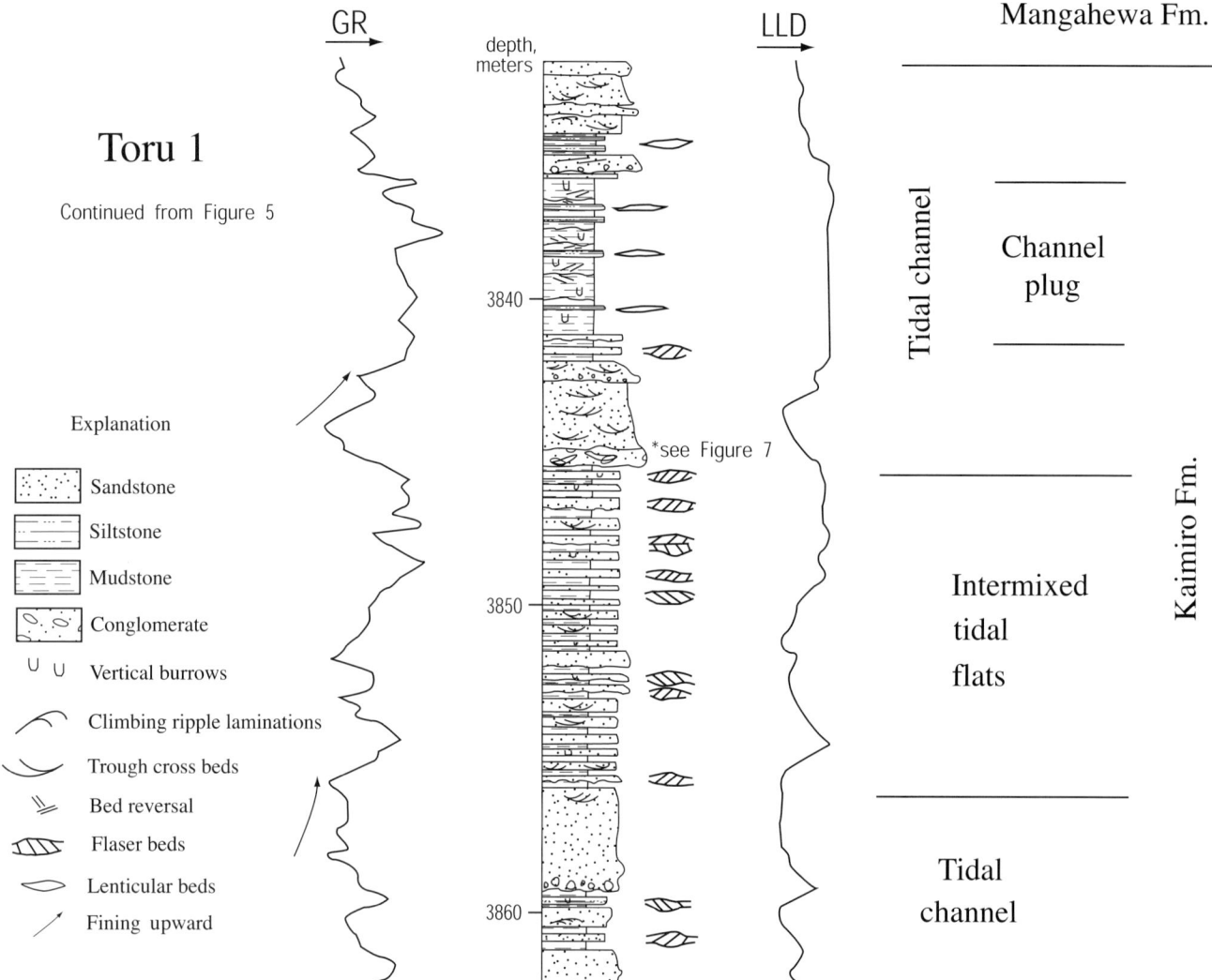

Figure 6. Vertical lithofacies profile and environmental interpretation of the study interval of the Kaimiro Formation in Toru 1. Related geophysical logs are also shown. Fm = Formation; GR = gamma ray; LLD = laterolog.

pattern (see Figure 5). The vertical pattern of these sandstones may be characterized by predominantly thick sandstone beds in the lower part (3830–3818 m [12,566–12,526 ft] depth) of the lithofacies, predominantly thin sandstone beds in the middle part (3808–3792 m [12,493–12,441 ft]), and mixed thin and thick sandstone beds in the upper part (3792–3769 m [12,441–12,365 ft]). The thick sandstone beds are commonly trough and planar cross-bedded, and the thin sandstone beds are mainly trough cross-bedded.

The sandstone beds in the uppermost part of the lithofacies are interbedded with alternating siltstone and mudstone (3780–3769 m [12,402–12,365 ft]). Flaser beds in the siltstone are bounded by mudstone drapes and are in bundles as much as 22 cm (8.7 in.) in thickness. These fine-grained rocks also contain lenticular beds that are 1–20 mm (0.04–0.79 in.) thick and occur in bundles as much as 7 cm (2.8 in.)

thick. Ripple laminations also occur in these rocks as climbing ripples in units as much as 26 cm (10.2 in.) thick. These sedimentary structures are common in the uppermost part of the lithofacies, where they cap the fining-upward sandstone (see Figure 5). The bundles of mudstone and siltstone with flaser, lenticular, and ripple laminations reflect heterolithic stratification (HTS) (see Figure 5).

The thin to thick sandstone beds of the Mangahewa Formation in Toru 1 are underlain mainly by interbedded sandstone, siltstone, and mudstone and subordinate conglomerate (Figure 6) in the upper part of the Kaimiro Formation. The sandstones are buff to light gray, medium to very coarse grained, quartzose (>75% subrounded-rounded quartz), moderately to well sorted, and gradational-, sharp-, and erosional-based. Grit-sized quartz and coal spars commonly mark erosional bases. The gradational bases are marked by

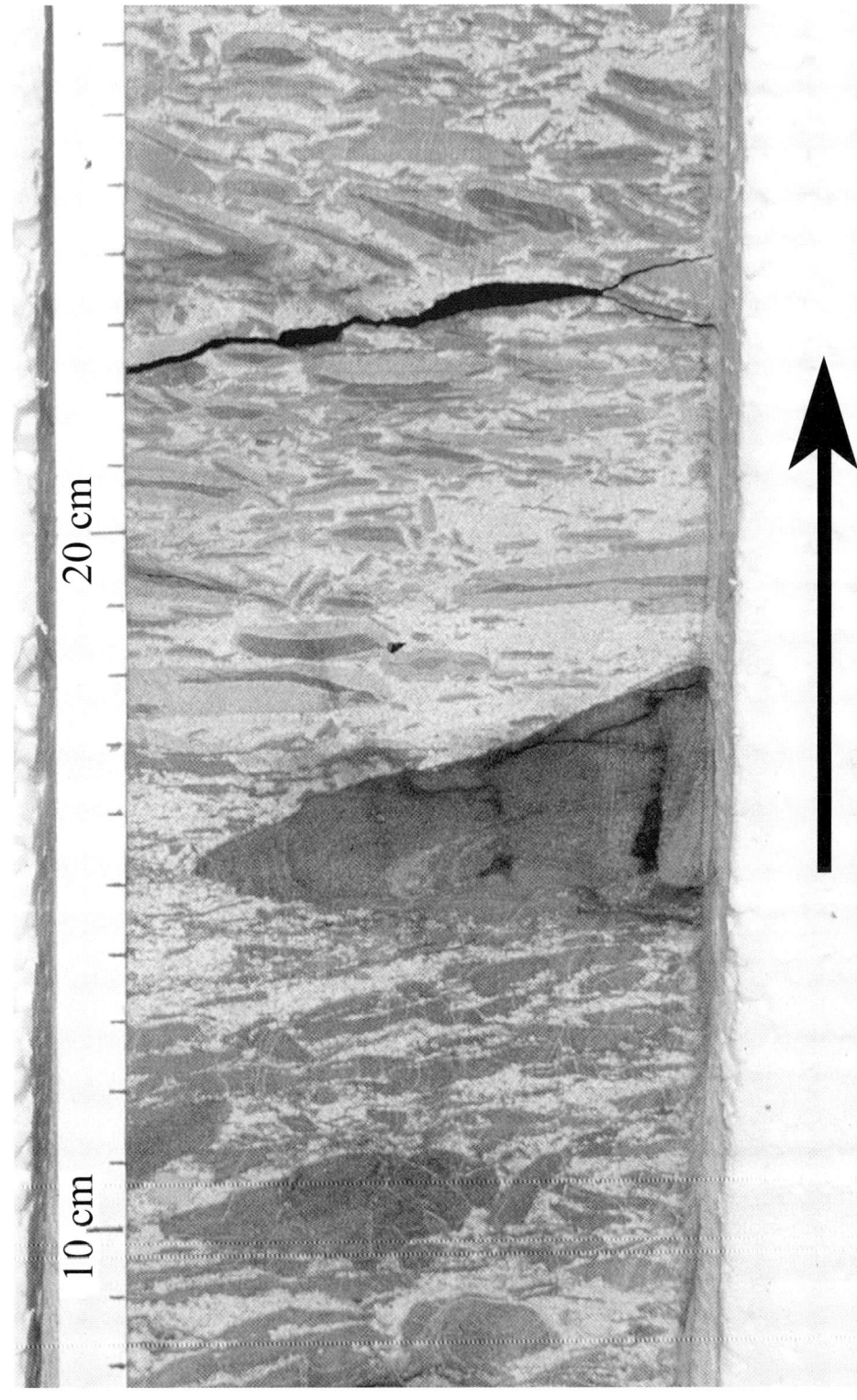

Figure 7. Bidirectional flat-pebble conglomerate in the Kaimiro Formation in Toru 1. See Figure 6 for the location of the conglomerate in the lithofacies succession. The core (depth 3845 m [12,615 ft]) is about 20 cm (8 in.) thick, as indicated by the scale on left of the photograph. Arrow indicates upsection.

and massive beds are sparse. Fining-upward, erosional-based sandstone and a basal, flat-pebble, mudstone conglomerate are observed at depths of 3843–3846 m (12,608–12,618 ft). The flat pebbles are imbricated with bidirectional orientations (Figure 7). Siltstone, very fine-grained sandstone, and mudstone are replete with flaser and lenticular bedding occurring in thin to thick cycles (2–43 cm [0.8–17 in.] in height). A thick mudstone bed (3836–3842 m [12,585–12,605 ft]) occurs with a gradational base and top. It exhibits flat lenses, lenticular laminations, and herringbone cross-stratification. These fine-grained rock types are commonly vertically burrowed.

Interpretation of Toru 1

In the study interval in Toru 1, the alluvial paleovalley represented by the sandstone-dominated lithofacies succeeded a tidally influenced coastal plain represented by the interbedded mudstone, siltstone, sandstone, and flat-pebble conglomerate lithofacies. More specifically, the multistory erosional-based and fining-upward sandstone beds represent deposits of a fluvial channel complex in an incised paleovalley. The stacked, thin to thick sandstone beds indicate

gradual change in grain size with underlying rocks; the sharp bases are marked by abrupt change in grain size. Sedimentary structures consist of abundant trough cross-beds (1–20 cm [0.4–7.9 in.] in height) and large foresets (3948 cm [1554 in.] in height). Herringbone cross-strata (24 cm [9.4 in.] in height) are common,

that a braided stream drained this alluvial paleovalley. The thick sandstone beds represent deposits in deep and large channels, and the thin sandstone beds represent deposits in minor shallow and small subchannels. The vertical pattern of sandstone beds suggests that the paleovalley was initially drained by large braided channels that evolved into small braided subchannels, with both channel systems infilled by gravel lags and sand bars. The evolution of the channel system continued with large channels mixed with small subchannels in the braidbelt. This sand-laden braided stream is similar to fluvial channels described by Flores and Sykes (1996) for the Brunner Coal Measures in the Buller coalfield, South Island, New Zealand.

The alternating mudstones and siltstones above the fluvial channel sandstone complex represent heterolithic sedimentary structures in a tidal environment (Knight and Dalrymple, 1975; de Boer et al., 1988; Nio and Yang, 1991). The flaser, lenticular, and ripple-laminated siltstone and mudstone reflect mudflat lithofacies. This lithofacies suggests that during the final braidbelt aggradation, tidal currents affected the small subchannels. This indicates progressive drowning of the paleovalley and landward migration of marine influence. The lithofacies in Toru 1 probably represent a landward equivalent of the Kapuni 13 well lithofacies (see Figure 2).

The fine-grained rocks below the fluvial channel sandstone complex represent deposits of a tide-influenced coastline. The thin to thick, sharp- and gradational-based sandstones, replete with trough cross-beds, large foresets, flaser and lenticular laminations, and vertical burrows, are interpreted as mixed mud-sand and sand tidal-flat lithofacies. The gradational-based, thick mudstone with herringbone stratification represents abandoned tidal-channel deposits. The fining-upward sandstone underlain by the flat-pebble conglomerate was deposited by ebb and flow currents across a tidal channel.

Thus, the Toru 1 study interval shows a tidal-flat and tidal-channel lithofacies in the lower part and alluvial paleovalley lithofacies in the upper part. The erosional contact between these lithofacies is a sequence boundary that defines the contact of the Kaimiro and Mangahewa Formations (see Figure 2). The alluvial paleovalley lithofacies is assigned to the lowermost part of the Mangahewa Formation, and the tidal-flat and channel lithofacies association is assigned to the uppermost part of the Kaimiro Formation. The paleovalley represents incising during lowering of sea level that was followed by braided-stream infill. The tide-influenced deposits in the uppermost part of the fluvial channel complex may indicate a paleoshoreline turnaround. The tidal-flat and channel lithofacies association indicates an earlier advance of the paleoshoreline prior to regression and incisement of the paleovalley.

Coastal Plain Lithofacies

The coastal plain lithofacies are best described from the upper part of the Kaimiro Formation and the lower part of the Mangahewa Formation at Kapuni field (see Figures 8, 9). These lithofacies are contemporaneous with and seaward of the alluvial paleovalley and tidal-flat and tidal-channel lithofacies described in Toru 1 (see Figures 2, 4).

Kapuni 8

The lithofacies in the upper part of the Kaimiro Formation consists of interbedded coal, carbonaceous shale, mudstone, siltstone, and sandstone (Figure 8). The sandstone is gray, medium grained, quartzose (>75% subrounded quartz grains), and moderately sorted. No core has been recovered from the base of the sandstone (4058–4062 m [13,314–13,326 ft]); however, the geophysical log (GR) shows a fining-upward sandstone. Sedimentary structures in the sandstone consist of abundant trough cross-beds, common planar cross-beds, and sparse massive beds. Vertical sets of these cross-beds, which are 7–30 cm (2.8–11.8 in.) in height, are commonly capped by climbing ripple laminations (2–20 cm [0.8–7.9 in.] in height). The trough cross-beds range from 3 to 17 cm (1.2 to 6.7 in.) in height. The planar cross-beds range from 3 to 14 cm (1.2 to 5.5 in.) in height. The massive beds are normally graded from fine to medium grained.

The mudstone is black, generally massive bedded, and contains scattered zones of flat lenses and small vertical burrows (1–3 mm [0.04–0.12 in.] in height) infilled with siltstone (4054–4058 m [13,301–13,314 ft]). Wormlike traces, 1.5 cm (0.6 in.) long, were observed. The mudstone contains common macerated plant fragments and exhibits sharp-to-gradational contact. Roots marked by vitrain and root fills (vertical silt infills lined with carbonaceous matter), occurring in zones from 2 to 12 cm (0.8 to 4.7 in.) thick, are abundant below coal beds. Mudcracks in a mudstone interbedded with sandy siltstone and coal beds were observed (see Figures 8, 10). The siltstone is gray, rooted, and burrowed; it displays a sharp to gradational contact and contains ripple and lenticular laminations.

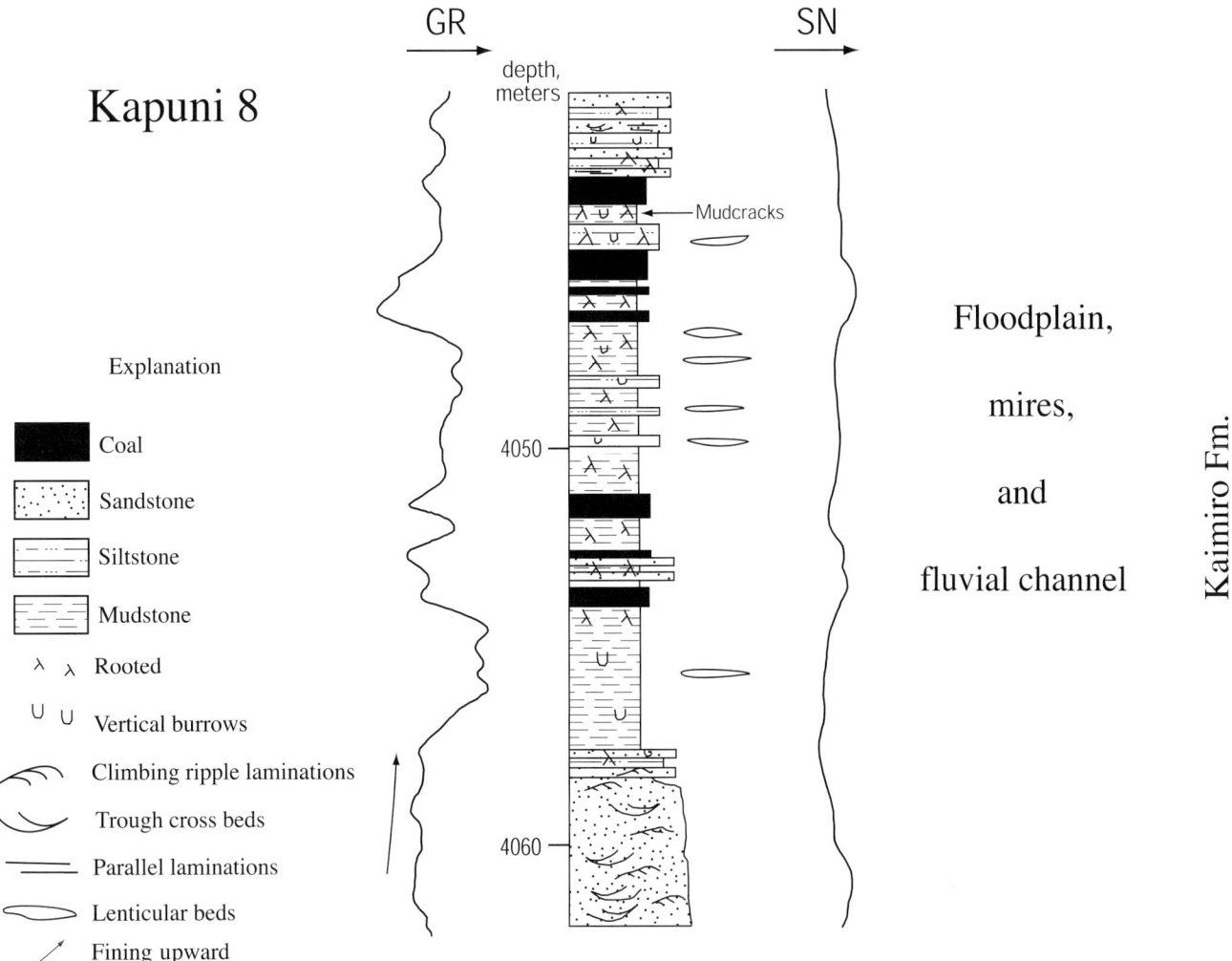

Figure 8. Vertical lithofacies profile and environmental interpretation of the study interval of the Kaimiro Formation in Kapuni 8. Related geophysical logs are also shown. Fm = Formation; GR = gamma ray; SN = sonic.

The coal beds are bright-banded, vitrain-rich, and range from 3 to 52 cm (1.2 to 20.5 in.) thick (4043–4054 m [13,264–13,301 ft] depth) (Figure 10). The coal grades into carbonaceous shale containing vitrain lenses or into a black, carbonaceous mudstone. Coal beds occur in two coal zones, each consisting of three to four beds separated by heavily rooted mudstone and siltstone. These coal zones, in turn, are separated by 12-m (39-ft)-thick heavily rooted mudstone with thin siltstone and carbonaceous shale beds.

Interpretation of Kapuni 8

The coal-bearing study interval of the lower part of the Mangahewa Formation in Kapuni 8 represents an alluvial lithofacies. The fining-upward, trough- and planar-cross-bedded sandstone in the lower part of the interval is a fluvial channel deposit. The stacked cross-bed sets capped by climbing ripple laminations

reflect deposits from migrating sandbars. The ripple- and lenticular-laminated, rooted siltstone and mudstone represent flood-plain deposits. The thick, heavily burrowed mudstone with zones of flat lenses is a flood-plain lacustrine deposit. The banded, vitrinite-rich coal beds accumulated in topogenous or low-lying wooded mires, perhaps in freshwater environments, and were subjected to frequent inundation by flood-plain sediments.

Kapuni 14

The lithofacies in the lower part of the Mangahewa Formation consist chiefly of sandstone and subordinate coal, mudstone, and siltstone (Figure 9). The sandstone is gray, fine to coarse grained, quartz-rich (>65% subrounded quartz grains), and moderately to well sorted. The basal contact of the sandstone is either erosional (marked by grit-size quartz and feldspar grains) or sharp. Sedimentary structures are

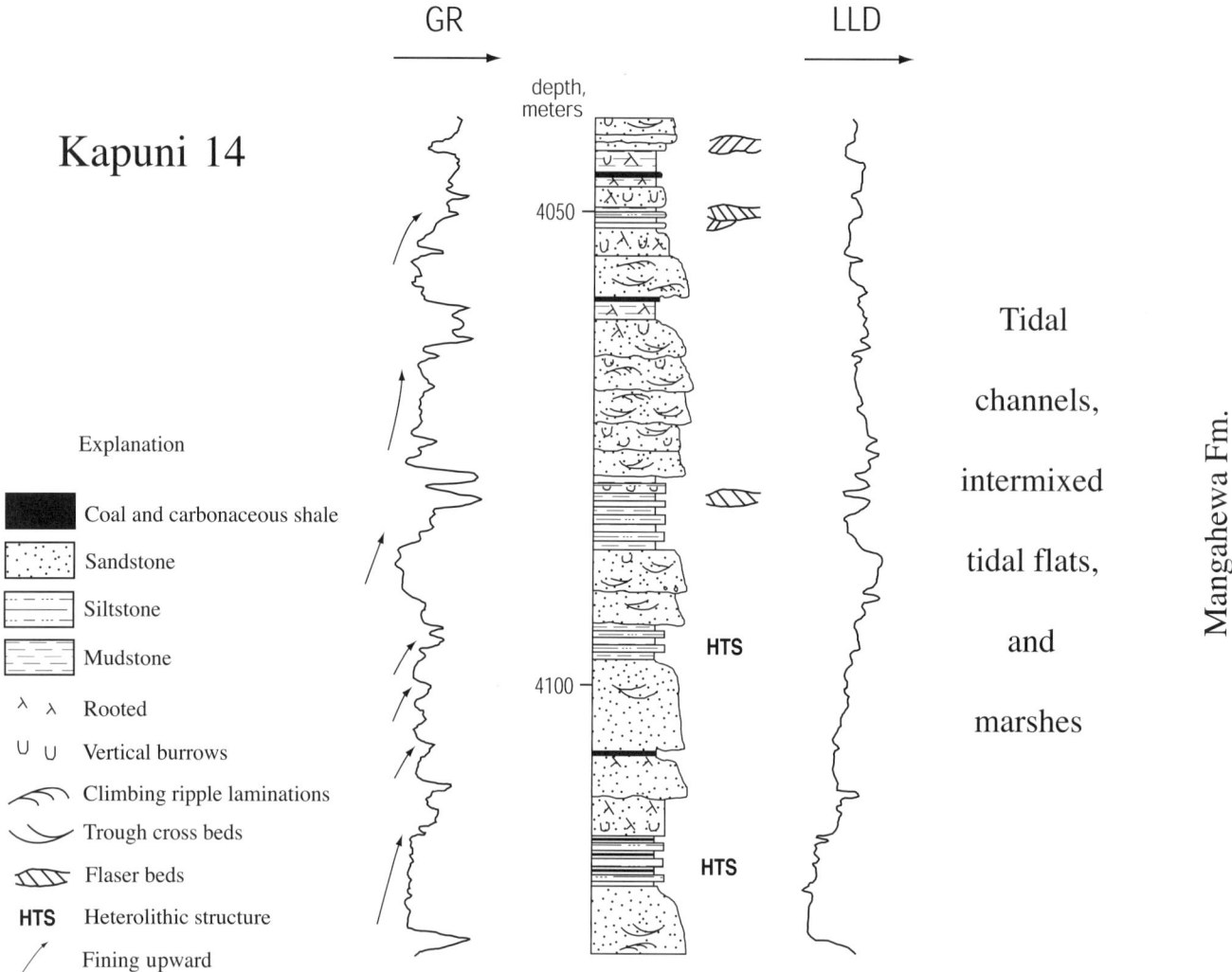

Figure 9. Vertical lithofacies profile and environmental interpretation of the study interval of the Mangahewa Formation in Kapuni 14. Related geophysical logs are also shown. Fm = Formation; GR = gamma ray; LLD = laterolog.

mainly trough cross-beds, ranging from 2 to 57 cm (0.8 to 22 in.) in height, and sparse subparallel lamination, ranging from 5 to 29 cm (2 to 11.4 in.) in height. These cross-beds are commonly bidirectional and are capped by climbing ripple laminations ranging from 7 to 28 cm (2.8 to 11 in.) in height. Flaser beds with mud-drape pairs (6–33 cm [2.4–13 in.] thick) are common. *Ophiomorpha* burrows are generally common but locally abundant and range from 0.2 to 50 mm (0.008 to 2 in.) long. *Teredo*-bored wood was also observed. Root marks and fills are common, both marked by vitrain lenses. Heterolithic stratification is common above the sandstone beds (4120–4080 m [13,517–13,386 ft] depth).

The sandstone is interbedded with mudstone, siltstone, and coal in the lower part of the Mangahewa Formation. The mudstone is black, commonly contains lenticular beds and flat lenses, and is vertically

burrowed and rooted. It exhibits sharp to gradational contacts. The siltstone is gray, contains ripple laminations, flaser and lenticular bedding, and is commonly rooted and burrowed. Coal beds are banded, vitrain-rich, and as much as 11 cm (4.3 in.) thick.

Interpretation of Kapuni 14

The study interval of the upper part of the Kaimiro Formation in Kapuni 14 represents a tide-influenced coastal plain lithofacies. The trough and planar cross-bedded sandstone beds that have an erosional base and that contain *Ophiomorpha* burrows and *Teredo*-bored wood and are capped by heterolithic stratification. Accordingly, this interval is interpreted as a tidal-channel deposit. The bidirectional cross-beds indicate ebb and flow tidal currents. The thin, sharp-based sandstone beds with flaser and mud-drape pairs are interpreted as tidal sand-flat deposits. The

Figure 10. Burrowed mudstone interbedded with a thin coal bed and flaser-bedded, burrowed siltstone of the coastal-plain lithofacies in the Kaimiro Formation. The core (Kapuni 8; depth 4044 m [13,268 ft]) is about 138 cm (54 in.) thick, as indicated by the scale on upper left corner of the photograph. Arrows indicate upsection. The cores are continuous from left to right upsection.

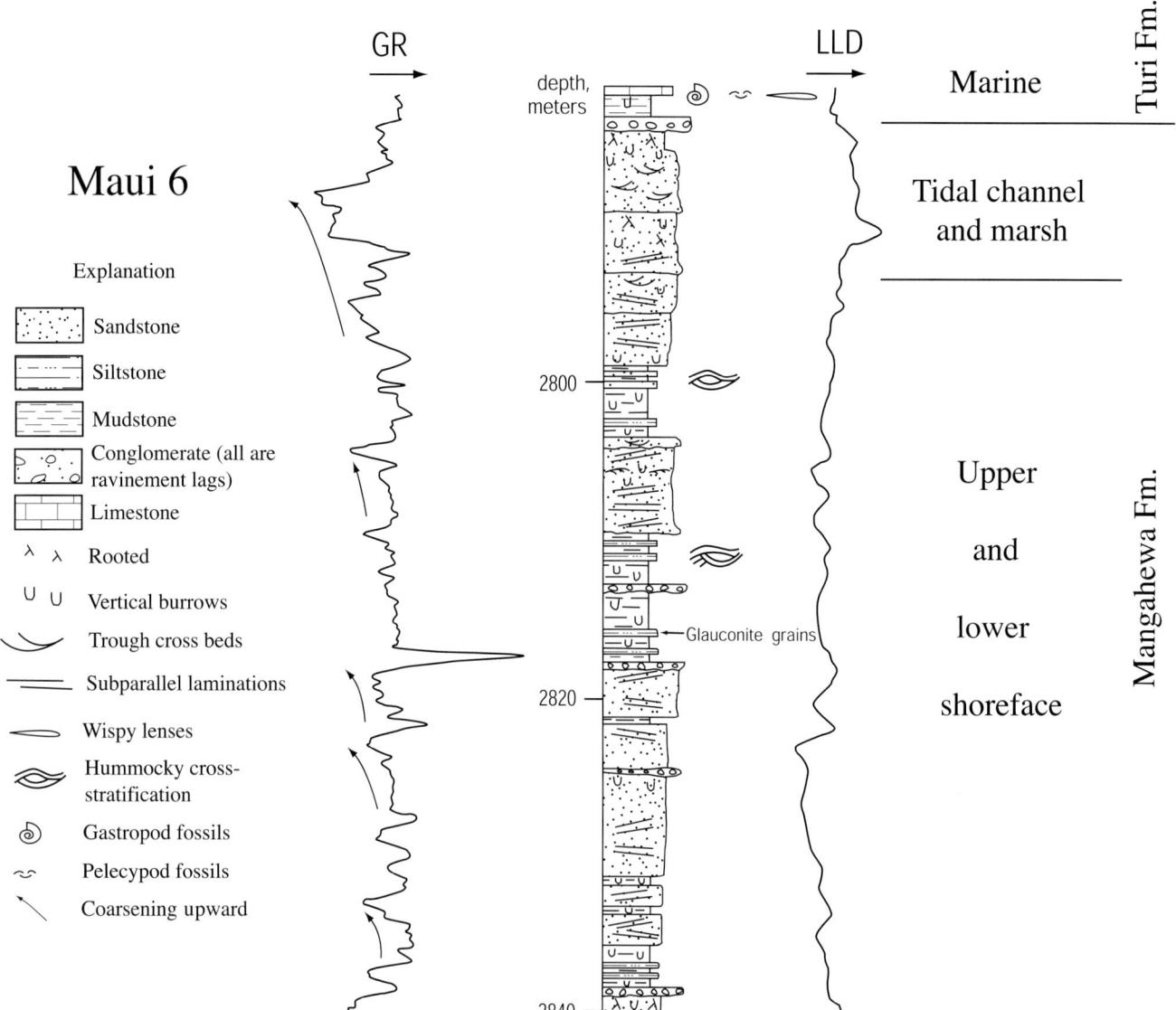

Figure 11. Vertical lithofacies profile and environmental interpretation of the study interval of the Mangahewa Formation in Maui 6. Related geophysical logs are also shown. Arrows indicate coarsening-upward cycles. Fm = Formation; GR = gamma ray; LLD = laterolog.

ripple-and lenticular-laminated, heavily rooted, and bioturbated siltstone and mudstone represent low marsh and mixed mud-silt tidal-flat deposits. The banded, vitrinite-rich coal beds apparently accumulated in high marshes.

Marine-shoreface Lithofacies

The marine-shoreface lithofacies are best described in the upper part of the Mangahewa Formation in Maui 6 (Figure 11). The immediate landward equivalent of the marine-shoreface lithofacies is best demonstrated by the Kaimiro Formation in Maui 5 and Maui A1G (Figures 12, 13). These landward lithofacies are contemporaneous with and seaward of the

tidal-flat and channel lithofacies described at Kapuni field (see Figure 2).

Maui 6

The lithofacies in the upper part of the Mangahewa Formation in Maui 6 consist of interbedded sandstone, mudstone, and siltstone. The sandstone is gray, coarsening-upward, fine to very coarse grained, quartzose (>75% subrounded quartz grains), and poor to moderately well sorted. The basal contact of the sandstone is gradational, sharp, and erosional. Sedimentary structures consist of abundant subparallel to parallel or flat laminations and hummocky cross-stratification (HCS) (see Figure 11). The subparallel

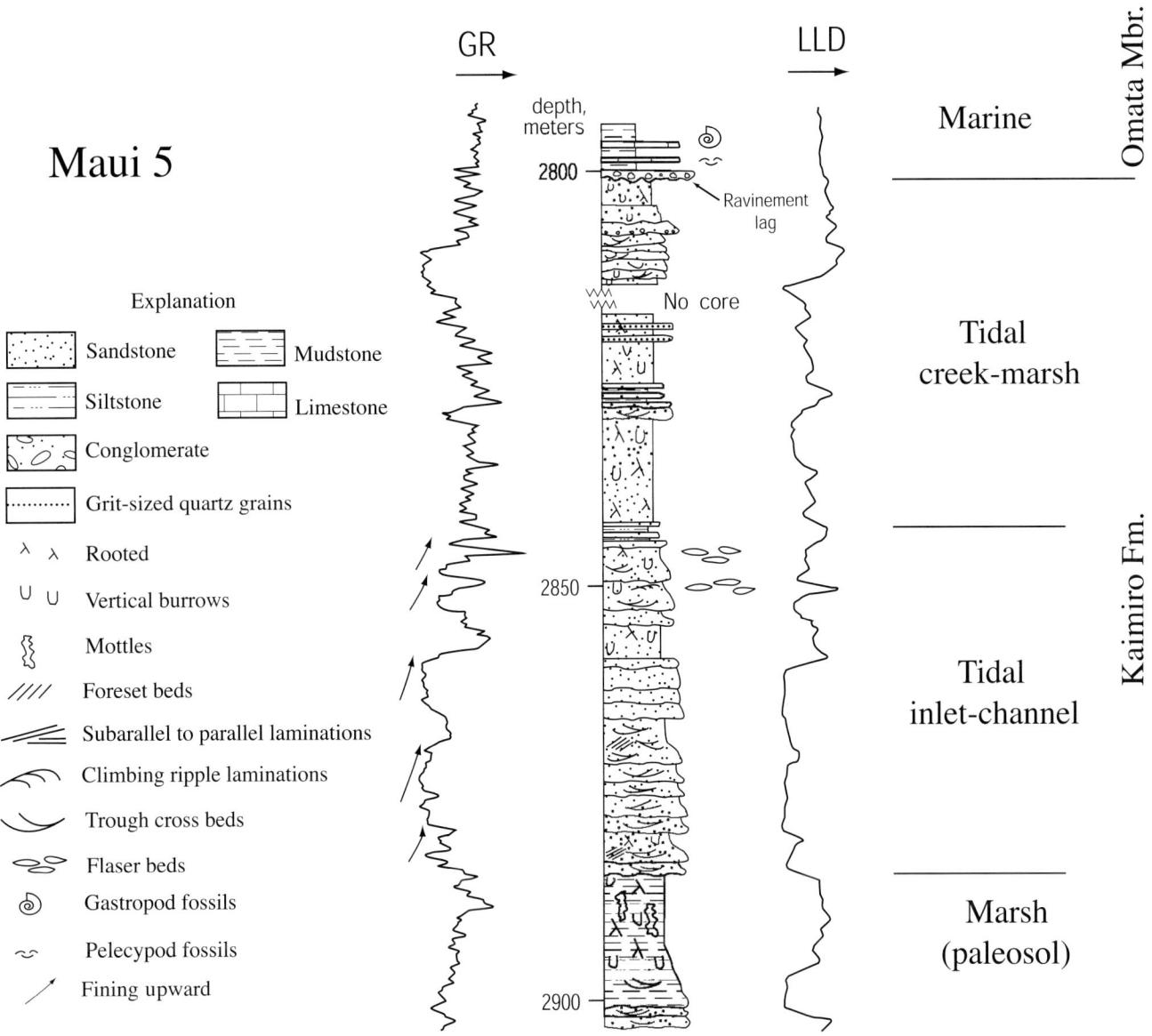

Figure 12. Vertical lithofacies profile and environmental interpretation of the study interval of the Mangahewa Formation in Maui 5. Related geophysical logs are also shown. Fm = Formation; Mbr = Member; GR = gamma ray; LLD = laterolog.

laminations are 2–55 cm (0.8–21.7 in.) in height and bidirectional. The HCS is 6–60 cm (2.4–23.6 in.) in height, is vertically burrowed, and, in places, has an erosional base (2800 m [9186 ft]). Trough cross-beds are not common, but can be as much as 14 cm (5.5 in.) in height. The subparallel laminations are either capped or alternating, with sets of wave ripple laminations that are 2–11 cm (0.8–4.3 in.) in height. Massive beds are commonly vertically burrowed. *Ophiomorpha* trace-fossil burrows are common to abundant and range from 3 to 9 mm (0.12 to 0.35 in.) in height (Figure 14). Ravinement lag conglomerates (2839, 2819,

and 2785 m [9314, 9249, and 9317 ft]) commonly cap the coarsening-upward sandstone (see Figure 11).

The mudstone is dark gray to black and contains common starved, wispy ripple laminations 1–2 cm (0.4–0.8 in.) in height, which are silty to sandy, normally graded, and burrowed. The mudstone contains sparse macerated plant fragments and exhibits common sharp and gradational contact, but in places, erosional contact. The mudstone contains sparse glauconite grains (see Figure 11). The siltstone is gray, contains ripple and lenticular laminations, is burrowed, and displays sharp to gradational contact.

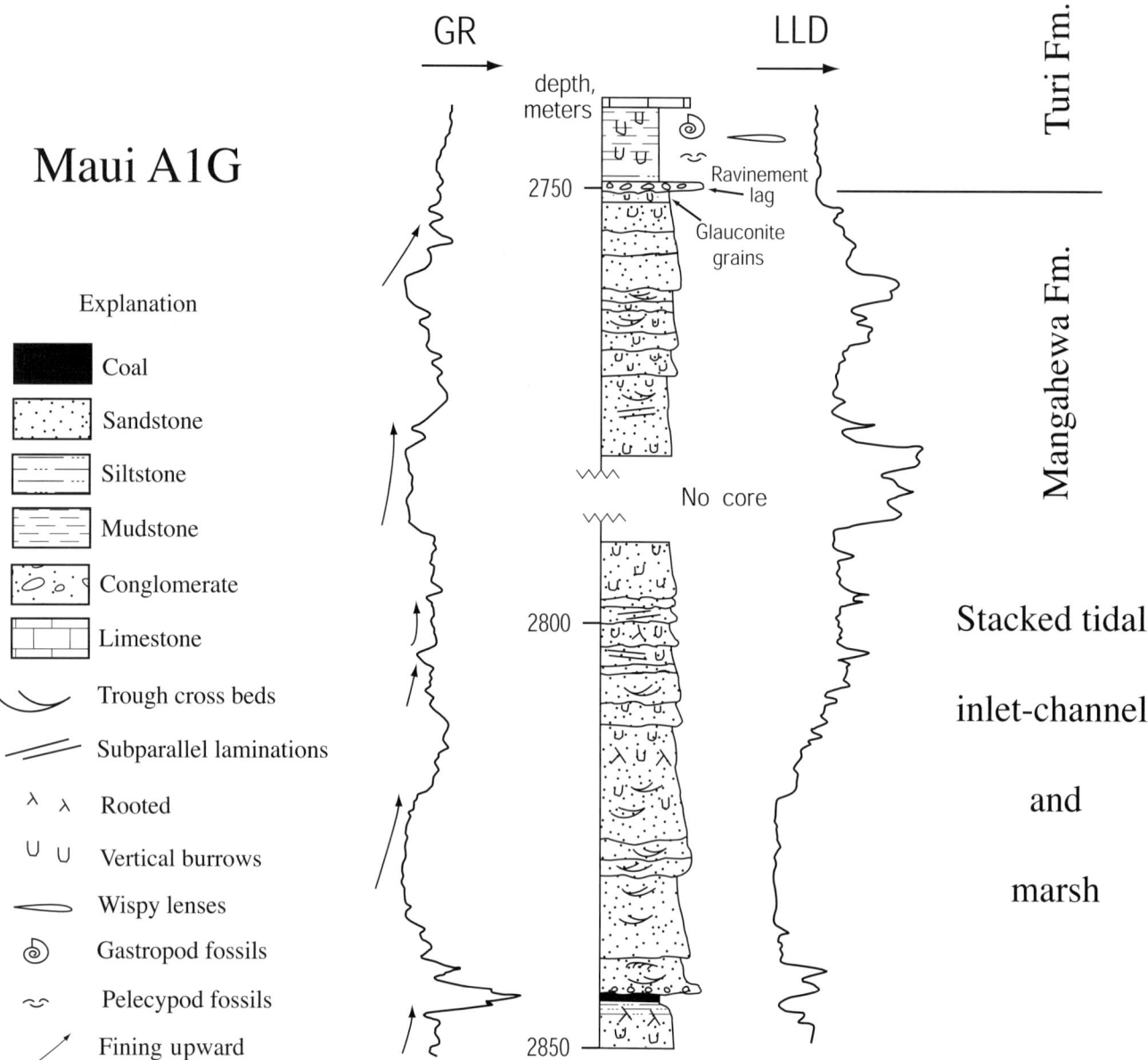

Figure 13. Vertical lithofacies profile and environmental interpretation of the study interval in the Kaimiro Formation of Maui A1G. Related geophysical logs are also shown. Fm = Formation; GR = gamma ray; LLD = laterolog.

Interpretation of Maui 6

The study interval of the Mangahewa Formation in Maui 6 represents a marine-shoreface lithofacies. The subparallel and parallel- or flat-laminated sandstone beds with wave ripple laminations, HCS, and *Ophiomorpha* burrows are characteristic of a shoreface deposit. The bidirectional, subparallel laminations indicate barred shoreface influenced by landward and seaward currents (Hunter et al., 1979). The HCS suggests storm-generated sand beds (Reading, 1978). The underlying mudstone and siltstone, which contain wispy ripple laminations, reflect deposition of suspended sediments during fair-weather periods. The mud-

stone with erosional base may represent the position of storm wave base, and the presence of trough cross-bedded, burrowed, rooted, and erosional-based sandstone (see Figure 11) indicates that the shoreface sandstone beds were incised by tidal channels. The tidal-channel sandstone is, in turn, capped by heavily bioturbated (*Ophiomorpha*) and rooted sandstone, which reflects reworking of the tidal deposits under a high marsh condition.

The mudstone, which contains starved ripple laminations, burrows, and sparse glauconite, is interpreted as marine lithofacies and is interbedded with the shoreface sandstone. The mudstone lithofacies

Figure 14. Marine-shoreface sandstone heavily bioturbated by *Ophiomorpha* trace fossils in the Mangahewa Formation. The core (Maui 6; depth 2850–2851 m [9350–9354 ft]) is more than 80 cm (31 in.) thick, as indicated by the scale in the center of the photograph. Arrows indicate upsection. The cores are continuous from left to right upsection.

commonly grade upward into the rippled, burrowed siltstone lithofacies, which, in turn, grades upward into the shoreface lithofacies. These lithofacies make up a coarsening-upward interval, and the stacked coarsening-upward intervals represent a parasequence set (Figure 11). This parasequence set is capped by ravinement (transgressive) lag conglomerates, which contain reworked deposits from marine flood events.

Maui 5

The lithofacies of the Kaimiro Formation in Maui 5 consist mainly of interbedded sandstone, mudstone, siltstone, and minor conglomerate (Figure 12). The sandstone is gray, fine to very coarse grained, quartz-ose (>70% subrounded-subangular quartz grains), and moderately to well sorted. Numerous sandstone beds represent fining-upward units. The contact of the sandstone is mainly erosional and subordinately sharp to gradational. Coal spars, very coarse grains of quartz sand, and mudstone to sandstone clast mark the erosional-based sandstone. Sedimentary structures consist of abundant subparallel to parallel lamina-tions and trough cross-beds (Figure 12). The subpar-allel to parallel laminations are 2–26 cm (0.8–10 in.) in height. Trough cross-beds are as much as 21 cm (8.3 in.) in height. A few large high-angle foreset lam-inations from 37 to 74 cm (15 to 29 in.) in height are present and show reactivation surfaces. *Ophiomorpha* trace-fossil burrows are locally abundant to common and range from 2 to 6 mm (0.08 to 0.24 in.) in height. *Teredo*-bored wood is found above erosional-based sandstones. Flasers, wave ripples, and climbing ripples are sparse. Thin sandstone beds, 20–50 cm (8–20 in.) thick, contain burrows and heterolithic stratification and alternate with burrowed mudstone beds. The conglomerate, which occurs at the base of some thin to thick sandstones, consists of coal spars and grit-size quartz grains.

Thin mudstone beds are dark gray to black, con-tain common wispy ripple laminations 1 cm (0.4 in.) in height, and are burrowed. The siltstone is gray, contains ripple laminations, is burrowed, and dis-plays gradational contact. These rock types are in-terbeded with the sandstone. Glaebules consisting of red mottles of mixed coarse grit to fine sand and mud are associated with abundant rootlets and vitrain lenses and form a zone (2900–2885 m [9514–9465 ft]) between sandstone beds in the lower part of the core section (Figure 12). They are defined by a 2.5-m (8-ft)-thick, blocky mudstone that consists of red spots or blotches and roots defined by vitrain lenses. The blotches are composed of hematite and mixed mud, fine sand, and quartz grains.

Interpretation of Maui 5

The study interval of the Kaimiro Formation in Maui 5 represents a tidal-channel and marsh-paleosol lithofacies. The erosional-based, subparallel and par-allel laminated, trough cross-bedded sandstone is in-terpreted as a tidal-channel deposit. The multiple stacking of these fining-upward units (Figure 12) indicates infilled sand cycles of small tidal inlets. The channel fill starts with grains of quartz sand or coal spars overlain by medium to fine sand. The presence of *Ophiomorpha* trace fossils and *Teredo*-bored wood suggests brackish-estuarine conditions in a tidal in-let. The rooted and intensely bioturbated sandstone represents abandoned tidal-channel deposits reworked by burrowing animals and inhabited by plants in a high marsh setting. The red glaebules or mottles indi-cate pedogenesis. The mudstone and siltstone, which contain wispy ripple laminations and burrows, repre-sent intertidal mud-flat lithofacies.

The red, rooted, and mottled mudstone is inter-preted as a well-developed paleosol in a high marsh environment. The top of the paleosol is an erosional surface, which probably represents local microero-sional relief formed by the overlying tidal inlet chan-nel. The paleosol formed in a high intertidal area, where there was no sedimentation or vertical aggrada-tion; thus, it represents a sedimentary hiatus (nondep-osition or erosion). This sediment-starved condition permitted prolonged pedogenesis of the intertidal mud. However, the paleosol is difficult to correlate to a distinct sequence boundary, which suggests that the paleosol may be a localized phenomenon.

Maui A1g

The lithofacies of the Mangahewa Formation in Maui A1G consist mainly of interbedded sandstone, mudstone, siltstone, and very minor coal (Figure 13). The sandstone is gray, quartzose (>75% subround-ed quartz grains), and moderately sorted. Numerous sandstone beds fine upward from fine grained to very coarse grained. The basal contact of the sandstone is mainly erosional, subordinately sharp to gradational, and is marked by gritty quartz fragments. Sedimen-tary structures consist of abundant trough cross-beds, which range from 4 to 70 cm (1.6 to 28 in.) in height (Figure 13). Subparallel laminations are com-mon and are 3–41 cm (1.2–16 in.) in height. The trough cross-beds and subparallel laminations are commonly capped by climbing ripple laminations. Foreset beds, 7–38 cm (2.8–15 in.) in height, contain reactivation surfaces and are capped by climbing ripple laminations, which are as much as 15 cm (6 in.) in height. *Ophiomorpha* trace-fossil burrows are locally common to abundant and range from 1.5 cm (0.6 in.) in diameter and 15 cm (6 in.) long. *Diplocriterion* burrows, 4–6 cm (1.6–2.4 in.) wide, are also common. A few *Teredo*-bored wood fragments are present in the erosional-based sandstone.

The lithofacies in the Kaimiro Formation consists mainly of silty to sandy flaser beds, which are abundant and range from 4 to 36 cm (1.6 to 14 in.) thick; they contain mud drapes that are moderately to intensely burrowed (Figure 15). The flaser beds contain alternating ripple laminations, 1–4 cm (0.4–1.6 in.) in height, and burrowed mud drapes, 1–1.5 cm (0.4–0.6 in) in height (Figure 15) and, in places, are so heavily bioturbated that only relict bedding is observed. Lenticular beds, which are silty to sandy, alternate with flaser beds and form cosets. The lenticular beds are rippled and burrowed and contain burrowed mud drapes. The silty to sandy flaser and lenticular beds alternate with thin to thick burrowed mudstone and form heterolithic stratification.

The mudstone is dark gray to black and contains flat lenses of siltstone, 1–5 mm (0.04–0.2 in.) thick, and lenticular ripple laminations, 1–1.5 cm (0.4–0.6 in.) in height, which is silty to sandy and burrowed. The mudstone commonly has both sharp and gradational contacts. The mudstone commonly is intercalated with gray siltstone that is flaser- and lenticular-bedded. Rootlets are common in mudstone below thin coal beds. The coal beds are typically 1–3 cm (0.4–1.2 in.) thick and contain vertical burrows in the upper part.

Interpretation of Maui A1g

The study interval of the Mangahewa and Kaimiro Formations at Maui A1G represents tidal-channel, tidal-flat, and marsh lithofacies. The erosional-based, trough, subparallel, and foreset cross-bedded sandstone is interpreted as a tidal-channel deposit. The stacked, fining-upward sandstone beds (Figure 13) represent small, amalgamated tidal inlets infilled by tidal sand waves. The foresets reflect tidal bundles of alternating foreset laminae separated by reactivation surfaces and climbing ripples. These types of foresets represent spring-neap tidal cycle deposits (Terwindt, 1981; Visser, 1980). The presence of *Ophiomorpha* and *Diplocriterion* trace fossils and *Teredo*-bored wood suggests an estuarine-beach setting for the tidal-channel complex.

The alternating mudstone and siltstone, which contains burrowed flaser and lenticular beds and burrowed mud drapes, represent intertidal-flat deposits. The alternating (heterolithic structure) flaser and lenticular cosets with thin burrowed mud drapes reflect deposition from ebb and flow tidal currents. The mud drapes represent slack-water deposition during ebb flow (Reineck and Wunderlich, 1968; de Raaf and Boersma, 1971; Reineck and Singh, 1980). The thin coal beds interbedded with the intertidal deposits represent brackish-water marsh deposits, as indicated by the burrows on top of the coal.

COAL BUILDUP

The lithofacies analyses and interpretations indicate that the Eocene Kapuni Group was deposited on a tide-influenced coastal plain bounded seaward by marine shoreface and on the hinterland by alluvial paleovalleys. Brackish-water marshes and freshwater mires in these tide-influenced coasts formed coal deposits as observed in Maui A1G, Kapuni 14, and Kapuni 8. These coal deposits are either interbedded with fluvial channel, flood-plain lacustrine, and flood-plain deposits (alluvial belt lithofacies), as in Kapuni 8, or intercalated with tidal-channel and intertidal deposits with paleosols, as in Kapuni 14, Maui 5, and Maui A1G. The coal beds are thicker in the freshwater mires of the alluvial belt lithofacies of the Kapuni Group than in the tidal-channel–intertidal lithofacies, which was deposited on the tidally influenced coastal plain.

Coal beds in the Eocene Kapuni Group (Kaimiro and Mangahewa Formations) accumulated in coastal mires between marine-shoreface environments, and are exemplified in the Maui field and alluvial paleovalley environments, and in Toru-1 (Figures 2, 5). Although the coal beds formed in alluvial mires and tidal marshes, a fundamental association exists between the occurrence of volumetrically significant coal beds and their position in the stratigraphic architecture of the Kapuni Group. For example, numerous studies in the past several decades have demonstrated that major coal beds occur in specific stratigraphic positions in the transgressive-regressive sequences of the Upper Cretaceous of the Rocky Mountain region of the United States (Sears et al., 1941; Weimer, 1960; Fassett and Hinds, 1971; Ryer, 1984; Flores and Cross, 1991). These studies have

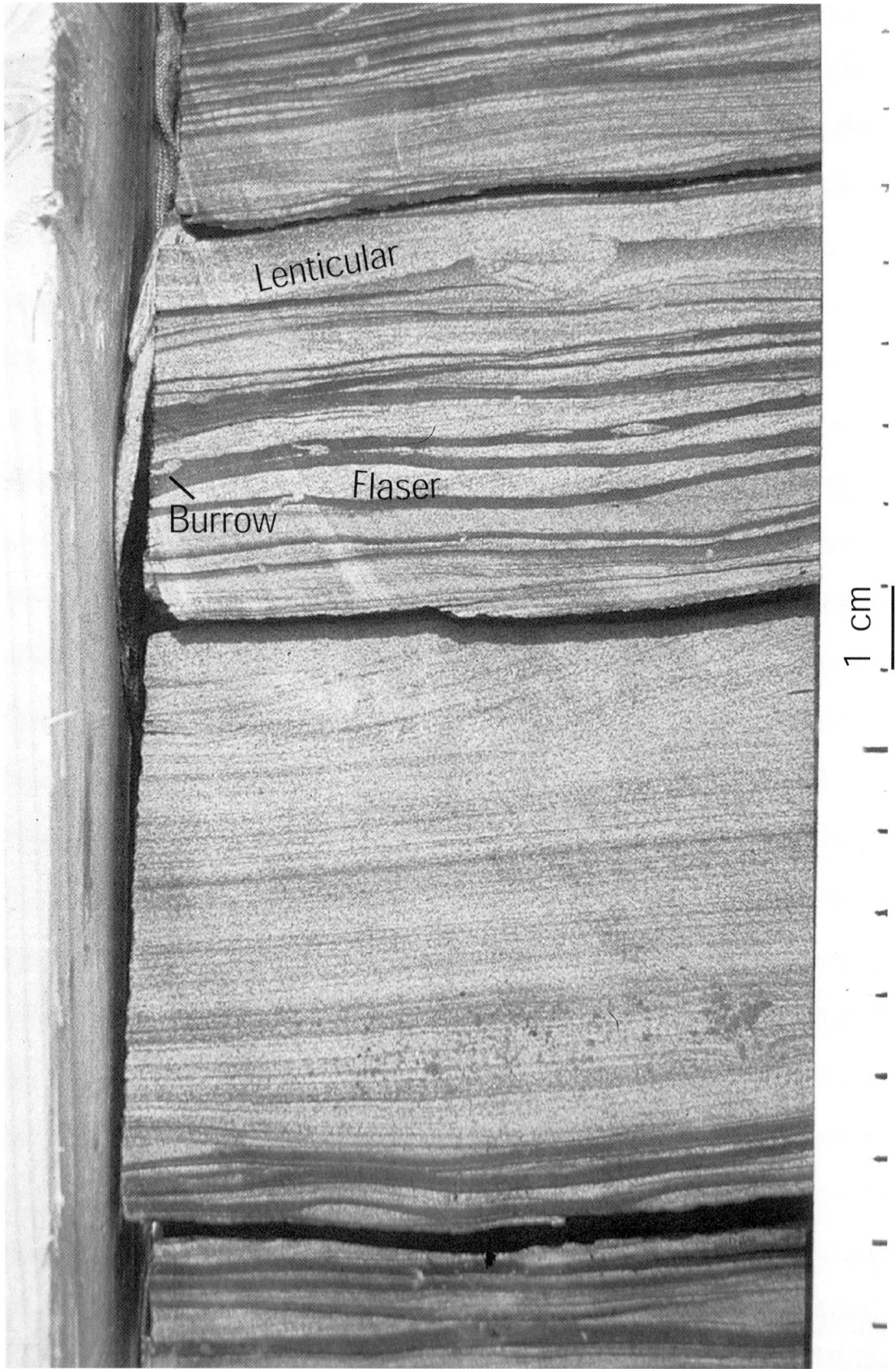

Figure 15. Flaser and lenticular beds of siltstones alternating with burrowed mud drapes and a trough cross-bedded and subparallel laminated sandy siltstone. The core (Maui 6; depth 2810 m [9219 ft]) is about 17 cm (6.7 in.) thick, as indicated by the scale on the right of the photograph.

shown that coal beds occur on top and landward of marine-shoreface lithofacies in progradational packages that make up transgressive-regressive sequences. More specifically, coal buildup preferentially occurs at stratigraphic positions where the marine-shoreface lithofacies of successive progradational events are stacked vertically (Flores and Cross, 1991). However, in the Kapuni Group, the coal buildup is related to marine-shoreface lithofacies that successively stepped landward as shown in Figure 2. In addition, minor wedges of these marine-shoreface lithofacies have retrograded farther landward and served as platforms for coal buildup (Figure 16). These retrogradations, which reflect minor landward movement of paleo-shorelines caused by fluctuations of sea level (Figure 2), are demonstrated by the marine-shoreface lithofacies in Maui 6 and by the geophysical and lithological logs of Kapuni 2 and Kapuni 13. The landward movement of the paleoshoreline by wave action may be reflected by transgressive surfaces and ravinement lag conglomerates atop the parasequence sets exemplified in Maui 6 (Figure 11). Landward-stepping parasequence sets result in proportionately more sediment accumulation in the marine-shoreface environments than in the coastal plain.

Control of Sea Level Fluctuations on Coal Buildup

The geophysical and lithological logs of Kapuni 2 and Kapuni 13, which are 0.75 km (0.47 mi) apart (Figure 16), show generalized lithofacies variations and the vertical and lateral distribution of coal beds in the Mangahewa Formation. The formation may be divided into three coal zones: lower coal zone (3620–3780 m [11,877–12,402 ft] in Kapuni 2 and 3680–3810 m [12,073–12,500 ft] in Kapuni 13), middle coal zone (3490–3500 m [11,450–11,483 ft] in Kapuni 2 and 3500–3550 m [11,483–11647 ft] in Kapuni 13), and upper coal zone (3050–3360 m [10,007–11,24 ft] in Kapuni 2 and 3090–3410 m [10,138–11,188 ft] in Kapuni 13). These coal zones are separated by coarsening-upward, sandstone-dominated intervals as much as 125 m (410 ft) thick. The lower coal zone consists of four coal beds ranging from 2 to 10 m (6.6 to 33 ft) thick. The middle coal zone contains two to three coal beds ranging from 2 to 5 m (6.6 to 16 ft) thick. The upper coal zone includes 28 to 33 coal beds ranging from 2 to 14 m (6.6 to 46 ft) thick. In these wells, the coal beds are stacked vertically and become more numerous and thicker toward the upper coal zone.

The presence of thin, coarsening-upward marine-shoreface lithofacies below the upper coal zones (Figure 16) and thick, coarsening-upward marine-shoreface lithofacies between the middle and lower coal zones and below the lower coal zone indicates the retrogradation of paleoshorelines. Pocknall and Beggs (1990) proposed for these Eocene rocks that the succession of fluvial, tidal, and marine-shoreface lithofacies might be attributed to third- and fourth-order fluctuations of sea level. In the Cretaceous Ferron Sandstone, Ryer (1984) indicated that lithofacies of transgressive and regressive phases of third-order cycles contain only stacked coal beds of the fourth-order cycles. In addition, Ryer indicated that coal buildup is best associated with the fourth-order cycle in which a one-to-one relationship with thick coal seams can be observed.

As shown in Figure 2, the coal buildup, best demonstrated by vertical stacking of coal beds, particularly in the Mangahewa Formation, is associated with the fluvial-tidal–marine-shoreface lithofacies related to the third-order fluctuation of sea level according to Haq et al. (1988). This order of fluctuation of sea level corresponds roughly to about 1.5-m.y. time interval, which occurs three times from the middle Eocene (46.5 m.y.) to late Eocene (41.0 m.y.). This time interval is between retrogradational or landward pinch-outs of marine-shoreface sandstone lithofacies, which also roughly correlates with the sea level rises shown in Figure 2. However, a variation in thickness and number of coal beds occurs in these third-order cycle buildups. The lower, middle, and upper coal zones bounded by the marine-shoreface sandstone lithofacies represent the third-order-cycle coal buildup. The occurrence of a few thin coal beds in the lower and middle coal zones separated by thick marine-shoreface lithofacies suggests that coal buildup in these coal zones was interrupted by rapid landward-stepping shoreface events. The buildup of thick and numerous coal beds in the upper coal zone represents prolonged accumulation of peat mires behind a very thick, vertically stacked marine-shoreface lithofacies, which permitted aggradation by alluvial belts, which served as platforms for freshwater mires in which thick coal beds accumulated.

A second-order-cycle (Haq et al., 1988) coal buildup is represented by the entire lower to upper coal zones of the Mangahewa Formation from 49- to 39-m.y. interval. The interval is bounded by major cycles of sea level drops marked by erosional surfaces or sequence boundaries (see Figure 2). During this time interval, coal buildup became well developed through time, as evidenced by the upward thickening of the coal beds.

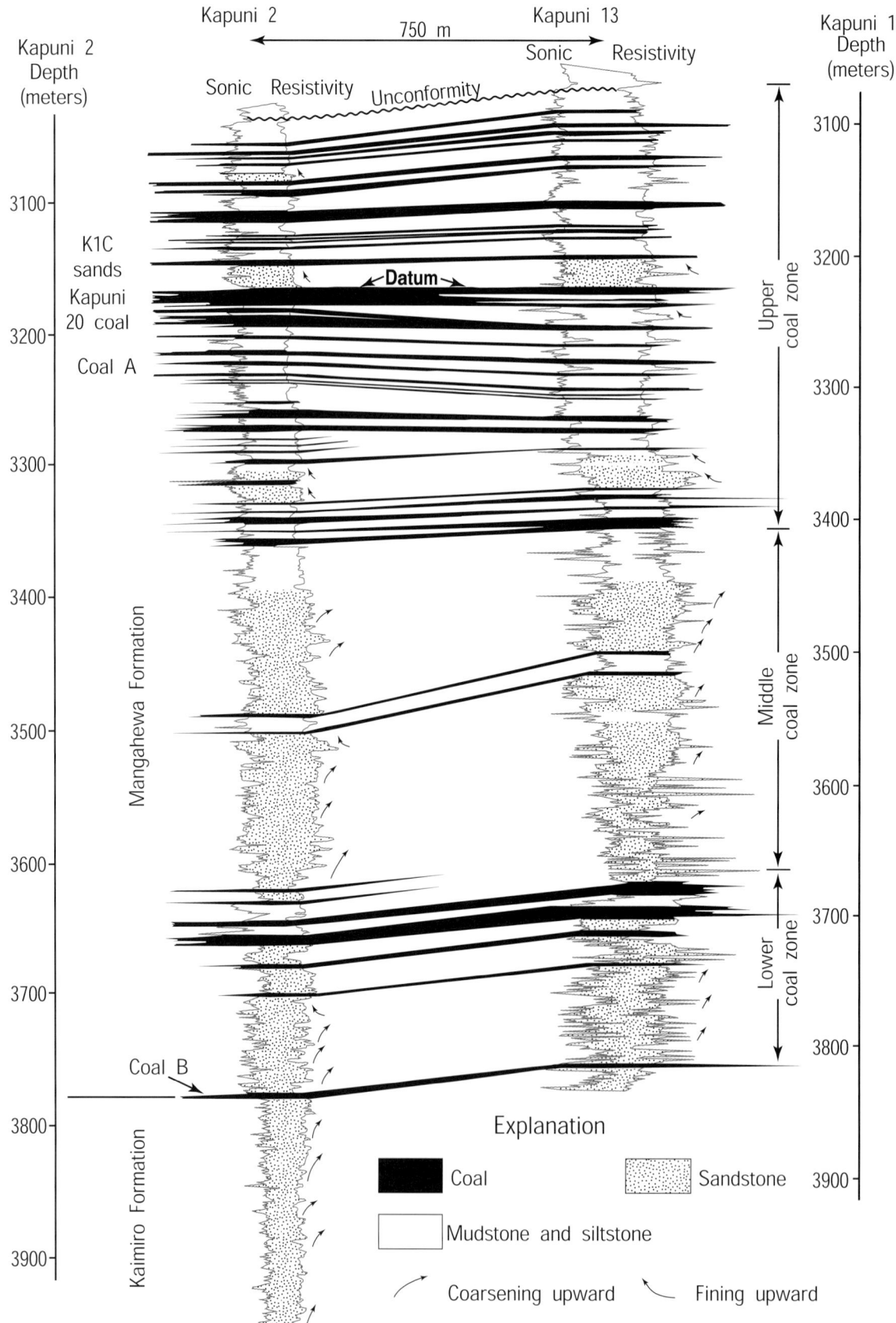

Figure 16. Vertical stacking of coal beds in the Mangahewa Formation in Kapuni 2 and Kapuni 13. The coal beds are interbedded with coarsening-upward and fining-upward complexes of sandstone, mudstone, and siltstone. Modified from Shell BP & Todd Oil Services (1984).

Control of Depositional Environments on Coal Buildup

The third- and fourth-order cycles of sea level fluctuation, combined with local depositional processes, affected coal buildup in the Eocene Kapuni Group. Depositional processes in the tidal and intertidal environments may have controlled the accumulation of individual coal beds in the fourth-order-cycle coal buildup. The Kapuni 20 coal bed, coal A, coal B, and intervening coal deposits between coarsening- and fining-upward sandstones (as shown in Figure 16) best demonstrate this buildup.

The Kapuni 20 is the thickest coal bed (as much as 14 m [46 ft]) in the upper coal zone in Kapuni 2 and Kapuni 13, and was mapped as a laterally extensive marker bed several kilometers in extent, based on three-dimensional seismic modeling studies by Voggenreiter (1992). However, as shown in Figure 16, the Kapuni 20 coal bed locally splits into three coal beds between Kapuni 2 and Kapuni 13, which is less than 0.75 km (0.47 mi) apart. This coal bed is overlain by a fining-upward fluvial channel sandstone (K1C sands; Figure 16). Bryant and Bartlett (1992) interpreted the Kapuni 20 coal bed as a raised mire deposit based on very low ash yield (less than 2%); Voggenreiter (1992) considered it to have formed in a mire juxtaposed with a meandering fluvial channel (Figure 17). This depositional environment probably represents a fluvial channel or alluvial lithofacies interspersed with the tidal and intertidal lithofacies (see lithofacies associations in Kapuni 6 and Kapuni 14). There are several fining-upward sandstones in the upper coal zone at Kapuni 2 and Kapuni 13 (Figure 16), which indicate the presence of the fluvial channel lithofacies (see Kapuni 8 for lithofacies descriptions and interpretations). In contrast, the coal beds in the lower and middle coal zones lacked the fining-upward fluvial channel lithofacies. The coal beds are intercalated with tidal and intertidal lithofacies (see Kapuni 14 for lithofacies descriptions and interpretations).

The occurrence of tidal and intertidal lithofacies successions in the lower and middle coal zones and fluvial channel lithofacies in the upper coal zone in Kapuni 2 and Kapuni 13 suggests that alluvial belts are more common in the tidal-influenced coastal plain of the upper Mangahewa Formation than in the lower and middle parts of the formation. It also suggests that alluvial aggradation in the coastal plain provided belts of freshwater mires, where thick coal accumulated. As stated earlier, Bryant and Bartlett (1992) suggested that the low ash yield (less than 2%) of the Kapuni 20 coal bed might reflect accumulation in a raised or domed mire. Indeed, if the coal-depositional environment is a raised mire, it may be similar to the domed mires of central Sumatra as described by R. M. Flores and T. A. Moore (2003, personal communication). In central Sumatra, the domed mires are formed on a tidally influenced coastal plain. Thick peats of these domed mires are more than 10 m (33 ft) thick and greater than 3500 km^2 (1351 mi^2) in areal extent and are laterally juxtaposed with meandering and anastomosed fluvial channels developed about 70–80 km (43–50 mi) from the shoreline.

Although peat deposits form in coastal areas where tidal ranges are high (Coleman et al., 1970), there are only a few thin coal beds of Eocene age that accumulated in tidally influenced coastal plains (Flores and Johnson, 1995; Flores and Sykes, 1996; Flores et al., 1996). The occurrence of very thin coal beds associated with tidal-channel and intertidal-flat lithofacies in Kapuni 14 and Maui A1G supports this concept. In tidally influenced areas of the coastal plain, the brackish-water nature of the marshes, constant influx of mud into the marshes, and erosion accompanying the daily ebb and flow of tidal currents prevented substantial accumulation of peat deposits. In addition, nonwoody vegetation in these marshes did not lend to accumulation of thick coal deposits. Further, burrowing animals intensively reworked any accumulation of organic deposits in the tidal environments.

Control of Basin Subsidence and Accommodation Space on Coal Buildup

Variations in subsidence rate of the Taranaki Basin may have had some control on the coal buildup from early to late Eocene. Based on the geohistory curves of several exploration wells in the Taranaki Basin (Hayward, 1987), King and Thrasher (1996) indicated generally rapid subsidence during the Late Cretaceous–Paleocene and waning subsidence during the Eocene. However, based on paleobathymetry and paleoshoreline transgressions, relatively rapid subsidence began in the late Eocene in the northwest part of the basin and continued eastward and southward.

The stacking of coal beds in the late Eocene Mangahewa Formation may have been controlled by creation of accommodation space caused by basin subsidence. Rapid basin subsidence at this time (King and Thrasher, 1996) may have allowed preservation and burial of successions of peat deposits, thus resulting in vertically stacked coal beds. In addition, this may have promoted vertical aggradation in alluvial belts,

Figure 17. Possible lithofacies relationship at the time that the top of the Kapuni coal bed was deposited, as represented by the combined seismic signal for the K1C sands and Kapuni 20 coal in the Kapuni field area. Modified from Voggenreiter (1992).

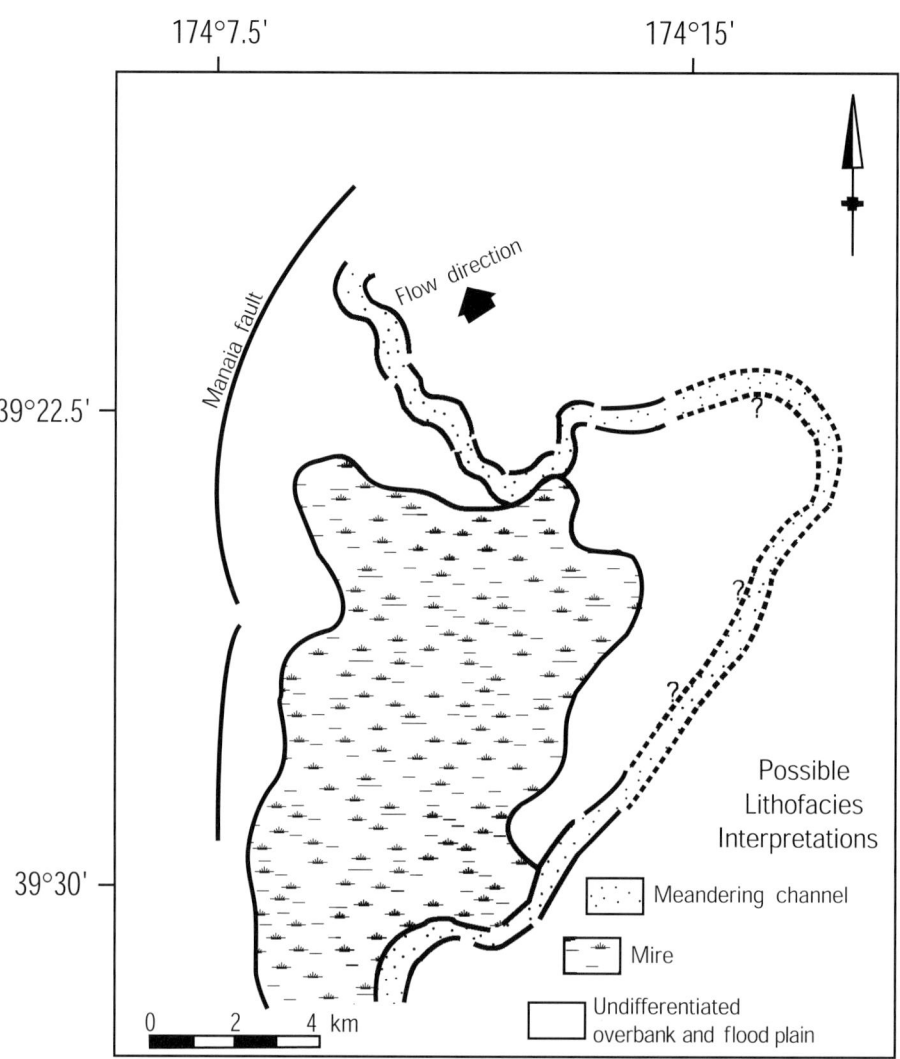

forming raised platforms where freshwater mires accumulated peat undisturbed by sediment floods.

During the early–late Eocene, sediment supply from erosion of the hinterlands to the south may have kept up with the intermittent paleoshoreline transgressions, which, in turn, narrowed the accommodation space. Thus, thinner and lesser numbers of coal beds were formed during the early–middle Eocene.

CONCLUSIONS

The coal buildup described here for the Eocene Kapuni Group in the Taranaki Basin, New Zealand, is generally similar to that in the Rocky Mountain region of the United States. The buildup is similar in that the coal beds are vertically stacked landward of marine-shoreface environments. Coarsening-upward sandstones of these marine-shoreface environments formed as landward-stepping parasequences. Buildup of these stacked coal beds is attributed to third- and fourth-order fluctuations of sea level during an overall marine transgression. In the Kapuni Group, vertical stacking of coal beds is variable during fourth-order cycles of sedimentation; however, the overall stacking is

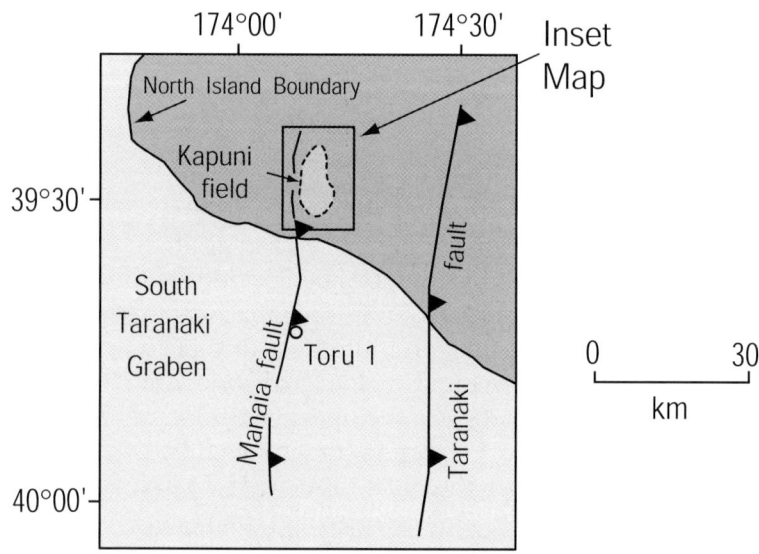

best developed during the third-order cycle of a transgressive phase. During the third-order cycle of sea level fluctuation, thick coal beds formed in raised mires near alluvial belts. Thinner coals formed in tidally affected areas of the coastal plain. The occurrence of thick coal beds and their stacking is attributed largely to interactions of sea level fluctuation, basin subsidence, and accommodation space, resulting in changes in depositional environments. The genetic relationship of the coal buildup should prove useful for targeting thick coal beds for future coal-bed methane or natural gas resource assessment in the Taranaki Basin.

ACKNOWLEDGMENTS

This study was made possible by travel financial support to R. M. Flores from the Institute of Geological and Nuclear Sciences Inc. (IGNS), Lower Hutt, New Zealand. Core work during 1992 was facilitated by John Simes of IGNS and the Ministry of Commerce, Wellington, New Zealand. The author deeply appreciates the assistance of J. M. Beggs, Greg Browne, P. R. King, and R. Sykes during this study. Special thanks goes to David Pocknall for allowing the author to stay in his home during the entire period of study. The manuscript benefited considerably from reviews by Ted Dyman (U.S. Geological Survey) and David Pocknall (BP, Houston).

REFERENCES CITED

Beggs, J. M., and D. T. Pocknall, 1992, Sequence stratigraphic controls on reservoir architecture, lower Eocene Kapuni Group, Maui field, Taranaki Basin, New Zealand: New Zealand Geological Survey Report G 162, 36 p.

Beggs, J. M., G. J. Wilson, J. I. Raine, and D. T. Pocknall, 1992, Biostratigraphic review of Cretaceous and early Cenozoic section in Taranaki Basin exploration wells: Institute of Geological and Nuclear Sciences Ltd., Client Report 1992/18, 28 p.

Berggren, W. A., D. V. Kent, C. C. Swisher, and M. P. Aubry, 1995, A revised Cenozoic geochronology and chronostratigraphy, in W. A. Berggren, D. V. Kent, M. P. Aubry, and J. Hardenbol, eds., Geochronology, times scales and global stratigraphic correlation: SEPM Special Publication 54, p. 129–212.

Bryant, I. D., and A. D. Bartlett, 1992, Kapuni 3-D reservoir model and reservoir simulation, in Proceedings of the 1991 New Zealand Oil Exploration Conference, Ministry of Commerce, p. 404–412.

Bryant, I. D., M. G. Marshall, C. W. Greenstreet, W. R.

Voggenreiter, J. M. Cohen, and J. F. Stroemmen, 1994, Integrated geological reservoir modelling of the Maui field, Taranaki Basin, New Zealand, in Ministry of Commerce Proceedings of the 1994 New Zealand Petroleum Conference, p. 256–280.

Chatellier, J. Y., and V. H. Hitchings, 1987, Geological investigations of cores from the C and D reservoirs in well Maui 7, Maui field, New Zealand: New Zealand Open-File Petroleum Report 1302, 81 p.

Coleman, J. M., S. M. Gagliano, and W. G. Smith, 1970, Sedimentation in a Malaysian high tide tropical delta, in J. P. Morgan, ed., Deltaic sedimentation: Modern and ancient: SEPM Special Publication 15, p. 185–197.

Cook, R., and R. Gregg, 1997, Overview of New Zealand's petroleum systems and potential: Oil & Gas Journal, May, p. 55–58.

de Boer, P. L., A. van Gelder, and S. D. Nio, 1988, Tide-influenced sedimentary environments and facies: Reidel Publishing Company, Dordrecht, The Netherlands, 342 p.

de Raaf, J. F. M., and J. R. Boersma, 1971, Tidal deposits and their sedimentary structures: Geologie en Minjnbouw, v. 50, p. 479–503.

Fassett, J. E., and J. S. Hinds, 1971, Geology and fuel resources of the Fruitland Formation: U.S. Geological Survey Professional Paper 676, 76 p.

Flores, R. M., and T. A. Cross, 1991, Cretaceous and Tertiary coals of the Rocky Mountains and Great Plains regions, in H. J. Gluskoter, D. D. Rice, and R. B. Taylor, eds., Economic geology, U. S.: Geological Society of America, The geology of North America, v. P-2, p. 547–571.

Flores, R. M., and S. Y. Johnson, 1995, Sedimentology and lithofacies of the Eocene Skookumchuck Formation in the Centralia coal mine, southwest Washington, in A. E. Fritsche, ed., Cenozoic paleogeography of the western United States— II: Pacific Section, SEPM, p. 274–290.

Flores, R. M., and R. Sykes, 1996, Depositional controls on coal distribution and quality in the Eocene Brunner coal measures, Buller coalfield, South Island, New Zealand: International Journal of Coal Geology, v. 29, p. 291–336.

Flores, R. M., J. M. Beggs, and P. R. King, 1993, Sedimentology of tide-dominated reservoir sandstones in the Eocene Kapuni Group, Taranaki Basin (abs.): AAPG Bulletin, v. 77, p. 102.

Flores, R. M., G. D. Stricker, and R. B. Stiles, 1996, Tidal influence on deposition and quality of coals in the Miocene Tyonek Formation, Beluga coalfield, Upper Cook Inlet, Alaska: U.S. Geological Survey Professional Paper 1574, p. 137–156.

Haq, B. U., J. Hardenbol, and P. R. Vail, 1987, Chronology of fluctuating sea levels since the Triassic: Science, v. 235, p. 1156–1167.

Haq, B. U., J. Hardenbol, and P. R. Vail, 1988, Mesozoic and Cenozoic chronostratigraphy and cycles of sea-level change, in C. K. Wilgus, B. S. Hastings, C. G. St. C. Kendall, H. W. Posamentier, C. A. Ross, and J. C. van

Wagoner, eds., Sea-level changes: An integrated approach: SEPM Special Publication 42, p. 71–108.

Hayward, B. W., 1987, Paleobathymetry and structural and tectonic history of Cenozoic drillhole sequences in Taranaki Basin: New Zealand Geological Survey Report PAL 122, 45 p.

Hitchings, V. H., 1986, Geological investigations of cores from the C1 and D1 reservoirs in well Maui-A1 (G), Maui field, New Zealand: New Zealand Open-File Petroleum Report 1036, 104 p.

Hitchings, V. H., 1987a, Geological investigations of cores from the C1 reservoir in well Maui-5, Maui field, New Zealand: New Zealand Open-File Petroleum Report 1279, 98 p.

Hitchings, V. H., 1987b, Geological investigations of cores from the C1 reservoir in well Maui-6, Maui field, New Zealand: New Zealand Open-File Petroleum Report 1279, 96 p.

Hunter, R. E., H. E. Clifton, and R. L. Phillips, 1979, Depositional processes, sedimentary structures and predicted vertical sequences in barred nearshore systems, southern Oregon coasts: Journal of Sedimentary Petrology, v. 49, p. 711–726.

King, P. R., 1991, Physiographic maps of the Taranaki Basin— Late Cretaceous to Recent: New Zealand Geological Survey Report G155, 32 p.

King, P. R., T. R. Naish, and G. P. Thrasher, 1991, Structural cross sections and selected palinspastic reconstructions of the Taranaki Basin: New Zealand Geological Survey Report G150, 49 p.

King, P. R., and G. P. Thrasher, 1996, Cretaceous–Cenozoic geology and petroleum systems of the Taranaki Basin: New Zealand Institute of Geological & Nuclear Sciences Monograph 13, 243 p.

Knight, R. J., and R. W. Dalrymple, 1975, Intertidal sediments from the south shore of Cobequid Bay, Bay of Fundy, Nova Scotia, Canada, in R. N. Ginsburg, ed., Tidal deposits— A casebook of recent examples and fossil counterparts: New York, Springer-Verlag, p. 47–56.

Nio, S.-D., and C. Yang, 1991, Diagnostic attributes of clastic tidal deposits— a review, in D. G., Smith, G. E. Reinson, B. A. Zaitlin, and R. A. Rahmani, eds., Clastic tidal sedimentology: Canadian Society of Petroleum Geology Memoir 16, p. 3–28.

Pocknall, D. T., and J. M. Beggs, 1990, Palynofacies as a tool for the interpretation of depositional environments in the Waikato and Taranaki Basins, New Zealand, in Proceedings of the 1989 New Zealand Oil Exploration Conference, Ministry of Commerce, p. 250–258.

Pocknall, D. T., J. M. Beggs, and I. D. Bryant, 1992, Sequence stratigraphic controls on architecture, lower Eocene Kapuni Group, Maui field, Taranaki Basin, New Zealand: Palynology, v. 16, p. 228.

Reading, H. G., 1978, Sedimentary Environments and facies: London, Blackwell Scientific Publications, 615 p.

Reineck, H.-E., and F. Wunderlich, 1968: Classification and origin of flaser and lenticular bedding: Sedimentology, v. 11, p. 99–104.

Reineck, H. E., and I. B. Singh, 1980, Depositional Sedimentary Environments: New York, Springer-Verlag, 549 p.

Ryer, T. A., 1984, Transgressive-regressive cycles and the occurrence of coal in some Upper Cretaceous strata of Utah, U.S.A., in R. A. Rahmani and R. M. Flores, eds., Sedimentology of coal and coal-bearing sequences: International Association of Sedimentologists Special Publication 7, p. 217–227.

Sears, J. D., C. B. Hunt, and T. A. Hendricks, 1941, Transgressive and regressive Cretaceous deposits in southern San Juan Basin, New Mexico: U.S. Geological Survey Professional Paper 193F, p. 110–121.

Shell BP & Todd Oil Services, 1984, Geological Summary Kapuni Deep-1, Petroleum Mining Licence 839 Taranaki, New Zealand: New Zealand Unpublished Open-File Petroleum Report 1024: Wellington, Crown Minerals, Ministry of Economic Development, unpaginated.

Smale, D., 1992, Provenance of sediments in Taranaki Basin: An assessment from heavy minerals, in Proceedings of the 1991 New Zealand Oil Exploration Conference, Ministry of Commerce, p. 245–254.

Smale, D., 1996, Petrographic summaries of Taranaki petroleum reports: Institute of Geological and Nuclear Sciences Ltd., Science Report 96/1, 56 p.

Smale, D., and A. C. Morton, 1987, Heavy mineral suites of core samples from the McKee Formation (Eocene–lower Oligocene), Taranaki: Implications for provenance and diagenesis: New Zealand Journal of Geology and Geophysics, v. 30, p. 299–306.

Terwindt, J. H. J., 1981, Origin and sequences of sedimentary structures in inshore mesotidal deposits of the North Sea, in S.-D. Nio, R. J. E. Shuttenhelm, and Tj. C. E. van Weering, eds., Holocene marine sedimentation in the North Sea: International Association of Sedimentologists Special Publication 5, p. 4–26.

Thrasher, G. P., 1992, Late Cretaceous source rocks of Taranaki Basin, in Proceedings of the 1991 New Zealand Oil Exploration Conference, Ministry of Commerce, p. 147–154.

Visser, M. J., 1980, Neap-spring cycles reflected in Holocene subtidal large-scale bedform deposits: A preliminary note: Geology, v. 8, p. 543–546.

Voggenreiter, W. R., 1992, Kapuni 3-D interpretation-imaging Eocene paleogeography, in Proceedings of the 1991 New Zealand Oil Exploration Conference, Ministry of Commerce, p. 287–298.

Weimer, R. J., 1960, Upper Cretaceous stratigraphy, Rocky Mountains area: AAPG Bulletin, v. 44, p. 1–20.

Kvale, E. P., M. Mastalerz, L. C. Furer, D. W. Engelhardt, C. B. Rexroad, and C. F. Eble, 2004, Atokan and early Desmoinesian coal-bearing parasequences in Indiana, U.S.A., *in* J. C. Pashin and R. A. Gastaldo, eds., Sequence stratigraphy, paleoclimate, and tectonics of coal-bearing strata: AAPG Studies in Geology 51, p. 71–88.

4

Atokan and Early Desmoinesian Coal-bearing Parasequences in Indiana, U.S.A.

Erik P. Kvale

Indiana Geological Survey, Indiana University, Bloomington, Indiana, U.S.A.

Maria Mastalerz

Indiana Geological Survey, Indiana University, Bloomington, Indiana, U.S.A.

Lloyd C. Furer

N799 Waubunsee Trail, Fort Atkinson, Wisconsin, U.S.A.

Donald W. Engelhardt (deceased)

Earth Sciences and Resources Institute, University of South Carolina, Columbia, South Carolina, U.S.A.

Carl B. Rexroad

Indiana Geological Survey, Indiana University, Bloomington, Indiana, U.S.A.

Cortland F. Eble

Kentucky Geological Survey, Lexington, Kentucky, U.S.A.

ABSTRACT

Pennsylvanian (Atokan–early Desmoinesian) parasequences in Indiana are thin (2–13 m; 5–40 ft) intervals that are composed of coal, siliciclastic, and carbonate-clastic units bounded by paleosols. Because the parasequences exhibit significant lateral and vertical lithologic variability and are so thin, they are difficult or impossible to discern on standard oil and gas geophysical logs. Therefore, in Indiana, regional correlations of this interval based primarily on geophysical logs and lithologic strip logs created from drill cuttings remain controversial. Detailed analyses of proprietary core from numerous locations in Daviess County in southwestern Indiana reveal that the most traceable of the parasequence facies in core are the paleosols which represent exposure surfaces that developed, in most cases, during apparent basinwide drops in relative sea level. Correlations are substantiated by detailed palynologic analyses of material collected from the bases of overlying marine-influenced flooding surfaces and by the use of thin, nearly continuous marker beds and the presence of certain biostratigraphically significant conodonts. Transgressive and regressive facies above the exposure surfaces are preserved with varying significance. The relative significance of the transgressive-regressive facies in a parasequence is, in part, related to the relative rates of changes in accommodation space and sea level. Detailed analyses of coal lithotypes and maceral compositions in two Atokan coal seams reveal that base-level rises during paleomire development were gradual in one and abrupt (catastrophic?) in another. Abrupt transgressions and the preservation of relatively thick transgressive sequences above the exposure surfaces were perhaps related to rates of mire collapse and compaction of the underlying peat (now coal) and soil (paleosol).

INTRODUCTION

From Weller (1930) and Wanless and Shepard (1936) to Heckel (1994), much has been written about depositional sequences in Desmoinesian and younger rocks (largely that interval above the Colchester Coal Member) in the Illinois basin (Figures 1A, 2). These upper Pennsylvanian rocks record transgressive-regressive parasequences and parasequence sets that are recognizable basinwide largely because of the presence of coal seams and limestone intervals that serve as marker beds. Most, if not all, of these Desmoinesian and younger units can be identified on standard oil and gas geophysical logs because they are relatively thick, and many of them exhibit a predictable succession of facies. Although there are lateral and vertical complexities in these sequences, and the nature of the depositional origins of certain of the facies in them are somewhat problematic, more or less repetitive sequences are acknowledged by most who study this interval.

This is not so for the older Pennsylvanian sediments in the Illinois basin, in which regional correlations are much more tenuous. In southwestern Indiana, near the eastern edge of the Pennsylvanian basin margin, 15 members have been named formally in the lower Pennsylvanian Mansfield, Brazil, and Staunton Formations, an interval as much as 130 m (400 ft) thick (Shaver et al., 1986). Each of these three formations is interpreted to thicken toward the center of the basin (Droste and Furer, 1995). Two of the members (top of Buffaloville/Minshall Coal Members and base of Lower Block Coal Member) define the boundaries between the formations. However, few of the 15 members have been convincingly correlated beyond their type areas. The most convincing of these correlations have relied largely on micropaleontologic data (e.g., Rexroad et al., 1997).

The reason that correlations in the lower part of the Pennsylvanian are so tenuous is, in part, that natural and man-made outcrops are few and far between, and relatively thick marker beds are absent. The Mansfield Formation has the best natural outcrops. Most of the outcrops are restricted, however, to the coarser-grained sandstone facies. Few cores from the Mansfield Formation exist, largely because there are few economically mineable coals in this

formation. Natural outcrops of fine-grained facies in the Mansfield Formation are few, deeply weathered, and heavily vegetated.

Overall, the Brazil and Staunton Formations are dominated by a greater relative percentage of fine-grained facies and are even less well exposed in natural outcrops than the Mansfield Formation. Core information through the Brazil and Staunton Formations remains largely the proprietary property of coal companies. Subsurface correlations of units in these two formations are difficult even where locally persistent coal or limestone deposits occur, because these local markers are generally thin and are typically difficult, if not impossible, to identify on most standard oil and gas geophysical logs. Cores of public record that exist in this interval are widely spaced. The result is that, whereas attempts have been made to regionally correlate members in the Mansfield, Brazil, and Staunton Formations (largely on the basis of oil and gas geophysical logs), these studies remain controversial because time-stratigraphic units cannot be traced readily and because those time-stratigraphic units that have been proposed are not convincing or confirmed easily by others.

This study demonstrates that, despite the inherent complexities that exist in the Atokan and lower Desmoinesian Pennsylvanian sections of the Illinois basin, reliable and reproducible regional correlations are possible if suitably long cores are available. This study made use of a fairly closely spaced series of Pennsylvanian cores, obtained from two coal companies, which were used to identify a series of transgressive-regressive sedimentation sequences in Daviess County, Indiana (Figure 1B, C). The correlation of these sequences was based, in part, on the recognition of persistent exposure surfaces that manifest themselves as paleosols (e.g., Figure 3A) that are overlain by coals in some places but not everywhere. Most of the paleosols above paleosol 11 (Figure 4) can be traced into a deep core hole that was recently drilled in Posey County, Indiana, approximately 100 km (60 mi) southwest of Daviess County, by the Indiana Geological Survey as part of a deep coal-quality investigation, and to another deep core hole drilled in southernmost Sullivan County approximately 32 km (20 mi) northwest of the Daviess County holes. These correlations were substantiated by the identification

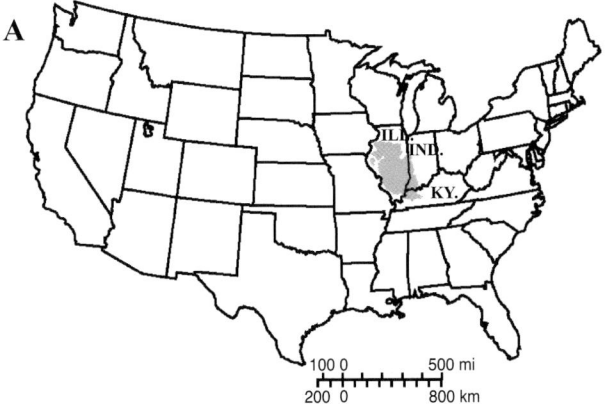

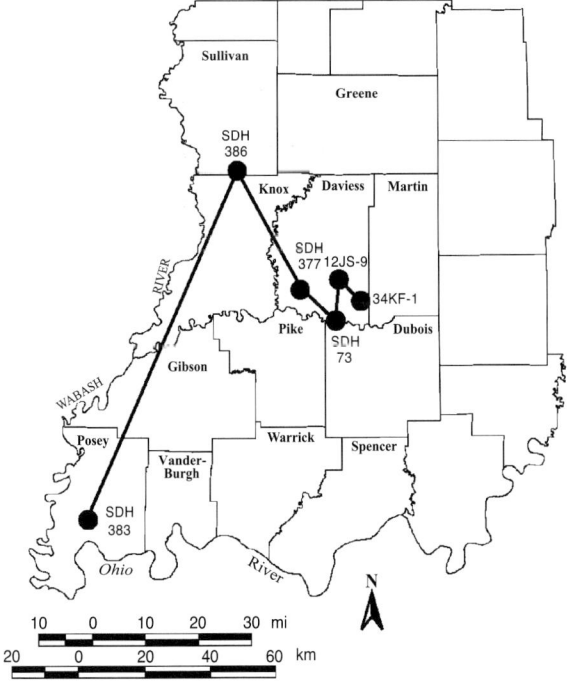

of certain relatively persistent thin beds, including a few coals and limestones, as well as by the presence of certain biostratigraphically significant palynomorphs and conodonts. This chapter documents the results of the investigation and discusses some of the important tectonic and sequence-stratigraphic implications of this study.

GEOLOGY OF THE STUDY INTERVAL

The Illinois basin is a spoon-shaped structural depression that extends 600 km (370 mi) in a general northwest-southeast direction and 320 km (200 mi) in a general east-west direction (Figure 1A). It is a cratonic basin that was located approximately 300 km (190 mi) north of the southern margin of the North American craton (Collinson et al., 1988). Paleozoic deposits fill and dip toward the center of the basin. In Indiana, dip of Pennsylvanian deposits generally ranges from 4 to 6 m/km (20 to 30 ft/mi) to the southwest. Pennsylvanian deposits reach a maximum thickness of 660 m (2000 ft) in Indiana (Droste et al., 2000).

The basal Pennsylvanian unit in Indiana is the Mansfield Formation (Figure 2), which is reported to achieve a maximum thickness of 220 m (800 ft) in sub-Pennsylvanian paleovalleys, of which only the upper 100 m (300 ft) is exposed along the outcrop belt in the southwestern part of the state (Droste and Furer, 1995; Droste et al., 2000). The Mansfield rests unconformably on Mississippian and older rocks and fills a major sub-Pennsylvanian drainage system that formed during late Mississippian into early Pennsylvanian time (Droste and Keller, 1989). The Mansfield, Brazil, and Staunton Formations are dominated by interbedded sandstone and mudstone. Thin coals and limestones and channel fills of mudstone, interbedded mudstone and sandstone, and sandstone are also present. The Mansfield Formation (particularly the lower half) characteristically contains more sandstone-filled channels than the next two younger formations. Many of the channels in the lower part of the Mansfield are filled with coarse- to medium-grained sandstone and contain quartz-clast conglomerates near their bases. Stacked channel successions are also more common in the Mansfield, and sandstone belts more than 30 m (90 ft) thick are present in both the outcrop and subsurface.

Figure 1. Location maps showing (A) distribution of Pennsylvanian sediments in the Illinois basin (shaded), (B) general study area (shaded), and (C) the line of cross section illustrated in Figure 4.

System	Series	Group	Formation	Member Unless Bed is Stated	
Pennsylvanian	Desmoinesian	Carbondale	Petersburg	Springfield Coal Mbr.	Foisomville Mbr.
				Stendal Ls.	
				Houchin Creek Coal	
			Linton	Survant Coal	
				Velpen Ls.	
				Mecca Shale	
				Colchester Coal	
				Coxville Ss.	
	Atokan	Raccoon Creek	Staunton	Seelyville Coal	
				Silverwood Ls.	
				Holland Ls.	
				Perth Ls.	
			Brazil	Minshall Coal	Buffaloville Coal
				Upper Block Coal	
				Lower Block Coal	
	Morrowan		Mansfield	Shady Lane Coal	
				Lead Creek Ls.	Ferdinand Bed
					Fulda Bed
				Mariah Hill Coal	
				Blue Creek Coal	
				Pinnick Coal	
				St. Meinrad Coal	
				French Lick Coal	

M i s s i s s i p p i a n

Figure 2. Stratigraphic column of part of the Pennsylvanian section in Indiana. The Pennsylvanian unconformably overlies the Mississippian. Modified from Shaver et al. (1986) and Tri-State Committee on Correlation of the Pennsylvanian System in the Illinois basin (2001). Ls = limestone.

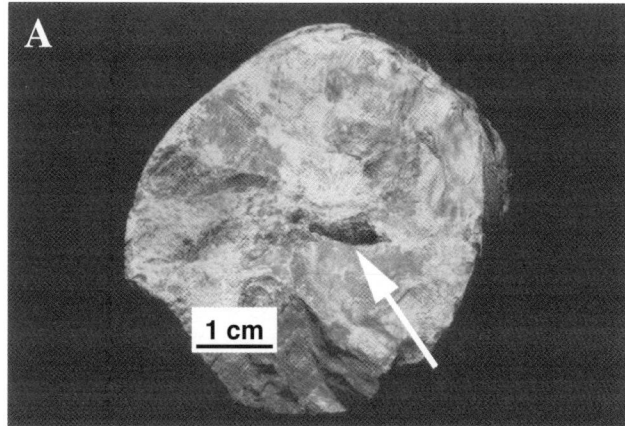

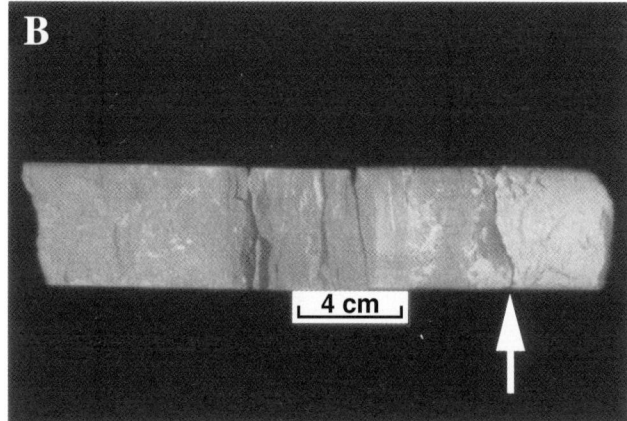

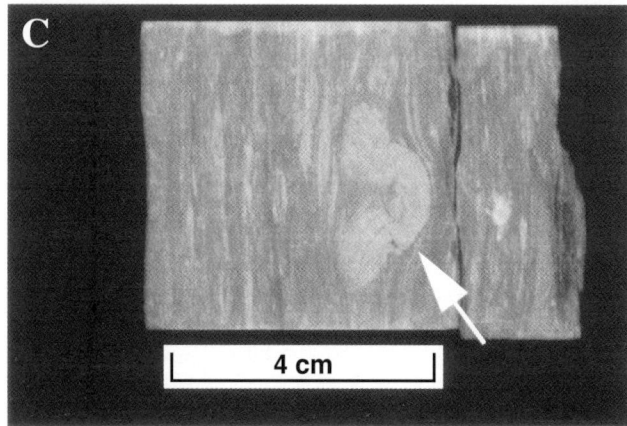

Figure 3. Photos of segments of core from (A) SDH 377 (depth of 87 m [286 ft]) showing rooting (arrow) and slickenslides preserved in a paleosol associated with exposure surface 16; (B) SDH 377 showing the erosive contact (arrow) between the paleosol illustrated in (A) (light gray material left of arrow) and the overlying marine shale and dark carbonate (right of arrow); and (C) showing *Conostichus*-like trace (arrow) in SDH 377 at the 83-m (273-ft) level. Top of the core is to the left on both (B) and (C).

Thick, coarse-grained sandstone-filled channels are sparsely present in the Brazil and Staunton Formations. Where channels do occur, they tend to be thin-

ner, generally less than 10 m (30 ft) thick, and typically contain abundant clay-draped bedforms and generally are not stacked. Coal seams and limestone beds are thicker and more laterally continuous upward in both the Brazil and Staunton Formations than in the Mansfield Formation. In general, coal seams are lower in sulfur content in the Brazil Formation than in the Staunton Formation.

METHODOLOGY

In 1995, the Indiana Geological Survey acquired 100 confidential coal company driller's logs and cores that were scattered across a 340-km^2 (132-mi^2) area. Each core or log included all or part of the upper portion of the Mansfield Formation and the Brazil and Staunton Formations. The confidential data were supplemented with existing Indiana Geological Survey core-hole data, mine outcrop descriptions, and samples obtained in working surface coal mines. The data were used to construct several regional and detailed cross sections in the study area. Correlations were attempted on a bed-by-bed basis. Detailed core descriptions and coal-quality analyses allowed for the recognition of time-stratigraphic markers that are beyond the resolution of standard oil and gas geophysical logs. The most reliable of the markers are the paleosols that are identifiable by the presence of rootlets in massive, weathered sandstones or clay-rich mudstones that typically exhibit slickenslides and sparse, small, iron-rich, centimeter-scale nodules (Figure 3A). Other markers in Daviess County include marine zones characterized by bioturbated mudstones and limestones, well-developed tidal rhythmites above exposure surface 13 (Figure 4), and specific sequences of coal lithotypes in certain coal seams. Paleosols, once identified, could be correlated easily throughout Daviess County using the marker horizons as guides. Once correlated, the paleosols were subsequently numbered for identification (Figure 4).

Estimates of relative water salinities existing during burial of the underlying mires were evaluated by looking at the coal characteristics beneath the transgressive surface and at the degree of bioturbation and the general lithologies of the rocks immediately overlying the transgressed surface. Higher sulfur contents in coals can be correlated with those seams that are overlain within a few tens of centimeters by either a marine limestone or mudstones that contain a diverse trace-fossil assemblage. Where coals were present, seam and bench channel samples were collected for proximate analysis. Lithotype compositions were described in detail for each coal seam if possible, and a lithotype stratigraphy for each seam was established. The lithotype stratigraphy proved to be very useful for correlating seams in Daviess County (Kvale et al., 1996). Coals encountered in coal company coring operations were not available to us for detailed megascopic descriptions, as they were collected by the companies for coal analyses immediately after drilling. To compensate for this, detailed examinations of the economic seams were carried out in coal mines located in the study area in Daviess County. Generally, those seams being mined are the coals present in the Brazil Formation and the lower part of the Staunton Formation. The investigation also included the only two publically accessible core holes (SDH 383 and SDH 386) that were drilled downdip from Daviess County and that penetrated the study interval.

Thirty-two samples, primarily of dark mudstone, were collected from five core holes in Daviess County for palynological analysis (Engelhardt, 2003). Most of these samples were collected just above the transgressive surfaces. An additional 14 core samples were collected from similar deposits present in the deep core hole in Posey County (SDH 383; Figure 4) (Engelhardt, 2003). Abundant and well-preserved palynomorphs were recovered from most of the core samples. All of the major vegetation types present in the coal mires of the Illinois basin were represented by the miospores recovered. A comparison of the miospores recovered and identified from the samples was made with the lower and middle Pennsylvanian palynomorphs reported from other studies of coals and shales of Indiana, Illinois, Iowa, and Kentucky. More than 150 different taxa were identified from the samples, with relative abundances noted (Engelhardt, 2003).

Marine intervals of black shale, limestone, and gray shale were collected for conodonts from SDH 383 in Posey County in the interval between exposure surface 11 and the top of the first limestone above exposure surface 24 and from SDH 377, SDH 386, 12JS-9, and 34KF-1 in Daviess County (Figure 4). The results of the conodont analysis were then used to evaluate the regional correlations that were based largely on the laterally persistent paleosols, lithofacies analysis, and interpretation of the palynomorphs.

BIOSTRATIGRAPHIC CORRELATIONS

Conodonts

The top limestone in SDH 383 (Figure 4A), approximately 8 m (25 ft) above paleosol 24, corresponds to the Velpen Limestone Member of the Linton Formation, which, together with the underlying black shale corresponding to the Mecca Quarry Shale Member,

A

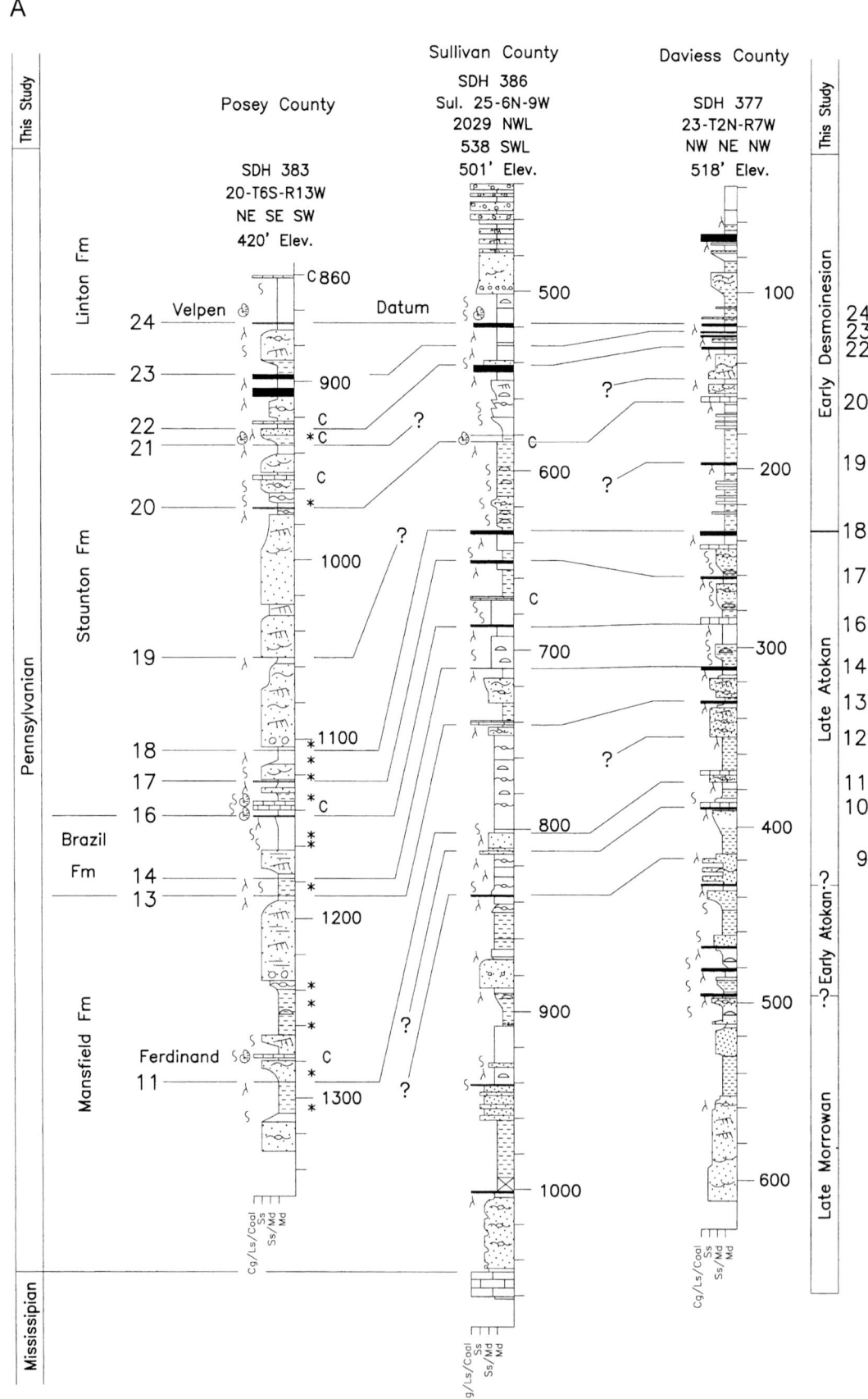

Figure 4. Detailed cross sections (A and B) showing correlation between selected cores and an explanation of symbols (C). Locations of cores are shown in Figure 1C.

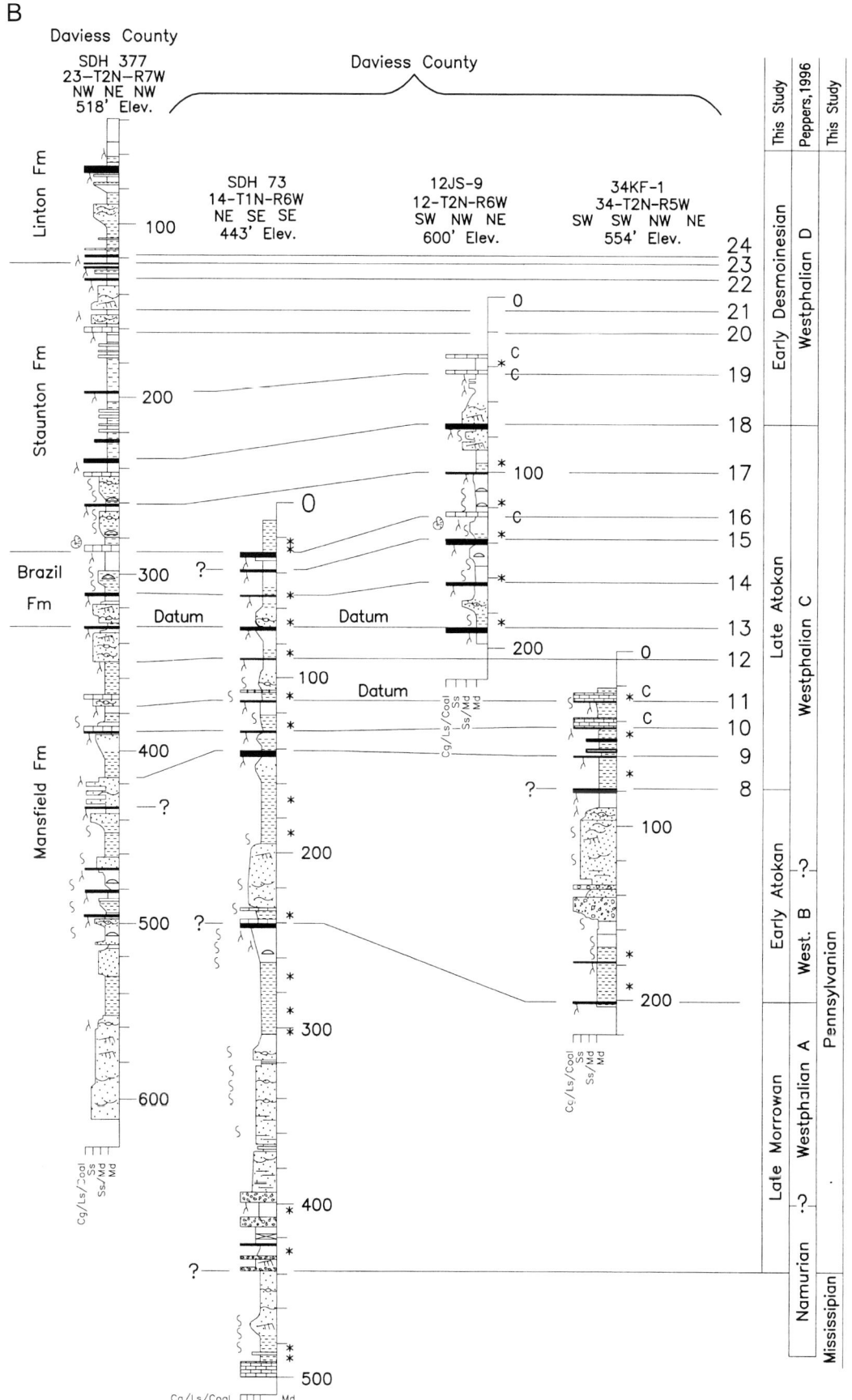

Figure 4. (cont.).

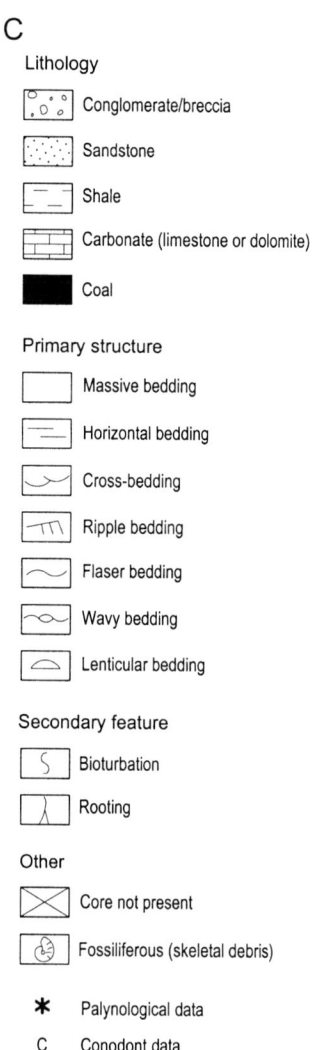

Figure 4. (cont.).

places a stratigraphic upper limit on this study. The lowest named Pennsylvanian limestones encountered in this study can be assigned to the Lead Creek Limestone Member of the Mansfield Formation, which locally can be subdivided into two beds. In SDH 383, the youngest of these beds, the Ferdinand Bed, is present a few tens of centimeters above paleosol 11 (Figures 2, 4A). In 34KF-1 (Figure 4B), both the Ferdinand (8.2–9 m [26.8–29.5 ft]) and the underlying Fulda beds (11.6–13.2 m [38.0–43.3 ft]) are identified.

Lead Creek conodonts are characterized by the presence of *Idiognathoides*, which is not found above the Lead Creek, and the lack of *Idiognathodus klapperi*, which is present in an unnamed limestone just below the Lead Creek. A feature unique to the Lead Creek in Indiana is also recognized by the presence of faceting on one side of the blade of some specimens of *Idiognathodus*, although it was found by

Stamm (1996, 2000, personal communication) in the slightly younger Stoney Fork Member of the Breathitt Formation of Kentucky and from other units in Ohio, Tennessee, and Wyoming. Faceting has also been reported by Grubbs (1984) from the upper Atoka Formation of Oklahoma and by van den Boogaard and Bless (1985) from the Aegiranum marine bed of the United Kingdom.

The next younger limestone encountered in this study occurs just above paleosol 16 in SDH 383 and 12JS-9 (Figure 4). This limestone contains *Neognathodus*. Although not abundant enough for mathematical certainty, this appears to be a very primitive form of *Neognathodus bothrops*. *Neognathodus* so far has proved to be the best genus for correlation by conodonts in the Desmoinesian when it is fully developed, as it is in even younger limestones in the cores. Thus, the limestone above paleosol 16 in SDH 383 and 12JS-9 appears to be slightly older than the Desmoinesian limestones that contain the well-developed *Neognathodus*.

The limestone that occurs in SDH 386 at the 204–205-m (671–672.1-ft) interval (Figure 4A, between paleosols 16 and 17) also occurs between the Lead Creek limestones and younger Desmoinesian limestones. This interval lacks *Idiognathoides* which disappears very close to the Atokan–Desmoinesian boundary and is important in the Lead Creek interval but absent from younger Desmoinesian limestones. However, it does contain a form of *Gnathodus* that is closely similar to those in the Lead Creek and markedly different from those in the oldest of the Desmoinesian limestones.

No limestones are present in SDH 383 in the position between paleosols 17 and 20, an interval that includes the Holland and Silverwood Limestone Members of the Staunton Formation of Indiana (Figure 2) and the Mitchellsville, Creal Springs, Stonefort, and Seahorne Limestone Members of the Tradewater Formation in Illinois. However, there are two limestones of nearly the same age in 12JS-9, in the middle part of the Staunton Formation, that are not present in SDH 383 or in 386. The lower of these, located just above paleosol 19 (Figure 4B; 12.7–13.4 m [41.8–44.0 ft]), correlates well with the Holland Limestone as recognized by an unusual curvature of the carina of *Neognathodus medexultimus*, a species at the same stage of development in the core as it is in the Holland on outcrop. There is no strong evidence for equating the upper limestone, which is very thin and sits directly on top of the uppermost coal in 12JS-9 (Figure 4B; 9 m [30 ft]), with the Silverwood.

The next younger limestones in SDH 383 are present at 290–290.6 and 281–282 m (951.9–953.4 and 920.8–926.0 ft). A marine shale occurs at 284–285 m (933.0–935.5 ft) (Figure 4A). The lower limestone is of nearly the same age as the Velpen and Mecca Quarry, but conodonts are not abundant enough for a precise age determination. The limestone that occurs at the 290–291-m (951.9–953.4-ft) interval between paleosols 20 and 21 in SDH 383 does appear to be equivalent to a fossiliferous calcareous shale that occurs in SDH 386 in the 176.9–177.8-m (580.6–581.3-ft) interval (Figure 4A, above paleosol 20). This unit in SDH 386 appears to be very close to the Holland Limestone in age, but may be slightly younger. It is much closer in age to the Holland Limestone than to the stratigraphically higher Velpen Limestone.

Conodonts from the top limestone in SDH 383 (Figure 4A, 262–262.7 m [859.8–861.9 ft], above paleosol 24) are uncommon, but the few present are compatible with the Velpen Limestone. Conodonts from immediately below this limestone compare with those from the Mecca Quarry Shale. *Neognathodus* suggests that the black mudstone in the core, just below the limestone, is slightly older than in outcrop, but that it is almost identical in age to the Oak Grove Limestone Member of the Carbondale Formation in Illinois, the unit which overlies the Mecca Quarry Shale in northwestern Illinois and which normally is correlated directly with the Velpen. This, in turn, suggests a general transgression from the west and south during this time.

Palynology

For this study, Pennsylvanian mid-continent series boundaries were identified primarily through the use of palynology. A chronostratigraphic correlation to European series and stages follows that of Peppers (1996).

The Desmoinesian–Atokan boundary was identified in most of the cores analyzed (Figure 4). An interpretation of early Desmoinesian was based largely on the presence or frequent occurrences of *Muraspora kosanke* and *Torispora securisi*, very abundant *Punctatisporites minutus*, and sparse *Thymospora obscura* (Engelhardt, 2003).

A late Atokan age was assigned to those rocks that contain common occurrences of *Cristatisporites indignabundus* and *Radiizonates difformis*, along with common *Dictyotriletes bireticulatus*, the presence of *Retispora staphlinii*, *Savitrisporites nux*, *Vestispora pseudoreticulata*, *Camptotriletes bucculentus*, and *Trilobates belli*. An early Atokan age was assigned to those rocks that

contained sparse *Sinuspores sinuatus*, *Triquitrites sculptilus*, and *V. pseudoreticulata*, few *C. indignabundus*, common to abundant *Florinites mediapudens*, abundant *Endosporites globiformis*, *Lycospora pellucida*, *D. bireticulatus*, and *Densosporites* spp. (Engelhardt, 2003).

A Morrowan age was determined by the presence of *Schulzaspora rara*, *Lycospora noctuina*, *Radiizonates striatus*, *Waltzispora prisca*, *Waltzispora polita*, *Vallatisporites* cf., and *Vallatisporites foveolatus*, *Densosporites irregularis*, *Densosporites variabilus*, and *Densosporites aculeatus*. In some samples *Stenozonosporites* Sp.1 Ravn was observed (Engelhardt, 2003).

Based on the above, the Atokan–Desmoinesian boundary is at paleosol 18 (Figure 4). The late–early Atokan boundary is in the Mansfield Formation at paleosol 8 (Figure 4). The Atokan–Morrowan boundary is situated at an unnumbered paleosol below surface 8.

PARASEQUENCE DESCRIPTION

In the study interval, successions of marine-influenced facies are bounded by exposure surfaces (paleosols) that formed during apparent, high-frequency, relative base-level drops. As such, each of these successions can be considered a parasequence (van Wagoner et al., 1990). Seventeen parasequences were identified in the Atokan and lower part of the early Desmoinesian succession in the Daviess County study (Figure 4). The associated exposure surfaces are numbered 8 to 24. Exposure surface 8 was arbitrarily assigned to the lowest laterally correlateable paleosol, with the anticipation that future work would allow for the identification of older and, as yet, undocumented regional exposure surfaces.

The parasequences range in thickness from less than 2 to nearly 13 m (5 to nearly 40 ft). They span an interval from the upper part of the Mansfield Formation through the Brazil and Staunton Formations and into the lower part of the Linton Formation. Each succession starts with a paleosol that, in most cores, is overlain by a coal. A mudstone facies (commonly a dark shale) typically overlies the coal or, if the coal is absent, the paleosol (Figure 3B). Typically, a mudstone then coarsens upward into, or is truncated by, a sandstone body or a thinly interbedded sandstone and mudstone or is overlain by a fossiliferous limestone. The sequences generally fine upward above the sandstone- or limestone-dominated facies. In some cores, scouring at the bases of the sandstone bodies has removed almost completely the underlying mudstone, but typically not the underlying exposure surface (e.g., Figure 4A; SDH 377, above 12; SDH 383, above surfaces 18 and 19).

The sandstones are all quartz arenites. These sandstones are very fine to medium grained in the Posey County core and very fine to fine grained in Daviess County. Sedimentary structures typically linked to tidal processes, such as well-developed tidal rhythmites and lenticular, wavy, and flaser bedding, dominate the siliciclastic intervals.

At least some invertebrate bioturbation was identified in most of the parasequences, although it tends to be sparse in most cores. The more common traces are typically horizontally to subhorizontally oriented tubes that include *Chondrites*, *Teichichnus*, and *Planolites*. *Zoophycos*, *Skolithos*, *Locheia*, and *Teichichnus* are less common, and *Rosselia*, *Conostichus*, and *Monocraterion* appear only in narrow bands that occur in sequences that also contain limestones. Intervals that contain *Rosselia*, *Conostichus*, and *Zoophycos* are considered to indicate sediments deposited under normal marine salinities. Coals that underlie sediments containing any of these traces, marine fossils, limestone beds, or a combination of these contain the highest amounts of sulfur (e.g., Mastalerz et al., 1997; Kvale and Mastalerz, 1998).

The facies overlying the paleosols, including coals, all reflect some marine component, although the water chemistry of a "marine" depositional event may have been essentially freshwater, such as above exposure surface 13 (Figure 4B). Immediately above exposure surface 13 in Daviess County, a coarsening-upward transgressive tidal-flat succession is present. This interval has been discussed in detail elsewhere (Kvale and Archer, 1990; Kvale and Mastalerz, 1998; Mastalerz et al., 1999). The laminae in the mudstone at the base of the transgressive phase are termed "tidal rhythmites," because the laminae thicken and thin vertically and record neap-spring tidal cycles in detail (Kvale and Archer, 1990; Kvale et al., 1999). Carbon-to-sulfur ratios, maceral types, and the dominance of terrestrial organic markers in the rhythmite facies reveal that the underlying coal (Lower Block coal) was transgressed by a freshwater tidal flat (Kvale and Mastalerz, 1998; Mastalerz et al., 1999). This is consistent with the low sulfur and ash contents in the Lower Block coal, which are typically below 1 and 6%, respectively. These tidal rhythmites appeared in every core and outcrop that contained the Lower Block coal examined during the course of this investigation in Daviess County, an area that extends over 400 km² (154 mi²) and, as a result, serves as a key marker for the Brazil Formation in Daviess County.

The rhythmite facies above the Lower Block coal, however, does not extend to the Posey or Sullivan county cores.

Limestones facies (packstone and wackestone) and heavily bioturbated facies are interpreted to represent the most fully marine phase of the transgressive cycle. Some sequences preserve both a relatively thick transgressive component and a relatively thick regressive facies. Other packages are dominated by regressive sediments. In several parasequences, transgressive and regressive facies cannot be distinguished. Coarsening-upward cycles apparently represent shoaling-upward events in some parasequences and deepening-upward in others (Figure 5).

All but two of the paleosols from intervals 11 and 24 (Figure 4) can be correlated from Daviess County to Posey County, and all but four can be correlated from Daviess County to Sullivan County. Those Daviess County paleosols that are not correlateable to one or both of the deeper core holes include 12 and 15 (absent in both SDH 386 and SDH 383) and 19 and 21 (absent in SDH 386 only).

The Posey County SDH 383 core is much coarser grained through the study interval than the cores examined in Daviess or Sullivan Counties (Figures 4, 6). Most of the thicker sandstones in SDH 383 are erosive based.

The presence of the sandstone bodies in the Posey County core may be related, in part, to syndepositional tectonic influence. A series of faults associated with the Wabash Valley fault system are located nearby. Furer (1996) has shown that at least some of these faults influenced sedimentation patterns in Indiana through much of the Paleozoic. Tectonic downwarping may have focused drainage through the SDH 383 area during the Atokan, resulting in coarser-channelized sediments being deposited.

SEQUENCE-STRATIGRAPHIC ASPECTS OF COAL SEAMS

Detailed analyses of two coal seams in the study interval suggest that drowning of coastal mires during transgressions was either gradual or very rapid. Relative rates of drowning of coastal mires can be deduced by noting vertical changes in coal lithotypes and maceral compositions, as well as in spore assemblages. In Indiana, the compositional changes in both the Lower Block Coal Member (above exposure surface 13) and the Upper Block Coal Member (above exposure surface 14) from Daviess County proved

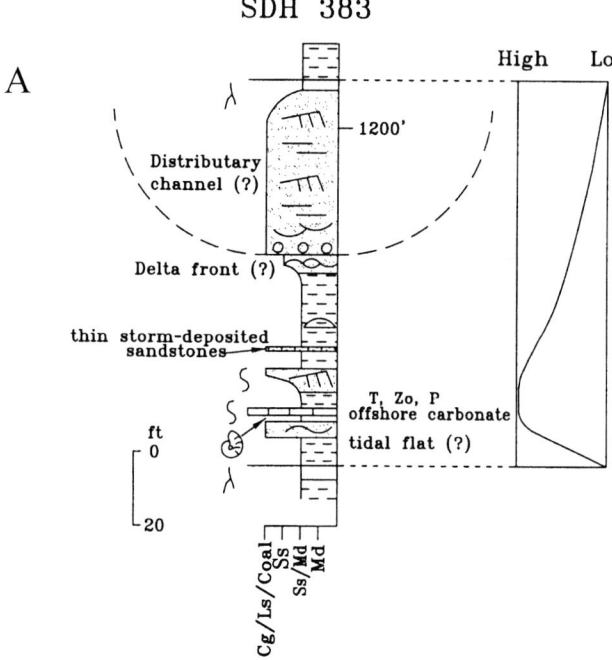

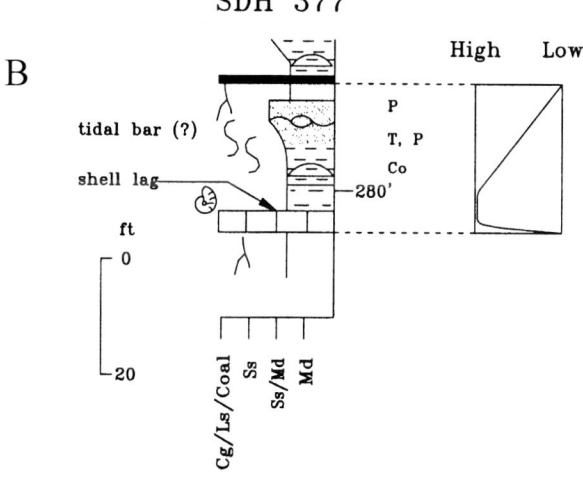

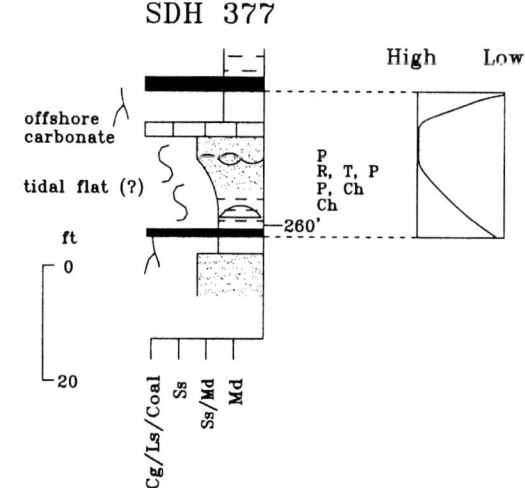

Figure 5. Interpretation of relative sea level curves for three sections of core illustrated in the cross section shown in Figure 4. Letters refer to the following trace fossils: *Chondrites* (Ch), *Teichichnus* (T), *Planolites* (P), *Zoophycos* (Zo), *Rosselia* (R), and *Conostichus* (Co).

to be very useful in helping us to understand the sequence-stratigraphic aspects of this succession.

Lower Block Coal Member

Variations in coal lithotypes, spore assemblages, and petrographic character of the Lower Block coal are systematic in the seam in Daviess County. Lithotypes of this seam are composed predominantly of vitrinite and the highest vitrinite content is recorded in the middle part of the seam (Figure 7A). Inertinite is the second most dominant maceral group, averaging 18.0%. It commonly increases toward the top and bottom thirds of the columns. Liptinite, composed primarily of sporinite, occurs in the lowest concentrations in the middle part of the seam and increases toward the top and the bottom of the column. Mineral matter is commonly elevated in the lowermost part of the seam and, in places, in the uppermost part of the seam. Sulfur content in the vertical section shows relative increases in the top and bottom portions (Figure 7A). Ash yield from lithotypes is relatively low (commonly below 6%) and does not appear to be related to sulfur content.

Spore assemblages in the Lower Block coal show that it was dominated by lycopod trees and tree ferns (Figure 7A). The highest concentration of lycopod tree spores occurs in the central part of the seam, corresponding to the brightest (vitrinite-dominated) portion of the seam. The concentration of lycopod trees decreased slightly by the end of deposition of the Lower Block paleomire. The contribution by other plant groups was minor.

Petrographic and palynological examinations suggest that the Lower Block coal was most likely a mesotrophic to perhaps oligotrophic, planar paleomire with a lycopod-dominated, or lycopod-tree fern codominant flora (Mastalerz et al., 2000). Thick paleosols commonly found below the Lower Block suggest prolonged exposure and a hiatus before peat deposition. Peat deposition appears to have been initiated by a regional rise of the ground-water table, likely induced by a relative sea level rise. Peat began to accumulate because of an associated increase in accommodation space. Hence, the boundary between the paleosol and the coal corresponds to a sequence-stratigraphic boundary referred to as a paludification

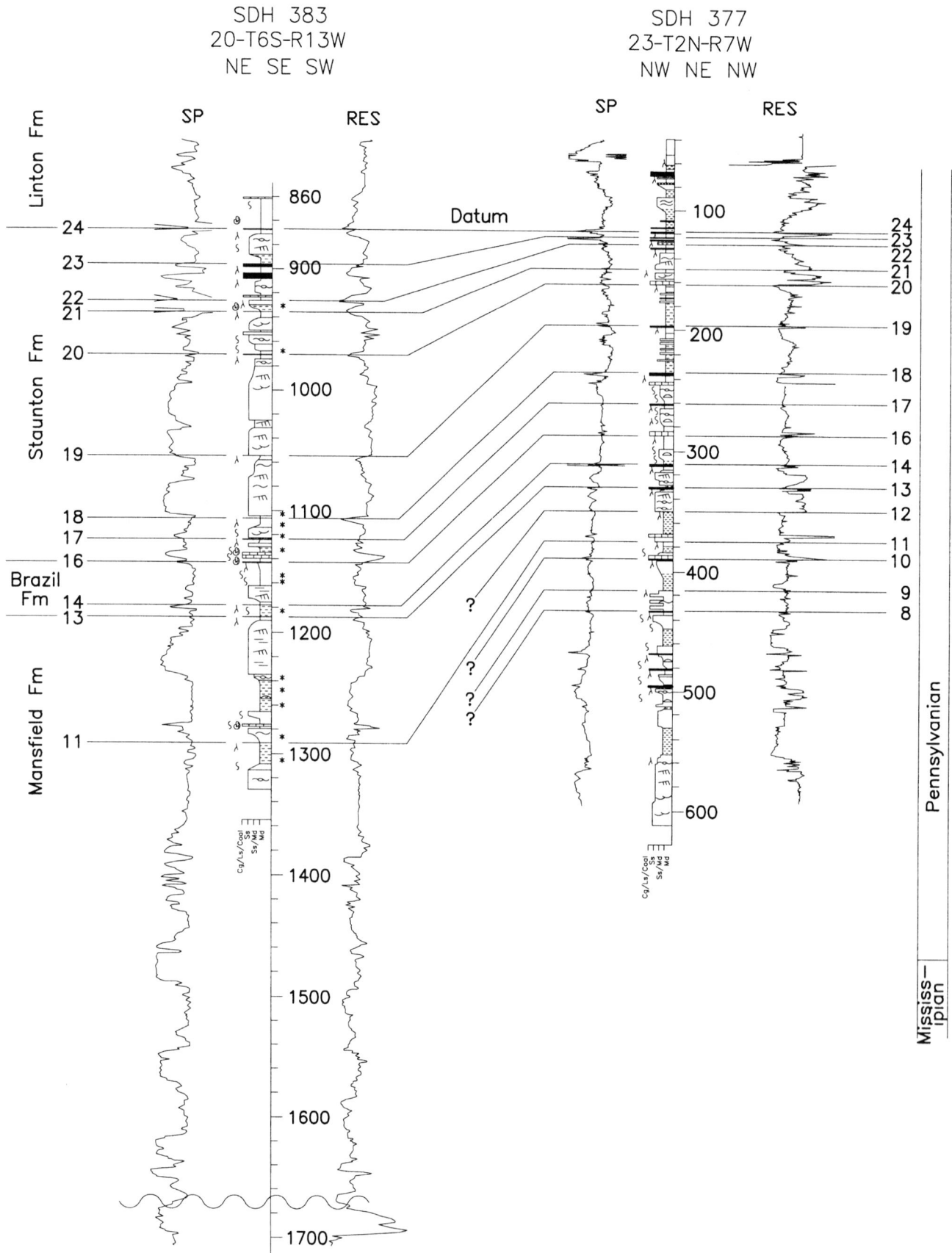

Figure 6. Graphic logs for two cores and their associated geophysical logs. Cores are located approximately 100 km (60 mi) apart.

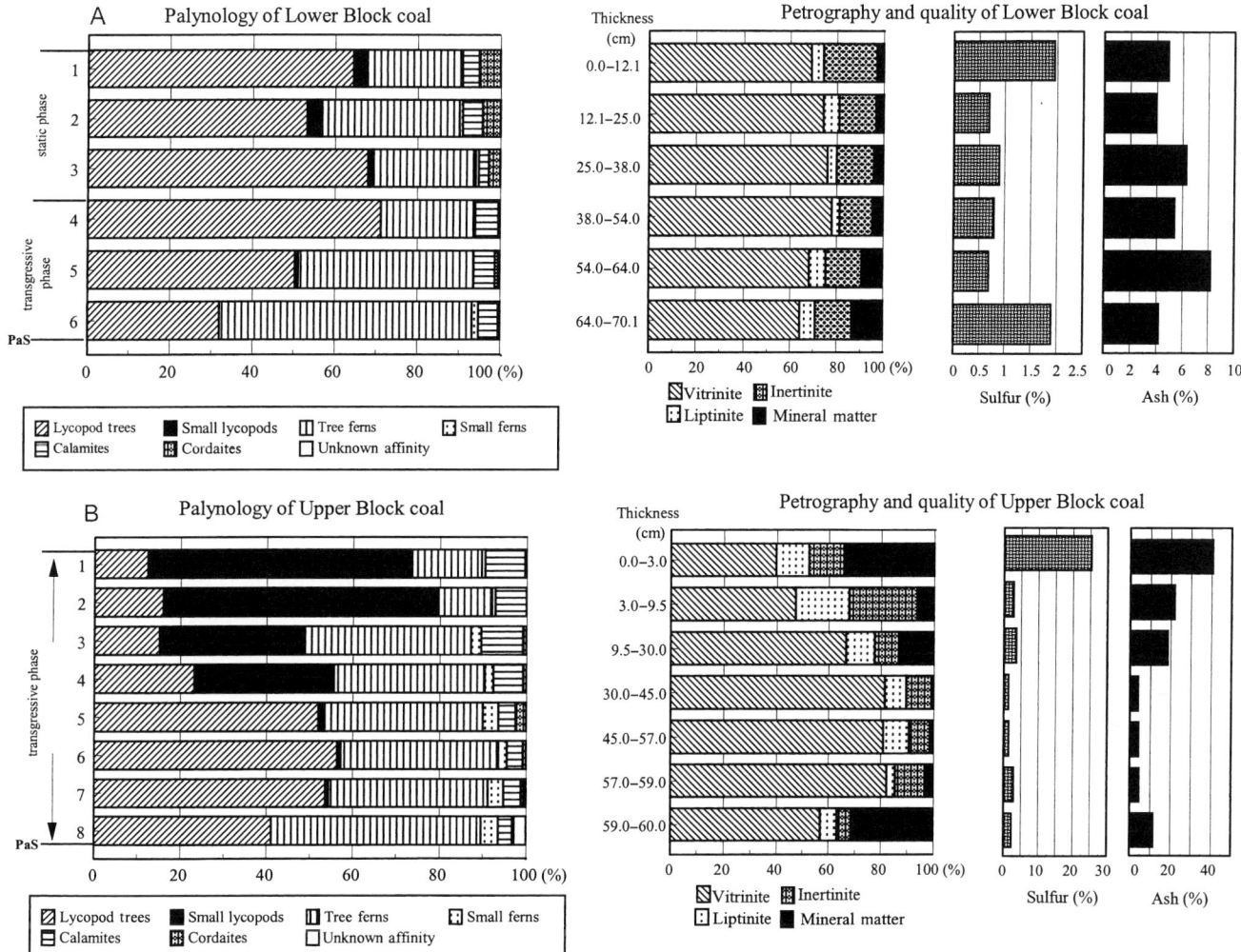

Figure 7. Palynology, petrography, and coal quality of the Lower Block coal (A) and the Upper Block coal (B). PaS = paludification surface.

surface (PaS) sensu Diessel et al. (2000). A brightening of the coal toward the middle portion of the seam, followed by a slight dulling toward the top of the seam and a corresponding change in the spore assemblages (lycopod trees dominant in the middle portion of the seam), suggests that the wettest (deepest) conditions were during accumulation of the middle portion of the seam, when lycopod trees flourished, as opposed to other less water-tolerant species (Mastalerz et al., 2000). This was followed by a period of very minor oscillations in water depths until a sudden transgressive event over the top of the peat by siliciclastics (Figure 8). Detailed geochemical studies of organic matter in these transgressive sediments indicate a freshwater chemistry for this rapidly formed tidal flat (Mastalerz et al., 1999).

Kvale and Mastalerz (1998) suggested that this rapid transgression was caused by the very rapid, perhaps catastrophic, collapse of the Lower Block peat

mire and the resulting very rapid generation of accommodation space and subsequent fill by tidal-flat facies. This interpretation is supported by the thick accumulation of tidal rhythmites with well-developed neap-spring cycles over the top of the Lower Block coal and the preservation of upright lycopod trees in growth positions in many of the Lower Block coal mines (Kvale and Archer, 1990; Kvale and Mastalerz, 1998). The generation of accommodation space by mire collapse is suggested by the documentation of rapid collapses of modern organic-rich soils (e.g., DeLaune et al., 1994). In the modern Mississippi River delta plain, the vertical accretion of plant material has locally kept up with modern sea level rise. In those areas where the accumulation of plant material fails to keep up with the concomitant rise in base level (triggered perhaps because of ecological stresses induced by salt-water intrusions, flooding, inadequate mineral sediment supply, or fires), so-called

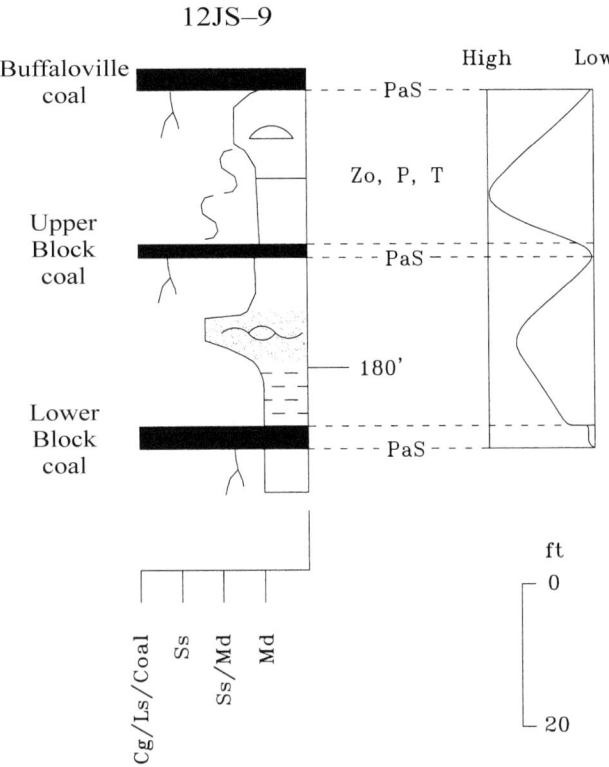

Figure 8. Detailed relative sea level curve for Lower Block to Buffaloville coal. Note the interpretation of a very abrupt increase in water depth and relatively thick transgressive succession above the Lower Block coal as opposed to the gradual increase in water depth and thinner transgressive interval above the Upper Block coal. See Figure 4C for symbols.

"marsh-loss hotspots" occur (DeLaune et al., 1994). In such cases, the vertical accretion of plant material (which locally averages 0.98 cm/yr over a 27-yr period) cannot keep pace with the submergence of the coastal plain, and plant mortality and peat collapse occur by autocompaction primarily through an increase in the decomposition rates of root tissue and/or the loss of root turgor (DeLaune et al., 1994). This causes depressions to quickly form in the marshes, which subsequently are drowned by flood waters or are transgressed by brackish to marine water. Submergence rates on the Mississippi delta plain as a combination of peat collapse, the consolidation of underlying sediments, and global sea level rise exceed 1 cm/yr (DeLaune et al., 1994). Elevation loss is documented to be greatest in marshes, where organic contents on the delta plain are highest and less in areas such as along levees, where vegetation persisted and soil organic contents were low. As noted by Kvale and Mastalerz (1998), subsidence rates noted by DeLaune et al. (1994) are two orders of magnitude

less than that interpreted to have occurred with the collapse of the Lower Block coal paleomire. This discrepancy could be explained by the presence of a much thicker peat deposit (Lower Block coal) and its potential for a much greater and perhaps more rapid collapse than that reported for the relatively thin and more inorganic-rich coastal marsh peats of the Mississippi delta. More work on modern tropical peats, however, needs to be done to further substantiate the potential importance of mire collapse for generating significant sediment accommodation space.

Additional triggers of rapid transgression above the Lower Block coal, such as a several-hundred-kilometer-scale, seismically induced subsidence, cannot be ruled out. Atwater et al. (2001) described paleosols and upright trees buried by well-developed tidal rhythmites over a broad area near the town of Portage in the Turnagain Arm of Cook Inlet in Alaska. They show that the rhythmites were deposited on intertidal flats that formed as a consequence of subsidence of supratidal marshes and forests, the result of the 1964 Alaska earthquake of magnitude 9.2. In their study area, the earthquake produced 1.5 m (5 ft) of tectonic subsidence and unknown amounts of nontectonic compaction, for a total subsidence of as much as 2.3 m (7.5 ft). However, no direct evidence of such a major tectonic event at the end of Lower Block coal deposition has been documented yet.

To summarize, the Lower Block coal mire originated because of an increase in accommodation space driven by a rise in relative sea level and the initiation of paludification. The lower part of the seam represents a transgressive phase, and the upper part represents an oscillating water depth phase. The seam terminated because of a very rapid acceleration in the generation of accommodation space perhaps driven by collapse of the peat mire. Deposition of the whole seam (70 cm [28 in.] thick) is estimated to have spanned about 4900 yr, assuming a rate of peat compaction of 7:1 (Taylor et al., 1998) and a peat growth rate in subtropical conditions of 1 mm (0.04 in.) per year (see Fisk, 1960; Anderson, 1983; Taylor et al., 1998). Consequently, the transgressive phase of the seam (lower 32 cm [13 in.]) would correspond roughly to 2200–2300 yr of continuous deposition.

Upper Block Coal Member

The Upper Block coal, like the Lower Block coal, shows very systematic variation in lithotypes, coal quality, and palynomorphs across Daviess County. Lithotypes in the lower part of the Upper Block coal are composed predominantly of vitrinite, except the lowermost part that has a vitrinite content below 60% (Figure 7B) because of an enrichment in mineral matter. The upper part of the seam is characterized by relatively more liptinite, inertinite, and mineral matter content, with a concomitant decrease of vitrinite. In addition, inertodetrinite (a component of the inertinite shown in Figure 7) is more abundant in the upper part than in the lower part. Locally, small *Botryococcus* fragments occur along bedding surfaces close to the top of the seam.

Coal quality of the lower part of the Upper Block differs from that of its upper part, mainly with regard to ash content, with the mean value of 6.4% compared to 25.9% in the upper part (Mastalerz et al., 2000). Sulfur content of the lower part is also lower than in the upper part. The vertical seam section presented in Figure 7B shows that the sulfur content can exceed 20% in the uppermost few centimeters of the seam.

In the Upper Block coal, spore assemblages in the lower part differ significantly from those in the upper part. The lower part is similar to the Lower Block coal, with lycopod tree and tree fern spores being dominant (Figure 7B). The upper part of the seam is dominated by spores from small lycopods, with a high input from tree ferns toward the middle of the seam.

Like the Lower Block coal, the Upper Block paleomire also developed initially with a tree fern- and lycopod-dominant flora, likely with a persistent water cover that deepened toward the middle of the seam (as inferred from a brightening-upward trend and an increase in lycopod tree spores). The conditions changed during deposition of the upper half of the Upper Block peat. Lycopod trees and tree ferns that dominated the early phase of deposition were replaced by small lycopods in the later phase. An increase in mineral matter content at the top of the coal and the appearance of *Botryococcus* algae suggest that the spores and other associated liptinite and inertinite were transported by water into the mire instead of being deposited in situ. An increase in both clastic input and transported organic matter indicates a further increase in the depth of the water cover and gradual drowning of the mire (Figure 8). The transgressing waters were at least brackish, as inferred from the increase in sulfur content in the upper part of the seam.

The Upper Block coal is underlain by a prominent paleosol, indicating, as in the case of the Lower Block, exposure before the initiation of peat deposition. Peat began to accumulate as a result of a rising ground-water table and the generation of accommodation space. The lower boundary of the peat is, as in the Lower Block, a PaS. The lower part of the seam registers a brightening-upward trend (increasing accommodation; transgressive phase with the rise of a freshwater ground-water table). The upper part of the seam reflects further transgression under the influence of at least brackish waters. Accommodation rates eventually surpassed rates of peat production, resulting in a gradual drowning of the mire. Thus, in this seam, there was a continual and gradual increase in accommodation space, unlike the Lower Block seam in which there was a "catastrophic" generation of accommodation space to end peat accumulation (Figure 8). Using the same compaction and peat growth rates as for the Lower Block, the deposition of the Upper Block seam would correspond to 4200 yr, nearly twice as long as the transgressive phase in the Lower Block seam.

INTERPRETATION OF PARASEQUENCES

The debate over what has driven the deposition of Pennsylvanian parasequences and parasequence sets in the Illinois basin has focused on several models: glacioeustatic, climatic, delta lobe shifting, tectonic, or a complex combination of these factors. The parasequences described herein lack internal consistency from interval to interval. Both transgressive and regressive phases of sedimentation are preserved in the intervals, but their relative significance and associated lithofacies within a package vary.

In Daviess County, the absence of obvious large-scale channels that could be interpreted as deltaic distributaries and the general character of the facies led Kvale and Mastalerz (1998) to suggest that each parasequence in Daviess County was deposited in a nearshore to marginal marine setting. Each parasequence appears to reflect deposits of a shifting mud-dominated coastal plain in which the bulk of the sediments were delivered via longshore currents. However, this interpretation is tempered by studies of modern mud-rich, tide-dominated tropical coastlines that reveal the difficulty of differentiating facies associated with active distributary channels from those of prograding, coastal tidal flats (e.g., Coleman et al., 1970; Cecil et al., 1993; Staub and Esterle, 1993). Channelization is more prevalent in the Posey County core in the deeper part of the basin. The apparent absence of significant channelization in the Daviess County portion of the study (upper Mansfield through Staunton) could be explained if the relative base-level changes were minor, resulting in minimal changes in gradients between cycles. This would also explain, at least in part, why the parasequences are so thin along the basin margins.

Although it is difficult to know for certain the duration of any or all of the depositional sequences discussed, it is possible to estimate, in a very general way, the interval of time represented by each parasequence. Kunk and Rice (1994) report a mean Ar-Ar age of 310.9 ± 0.8 Ma for the Fire Clay tonstein (weathered volcanic ash) in the Appalachian basin. This unit can be correlated biostratigraphically with the Trace Creek Shale Member of the Atoka Formation in Arkansas and is very close to the stratigraphic position of the Westphalian BC boundary in Western Europe (Kunk and Rice, 1994). This study puts the early–late Atokan boundary at paleosol 8 and the Westphalian BC boundary below that (Figure 4B).

The Westphalian AB boundary is considered by Peppers (1996) to be equivalent to the Atokan–Morrowan boundary in the United States. Spore data from this study indicate that an Atokan–Morrowan boundary is present near the base of 34KF-1 (by correlation to a nearby core not shown in Figure 4 but reported in Englehardt, 2001) and in the SDH 73 core.

Peppers (1996) correlated the Westphalian CD boundary with the Atokan–Desmoinesian boundary in the Illinois basin. This would mean that the Westphalian CD boundary (based on palynomorphs) is above paleosol 18. Menning et al. (2000) reported the Westphalian CD boundary to be at approximately 308 Ma. As such, it is unlikely that the Atokan spans much more than 3 m.y. Hence, it appears unlikely that the entire study interval from exposure surface 8 to 18 took more than 3 m.y. to be deposited. If this assumption is correct, then each parasequence perhaps averaged from 200,000 to 300,000 yr/parasequence. Such high-frequency periods of deposition are more likely to be considered glacial eustatic in origin instead of tectonic. A glacial eustatic origin for the upper Morrowan, Atokan, and lower Desmoinesian sediments in the study interval is consistent with the generally accepted view of an established Pennsylvanian icehouse world for this period (Caputo and Crowell, 1985; Veevers and Powell, 1987).

Chesnut (1994) reported similar high-frequency transgressive-regressive cycles in the lower and middle Pennsylvanian of the central Appalachian basin and interpreted them as glacial-eustatic in origin. He reported coal-clastic transgressive-regressive cycles in the Breathitt Group, with average durations of 0.4 m.y. (ranges of 0.2–0.7 m.y.) in larger-scale transgressive-regressive cycles that exhibited average durations of 2.5 m.y. The upper part of the Breathitt Group is largely equivalent to the interval reported herein. As such, it does appear that high-frequency transgressive-regressive sequences were influencing sedimentation in both the Illinois basin and the central Appalachian basin during the early Atokan into early Desmoinesian. It is not clear, however, if there was a uniform process driving the Illinois basin and central Appalachian sequences, as individual depositional packages have not yet been correlated from one basin to another.

TECTONIC IMPLICATIONS

The recognition of correlatable parasequences over a 60-km (37-mi) portion of Indiana provides information about tectonic subsidence of the Illinois basin. There is an apparent basinward thickening of the stratigraphic interval between paleosols 11 and 13 (Figures 4, 6). Stratigraphic thicknesses remain fairly uniform from paleosols 13 to 18, with a slight thickening updip from Posey County into the Sullivan and Daviess County wells. Between surfaces 18 and 24, there is again a thickening basinward. This suggests that the deeper parts of the basin were subsiding more during the early part of the late Atokan and again during the early Desmoinesian. The Posey, Sullivan, and Daviess County portions of the basin appear to have subsided more uniformly during the later part of the late Atokan. Some of the apparent basin subsidence, as inferred from SDH 383, however, may have been generated by local tectonic downwarping such as along fault blocks. Until more core data become available from the deeper parts of the basin, it will not be clear if this subsidence was local or basinwide.

CONCLUSIONS

Pennsylvanian parasequences have been documented in the Illinois basin for the Atokan and lower part of the Desmoinesian. The parasequences are bounded by exposure surfaces (paleosols) that can be correlated from subsurface core data over a linear distance of more than 100 km (60 mi). These intervals were deposited with an average frequency on the order of 10^5 yr. Such parasequences were previously undocumented in the Illinois basin. A preliminary analysis of the spatial thickness variability of the parasequences suggests that the deeper portions of the Illinois basin subsided more rapidly than the eastern margin during the early part of the late Atokan and again during the early Desmoinesian. Basin subsidence appears to have been more uniform during the later stages of the late Atokan. However, more regional subsurface data are necessary to confirm this.

ACKNOWLEDGMENTS

The authors are grateful to Solar Sources, Inc., and Black Beauty Coal Company for access to the boreholes, and to the Indiana Department of Commerce and the Electric Power Research Institute for their support during data collection. We also thank John Nelson and Jim Staub for very thorough and helpful reviews and the former director of the Indiana Geological Survey, Norman C. Hester, and the current director, John C. Steinmetz, for supporting and encouraging this research.

REFERENCES CITED

Anderson, J. A. R., 1983, Tropical peat swamps of western Malaysia, in A. J. P. Gore, ed., Ecosystems of the world, 4B Mires: Swamp, bog, fen and moor, regional studies: Amsterdam, Elsevier, p. 181–199.

Atwater, B. F., D. K. Yamaguchi, S. Bondevik, W. A. Barnhardt, L. J. Amidon, B. E. Benson, G. Skjerdal, J. A. Shulene, and F. Nanayama, 2001, Rapid resetting of an estuarine recorder of the 1964 Alaska earthquake: Geological Society of America Bulletin, v. 113, p. 1193–1204.

Caputo, M. V., and J. C. Crowell, 1985, Migration of glacial centers across Gondwana during Paleozoic Era: Geological Society of America Bulletin, v. 96, p. 1020–1036.

Cecil, C. B., F. T. Dulong, J. C. Cobb, and Supardi, 1993, Allogenic and autogenic controls on sedimentation in the central Sumatra basin as an analogue to Pennsylvanian coal-bearing strata in the Appalachian basin, in J. C. Cobb and C. B. Cecil, eds., Modern and ancient coal-forming environments: Geological Society of America Special Paper 286, p. 3–22.

Chesnut, D. R., 1994, Eustatic and tectonic control of deposition of the lower and middle Pennsylvanian strata of the central Applachian basin, in J. M. Dennison and F. R. Ettensohn, eds., Tectonic and eustatic controls on sedimentary cycles: SEPM Concepts in Sedimentology and Paleontology, v. 4, p. 51–64.

Coleman, J. M., S. M. Gagliano, and W. G. Smith, 1970, Sedimentation in a Malaysian high tide tropical delta, in J. P. Morgan, ed., Deltaic sedimentation modern and ancient: SEPM Special Publication 15, p. 185–197.

Collinson, C., M. Sargent, and J. R. Jennings, 1988, Illinois basin region, in L. L. Sloss, ed., Sedimentary cover—North America craton: Geological Society of America, The geology of North America, v. D-2, p. 383–426.

DeLaune, R. D., J. A. Nyman, and W. H. Patrick Jr., 1994, Peat collapse, ponding, and wetland loss in a rapidly submerging coastal marsh: Journal of Coastal Research, v. 10, p. 1021–1030.

Diessel, C., R. Boyd, J. Wadsworth, D. Leckie, and G. Chalmers, 2000, On balanced and unbalanced accommodation/peat accumulation ratios in the Cretaceous coals from Gates Formation, Western Canada, and their sequence-stratigraphic significance: International Journal of Coal Geology, v. 43, p. 143–186.

Droste, J. B., and L. C. Furer, 1995, Early Pennsylvanian stratigraphy and the influence of sub-Pennsylvanian topography in the subsurface of Indiana: Indiana Geological Survey Special Report 58, 11 p.

Droste, J. B., and S. J. Keller, 1989, Development of the Mississippian–Pennsylvanian unconformity in Indiana: Indiana Geological Survey Occasional Paper 55, 11 p.

Droste, J. B., L. C. Furer, and A. S. Horowitz, 2000, Patterns of deposition during the early Pennsylvanian (Morrowan) in the Illinois basin: Indiana Geological Survey Special Report 62, 16 p.

Engelhardt, D. W., 2003, Palynologic analysis of lower Pennsylvanian core hole samples from Daviess and Posey Counties, Indiana: Indiana Geological Survey OFS 03-17, 37 p.

Fisk, H. N., 1960, Recent Mississippi River sedimentation and peat accumulation, in IV International Congress of Carboniferous Stratigraphy and Geology, Heerlen 1958 Proceedings, Compte Rendu, v. 1, p. 187–199.

Furer, L. C., 1996, Basement tectonics in the southeastern part of the Illinois basin and its effect on Paleozoic sedimentation, in B. A. van der Pluijm and P. A. Catacosinos, eds., Basement and basins of eastern North America: Geological Society of America Special Paper 308, p. 109–126.

Grubbs, R. K., 1984, Conodont platform elements from the Wapanucka and Atoka Formations (Morrowan–Atokan) of the Mill Creek syncline, central Arbuckle Mountains, Oklahoma, in P. K. Sutherland and W. L. Manger, eds., The Atokan series (Pennsylvanian) and its boundaries: Oklahoma Geological Survey Bulletin, v. 136, p. 65–79.

Heckel, P. H., 1994, Evaluation of evidence for glacioeustatic control over marine Pennsylvanian cyclothems in North America and consideration of possible tectonic effects, in J. M. Dennison and F. R. Ettensohn, eds., Tectonic and eustatic controls on sedimentary cycles: SEPM Concepts in Sedimentology and Paleontology, v. 4, p. 65–87.

Kunk, M. J., and C. L. Rice, 1994, High-precision ^{40}Ar/^{39}Ar age spectrum dating of sanidine from the middle Pennsylvanian Fire Clay tonstein of the Appalachian basin, in C. L. Rice, ed., Elements of Pennsylvanian stratigraphy, central Appalachian basin: Geological Society of America Special Paper 294, p. 105–113.

Kvale, E. P., and A. W. Archer, 1990, Tidal deposits associated with low-sulfur coals, Brazil Formation (lower Pennsylvanian), Indiana: Journal of Sedimentary Petrology, v. 60, p. 563–574.

Kvale E. P., and M. Mastalerz, 1998, Evidence of fresh water tidal deposits, in C. Alexander, R. A. Davis Jr., and J. H. Vernon, eds., Tidalites: Processes and products: SEPM Special Publication 61, p. 95–107.

Kvale, E. P., L. C. Furer, and M. Mastalerz, 1996, Exploration models for selected economic coals in southeastern Daviess and southwestern Martin Counties: Indiana Geological Survey Open-File Report 96-2, 28 p.

Kvale, E. P., H. W. Johnson, C. P. Sonett, A. W. Archer, and A. Zawistoski, 1999, Calculating lunar retreat rates using tidal rhythmites: Journal of Sedimentary Research, v. 69, p. 1154–1168.

Mastalerz, M., B. A. Stankiewicz, G. Salmon, E. P. Kvale, and C. L. Millard, 1997, Organic geochemical study of sequences overlying coal seams; example from the Mansfield Formation (lower Pennsylvanian), Indiana: International Journal of Coal Geology, v. 33, p. 275–299.

Mastalerz, M., E. P. Kvale, B. A. Stankiewicz, and K. Portle, 1999, Organic geochemistry in Pennsylvanian tidally-influenced sediments from Indiana: Organic Geochemistry, v. 30, p. 57–73.

Mastalerz, M., P. Padgett Alano, and C. Eble, 2000, Block coals from Indiana: Inferences on changing depositional environment: International Journal of Coal Geology, v. 43, p. 211–227.

Menning, M., D. Weyer, G. Drosdzewski, H. W. J. van Amerom, and I. Wendt, 2000, A Carboniferous time scale 2000: Discussion and use of geological parameters as time indicators from central and western Europe: Geologische Jahrbuch, v. A156, p. 3–44.

Peppers, R. A., 1996, Palynological correlations of major Pennsylvanian (middle and Upper Carboniferous) chronostratigraphic boundaries in the Illinois and other coal basins: Geological Society of America Memoir 188, 111 p.

Rexroad, C. B., L. M. Brown, J. Devera, and R. J. Suman, 1997, Conodont biostratigraphy and paleoecology of the Perth Limestone Member, Staunton Formation (Pennsylvanian) of the Illinois basin, U.S.A., in H. Szaniawski, ed., Proceedings of the Sixth European Conodont Symposium (ECOS VI): Palaeontologia Polonica, v. 58, p. 247–256.

Shaver, R. H. et al., 1986, Compendium of Paleozoic rock-unit stratigraphy in Indiana— A revision: Indiana Department of Natural Resources: Geological Survey Bulletin, v. 59, 203 p.

Stamm, R. G., 1996, Reversals of misfortune; a new species of Idiognathodus (Conodonta) based on functional surface morphology, in J. E. Repetski, ed., Sixth North American Paleontological Convention (abs.): Paleontological Society Special Publication 8, p. 369.

Staub, J. R., and J. S. Esterle, 1993, Provenance and sediment dispersal in the Rajang River delta/coastal plain system, Sarawak, East Malaysia: Sedimentary Geology, v. 85, p. 191–201.

Taylor, G. H., M. Teichmüller, A. Davis, C. F. K. Diessel, R. Littke, and P. Robert, 1998, Organic petrology: Berlin, Gebrüder Borntraeger, 704 p.

Tri-State Committee on Correlation of the Pennsylvanian System in the Illinois basin, 2001, Toward a more uniform stratigraphic nomenclature for rock units (formations and groups) of the Pennsylvanian system in the Illinois basin: Illinois Basin Consortium Study 5: Bloomington, Indiana, Indiana Geological Survey, 26 p.

van den Boogaard, M., and M. J. M. Bless, 1985, Some conodont faunas from the Aegiranum marine band, in Proceedings of the Koninklijke Nederlandse Akademie van Wetenschappen. Series B: Palaeontology, Geology, Physics, Chemistry, Anthropology, v. 88, p. 133–154.

van Wagoner, J. C., R. M. Mitchum Jr., K. M. Campion, and V. D. Rahmanian, 1990, Siliciclastic sequence stratigraphy in well logs, cores and outcrops: Concepts for high resolution correlation of time and facies: AAPG Methods in Exploration Series 7, 64 p.

Veevers, J. J., and C. M. Powell, 1987, Late Paleozoic glacial episodes in Gondwanaland reflected in transgressive-regressive depositional sequences in Euramerica: Geological Society of America Bulletin, v. 98, p. 475–487.

Wanless, H. R., and F. P. Shepard, 1936, Sea level and climatic changes related to late Paleozoic cycles: Geological Society of America Bulletin, v. 47, p. 1177–1206.

Weller, J. M., 1930, Cyclical sedimentation of the Pennsylvanian period and its significance: Journal of Geology, v. 38, p. 97–135.

Greb, S. F., D. R. Chesnut Jr., and C. F. Eble, 2004, Temporal changes in
coal-bearing depositional sequences (lower and middle Pennsylvanian) of
the central Appalachian basin, U.S.A., *in* J. C. Pashin and R. A. Gastaldo,
eds., Sequence stratigraphy, paleoclimate, and tectonics of coal-bearing
strata: AAPG Studies in Geology 51, p. 89–120.

Temporal Changes in Coal-bearing Depositional Sequences (Lower and Middle Pennsylvanian) of the Central Appalachian Basin, U.S.A.

Stephen F. Greb
Kentucky Geological Survey, University of Kentucky, Lexington, Kentucky, U.S.A.

Donald R. Chesnut Jr.
Kentucky Geological Survey, University of Kentucky, Lexington, Kentucky, U.S.A.

Cortland F. Eble
Kentucky Geological Survey, University of Kentucky, Lexington, Kentucky, U.S.A.

ABSTRACT

Middle Pennsylvanian coal-measure sequences of the eastern Kentucky coal field, central Appalachian basin, occur in ordered groupings of five to six fourth-order coal-clastic cycles, between third-order marine flooding surfaces. Lower Pennsylvanian coal measures also are present, but are laterally truncated by at least four, 60–80-km (37–50-mi)-wide belts of quartz-pebble-bearing quartzarenites that were deposited in a longitudinal drainage system. Successive quartzarenite belts are truncated updip by the next youngest belt. Each belt consists of at least a pair of vertically stacked, composite sandstones separated by a coal bed and estuarine or marine shale facies. Although less marine than their middle Pennsylvanian counterparts, the base of these lower Pennsylvanian midformation shales also represents marine flooding surfaces, or the updip equivalents of flooding surfaces. Therefore, lower Pennsylvanian third-order genetic sequences can be defined that include both marginward quartzarenites and basinward coal-measure facies.

Changes in foreland-basin subsidence, sedimentation patterns, climate, and marine influences affected depositional sequences from the early to middle Pennsylvanian. The westward shift of the longitudinal drainage belt was accompanied by a westward shift in basinward coal measures, resulting in increasingly more extensive coals with time. Increasing expanse and uniformity of coal measures was accompanied by decreasing foreland accommodation. In each third-order sequence, the greatest accommodation appears to occur in the regressive parts of the brackish to marine shales that bound each sequence. The greatest spatial changes in sequence thickness occur across the northern hinge line of the basin and along the basinward limit of successive quartzarenite belts. Foreland-basin subsidence influenced the stacking of successive lower Pennsylvanian quartzarenites, the westward overlap of successive quartzarenite belts, basinward increases in the number of coal beds, development of coal zones in third-order sequences, and basinward increases in the thickness of coal beds.

INTRODUCTION

The eastern Kentucky coal field is one of the leading producers of coal in the United States, producing between 120 and 173 million tons of coal annually from more than 30 lower and middle Pennsylvanian coal beds (data from 1980 to present, Kentucky Coal Association). Chesnut (1992, 1996) constructed a series of strike and dip cross sections across the coal field into surrounding parts of the central Appalachian basin to develop a regional geologic framework for Carboniferous strata (Figure 1). These sections were used to confirm the occurrence of a middle Carboniferous unconformity across much of the basin, and as the basis for a revision of the stratigraphic nomenclature in eastern Kentucky (Figure 2). Boundaries of new formations were mostly based on regionally extensive marine horizons.

The Grundy dip section (Chesnut, 1992, 1996) extends from the northern part of the eastern Kentucky coal field, where the Pennsylvanian section is thin, into the deepest part of the preserved central Appalachian basin (the Pocahontas basin) in the southwestern Virginia coal field (Figure 1). The section was constructed from hundreds of downhole geophysical logs and drillers' records, coal exploration borehole records, and surface descriptions from 21 published 7.5-min geological quadrangle maps. Simplified versions of the cross section were previously published for the purpose of defining a stratigraphic framework and interpreting the broad relationship between lower Pennsylvanian quartzarenites and lateral facies (Chesnut, 1992, 1996). Herein, coal beds and small-scale sandstone bodies not illustrated on previously published versions of the section are illustrated. An additional 25 borehole descriptions were also added for the southwestern Virginia portion of the section. Subsections of the Grundy section, at the scale of individual formations, are used to illustrate temporal and spatial changes in the development of depositional sequences and coal beds. Similar changes were noted on each of Chesnut's (1992, 1996) dip sections, but the Grundy section was based on the greatest number of subsurface records (average spacing of one well per 2.25 km [1.4 mi]) and allows for more detailed examination of

changes at the scale of individual genetic sequences (third order) and coal-clastic cycles (fourth order).

Regional Tectonics

The central Appalachian basin is a subbasin of the Appalachian foreland basin (Figure 3A). The central Appalachian basin initially formed above a part of the Rome trough, a possible Precambrian aulacogen. The basin was subsequently enlarged in response to collisional tectonics along the eastern margin of North America during the Taconic, Acadian, and Alleghanian–Hercynian orogenies (Tankard, 1986; Chesnut, 1991b; Thomas, 1995). The northern boundary of the Rome trough is defined at the surface by the Kentucky River fault system (Figure 3B). The southern boundary in the eastern part of the coalfield corresponds in part to the axis of a post-Pennsylvanian downwarping, the eastern Kentucky syncline (Figure 3B). South of the Kentucky River fault system is another subparallel, east–west-oriented structural trend, formed by the Irvine-Paint Creek fault system and the Paintsville anticline (Figure 3B). Faults in the Kentucky River and Irvine-Paint Creek fault systems are downthrown to the south. The Pine Mountain thrust fault (Figure 3B), which is post-middle Pennsylvanian, represents the westernmost edge of Alleghanian deformation. The Grundy dip section is located northeast of the terminus of the thrust fault. Southeast of the thrust, lower Pennsylvanian strata thicken considerably into southwestern Virginia, into a subbasin called the Pocahontas basin (Figure 3A, B).

Structural controls on Pennsylvanian sedimentation are widely recognized in the basin. Regional downwarping of the foreland basin produced southeastward thickening of strata and an increase in the total number of coal beds basinward (Wanless, 1975). Stratigraphic thickness changes measured from surface outcrops across the Kentucky River and Irvine-Paint Creek fault systems have been used to infer that the structures acted as a hinge line on the northern margin of the basin during the middle Pennsylvanian (Ferm and Cavaroc, 1969; Donaldson, 1974; Horne, 1979; Powell, 1979; Donaldson and Shumaker, 1981; Donaldson and Eble, 1991). Paleochannel and paleovalley trends have been shown to parallel several

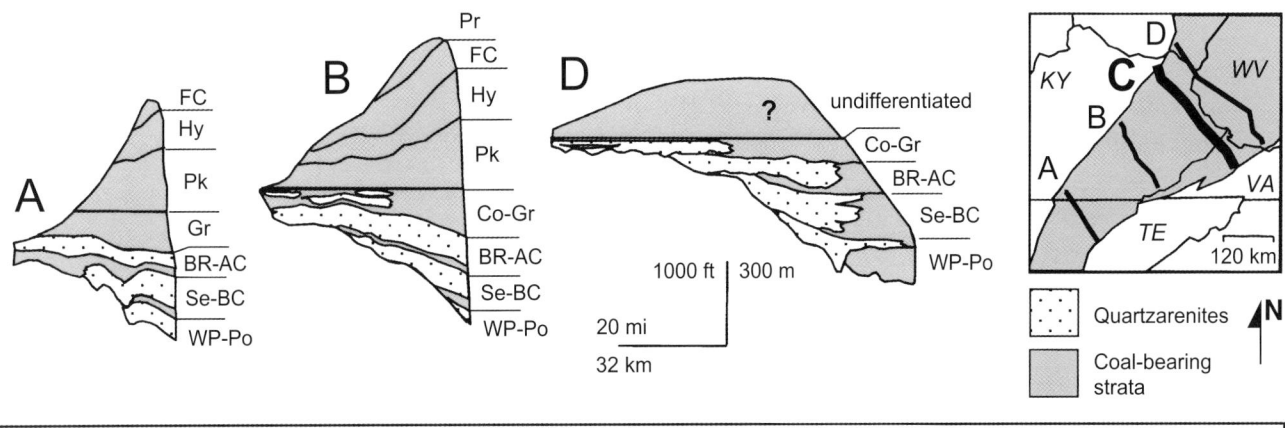

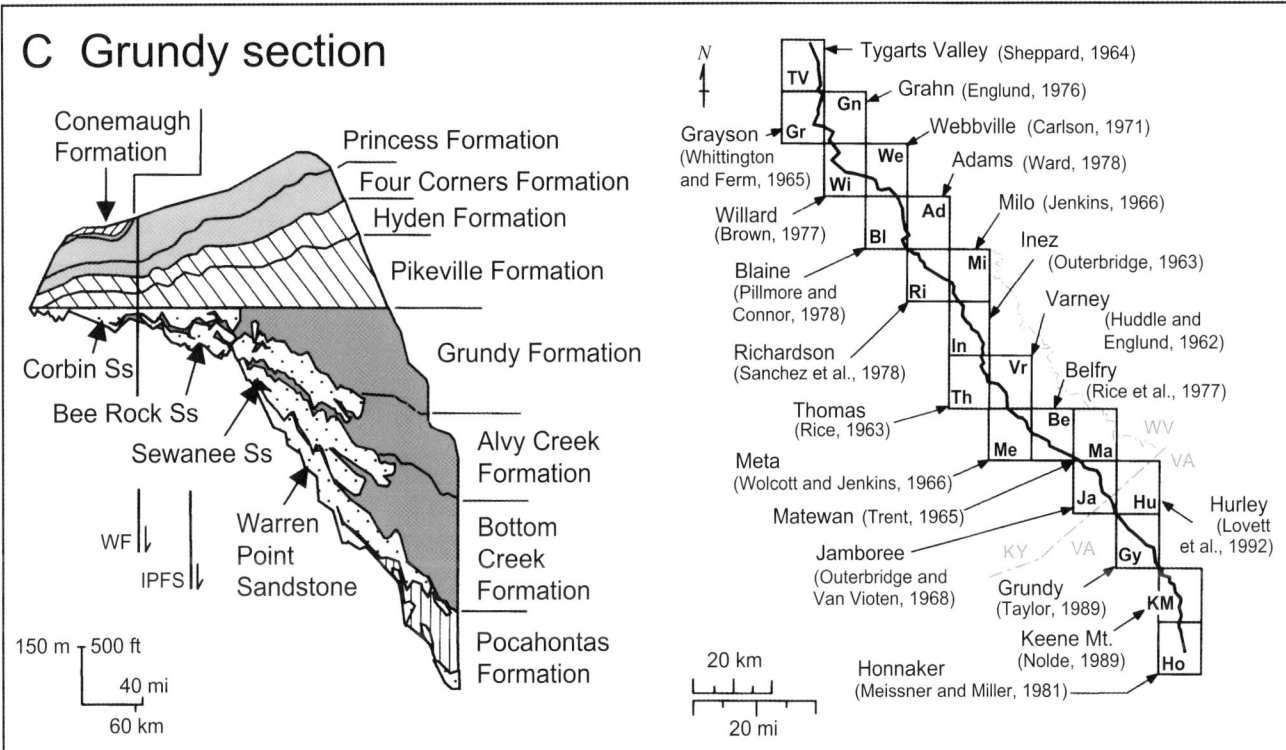

Figure 1. Cross sections of Pennsylvanian strata through the central Appalachian basin. Datum is the base of the Betsie Shale Member of the Pikeville Formation (from figure 2, Chesnut, 1996). The Grundy section is highlighted to show division of depositional sequences analyzed in this report. The location map for the Grundy section shows 7.5-min geologic quadrangle map references and abbreviations used in Figures 5–12.

major basin structures (Horne et al., 1978; Padgett and Ehrlich, 1979; Greb and Chesnut, 1989, 1996; Barnhill, 1994; Greb et al., 1999). Other studies have documented coal-bed thickness changes across smaller structures (Ferm and Weisenfluh, 1989; Hower et al., 1991, 1992; Staub, 1991; Weisenfluh and Ferm, 1991; Greb and Weisenfluh, 1996; Greb et al., 1999).

Regional Stratigraphy

In general, the Pennsylvanian section in the eastern Kentucky coal field can be divided into (1) a lower Pennsylvanian section in which coal-bearing strata

are juxtaposed against thick, quartz-pebble-bearing quartzarenites; (2) a middle Pennsylvanian section of more spatially continuous coals lacking lateral quartzarenites; and (3) an areally restricted upper Pennsylvanian section characterized by thinner coals and thicker claystones (Figures 1, 2). With some variation, lower Pennsylvanian quartzarenites were mapped as members of the Lee Formation during the U.S. Geological Survey-Kentucky Geological Survey joint mapping program, whereas the remaining coal-bearing strata were assigned to the Breathitt Formation (Figure 2). Chesnut (1992, 1996) elevated the Breathitt to group status and divided the coal measures into

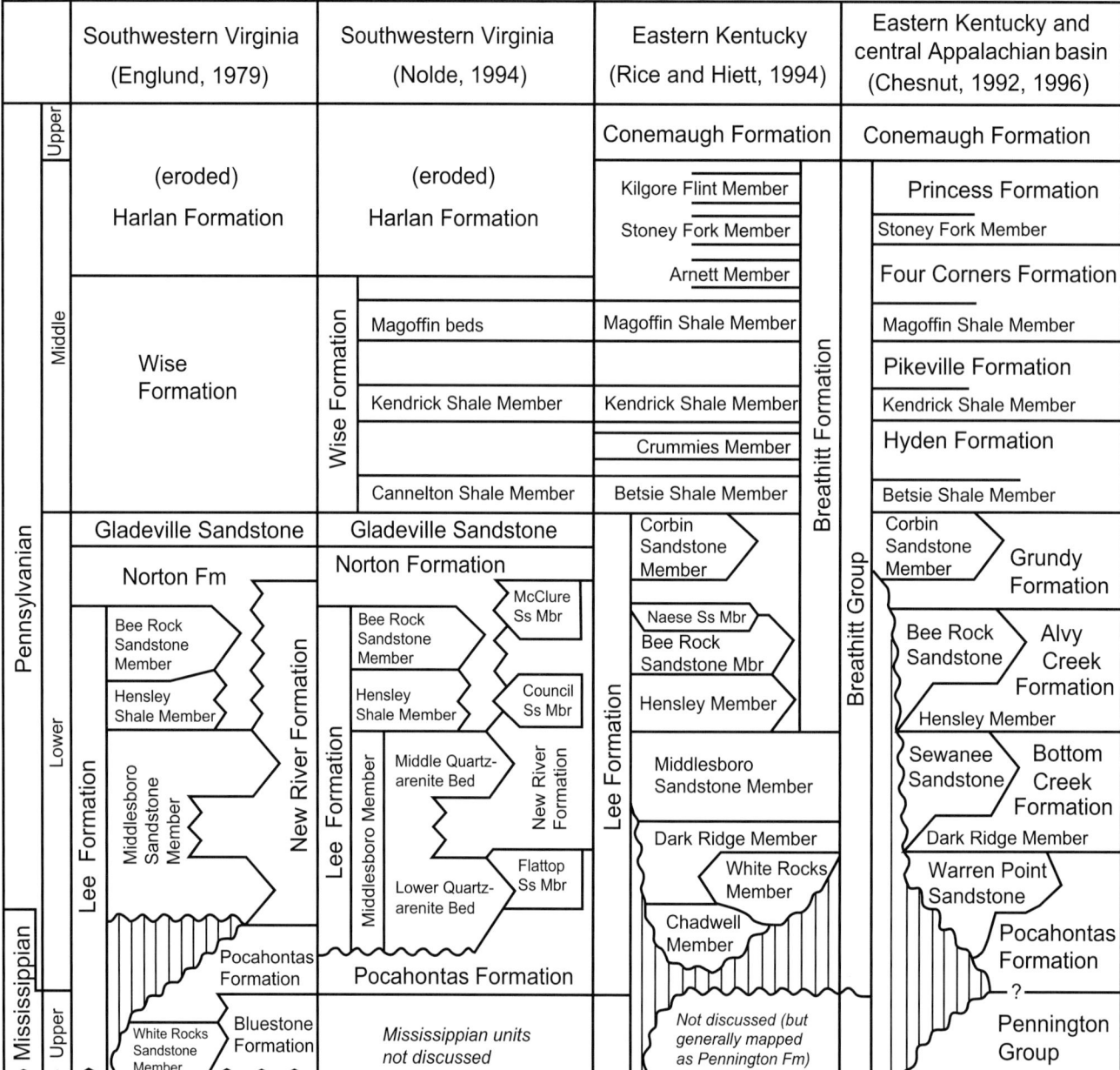

Figure 2. Lithostratigraphic correlation of Carboniferous units in the central Appalachian basin (from figure 5, Chesnut, 1996). Fm = Formation, Mbr = Member, Ss = Sandstone.

formations based on previously described, widespread, marine shale members (Figure 2). Many of the same marine units also have been used to revise correlations in surrounding states (Blake et al., 1994) and as key marker horizons for correlating coal beds between marine zones (Rice and Hiett, 1994). For simplicity, the nomenclature of Chesnut (1992, 1996) is used for both Kentucky and Virginia strata in this report.

Sandstone Composition

Lower Pennsylvanian strata on the western margin of the basin are dominated by thick, cliff-forming quartzarenites, whereas equivalent strata in basinward areas contain litharenites, sublitharenites, and quartzarenites (Davis and Ehrlich, 1974; Englund, 1974). Quartzarenites in the eastern Kentucky coal field generally exhibit south or southwest paleocurrents and contain extrabasinal quartz pebbles (Bement, 1976; Rice, 1984; Barnhill, 1994; Churnet, 1996; Greb and Chesnut, 1996). Litharenites and sublitharenites in lower Pennsylvanian coal-bearing strata, as well as litharenites in middle Pennsylvanian strata, exhibit mostly west or northwest paleocurrent modes and intrabasinal clasts (Donaldson, 1974; Ferm, 1974; Englund,

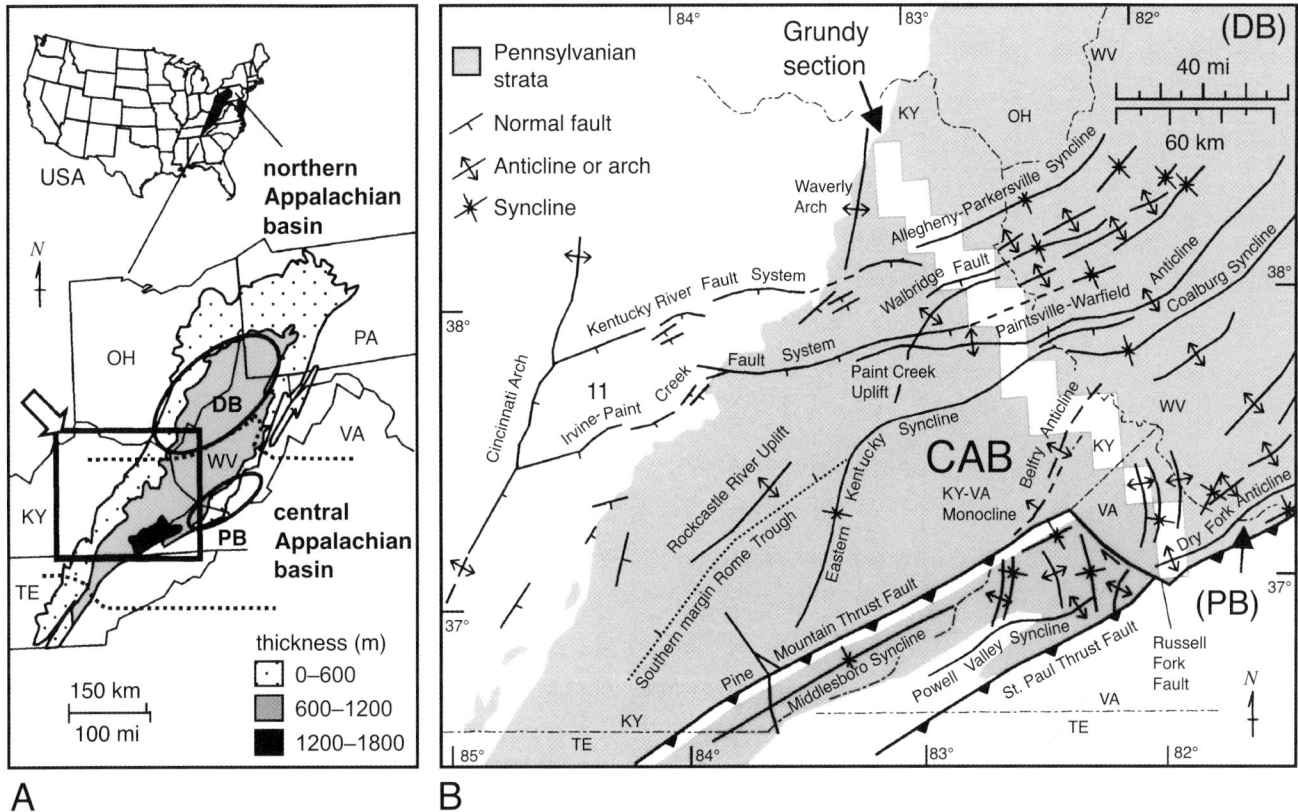

Figure 3. (A) Location of the eastern Kentucky coalfield of the central Appalachian basin showing thickness of Pennsylvanian strata. KY = Kentucky, OH = Ohio, PA = Pennsylvania, TE = Tennessee, VA = Virginia, WV = West Virginia, DB = Dunkard basin, PB = Pocahontas basin, (B) Map of the eastern Kentucky coalfield and surrounding parts of the central Appalachian basin showing major structural features and the position of Chesnut's (1992, 1996) Grundy dip section.

1979; Rice et al., 1979; Houseknecht, 1980; Rice and Schwietering, 1988). Stratigraphic changes in the composition of rock fragments in the lithic arenites were interpreted by Davis and Ehrlich (1974) as the breaching and erosion of a plutonic sequence southeast of the basin.

There has been significant debate concerning the source areas and depositional environments of Lee-type quartzarenites. Similar quartzarenites have been interpreted as beach barriers (Ferm et al., 1971; Ferm, 1974; Horne et al., 1971; Englund, 1974, 1979), fluvial bed-load (braided) systems (Rice, 1984; Rice and Schwietering, 1988; Churnet, 1996), and tidal straits (Cecil and Englund, 1989; Englund and Thomas, 1990). Greb and Chesnut (1996) inferred dominantly fluvial deposition for Lee-type sandstones in eastern Kentucky with the development of local tidal-estuarine sedimentation toward the tops of sandstone units. Relevant to this study is the lateral relationship with basinward litharenites. In some reports (e.g., Englund and Delaney, 1966a), the relationship is interpreted as intertonguing. Data from other reports (e.g., Miller, 1974) suggest that the quartzarenites do not inter-

tongue with laterally adjacent coal-bearing strata across their entire extent.

Depositional Sequences

Repetitive, coal-bearing depositional sequences in the Appalachian basin were termed cyclothems by Wanless and Shepard (1936). Cyclothems and similar scales of sedimentary cycles in the basin have been inferred by many researchers (e.g., Busch and Rollins, 1984) to be controlled by waxing and waning Gondwana ice sheets, i.e., glacial-eustatic controls. Late Pennsylvanian eustatic oscillations may also have created lateral shifts in maritime-influenced paleoclimates along an Appalachian coastline, which would have influenced peat accumulation and cyclothem development (Heckel, 1995). Climatic controls have been inferred as a primary control on Appalachian basin cyclothems, as well as analogs for extensive Pennsylvanian coals in the basin (Cecil et al., 1985; Cecil, 1990).

Temporal changes in Carboniferous sedimentation also have been interpreted as a depositional continuum related to advancing deltas from the Alleghanian

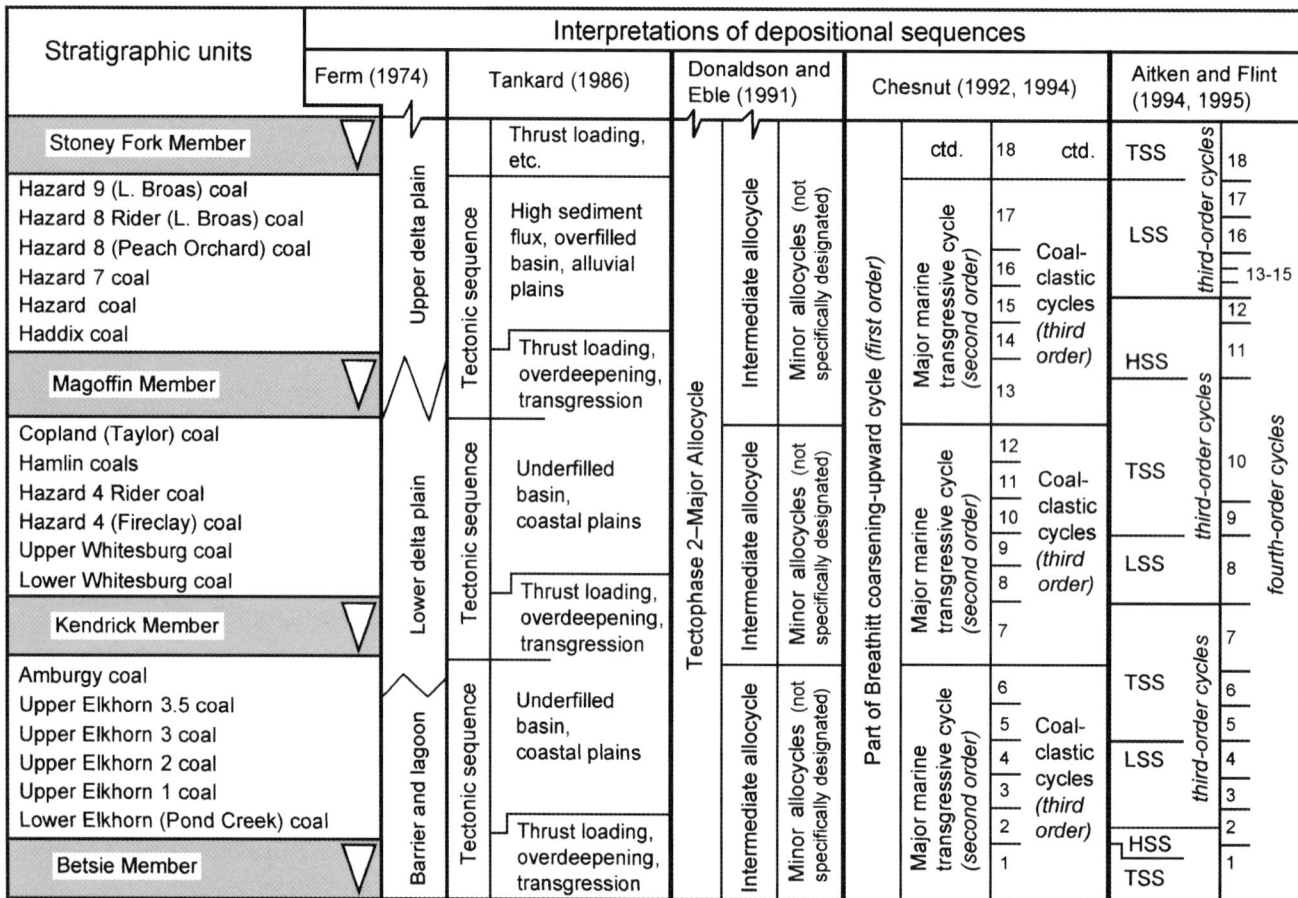

Figure 4. Comparison of controls on stratigraphic framework and depositional sequences in middle Pennsylvanian strata of the eastern Kentucky coalfield. HSS = Highstand sequence set, LSS = lowstand sequence set, TSS = transgressive sequence set.

orogeny (Ferm and Cavaroc, 1969; Ferm et al., 1971; Horne et al., 1971; Ferm, 1974, 1979). In this model, Carboniferous strata were deposited as a depositional continuum, with no unconformity between Mississippian and Pennsylvanian strata, and a general progression from marine to barrier and lagoon, to lower and upper delta-plain, and finally to alluvial-plain environments (shown in part in Figure 4). Lateral variation in so-called cyclothems and larger-scale depositional packages were interpreted to have been controlled by autocyclic delta-lobe switching (Ferm et al., 1971; Horne et al., 1978).

Weller (1930, 1964) advocated tectonic controls for Appalachian basin cyclothems. Quinlan and Beaumont (1994) interpreted flexural downwarping in the foreland basin and migration of forebulges cratonward as the mechanism for Carboniferous tectonics. Tankard (1986) expanded on the viscoelastic model and interpreted major middle Pennsylvanian marine horizons, including the Betsie, Kendrick, and Magoffin shales, as the result of foreland deepening following successive thrust loads. Coal-bearing strata between major marine zones were deposited in either underfilled or overfilled conditions that were related to the relative influx of sediment and accommodation space during the relaxation period that followed each deepening event (Figure 4). Klein and Willard (1989) inferred direct tectonic influences on Appalachian cyclothems and noted that glacial-eustatic changes were ultimately the result of Carboniferous tectonics. Donaldson and Eble (1991) interpreted three tectophases in the Appalachian basin, generally corresponding to the lower, middle, and upper Pennsylvanian series. They inferred tectophases to be composed of smaller allocycles bounded by the widespread marine zones, the same as those noted by Tankard (1986). Durations for the transgressive-regressive allocycles were calculated to be 2.5–3.0 m.y. (Donaldson and Eble, 1991; Chesnut, 1992, 1994), which did not match known Milankovitch oscillations; therefore, the allocycles were inferred to be controlled by tectonics or an unknown orbital mechanism.

Chesnut's (1992, 1996) revision of Pennsylvanian stratigraphy noted three general depositional trends

(Figure 4). The first was an upward-coarsening trend in the Breathitt Group, inferred to reflect increasing intensity and proximity of the Alleghanian orogeny. The second trend was the relatively equal spacing of seven widespread marine to marginal marine shales (lower and middle Pennsylvanian) and the six transgressive-regressive allocycles between them. Each major transgressive-regressive cycle contains five to seven coal-clastic cycles defined as the interval between the tops of regionally extensive coal beds, which is the scale of a cyclothem. Coal-clastic cycles were interpreted to have an average periodicity of 0.4 m.y., based on the inferred duration of marine-transgression cycles (Chesnut, 1991b, 1992, 1994). Hence, coal-clastic cycles were inferred to have been controlled by eccentricity cycles and glacial-eustatic oscillations. Donaldson and Eble (1991) termed this scale of cycle a minor allocycle and calculated a duration of 0.3–0.7 m.y., also inferring eustatic control, but stressing the significance of sediment-supply shifts, tectonics, and paleoclimates on differences between the changing appearance of minor allocycles in lower to upper Pennsylvanian strata.

Sequence-stratigraphic principles were applied to the middle Pennsylvanian, coal-bearing part of the Breathitt Group by Aitken and Flint (1994, 1995). They considered the coal-clastic cycles of Chesnut (1991b, 1992, 1994) to be fourth-order sequences, although there were some differences in designations (informally numbered for comparison in Figure 4). These sequences stack into lowstand, transgressive, and highstand sequence sets and are grouped into third-order composite sequences. For the most part, an attempt was made to place sequence set boundaries at the base of sandstones (interpreted as paleochannels and paleovalleys) of coarser grain size than sandstones or shales in the coarsening-upward, progradational part of the underlying highstand sequence set. Where scour-based sandstones were missing along the set boundary, the assumption was made that the base of the lowstand set must have been an interfluve and was traced along a paleosol, commonly at the base of a coal (Aitken and Flint, 1994, 1995).

PURPOSE

Stratigraphic, sedimentary, and coal data were used in association with a detailed dip section to compare temporal and spatial variability in lower and middle Pennsylvanian genetic sequences (third order) and coal-clastic cycles (fourth order) in the eastern Kentucky coal field and southwestern Virginia. Sections are illustrated for each formation, at the scale of third-order sequences, to examine spatial and temporal influences on peat accumulation and sedimentation relative to foreland-basin structure.

DEPOSITIONAL SEQUENCES

Warren Point and Pocahontas Formations

Figure 5 is a cross section of the strata between the top of the Warren Point Sandstone and the laterally equivalent Pocahontas Formation. Along the Grundy line of section, the Warren Point consists of two to three stacked, quartz-pebble-bearing quartzarenites. Regional cross sections reveal sequential truncation of lowermost Pennsylvanian (Pocohontas Formation) and Mississippian (Pennington Group and underlying units) rocks by the Warren Point Sandstone toward the western margin of the basin (Chesnut, 1992). Where the Pocohontas Formation is truncated, the base of the Warren Point is an apparent Mississippian–Pennsylvanian unconformity (Chesnut, 1992, 1994, 1996). Basinward, the presence of a systemic unconformity is debatable. Englund and Delaney (1966a) and Chesnut (1992, 1996) inferred an unconformity updip and a conformable sequence in the deepest part of the basin. Beuthin (1997) noted a thick paleosol toward the base of Pennsylvanian strata in the deepest part of the basin and inferred a basinwide unconformity. Along the Grundy section, there are not enough core data to determine if there is a thick paleosol toward the base of the Pocahontas; the systemic boundary is therefore shown as a dashed line in Figure 5.

The petrography and extent of thick sandstones in the lower part of the Pocahontas Formation in the southeastern portion of the section is uncertain because most of the borehole data do not extend beneath the Pocahontas 3 coal bed (po3 in Figure 5). Quartzarenites on the northwestern half of the cross section form an elongate, northeast-southwest–oriented belt at least 64 km (40 mi) wide (Chesnut, 1992, 1996). The uppermost sandstone in the elongate belt is truncated marginward by the overlying Sewanee Sandstone near the axis of the eastern Kentucky syncline (EKS in Figure 5). Downstructural dip, the upper part of the upper sandstone in the belt appears to be gradational, with sublitharenites and litharenites beneath the Pocahontas 8 (po8) coal, which correlates to the Flat Gap Sandstone Member of the Pocahontas Formation. This sandstone thickens from 9 to more than 30 m (30 to more than 100 ft) toward the sand belt. The top of the formation is the top of

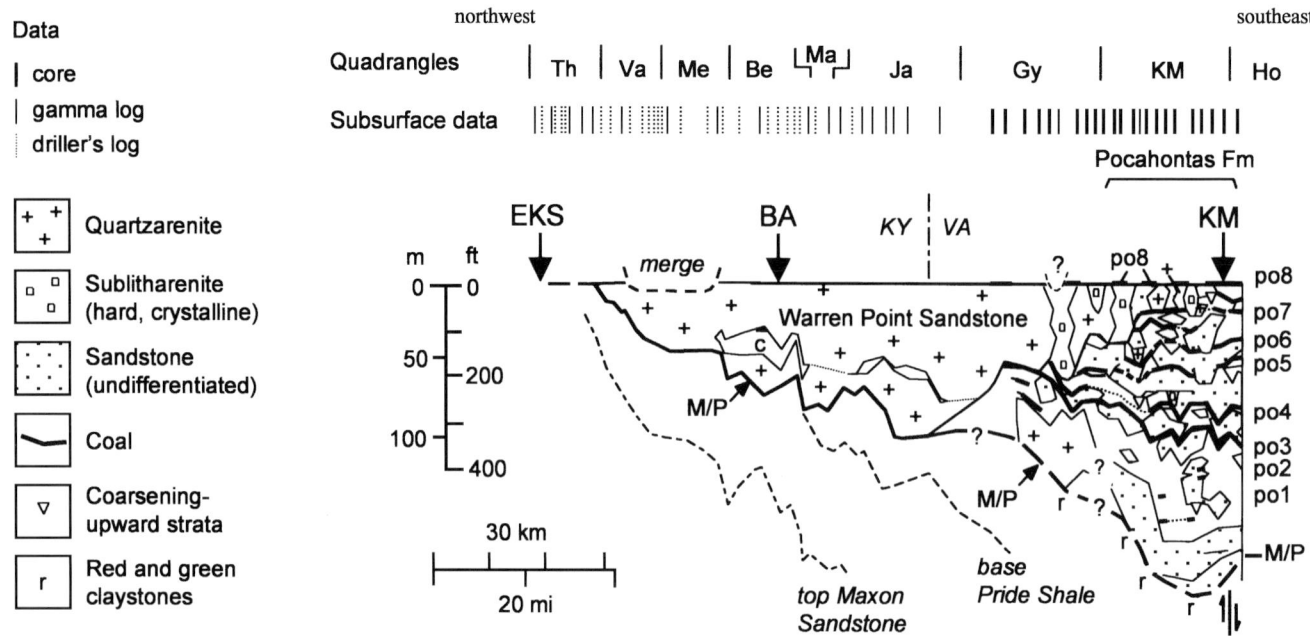

Figure 5. Warren Point Sandstone and Pocahontas Formation along the Grundy section (Chesnut, 1992). Datum redrawn to the top of the Warren Point Sandstone, Pocahontas 8 coal (po8), or base of the Dark Ridge Member. See Figure 2 for location of section. Coal-bed abbreviations explained in text (c = unnamed coal). Major structures located by arrows above upper datum and explained in text. M/P = Mississippian–Pennsylvanian contact. Dashed where uncertain. Vertical exaggeration is approximately 240× for Figures 5–12.

the Flat Gap Sandstone or base of the po8 coal bed. Miller (1974) indicated that the Flat Gap Sandstone was truncated by the Warren Point Sandstone (Middlesboro Member), although that relationship is not obvious along the Grundy dip section.

The Pocahontas Formation is 213 m (700 ft) thick to the southeast, as shown in Figure 5, but thins and is truncated to the northwest. Eight named coal beds are recognized in the Pocahontas Formation, including all of the coals between the Pocahontas 1 (po1) and Pocahontas 8 (po8) coals. The Pocahontas 3 is the most extensive and economically important coal bed in the formation (Miller, 1974; Arkle et al., 1979; Englund, 1979; Nolde, 1994). The Pocahontas 3 and higher coals are truncated laterally beneath the Warren Point Sandstone. Nolde (1989) indicated interfingering of the Warren Point (lower quartzarenite) between the Flat Gap Sandstone and the Pocahontas 6 coal, but later, Nolde (1994) indicated that the Warren Point truncated only the Pocahontas 3 to Pocahontas 5 (po5 in Figure 5) coals and inferred that the quartzarenites between the Pocahontas 6 (po6 in Figure 5) and Pocahontas 7 coals (po7 in Figure 5) were not equivalent to the Warren Point Sandstone (Middlesboro Member). Along the Grundy section, a thick sandstone between the Pocahontas 5 and 6 beds could be inferred to thicken toward the Warren Point and

be laterally intertonguing, or it could be interpreted as a separate sandstone truncated by the Warren Point (Figure 5).

Southeastward expansion of the Pocahontas stratigraphic interval is accompanied by the development of multiple benches, or multiple coal beds in the Pocahontas 6 zone, and perhaps in the Pocahontas 4 coal bed (Figure 5). Although not obvious in Figure 5 because of scale, most Pocahontas coal beds thicken toward the southeast (Miller, 1974; Englund and Thomas, 1991). None of these coal beds extend northwest into the eastern Kentucky coalfield. In addition, not all of the fourth-order (coal-clastic) cycles thicken evenly to the southeast. Lateral thickness changes from northwest to southeast are greater in the stratigraphic interval beneath the Pocahontas 3 coal bed than in the interval above it. Thickness also varies substantially between the Pocahontas 4 and 5 coal beds because of intervening sandstones (Figure 5). In contrast, the stratigraphic thickness of the intervals between coal beds above the level of the Pocahontas 5 only increases slightly to the southeast (Figure 5).

Sewanee and Bottom Creek Formations

Figure 6 is a cross section through the Sewanee Sandstone and Bottom Creek Formation. The Sewanee Sandstone overlaps the underlying Warren Point

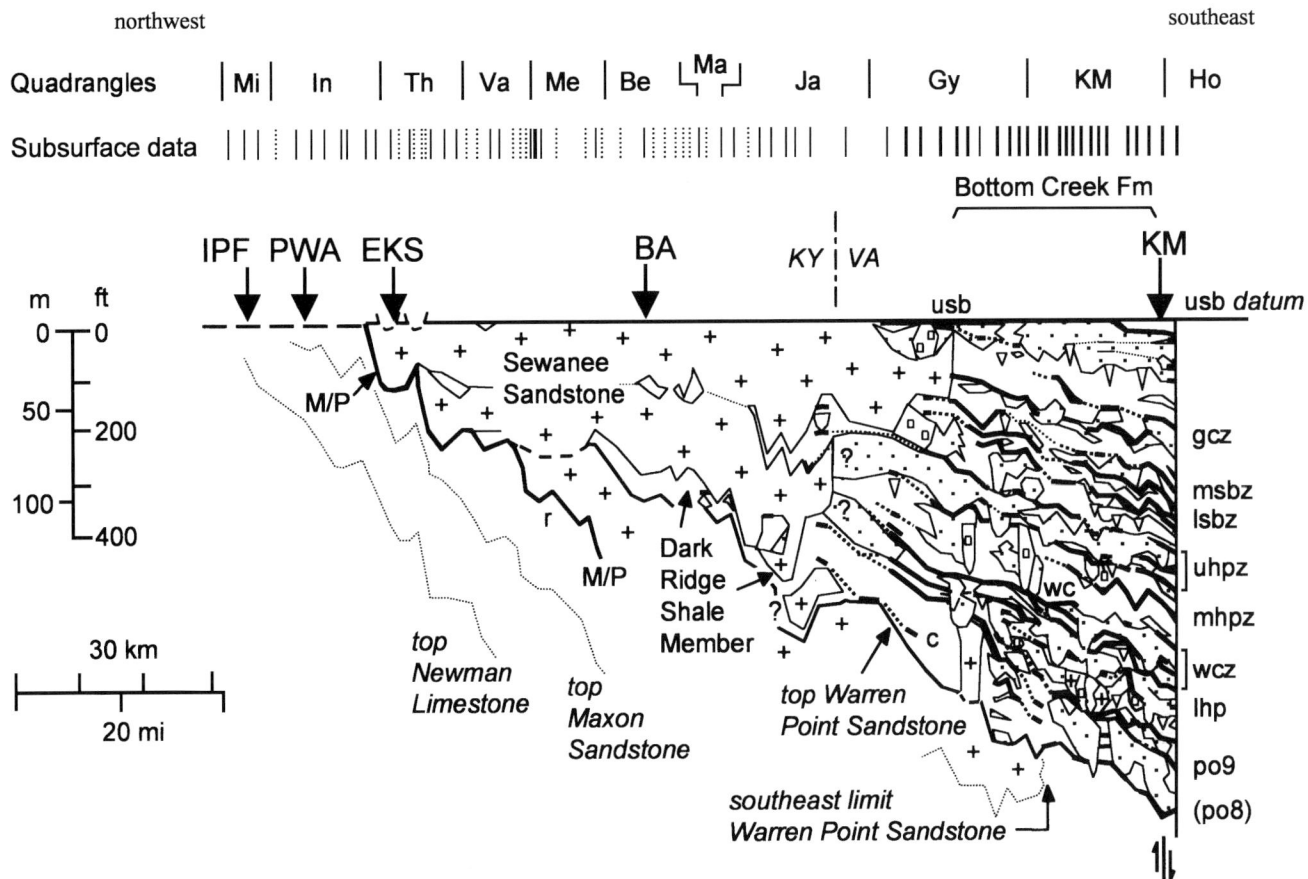

Figure 6. Sewanee Sandstone and Bottom Creek Formation along the Grundy section. Datum redrawn to the top of the Sewanee Sandstone, the upper Seaboard coal (usb), or the base of the Hensley Member. See Figure 2 for location of section and Figure 5 for legend. Coal-bed abbreviations explained in text. Major structures located by arrows above upper datum and explained in text. M/P = Mississippian–Pennsylvanian unconformity.

Sandstone. The shale between the Sewanee Sandstone and underlying Warren Point Sandstone is called the Dark Ridge Member (Rice et al., 1979; Chesnut, 1992; Rice and Hiett, 1994). This unit locally contains marine fossils (Chesnut, 1991a) and may be equivalent to a coal-barren interval of strata toward the base of the Bottom Creek Formation in Virginia. A gray shale with marine and brackish-water fauna (Henry and Gordon, 1979) has been noted above the Pocahontas 8 coal bed (Englund, 1974, 1979); it could be equivalent to the Dark Ridge Member.

The Sewanee Sandstone is another quartz-pebble-bearing quartzarenite (Churnet, 1996), which occurs in a belt approximately 80 km (50 mi) wide (Chesnut, 1992, 1996). On the Grundy section, the Sewanee Sandstone consists of at least two thick (>25 m; 75 ft) superimposed sandstone units (dotted line in quartzarenite in Figure 6). Two to three stacked units occur in each of Chesnut's (1992, 1996) other sections. The Sewanee Sandstone appears to pinch out against the middle Carboniferous unconformity near the eastern

Kentucky syncline (Figure 6). This position is just updip structurally from the point where the lower Sewanee Sandstone overlaps the underlying Warren Point Sandstone. Along this part of the Grundy section, the lower part of the Sewanee Sandstone cannot be differentiated from the top of the Warren Point Sandstone. Truncation is inferred along the dashed line in Figure 6. Basinward, both the upper and lower Sewanee terminate just east of the Kentucky–Virginia state line (KY-VA in Figure 6).

The lower part of the Sewanee Sandstone belt is laterally equivalent to a persistent sandstone that is capped by the upper Horsepen coal zone (uhcz in Figure 6). Another west-thickening sandstone occurs between coal beds in the War Creek coal zone (wcz in Figure 6) and middle Horsepen coal zone (mhpz in Figure 6). This sandstone may be transitional to the Sewanee or truncated by the Sewanee. The War Creek coal bed is, according to Nolde (1994), persistent between the lower and middle quartzarenites of southwest Virginia, which would be the upper part of the

Warren Point and lower part of the Sewanee herein. Most of the coals between the War Creek coal zone and Pocahontas 8 (po8) coal appear to pinch out prior to the southeastern terminus of the lower Sewanee Sandstone, although the Dark Ridge Shale continues eastward (Figure 6). A single coal bed, called the Cumberland Gap coal, has been noted in some boreholes in the Dark Ridge Member in Kentucky (Rice, 1984; Rice and Hiett, 1994).

The base of the upper part of the Sewanee Sandstone belt is separated from the lower part of the Sewanee Sandstone by a thick shale near the Kentucky–Virginia border (Figure 6). This shale may be equivalent to the thick, persistent shale and coarsening-upward interval between the upper Horsepen and lower Seaboard (lsb) coals to the southeast. The upper part of the upper Sewanee appears to be transitional, with a northwest-thickening sandstone beneath the upper Seaboard coal (usb) at the top of the Bottom Creek Formation (Figure 6). The sandstone beneath the upper Seaboard coal caps a thick coarsening-upward interval above the Greasy Creek coal zone (gcz in Figure 6). The upper part of the Sewanee Sandstone truncates coals in the Greasy Creek and middle Seaboard zones.

Each of the coals in the Bottom Creek Formation appears to develop into a zone of two to three coal beds to the southeast (coal labels ending in "z" in Figure 6), which makes correlation of individual beds difficult. The lateral thickness change between major coal beds is greatest in the lower part of the formation, and least between the Greasy Creek (gc) and upper Horsepen coals. As with coals in the underlying Pocahontas Formation, coal beds thicken to the southeast (Miller, 1974; Nolde, 1994).

Bee Rock and Alvy Creek Formations

Figure 7 is a cross section through the Bee Rock Sandstone and the laterally equivalent Alvy Creek Formation. The Bee Rock Sandstone overlaps the underlying Sewanee Sandstone between the eastern Kentucky syncline and Paintsville-Warfield anticline (PWA) (Figure 7). The Bee Rock is truncated updip by the Corbin Sandstone near the Irvine-Paint Creek fault system (IPF in Figure 7). The southeastern terminus of the Bee Rock Sandstone in this cross section is near the Kentucky–Virginia state line, which is northwest of the terminus of underlying quartzarenites.

The shale between the Bee Rock Sandstone and the Sewanee Sandstone is called the Hensley Member (previously of the Lee Formation; Englund and Delaney, 1966a). Much of the Hensley Member has been equated to marine-fossil-bearing strata above the middle Carboniferous unconformity along the western outcrop margin of the basin (Chesnut, 1991a). In southeastern Kentucky, a thin coal called the Tunnel coal bed (not labeled along the Grundy section) occurs locally near the top of the Hensley Member.

As with the underlying quartzarenites, the Bee Rock Sandstone occurs in a 91-km (57-mi)-wide belt, which is formed from at least two distinct, scour-based sandstones (Chesnut, 1992, 1996). Along the southwestern margin of the basin, the lower sandstone is mapped as the Rockcastle Sandstone Member, and the upper sandstone is mapped as an unnamed unit, but is informally called the Pine Creek sandstone. Quartz pebbles are abundant in the Rockcastle and occur only toward the base of the Pine Creek sandstone. Correlations have not been made to the northeast, so the nomenclature is not extended from the southwestern outcrop belt to the Grundy line of section. Both the Rockcastle and Pine Creek sandstones are multistory, cross-bedded, coarse-grained units with southwest paleocurrents (Wizevitch, 1992; Greb and Chesnut, 1996).

Along the southwestern margin of the coal field, the Rockcastle and Pine Creek sandstones are separated by the Barren Fork coal and an unnamed shale with a brackish-water fauna (Chesnut, 1991a; Greb and Chesnut, 1992). The Barren Fork coal contains a miospore assemblage that marks a change in spores between older and younger coal beds. The Lee coal in southwestern Virginia (Eble 1996), which occurs locally beneath the Tiller coal (ti in Figure 7), is the oldest coal with a similar assemblage in the Alvy Creek Formation. The overlying Jawbone coal contains the same spore assemblage. The coarsening-upward interval shown above the Jawbone coal in Figure 7 (jw) may be equivalent to the estuarine horizon developed above the Barren Fork coal on the southwestern margin of the basin. The Jawbone is reported to be the most extensive coal bed in this part of the section in southwestern Virginia (Miller, 1974) and appears to extend between the upper and lower sandstones of the Bee Rock Sandstone belt (Figure 7).

The lower part of the Bee Rock Sandstone belt is equivalent to the interval beneath the Jawbone coal and the top of the Bottom Creek Formation (Figure 7). The upper part of the lower Bee Rock may be transitional, with a thick sandstone beneath the Jawbone coal bed called the Council Sandstone Member (Css in Figure 7). Thickness increases sharply between the Council Sandstone and the lower part of the Bee Rock Sandstone. As with most extensive sandstones

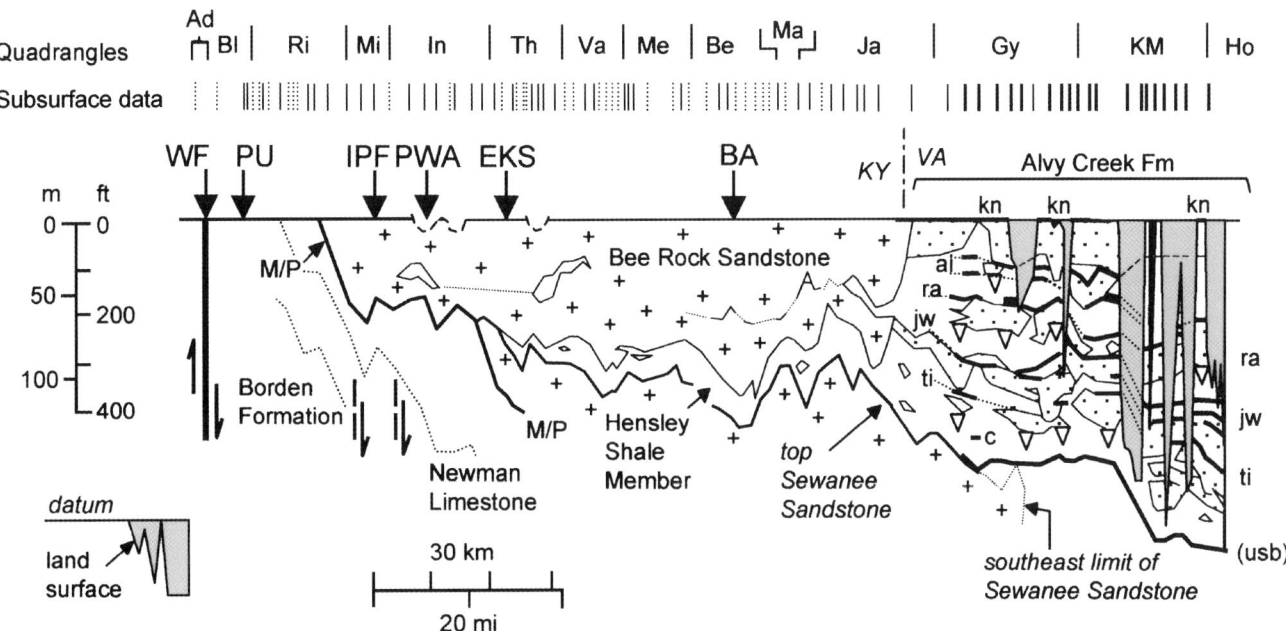

Figure 7. Bee Rock Sandstone and Alvy Creek Formation along the Grundy section. Datum redrawn to the top of the Bee Rock Sandstone, the Halsey Rough-Kennedy coal (kn), or the base of the Dave Branch Member. See Figure 2 for location of section and Figure 5 for legend. Coal-bed abbreviations explained in text. Major structures located by arrows above upper datum and explained in text. M/P = Mississippian–Pennsylvanian unconformity.

in this interval, the Council Sandstone thickens to the southeast.

The upper part of the Bee Rock Sandstone belt is capped by the Kennedy coal bed (kn in Figure 7). To the southeast, the Kennedy coal bed is underlain by the McClure Sandstone (Mss in Figure 7) of the lateral Alvy Creek Formation. The McClure is generally not a quartzarenite (Nolde, 1994; Whitlock, 1994), but locally is reported to contain sublitharenites transitional to the quartzarenites of the Bee Rock (Englund, 1979; Nolde, 1994). Hence, the McClure has been interpreted as both grading into (Englund, 1979; Nolde, 1994) and being truncated by the Bee Rock Sandstone (Miller, 1974; Nolde, 1994). Sporadic quartz pebbles (as are found in the quartzarenites) were noted by Nolde (1989) above the scoured base of the sandstone in the Keen Mountain quadrangle (KM in Figure 7). Along the Grundy section, the McClure Sandstone is less than half the thickness of the upper part of the Bee Rock Sandstone. Therefore, the McClure is interpreted to be mostly truncated by, albeit juxtaposed against, the Bee Rock Sandstone. If any part of the McClure is intertonguing, it is only with the uppermost part of the Bee Rock Sandstone.

All coal-clastic cycles increase in thickness to the southeast, but the interval between the Tiller coal and the base of the formation (above the usb in Fig-

ure 7), coal thickness changes the most, doubling to the southeast. Coal-clastic intervals in the upper part of the formation also thicken to the southeast, but to a lesser degree. Each of the coals beneath the McClure Sandstone thickens toward the southeast (Miller, 1974; Englund and Thomas, 1991). The Raven coal (ra in Figure 7) becomes a zone of multiple coal beds to the southeast.

Grundy Formation and Corbin Sandstone Member

Figure 8 is a cross section through the Grundy Formation, including the Corbin Sandstone Member. The Corbin is exposed near the surface along the western margin of the eastern Kentucky coal field and has been well studied (Potter and Siever, 1956; Rice, 1984; Wizevitch, 1992; Barnhill, 1994). It occurs in a belt 64 km (40 mi) wide, west of the Bee Rock Sandstone (Chesnut, 1992; Greb and Chesnut, 1996). The Corbin Sandstone Member is extensive across the northwestern part of the Grundy section and terminates near the axis of the eastern Kentucky syncline (Figure 8), close to the point at which the Sewanee Sandstone is truncated by the Bee Rock Sandstone (Figure 1).

The Corbin Sandstone Member also occurs as at least two vertically stacked sandstone units. The lower

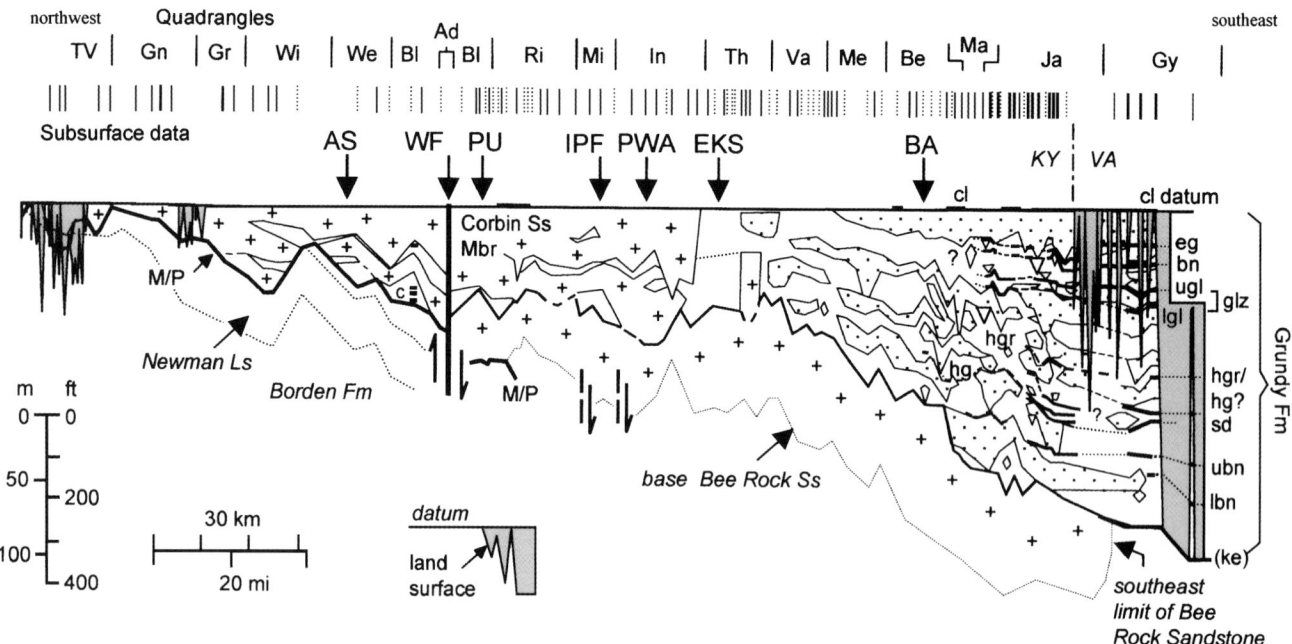

Figure 8. Corbin Sandstone Member and Grundy Formation along the Grundy section. Datum is the base of the Betsie Member or the top of the Manchester-Lily-Zachariah-Clintwood coal bed (cl). See Figure 2 for location of section and Figure 5 for legend. Some outcrop correlations inferred from geologic quadrangle maps referenced in Figure 2. Coal-bed abbreviations explained in text. Major structures located by arrows above upper datum and explained in text. M/P = Mississippian–Pennsylvanian unconformity.

part of the Corbin Sandstone may be equivalent to a unit in southeastern Kentucky informally called the Hazel Patch sandstone. The Hazel Patch is quartzose, but is commonly bioturbated and more akin to shore-face and estuarine deposits than to the fluvial facies that characterize the other thick quartzarenites (Greb and Chesnut, 1989, 1994). More work is needed to determine the extent of the bioturbated facies in the lower sandstone. The unit most commonly mapped as Corbin Sandstone is a multistory, cross-bedded, quartz-pebble-bearing quartzarenite (Rice, 1984; Chesnut, 1992, 1996; Wizevitch, 1992; Barnhill, 1994).

The lower sandstone in Figure 8 terminates along the eastern Kentucky syncline and merges with the upper, quartz-pebble-bearing part of the Corbin Sandstone south of the Walbridge fault (WF). In plan view, the Corbin Sandstone belt trends east-west between the Irvine-Paint Creek fault system, and the Walbridge fault then continues south as a belt parallel to the strike axis of the basin, west of the Grundy line of section (Barnhill, 1994; Chesnut, 1992; Greb and Chesnut, 1996). Thinning of the Corbin Sandstone north of the fault makes differentiation between the lower portion of the Corbin and the underlying upper Bee Rock Sandstone problematic. Herein, only the sandstones in the Corbin and laterally equivalent

Grundy Formation are interpreted to extend north of the fault. This is supported by palynology from samples of coal beds beneath the Corbin Sandstone and north of the Walbridge fault, which contain spore assemblages that place them stratigraphically above the Bee Rock Sandstone.

The Corbin Sandstone Member is separated from the underlying Bee Rock Sandstone by a gray shale. On the southwestern margin of the eastern Kentucky coal field, a dark-gray shale with sparse marine fauna and a basal ravinement, informally called the Dave Branch shale (Chesnut, 1991a, 1996; Greb and Chesnut, 1996), occurs at this horizon. Southeastward, there are two possible correlatives. Recent work in West Virginia indicates that the Nuttall Sandstone, a quartz-arenite considered to be an equivalent of the Bee Rock Sandstone, is overlain by the Gilbert coal bed, which in turn is overlain by a widespread, as of yet, unnamed marine zone (Blake et al., 2003). Dark shales with brackish to marine body fossils also occur above the lower Banner coal.

Across parts of northeastern Kentucky, the base of the Corbin Sandstone is the middle Carboniferous unconformity (Chesnut, 1992). In these areas, the Corbin Sandstone may truncate underlying strata to the level of the lower Mississippian Borden Formation

(Figure 8). North of the terminus of the Corbin Sandstone, sublitharenites of the Breathitt Group overlie the unconformity; whether any part of these are equivalent to the Alvy Creek Formation is uncertain. Along the AA Highway in northeastern Kentucky, a possible equivalent to the Betsie Shale Member occurs 10–12 m (33–40 ft) above the unconformity (Bennington, 1992; Martino and Rice, 1992a, b). In some areas, the interval is thinner, and a thick flint clay called the Olive Hill flint clay rests directly on the unconformity (see Englund, 1976, for example).

Along the line of section, the Grundy Formation thickens from 69 m (225 ft) at the terminus of the Corbin Sandstone to 229 m (750 ft) in Virginia (Figure 8). The Grundy Formation is the oldest formation in which significant coal resources occur in the eastern Kentucky coal field (Figure 2) and includes 8–12 coal beds. Coals include the Clintwood (cl), Eagle (eg), Glamorgan or Mason (gl), Hagy rider (hgr), Hagy (hg), Splashdam (sp), upper Banner (ub), and lower Banner (lb) (Figure 8). To the north and west, the coals thin from multiple coal beds (commonly termed upper, middle, and lower) in zones to single coal beds, to thin single coals. Figure 8 shows only a single Glamorgan coal bed and possibly a coal in the Hagy coal zone, continuing north as the sequence thins. Most coals appear to pinch out prior to being truncated updip by the Corbin Sandstone.

The Kennedy (beneath the base of the formation), lower Banner, Splashdam, and Hagy coal beds are overlain by thick, persistent coarsening-upward units. Shales at the base of these coarsening-upward units locally contain marine brachiopods near their bases (Nolde, 1989, 1994; Chesnut, 1991a). Micritic limestone concretions occur in the shales above the Kennedy and lower Banner coals (Nolde, 1989). If the Splashdam or Hagy coals are equivalent to the Grayhawk coal bed, then the overlying shale represents another widespread marine unit. Clearly, more work is needed in correlating this interval.

Along the Grundy section, the lowest coal in the section, called the Big Fork coal (bf in Figure 8), is laterally restricted to an area above an unnamed anticline in the Grundy quadrangle (Gy in Figure 8) (Taylor, 1989). The interval between the Splashdam and Kennedy (top of the Alvy Creek Formation) coal beds shows the greatest lateral change in thickness to the southeast, compared to overlying intervals between coals in the section (Figure 8). The largest lateral increase in thickness appears to occur just southeast of the terminus of the eastern margin of the underlying Bee Rock Sandstone (Figure 8). The interval beneath the Splashdam contains the upper and lower Banner coals, which pinch out north of the Jamboree quadrangle (Ja in Figure 8) in Kentucky.

The next largest increase in basinward thickness change occurs in the coarsening-upward unit above the Hagy rider coal (Figure 8). Above that stratigraphic level, the lateral thickness changes between coals in the upper part of the formation are less than those in the lower part of the formation. The uppermost coal, called the Clintwood coal (Figure 8), is the only coal bed in this formation that is nearly continuous across the eastern Kentucky coal field. The Clintwood (and equivalents) exhibits irregular thickness but is generally a single-benched coal bed above the Corbin Sandstone and a multiple-benched coal or coal zone basinward, where the Corbin is absent (Figure 8). The Lyons (ly), Dorchester (do), and Norton (no) coal beds of southwestern Virginia are equivalent to the Eagle, Blair (bl), and Glamorgan coals updip in Kentucky (Figure 8) (Lovett et al., 1992; Nolde, 1994). Zoning in the lower two coals to the southeast may complicate correlations.

Pikeville Formation

A cross section through the Pikeville Formation is shown in Figure 9. The base of the unit is the Betsie Shale Member, a regionally extensive marine zone (Rice et al., 1987; Chesnut, 1991a, 1992, 1996; Blake et al., 1994; Martino, 1996). Although extensive, the shale was not recognized regionally when 7.5-min geological quadrangles were mapped. The shale is more than 50 m (164 ft) thick in the southeast and thins to less than 5 m (16 ft) north of the Kentucky River fault system. Where thin, it is difficult to differentiate and may be truncated locally by overlying sandstones. The base of the Betsie Member has been assigned recently as the stratigraphic break between lower and middle Pennsylvanian strata, based on palynomorphs and plant megaflora (Eble, 1996; Blake et al., 2003).

The Pikeville Formation is not interrupted by thick quartzarenites. Coal beds are more widespread than in older strata. Figure 9 shows the Pikeville Formation thickening from less than 15 m (50 ft) in the northwest to more than 152 m (500 ft) in the southeast. It may reach a thickness of 214 m (700 ft) in the Jamboree quadrangle (Outerbridge and Van Vloten, 1968). Only two coals are preserved in the northernmost part of the section (Figure 9): the Gun Creek (gu) and Tom Cooper (tc) (Shepard, 1964). The formation thickens abruptly near the Tygarts Valley-Grahn quadrangle border (TV and Gn in Figure 9), and two additional

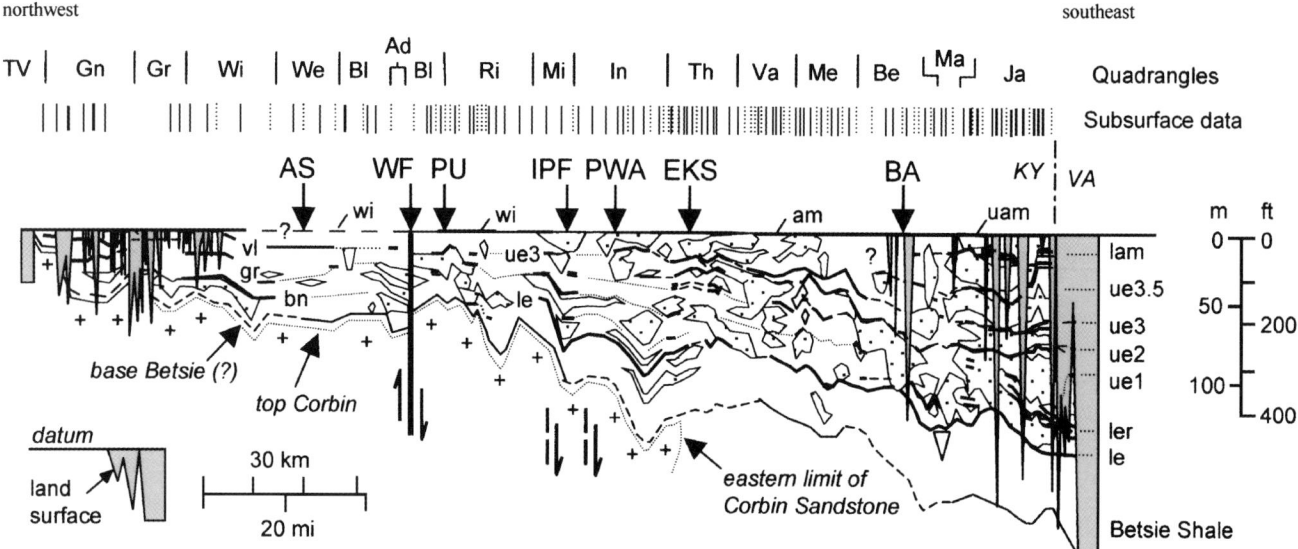

Figure 9. Pikeville Formation along the Grundy section. Datum is the base of the Kendrick Shale Member or the top of the Amburgy coal bed (am or uam where upper and lower Amburgy coal beds occur). See Figure 2 for location of section and Figure 5 for legend. Some outcrop correlations inferred from geologic quadrangle maps referenced in Figure 2. Coal-bed abbreviations explained in text. Major structures located by arrows above upper datum and explained in text.

coal beds are recognized; the Grassy (gr) and Bruin (br) coal beds. Down-structural dip, the Tom Cooper is equivalent to the Van Lear (vl), or upper Elkhorn 3 coal (ue3); the Grassy coal is equivalent to the upper Elkhorn 1 coal (ue1); and the Bruin coal is equivalent to the lower Elkhorn-Pond Creek coal bed (le) (Rice and Hiett, 1994). The Gun Creek coal is equivalent to the Williamson coal (wi), which is equivalent to the Amburgy coal (am) (Rice and Hiett, 1994). North of the Walbridge fault, the Williamson coal is less than 1 m (3.3 ft) thick (Brown, 1977; Pillmore and Connor, 1978; Ward, 1978) and may be cut out by sandstones at the base of the Kendrick Shale (Englund and DeLaney, 1966b). Because the Williamson coal is thin or missing, and the Kendrick thins and is locally replaced by sandstones, the top of the formation is approximated in northeastern Kentucky. On the cross section, the formation contact was drawn 10–12 m (33–39 ft) above the Van Lear coal, which is very extensive (Figure 9).

The thickness of the Pikeville Formation changes little north of the Paint Creek uplift (PU in Figure 9) and then increases sharply between the Paint Creek uplift and the eastern Kentucky syncline (Figure 9). The increase in thickness is accompanied by the addition southeastward of the Little Blue Gem coal bed, the lower Elkhorn rider (ler), the upper Elkhorn 2 coal bed (ue2), and the upper Elkhorn 3.5 or Darby coal bed (ue3.5). Several of these coals begin as single beds

in the northwest and become zones of multiple beds in the southeast. The upper Elkhorn 3 coal occurs downdip as a zone with as many as four individual beds (Figure 9). Most coals increase in thickness to the southeast.

Hyden Formation

A cross section through the Hyden Formation is shown in Figure 10. The Kendrick Shale at the base of the unit is an extensive marine stratigraphic marker horizon (Chesnut, 1991a, 1992). At the northern end of the Grundy section, the Kendrick Shale is less than 3 m (10 ft) thick, is locally truncated by sandstones (Sheppard, 1964; Englund and DeLaney, 1966b), and is not generally mapped. At its southern limit, the shale is as much as 15 m (50 ft) thick and is extensively mapped. Overlying units in the Hyden Formation are relatively uniform in thickness southward from the Willard quadrangle (Wi in Figure 10) area to their southernmost exposure in the Jamboree quadrangle (Figure 10), where the formation is 140 m (460 ft) thick.

The Whitesburg (wh) and Fire Clay (fc) coal beds extend along nearly the entire Grundy section. At the northern limit of the cross section, the Whitesburg and Fire Clay coals occur as two to three closely spaced, very thin (a few centimeters) coal beds, mapped together on several geologic quadrangle maps (see Whittington and Ferm, 1967, for example). The

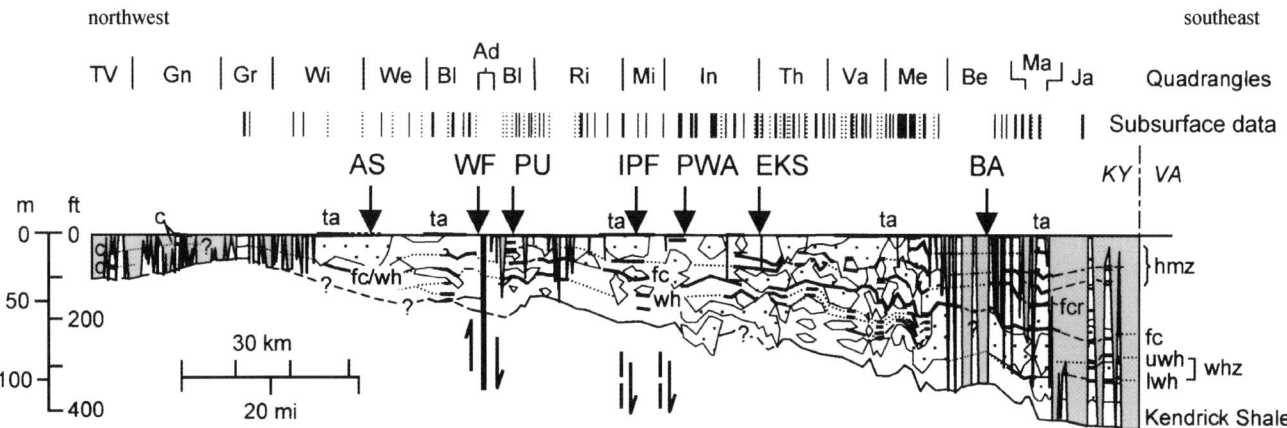

Figure 10. Hyden Formation along the Grundy section. Datum is the base of the Magoffin Member or the top of the Taylor (Copland) coal bed (ta). See Figure 2 for location of cross section and Figure 5 for legend. Some outcrop correlations inferred from geologic quadrangle maps referenced in Figure 2. Coal-bed abbreviations explained in text. Major structures located by arrows above upper datum and explained in text.

Fire Clay coal is distinctive to the south, where it contains a tonstein dated at 311 ± 1 m.y. (Lyons et al., 1992). Also to the south, the interval between the two coal beds increases to 30 m (100 ft), and the Whitesburg coal develops into a zone of at least two distinct coal beds, the lower and upper Whitesburg coal beds, which themselves split into multiple benches or beds. Additional coal beds, including the Fire Clay rider (fcr), Hamlin (hm), and Taylor (ta), develop southward between the Fire Clay coal bed and the base of the Magoffin Shale at the top of the Pikeville Formation. The Fire Clay rider and Hamlin coals cannot be traced more than a few kilometers north of the Walbridge fault (Figure 10). Southward, the Hamlin is a zone of two to three coal beds. Likewise, the Taylor coal pinches out north of the Willard quadrangle (WI in Figure 10), occurs as a single bed north of the Paintsville-Warfield anticline (PWA in Figure 10), and occurs as two distinct beds in a zone to the southeast in the Jamboree quadrangle (Outerbridge and Van Vloten, 1968). In each coal-clastic cycle or coal zone, additional local leader and rider coal beds may occur. For example, four rider coal beds are locally developed between the Fire Clay and Fire Clay rider coal beds (Greb et al., 1999).

Four Corners Formation

A cross section through the Four Corners Formation is shown in Figure 11. At its northern limit on the Grundy section, the formation is less than 30 m (100 ft) thick and thickens to nearly 122 m (400 ft) to the southeast. The formation occurs above drainage (the lowest level of stream valleys) across much of the coal field, so that units in the formation are widely mapped on geologic quadrangle maps. The Four Corners Formation thickens slightly in the Webbville quadrangle (We in Figure 11), just north of the axis of the Allegheny synclinorium (AS in Figure 11), but then increases in thickness at a nearly constant rate southward to its outcrop limit in the Jamboree quadrangle (Figure 11). The base of the formation is the Magoffin Shale Member, which may be the most widespread Pennsylvanian marine-fossil-bearing unit in the central Appalachian basin (Outerbridge, 1976; Chesnut, 1981, 1992; Martino, 1992, 1994; Bennington, 1996). The Magoffin is thin and may be truncated locally by sandstones on the northern end of the section, but thickens dramatically south of the Irvine-Paint Creek fault system and Paintsville-Warfield anticline toward the southeast (Figure 11).

The Four Corners Formation contains several coal beds. The coal with the widest occurrence on the Grundy section is the Princess 3 (pr3 in Figure 11), which is equivalent to the Peach Orchard zone. Southward, the Peach Orchard zone contains at least two major coal beds named the Hazard 7 and 8 (not shown in Figure 11). Significant increases in the thickness of the interval between the base of the Peach Orchard and the base of the Magoffin southeastward are accompanied by the addition of the Haddix (hx) and MHazard (hz) coal beds. The Haddix coal bed pinches out updip near the Paintsville-Warfield anticline, whereas the Hazard coal bed pinches out near an abrupt thickening of the formation near the Allegheny synclinorium (Figure 11). Both coals occur in zones, so that it may be difficult to correlate individual beds. The top of the formation is marked by the top of the Hindman coal, also called the Hazard 9 or lower

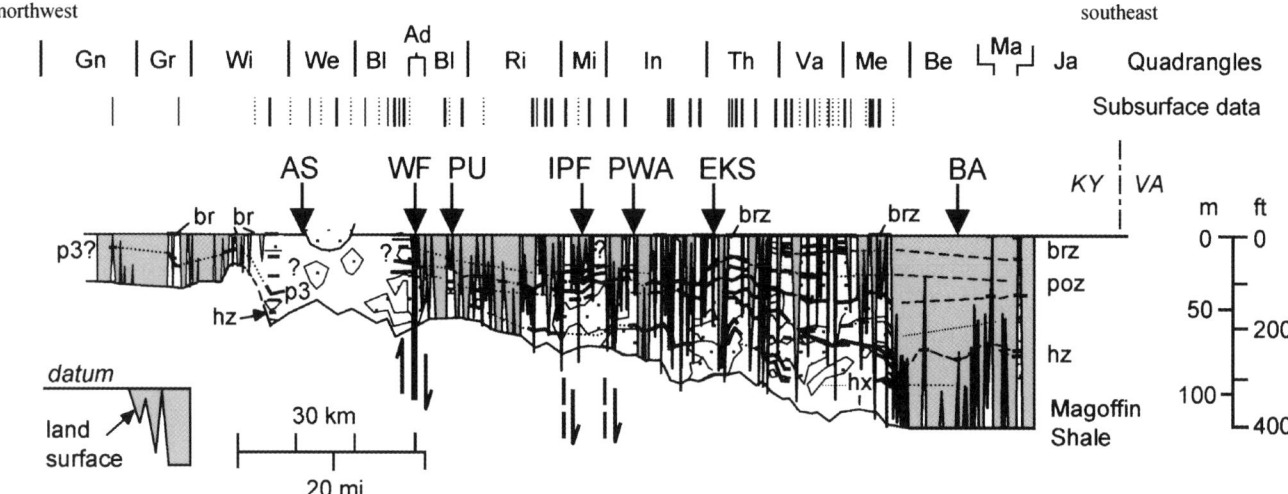

Figure 11. Four Corners Formation along the Grundy section. Datum is the base of the Stoney Fork Member or the top of the Hindman (Hazard 8) coal bed, which correlates between the lower (lbr) and upper Broas (ubr) coal beds. See Figure 2 for location of section and Figure 5 for legend. Some outcrop correlations inferred from geologic quadrangle maps referenced in Figure 2. Coal-bed abbreviations explained in text. Major structures located by arrows above upper datum and explained in text.

Broas (lbr in Figure 11) coal. The Stoney Fork Member, a regionally extensive, marine fossil-bearing shale and limestone, is well developed above the Hazard 9 coal southward and between the correlative lower and upper Broas coals. Northward, as the Broas interval thins, the member is more difficult to differentiate and may be absent. Where the member cannot be differentiated, the formation boundary is placed between the two coals.

The Princess Formation

The Princess Formation includes the stratigraphic interval between the base of the Stoney Fork Member and the base of the Conemaugh Formation (Figure 12). The Conemaugh Formation marks the base of the upper Pennsylvanian section in the basin (Figure 2). The entire Princess Formation is preserved only in the northeastern part of the eastern Kentucky coal field, but is extensive northeastward into West Virginia. North of the Walbridge fault, the formation is well exposed above drainage and shows little change in thickness. The interval increases in thickness significantly between the Princess 5B (pr5B) and the Stoney Fork Member across the fault. Southward, the Princess 5 coal is considered equivalent to the Richardson coal bed (ri) (Rice and Hiett, 1994). Between the Walbridge fault and the Paintsville-Warfield anticline, the thickness of the interval between the Richardson coal bed and the Stoney Fork Member does not change significantly, but then thickens more near the axis of the eastern Kentucky syncline (Figure 12).

The Princess Formation contains the upper Broas (ubr), or Tiptop; Princess 5, or Richardson; Princess 5A (pr5A); Princess 5B; Princess 6 (pr6); Princess 7 (pr7); and Princess 8 (pr8) coal beds (Figure 12). Of these, the Princess 5 (Richardson) coal bed is perhaps the most extensive. Northward, the coals are closely spaced, and correlation of individual coal beds can be difficult.

The Stoney Fork Member (Lost Creek Limestone) has a patchy distribution and may be truncated by sandstones, which are common at the base of the Princess Formation. Each of the major coal-clastic cycles is locally overlain by shale with marine fossils (summarized in Chesnut, 1991a). The Main Block Ore (a persistent sideritic bed overlain by dark shale) occurs below the Princess 5 coal and may be difficult to differentiate from the Stoney Fork Member, although their distribution is mutually exclusive as mapped on geologic quadrangle maps. The Kilgore-Flint Ridge Flint occurs above the Princess 5 coal bed. The Obryan Member, which is equivalent to the Vanport Limestone of Kentucky (Rice et al., 1979), occurs above the Princess 5B coal bed. These units are not shown in Figure 12.

The Princess Formation differs from other formations in the Breathitt Group in that it contains numerous flint clays between coal beds. The Kilgore-Flint Ridge Flint occurs above the Princess 5 coal and is equivalent to the Kanawha Black Flint of West Virginia. This flint clay contains abundant sponge spicules and may locally fill abandoned scours (Cavaroc

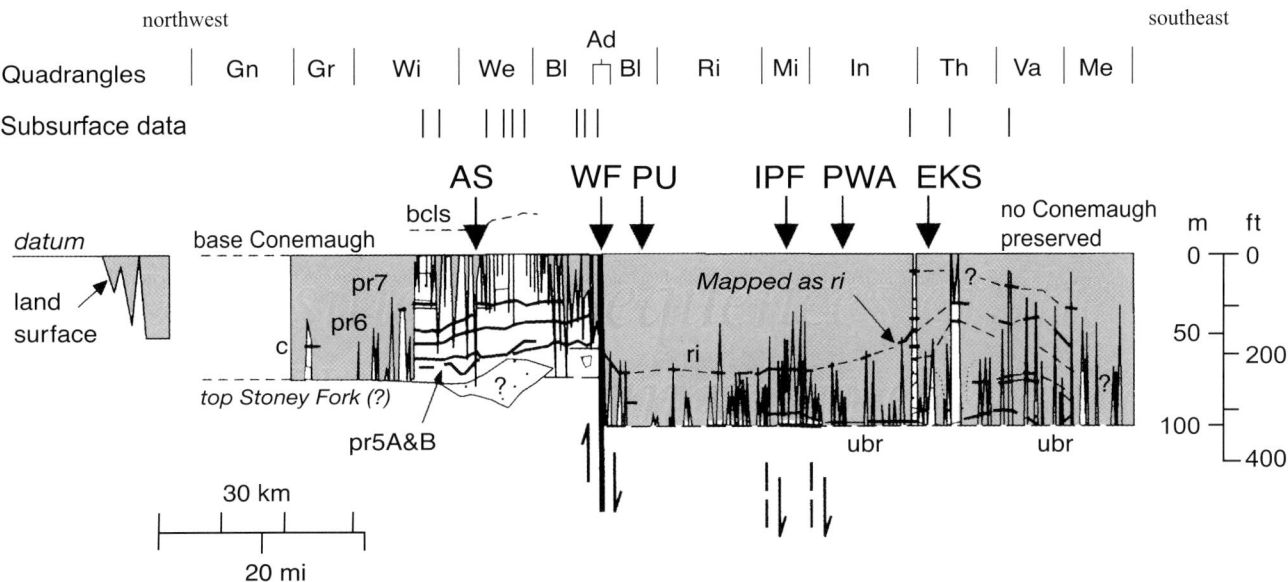

Figure 12. Princess Formation along the Grundy section. Datum is the base of the Conemaugh Formation, where the Conemaugh is preserved; otherwise, the datum is the base of the Stoney Fork Member. See Figure 2 for location of section and Figure 5 for legend. Some outcrop correlations inferred from geologic quadrangle maps referenced in Figure 2. Coal-bed abbreviations explained in text. Major structures located by arrows above upper datum and explained in text. In the Conemaugh Formation, bcls = Brush Creek Limestone.

and Ferm, 1968; Rice et al., 1994a). The Hitchins clay bed is a relatively thick flint clay between the Laurel coal (locally developed above the Princess 5B) or Obryan Member and Princess 6 coal (Rice et al., 1994a). This clay is a complex of at least five separate units, one of which has volcanic origin (Outerbridge et al., 1990). The Hitchins clay bed may be equivalent to the Lawrence and Oak Hill clays of Ohio and the Elk fire clay of West Virginia (Rice et al., 1994a). Together, these beds form a widespread clay unit at the northwestern margin of the basin.

Middle Pennsylvanian Marine Zones

Figure 13 shows the distribution of selected fauna from the middle Pennsylvanian of eastern Kentucky to illustrate relative marine influences in each formation. Possible lower Pennsylvanian marine zones are not illustrated because of limited sampling and uncertainty of some correlations and because most exhibit limited faunal diversity. Middle Pennsylvanian formation-bounding shales show the greatest diversity in marine fauna. The Pikeville Formation contains several extensive marine horizons. The Crummies Shale (Rice et al., 1994b) occurs above the lower Elkhorn-Pond Creek rider coal, the Dwale Shale (see Chesnut, 1991a) occurs in the upper Elkhorn 3 coal zone, and the Elkins Fork Shale (Morse, 1931) overlies the uppermost beds of the upper Elkhorn 3 (or 3.5)

coal zone. The Elkins Fork Shale is better developed than the overlying Kendrick Shale in the northern parts of the eastern Kentucky coal field, although it is less regionally widespread than the Kendrick.

Chesnut (1981) noted that marine to marginal-marine strata occur above all the coals in the eastern Kentucky coal field across at least part of their occurrence, but that faunal diversity and areal extent are varied. For example, in the Hyden Formation, marine and restricted-marine fauna have been noted above the upper and lower Whitesburg coal beds, the Fire Clay, Fire Clay rider, and the upper and lower Hamlin coal beds (Figure 13). The diversity of marine and brackish fauna from shales in this interval is cyclic relative to the formation as a whole; the least marine facies occur above the Fire Clay coal (Figure 13). The overall geographic distribution of the units mirrors diversity, with the most extensive marine units being the Kendrick and Magoffin Shales, and the least extensive being the marginally marine strata above the Fire Clay coal bed (Figure 14).

Middle Pennsylvanian Coal Resources

In eastern Kentucky, lower Pennsylvanian coals are mostly truncated by quartzarenites or occur in the deeper subsurface, so they are generally not significant resources. Middle Pennsylvanian coals constitute the bulk of eastern Kentucky's resources. In general,

Figure 13. Temporal distribution of selected invertebrate fauna, showing relative marine influences of major middle Pennsylvanian marine units and coal-clastic cycles (based on taxonomic distribution lists of Chesnut, 1991a). Shaded boxes indicate occurrence. Numbers indicate species collected during geologic mapping and other investigations. Dashed lines indicate units in which collections were not differentiated between adjacent units. Inferred maximum flooding surfaces = mfs. See text for coal-bed abbreviations.

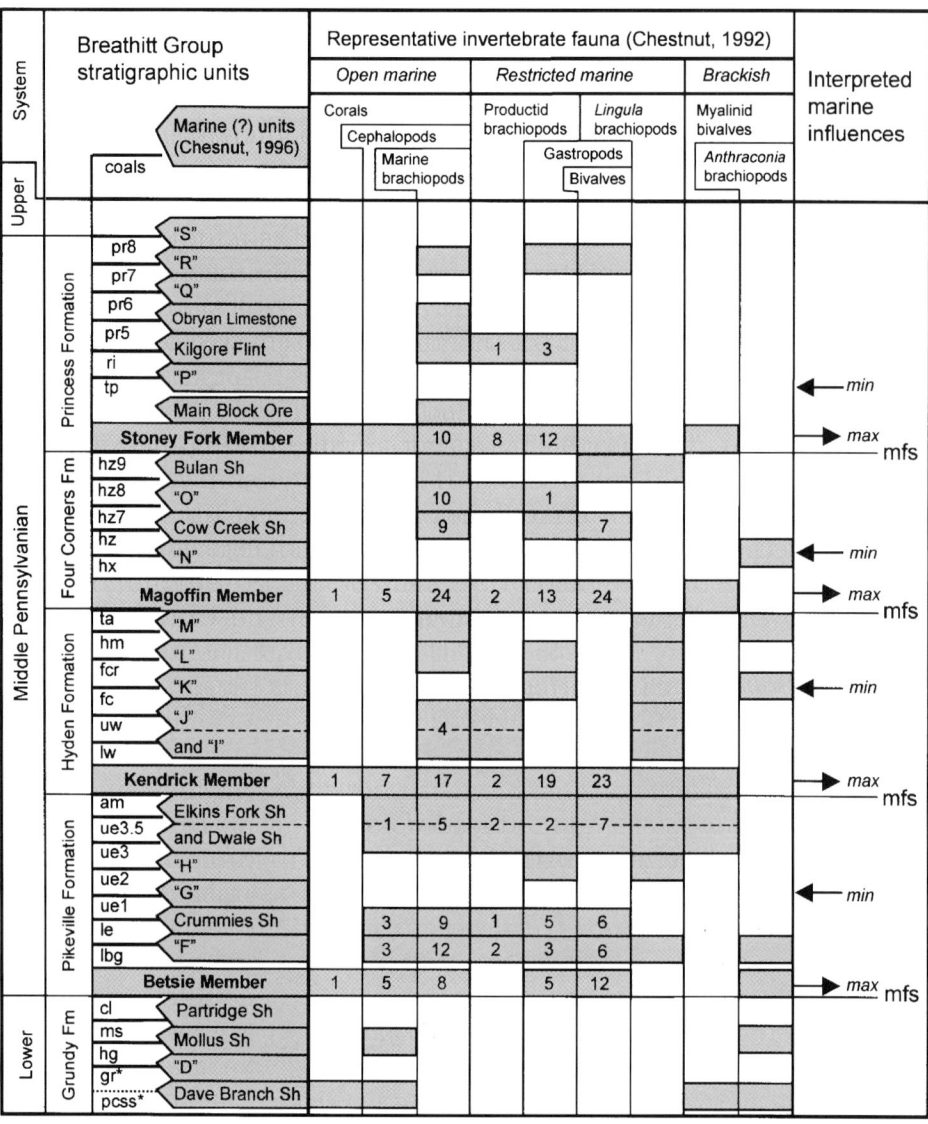

coal resources decrease from the Betsie Member upward to the Stoney Fork Member (Figure 15); this is partly a function of surface outcrop area and accessibility (Figures 9–12). Between the Betsie and Stoney Fork Members, resources occur in crude patterns, which are not strictly a function of surface area distribution. The greatest individual bed resources in the Pikeville Formation are toward the middle of the formation, but significant resources are also near the top and bottom of the formation (Figure 15). The Four Corners and Hyden Formations show a general increase in resources toward the middle of the formations, decreasing toward the bounding marine units. This pattern was previously noted by Cobb and Chesnut (1989), although it appears to be more complicated than previously assumed. For at least the Hyden Formation, the temporal distribution of resources is the inverse of the trend of marine species diversity and extent of overlying shales (Figures 13, 14).

Middle Pennsylvanian Coal Quality

Figure 16 shows mean coal quality (dry sulfur content and ash yield) for middle Pennsylvanian coals in eastern Kentucky. Not all beds have equal sampling. More work is needed in sampling some of the thinner coals for statistical evaluation of coal quality, but pre-liminary results are shown here for comparison with stratigraphic and resource data. Total average sulfur values for the means of the coal beds for each formation are 1.3% for the Pikeville, 1.7% for the Hyden, 1.5% for the Four Corners, and 2.3%, for the Princess. Hence, sulfur content increases overall from the Pikeville to Princess Formations (Figure 16). There is a slight increase in ash yield and variability in ash yield in coals younger than the Kendrick Shale as compared to coals older than the Kendrick.

In formations, temporal patterns in coal quality are variable, but in general, the highest sulfur occurs toward the top or bottom of each formation (Figure 16). In the Pikeville Formation, ash and sulfur increase upward to the upper Elkhorn 3 coal bed, but then decrease again toward the top of the formation. In the Hyden Formation, there is a general cyclic trend

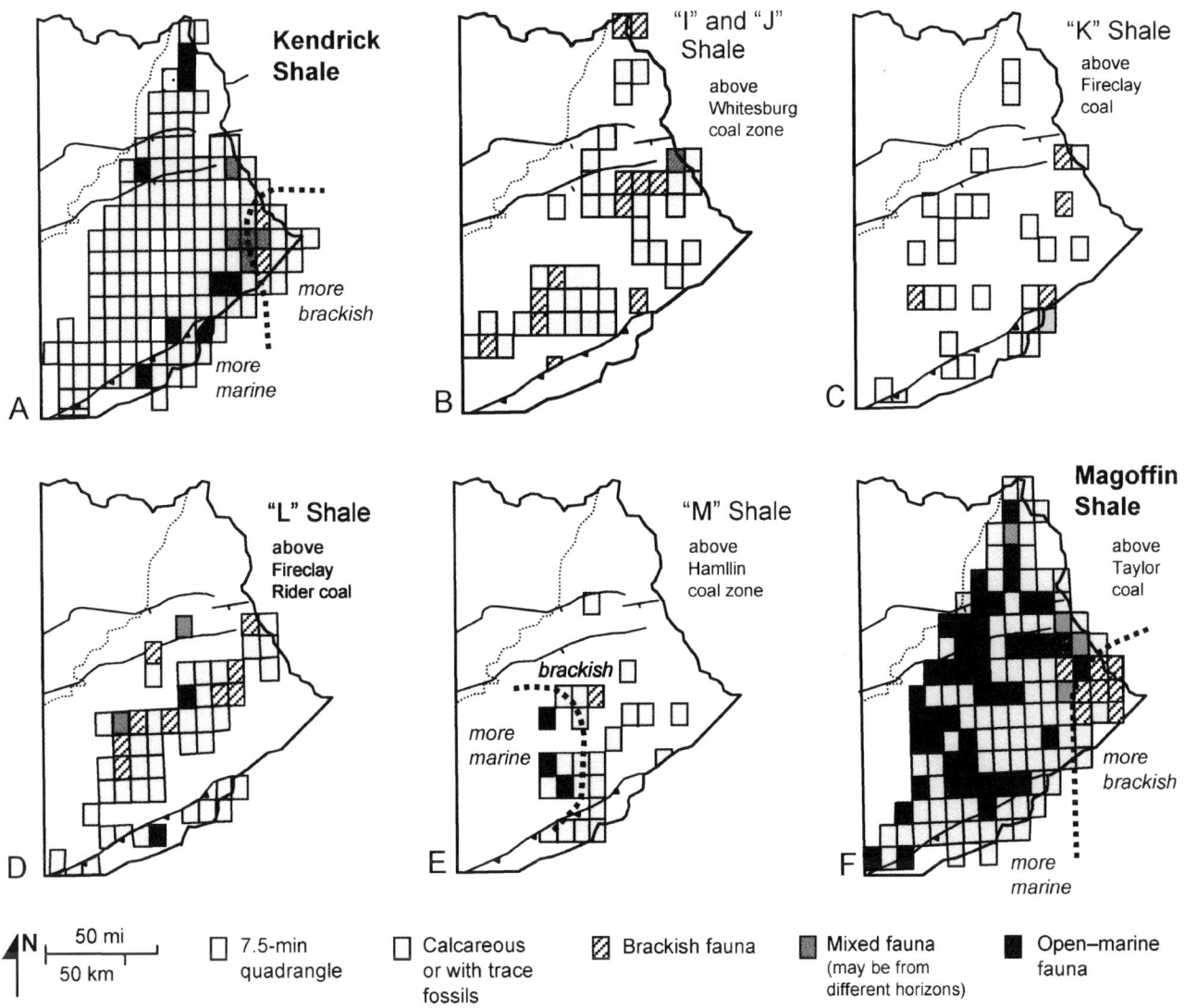

Figure 14. Regional maps of the eastern Kentucky coalfield, showing the general distribution of marine and brackish-water fauna from shales in the Hyden Formation. (Data from Chestnut, 1991a).

in quality; the highest mean sulfur contents are at the base and top, and lowest mean sulfur values are toward the center of the formation, with the exception of the Fire Clay rider coal bed (Figure 16). This trend is similar to the trend for resources (Figure 15) and marine influences (Figures 13, 14). In the Four Corners and Princess Formations, sulfur and ash content appear to increase from the lowest to uppermost coal bed.

DISCUSSION

Spatial and temporal comparisons of Pennsylvanian strata thickness, sedimentary patterns, coal distribution, coal thickness, and coal quality provide an opportunity to analyze the changing role of tectonics, eustasy, and sediment supply on the formation of coal measures in a foreland basin. These varied criteria can be compared with apparent accommodation space within and between formations, as well as coal-clastic cycles. Accommodation space is the space made available for potential sediment accumulation below base level (Jervey, 1988). Accommodation space is affected by the interdependent contributions of eustasy, tectonics, sediment supply, and climate.

Tectonic Accommodation

The lateral basinward thickening of coal-bearing formations shown in Figures 5–13 indicates increasing accommodation from the northwestern margin

Figure 15. Temporal distribution of resources for middle Pennsylvanian coal beds in the eastern Kentucky coalfield. *Total includes half of undifferentiated Broas coal. **Total includes half of undifferentiated Peach Orchard coal. ***Total includes half of undifferentiated Whitesburg coal. Coal resource data from Brant (1983a, 1983b) and Brant et al. (1983a, b, c, d).

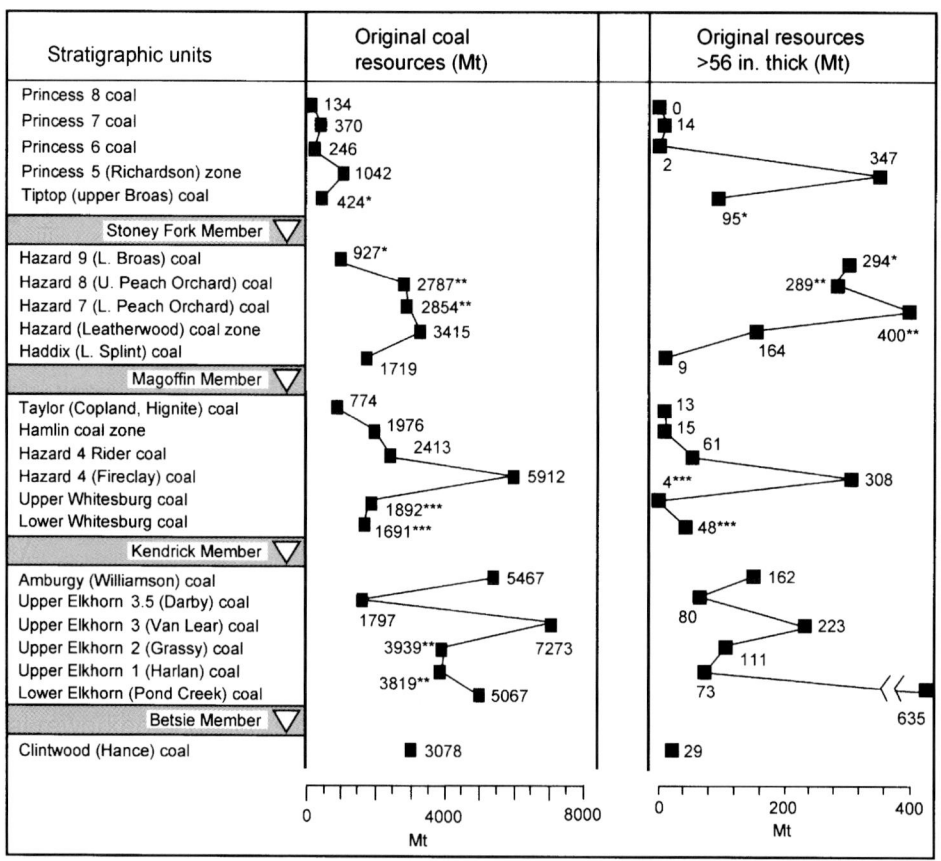

of the basin to the southeast. This reflects increased tectonic subsidence basinward. Figure 17A shows the relative thickness change for each formation along the Grundy dip section. The top of each formation is used as a datum, and then each formation is superimposed on the others for comparison. The greatest basinward thickness change occurs in the Sewanee-Bottom Creek (se-BC) and Warren Point-Pocahontas (WP-Po) intervals. Basinward, thickness decreases for each formation above the Sewanee Sandstone. Interestingly, if the quartzarenite belts are split into pairs of composite sandstones along the thick, midformation shale that occurs in each formation, then each of the sandstones in each pair shows decreasing thickness change to the level of the Corbin Sandstone (Figure 17B). Formation thickness continues to decrease into the middle Pennsylvanian. Decreasing basinward thickness from the lower to upper middle Pennsylvanian is interpreted as a temporal decrease in the space provided by foreland tectonic subsidence. Hence, increasing progradation of sediment onto the northwestern margin of the basin accompanied decreasing tectonic subsidence in the foreland basin.

Middle Pennsylvanian Coal-measure Sequences

As mentioned previously and shown in Figure 4, the middle Pennsylvanian coal measures have been placed in three stratigraphic sequences based on sequence-stratigraphic methods (Aitken and Flint, 1994, 1995). Because Chesnut (1992, 1996) defined formations on the basis of widespread marine-fossil-bearing

shales, the three middle Pennsylvanian formations generally conform to intervals between major marine flooding events, interpreted as maximum marine flooding surfaces. The formations, therefore, are similar to genetic sequences as defined by Galloway (1989), which are defined as stratal packages bounded by maximum flooding surfaces. The concepts of sequence stratigraphy and genetic stratigraphy both attempt to develop depositional frameworks for sedimentary sequences, but differ in their use of bounding surfaces and terminology.

The marine shales of the middle Pennsylvanian are the most widespread stratigraphic units in the basin and are easier to define consistently on subsurface logs than are the lowstand truncation surfaces and subaerial surfaces needed for sequence-stratigraphic analysis. The major marine shales can be differentiated south of the Walbridge fault as thick, coal-barren units that generally coarsen upward in core and on gamma-ray logs. North of the fault, some of the marine zones may be truncated by "incised" sandstones, but in northeastern Kentucky, decreasing data density complicates correlations, regardless of the method used. Many borehole records also refer to marine fossil accumulations, such as those that occur at the base of the

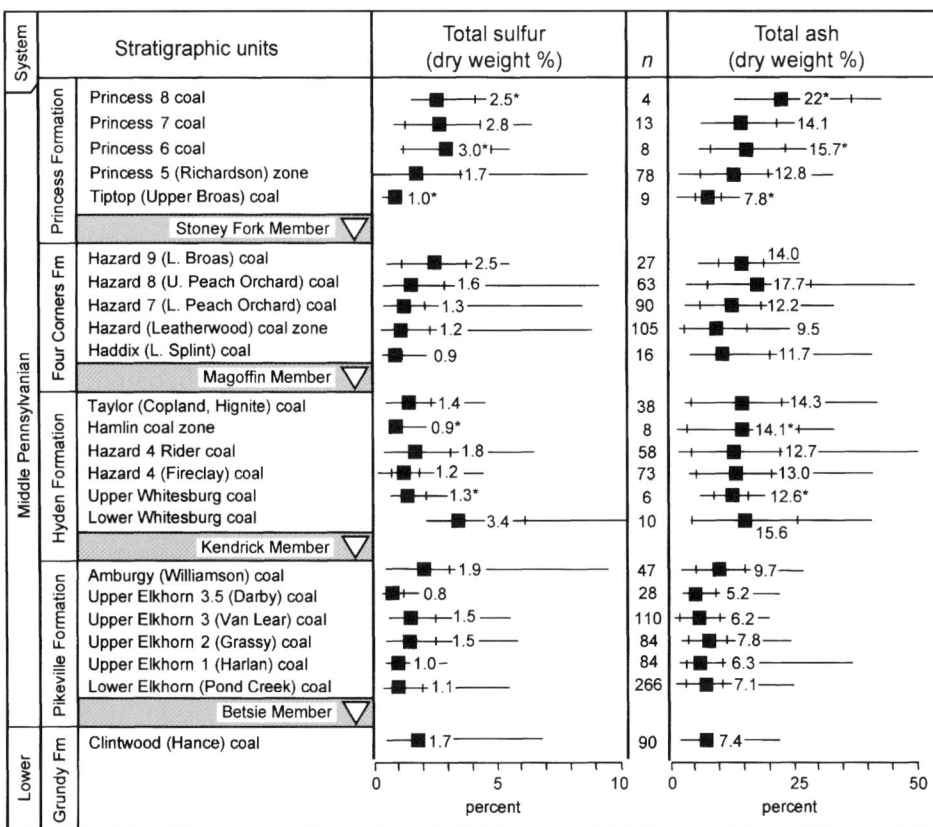

Figure 16. Temporal distribution of coal quality for middle Pennsylvanian coal beds in the eastern Kentucky coalfield. Numbers indicate mean for each bed. Lines extend from minimum to maximum values. Vertical tick marks on lines encompass standard deviation about mean. Coal-quality data from Kentucky Geological Survey database. Note that some beds have few data (column *n*).

base of the overlying shale member, but generally in a short distance of the base of the overlying shale member. Overall, maximum flooding occurred during deposition of the Pikeville Formation, as indicated by the relative distribution of fauna above coals in the Pikeville genetic sequence versus other genetic sequences (Figure 13). The greatest single transgression occurred at the base of the Magoffin Member (Figures 13, 14). If the relative degree of marine influences are used as a proxy for degree of offlap and onlap, then relative marine influences are vertically (i.e., temporally) symmetrical in the Hyden sequence (Figures 13, 14), but asymmetrical in other middle Pennsylvanian sequences (Figure 13). Thick coal distribution (Figure 15) and ash and sulfur contents (Figure 16) in each third-order sequence crudely parallel the trends of relative marine influence in these middle Pennsylvanian sequences.

major marine shale units, which are associated with marine flooding surfaces. Defining lowstand surfaces in middle Pennsylvanian strata using subsurface data is problematic, because distinguishing scours at the base of lowstand sandstones and lowstand sequence sets from scours at the base of progradational sandy channels in underlying highstand systems tracts or highstand sequence sets can be difficult. In addition where "scour-based" sandstones are absent, the lowstand surfaces must be inferred to be interfluves between channels, although subsurface paleosol information is typically not available to confirm this. Placing the interfluve at the base of the first coal also can be problematic, because this may not define the first paleosol surface. These problems are compounded in the eastern Kentucky coal field because subsurface geophysical data commonly lack density logs for the Pennsylvanian part of the section.

Each of Chesnut's (1992, 1996) middle Pennsylvanian formations generally conforms to a genetic sequence. The exception would be that the formation boundaries are placed at the top of the uppermost coal or base of the shale member in the overlying formation. The basal boundaries of the genetic sequences would be the maximum flooding surfaces, which are not necessarily at the top of the coal or the

Lower Pennsylvanian Coal-measure Sequences

Each of Chesnut's (1992, 1996) four lower Pennsylvanian coal-bearing formations (Pocahontas, Bottom Creek, Alvy Creek, and Grundy) are capped by coal beds overlain by widespread shales and coarsening-upward intervals (Figures 5–8). These shales are lithologically similar to the major middle Pennsylvanian marine zones. The upper three lower Pennsylvanian formation-capping shales contain at least marginal marine fauna at the base. Assuming these thick, coarsening-upward intervals represent a recurrent facies, the lower Pennsylvanian shales and coarsening-upward intervals also can be inferred to represent regional transgressions (although less open-marine than

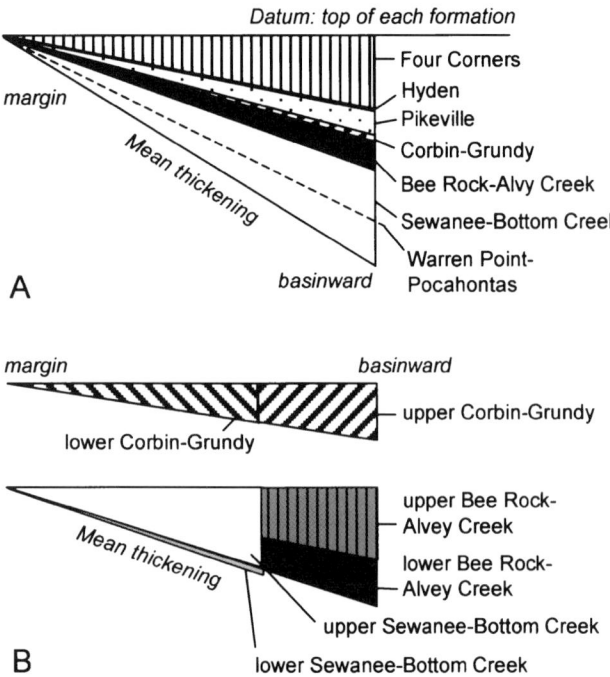

Figure 17. Graphical comparison of relative basinward thickening for (A) each of the formations shown in Figures 5–11 (the Princess Formation is not shown because it is not laterally extensive and the upper boundaries are not present along much of the Grundy section), and (B) each of the paired sandstones in the lower Pennsylvanian quartzarenite belts. Graphs were constructed by taking the average thickness change from the margin-ward limit of each formation to its basinward limit, aligning the margin for each formation, and then using the top of each formation as a datum.

their younger counterparts), each followed by progradation. Marine fauna may be lacking at the base of these lower Pennsylvanian shales because (1) the overall extent of marine influences was less than in the middle Pennsylvanian, and (2) greater early Pennsylvanian sedimentation and freshwater influx diluted marine influences.

If the extensive lower Pennsylvanian shales are used to divide sequences similar to the way the middle Pennsylvanian is divided, then the lower Pennsylvanian section can be divided into at least four genetic sequences (Figure 18). More work is needed to determine the placement and extent of possible flooding surfaces at the top of the Pocahontas and Bottom Creek Formations. Herein, the genetic-sequence boundaries are placed at the formation boundaries. Some lower Pennsylvanian sequence boundaries, however, might only represent the updip equivalents of marine flooding surfaces. As such, the sequence boundary might be placed in estuarine facies (Korus and

Erikkson, 2002; Greb et al., 2001) or in coals at the base of the shales.

Each lower Pennsylvanian coal-measure formation of Chesnut (1992, 1996) also contains a coarsening-upward shale and persistent sandstone toward the middle of the formation (Figures 6–8). Midformation sandstones are juxtaposed laterally against the lower sandstone of the lateral quartzarenite belt. In many cases, the midformation shale continues between the lower and upper quartzarenite for some distance before being truncated by the upper quartzarenite in each belt. These midformation shales and sandstones tend to split each of Chesnut's (1992, 1996) lower Pennsylvanian coal-bearing formations into two bundles of three to five coal-clastic cycles. The greatest lateral rate of thickness increase in each formation repeatedly occurs in the midformation and sequence-bounding shales. The fact that the most open-marine part of the section is thickest could be used to argue for eustatic controls on deposition. However, each of the sequence-bounding shales thickens basinward (to the southeast), although the greatest marine influences (Figure 14) were from the southwest, suggesting tectonic influences on deposition. If the bases of midformation shales are used to define marine flooding surfaces or their updip equivalents, then the lower Pennsylvanian section can be divided into at least eight genetic sequences.

Lower Pennsylvanian Quartzarenite Cycles

There are several repetitive patterns in the quartzarenite formations. Each of the major quartzarenites occurs in broad belts, northwest of coal-measure formations. Each belt thickens toward the southeast but terminates near its thickest extent, which is always updip of laterally thicker coal-measure facies. The maximum thickness of each quartzarenite belt is generally half of the maximum thickness of equivalent downdip coal-measure sequences. Each successive sand belt overlaps the underlying belt to the west, upstructural dip (Figures 1, 5, 7, 8).

Outcrop analyses of individual sandstones in the quartzarenite belts show that each belt consists of multistory, sheet-form, and channel-form elements (Bement, 1976; Rice, 1984; Wizevitch, 1992; Churnet, 1996; Greb and Chesnut, 1996). The pair of composite sandstones in each quartzarenite belt occurs along similar trends, although they are separated by shale and, in some cases, by a coal bed. The lower sandstone of the lower three quartzarenite belts (Warren Point, Sewanee, and Bee Rock) is truncated updip by the upper sandstone in each pair (Figures 5, 7, 8).

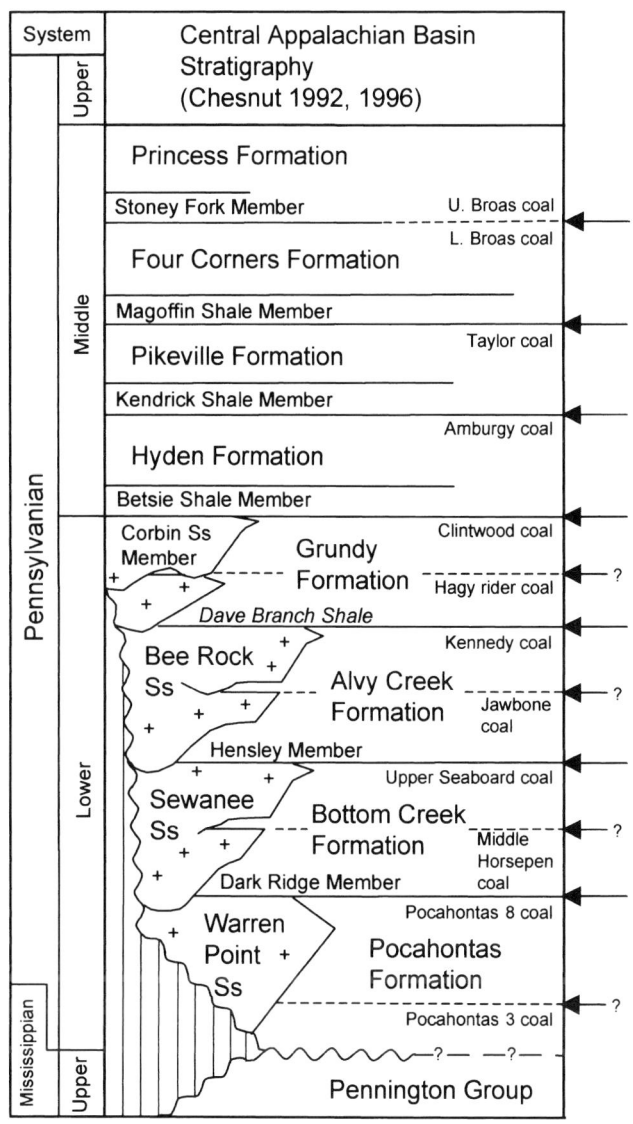

Figure 18. Inferred genetic-sequence boundaries (arrows) of lower and middle Pennsylvanian strata in the central Appalachian basin. Possible midformation sequence boundaries (arrows with question marks) in the lower Pennsylvanian require further investigation.

The upper sandstone in each quartzarenite pair also extends slightly farther to the southeast than the lower sandstone does. Maximum flooding surfaces occur above the upper sandstones in the Corbin and Bee Rock sandstone belts, whereas the intervening midformation shale between the upper and lower sandstone in each pair is commonly less developed. The midformation shales that divide the upper and lower sandstones in the Corbin and Bee Rock exhibit patchy estuarine to marginal marine facies, but whether this pattern occurs in the midformation shales between the paired sandstones in the Sewanee and Warren Point Sandstones is uncertain (Figures 5, 6).

Some investigations of lower Pennsylvanian quartzarenites (e.g., Englund and Delaney, 1966a) have inferred lateral intertonguing of the quartzarenites with the structurally downdip coal measures (Figure 19A). In the beach-barrier model (e.g., Ferm 1974), the quartzarenites are interpreted to intertongue throughout much of their extent with back-barrier, lagoonal, and lower-delta-plain coal facies. If the quartzarenites are inferred to be fluvial facies (see Rice, 1984, for example), then intertonguing can be interpreted in several ways. Fluvial, transverse drainage systems can be interpreted as flowing into the longitudinal quartzarenite belts in each fourth-order coal-clastic cycle (Figure 19B). Several sandstones in the Sewannee-Bottom Creek Internal (Figure 6) could be interpreted this way.

Quartz pebbles are concentrated only in those sandstones with southward paleocurrents. This suggests two source areas for lower Pennsylvanian sandstones. Greb and Chesnut (1996) inferred a lowstand surface at the base of each of the major quartzarenites that crop out on the western margin of the basin. This could be misinterpreted as incision of the transverse drainages by a single large paleovalley (Figure 19C). Each quartzarenite belt, however, represents composite valleys, with at least a pair of composite sandstones separated by a midformation shale, which in two of the sand belts contain brackish fauna (Figures 5–8). It is this scale of sandstone that was described on the western margin of the basin. Because the midformation shales are continuous with downstructural-dip coal-measure sequences, each quartzarenite belt could be interpreted with at least two third-order lowstand surfaces (Figure 19D). In this scenario, intertonguing of the quartzarenites and laterally equivalent facies occurs mostly in the upper parts of each of the paired units in each belt. Much of the record for fourth-order transitions lower in each of the paired units was removed by subsequent erosion in each of the paired units that comprise the quartzarenite belts. Because the quartzarenites represent only half of the thickness of downstructural dip coal measures, amalgamation and incision of fourth-order transverse and longitudinal channels by third-order longitudinal channels and valleys in the main channel belt seems likely.

If the base of midformation shales in each quartzarenite sand belt represents genetic-sequence boundaries (as shown by dashed lines in Figure 18), then each of the quartzarenites as defined by Chesnut (1992, 1996) could represent two stacked genetic sequences (Figure 19E). In a sequence-stratigraphic

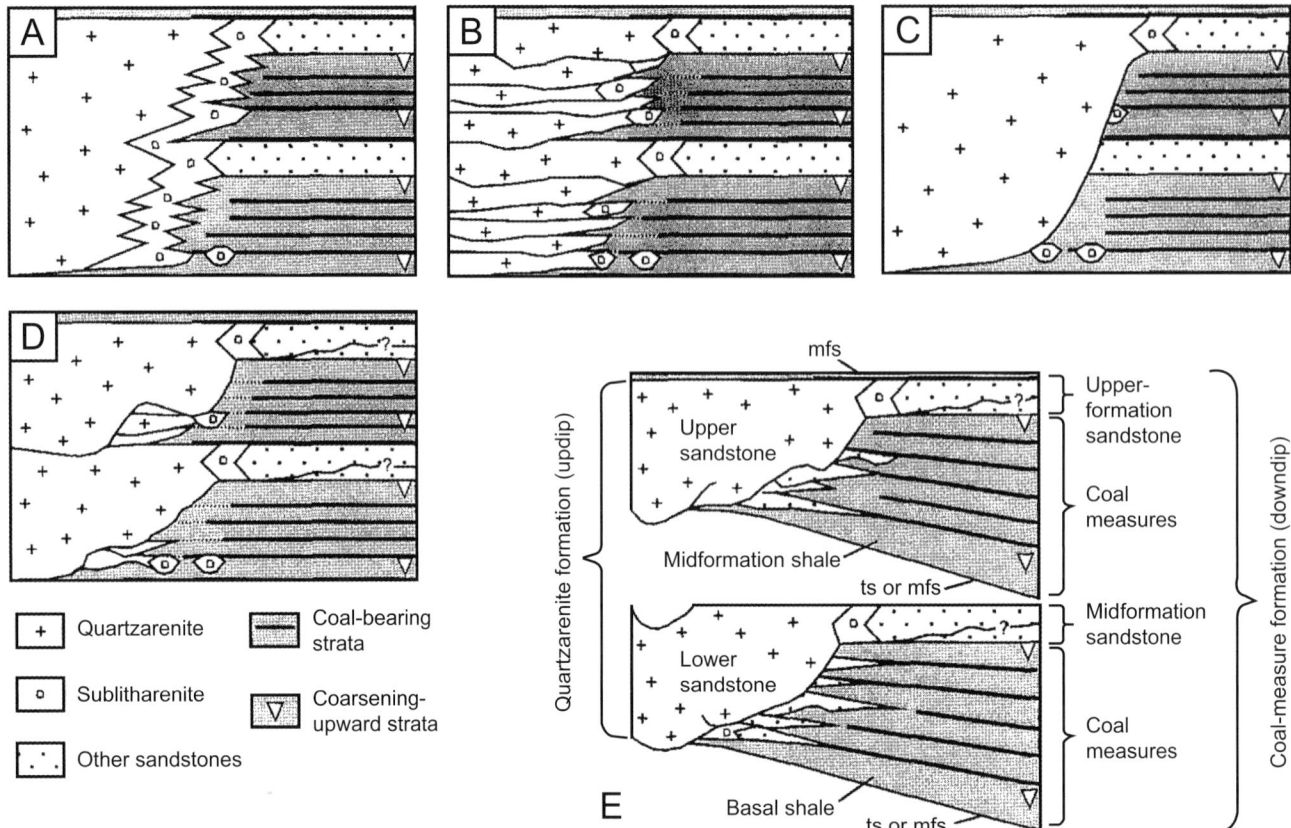

Figure 19. Four possible interpretations (A–D) for the juxtaposition of quartzarenites and lateral coal measures in the lower Pennsylvanian of the central Appalachian basin. (E) Interpretation of two genetic sequences for each of Chesnut's (1992) formations. mfs = maximum flooding surface.

approach, a third-order lowstand surface could be inferred to extend from the base of the quartzarenites laterally (1) to the top of the upper- and midformation-break sandstones, or a paleosol on the interfluve above those sandstones, effectively cutting through the upper- and midformation sandstones, or (2) along the base of the upper- and midformation sandstones of laterally equivalent coal measures, or (3) in the upper- and midformation sandstones as a third-order scour cutting through a fourth-order sandstone. For example, the Council Sandstone Member (Css in Figure 7) is a midformation sandstone that appears to be laterally transitional with the lower sandstone in the Bee Rock quartzarenite belt. Sporadic quartz pebbles at the base of the Council Sandstone might indicate that the basal scour surfaces of the quartzarenites and the laterally equivalent sandstones are the same. Miller (1974) also noted quartz pebbles isolated near the base of the McClure Sandstone (Figure 8), suggesting a similar scenario for the upper sandstone in each belt. The transition between quartzarenites and lateral sublitharenites to litharenites at the tops of

each quartzarenite sequence represents reworking of longitudinal valley sediments; as base level rose, rivers were converted to estuaries, and the interfluves were topped. More work is needed in determining the relation of major scour surfaces and possible paleosols in the upper- and midformation sandstone units. This will require detailed examination of sandstone architecture and scour surfaces along the transition zone between the quartzarenite belts and transverse sandstones in coal measures downdip.

Smaller, relatively local quartzarenites, and more commonly, sublitharenites, do occur in downdip coal measures lateral to the Warren Point and Sewanee Sandstones, but they are thinner and much less extensive than the quartzarenite belts (Figures 5, 6). In some areas, they are juxtaposed against the quartzarenite belts such that these may represent fourth-order intertonguing of the transverse and longitudinal drainages. Along the Grundy section, however, most of these more isolated quartzarenites do not interfinger with the main belts, but occur along distinct stratigraphic horizons separated from the main belt (Figure 8). Laterally isolated quartzarenites locally occur at the upper- and midformation level of each formation. Mid- and upper-formation quartzarenites that are separated from the main quartzarenite belt may be related to reworking during transgression. If fluvial in origin, some of these isolated quartzarenites also represent disconnected paleovalleys or paleochannels lateral to the main belts. Isolated quartzarenites also occur locally at the top of the stratigraphic level marked by the coarsening-upward unit that caps each of the belt. If the base of the shale and coarsening-upward unit marks transgression and a maximum flooding surface, then the second type of isolated quartzarenites records reworking of the preceding quartzarenites at highstand. Individual trends of isolated sandstones need to be resolved as belt-parallel and part of the longitudinal drainage system or belt-perpendicular and part of the transverse network to interpret their origin. Koros and Erikkson (2002) recently documented transverse drainages along the belt margin in part of West Virginia.

The vertical stacking of paired composite quartzarenites in a longitudinal drainage system, each 60–80 km (37–50 mi) wide, suggests tectonic control on sand belt position. Quartzarenites were stacked vertically in the belt, but shifted cratonward, between each belt. The cratonward shift in each belt is interpreted to have been controlled by progradation of the transverse sediment wedge, effectively pushing potential accommodation for the longitudinal sand-belt margin northwestward, in combination with migration of a forebulge, or progressive tilting of the craton because of continued continent-margin tectonics. In general, each successive lower Pennsylvanian sand belt shows a decrease in the amount of accommodation space generated. The greatest change is in the Sewanee Sandstone, and the least is in the Corbin Sandstone (Figure 17A, B). This thickness change was controlled by a temporal decrease in foreland subsidence. The increasingly marginward position of each sand belt also resulted in spatially limited accommodation relative to the foreland basin. Parts of the Corbin Sandstone and isolated paleovalleys on the western margin become bedrock confined, where they are incised into Mississippian, instead of Pennsylvanian, strata (Archer and Greb, 1995; Greb and Chesnut, 1996). An overall decrease in the width of the sand belts from the Bee Rock to the Corbin may also reflect decreased sediment flux from the northeast source area. This source completely ceased contributing clastics to the basin in the middle Pennsylvanian.

Comparison of Sequences

The major difference between lower and middle Pennsylvanian third-order genetic sequences is the longitudinal drainage system that deposited the lower Pennsylvanian quartzarenite belts. The end of quartzarenite deposition in the longitudinal belts indicates a loss of the northeastern source and a probable change in paleoslope at the early–middle Pennsylvanian boundary. Progressively increasing areal extent of marine conditions with each maximum marine flooding event from the early to late middle Pennsylvanian mirrors the temporally decreasing tectonic accommodation in the foreland. This seems opposite of what would be expected. In many basins, seas fill in the areas of greatest subsidence (and accommodation), such that increased marine influences tend to parallel increased accommodation. In the central Appalachian basin, however, clastic wedges from the southeast prograded to the northwest, progressively overlapping the basin margin. During the early Pennsylvanian, a southwest-flowing trunk stream system, oriented perpendicular to the clastic wedges, also transported large volumes of material from sources northeast of the basin (Archer and Greb, 1995; Greb and Chesnut, 1996). Only after the northeastern sources were effectively shut off near the early to middle Pennsylvanian boundary were marine influences persistent through several fourth-order coal-clastic cycles. Hence, increasing extent of marine influence also may be related to decreasing sedimentation rate.

Coal-clastic Cycles

Both the lower and middle Pennsylvanian genetic coal-bearing sequences contain multiple coal-clastic cycles. Middle Pennsylvanian coal-clastic cycles were previously interpreted as fourth-order sequences (Aitken and Flint, 1994, 1995). Dividing middle Pennsylvanian strata into three genetic sequences along Chesnut's (1992, 1996) formation boundaries results in sequences that each contain five to six coal zones. If the lower Pennsylvanian is divided into four genetic sequences along Chesnut's (1992, 1996) formation boundaries, then each sequence contains 6–12 coal zones. If the lower Pennsylvanian is divided into eight genetic sequences along the midformation shales, then each sequence contains three to six coal zones. The similarity of the latter number of coal zones to the number of coal zones in the middle Pennsylvanian sequences may lend support to the midformation breaks representing genetic-sequence boundaries if there was a temporally cyclic, eustatic control on coal accumulation.

Although the periodicity of fourth-order coal-clastic cycles may indicate eustatic oscillations, most of the lower and middle Pennsylvanian coal-clastic cycles thicken toward the southeast and the basin axis and locally exhibit thickness changes across major structural features (Figures 5–13), suggesting major tectonic and sedimentation influences on thickness and updip preservation. The most uniform fourth-order coal-clastic cycles in the eastern Kentucky coal field occur in the upper middle Pennsylvanian. Even more regionally uniform "classic" cyclothems occur in upper Pennsylvanian strata (mostly absent in eastern Kentucky) to the northeast in Ohio, West Virginia, and Pennsylvania (Donaldson and Eble, 1991; Heckel, 1995). Regional uniformity in the thickness and lateral distribution of facies in upper Pennsylvanian coal-clastic cycles suggest the gradual dominance of eustatic influences over tectonic and sedimentologic influences with time.

Major palynological changes occur between the Princess 5 (Richardson) and overlying coal beds in the Princess sequence. Younger coals thicken and are more laterally continuous northward into the northern Appalachian basin, instead of thickening to the southeast as with early and middle Pennsylvanian strata (Donaldson et al., 1985; Donaldson and Eble, 1991). The palynologic change may record a climatic change that accompanied a switch in basin depocenters from the central Appalachian basin to the northern Appalachian basin (Cecil et al., 1985; Donaldson et al., 1985; Donaldson and Eble, 1991). Along the Grundy section, a dramatic thickness change in the interval between the Stoney Fork Shale and Richardson coal occurs across the Walbridge fault (Figure 12). Underlying strata were showing a trend toward decreasing accommodation across the fault from the lower Pennsylvanian to the Princess 5 coal (Figures 5–11). The sudden increase in accommodation indicates renewed movement along the hinge line, which accompanied or preceded the shift in basin depocenters. The overlying Princess 5A and 5B coals are significantly more widespread north of the fault. The Hitchins clay bed (which occurs above the Princess 5B coal) may also record the continuation of this shift, with widespread, extensive pedogenesis occurring on the southwestern margins of the northern Appalachian basin.

Coal Distribution and Quality

At the bed level, major coals (and the peats from which they were derived) were initiated by regional base-level rise and ponding on exposed coastal plains generally accepted to be controlled by eustatic modulation, as discussed previously. Coals in the basin, however, do not thicken toward the inferred paleo-coastline to the south and west (see direction of marine influence in Figure 14); instead, they thicken basinward. In the Fire Clay coal, subtle structures directly affect thickness and cause rectangular peat thickness and rectilinear fluvial drainage patterns (Greb et al., 1999), which implies strong structural influence on peat accumulation. Rectangular coal distributions and paleochannel patterns have been noted for several lower and middle Pennsylvanian coal intervals in the basin (Padgett and Ehrlich, 1979; Staub, 1991; Weisenfluh and Ferm, 1991).

In the middle Pennsylvanian Pikeville, Hyden, and Four Corners sequences, generally, three coals or fewer are continuous onto the northern margin of the basin, and many of the coals thin toward major structural elements in the basin, especially the Walbridge fault and Paintsville-Warfield anticline (Figures 5–13). Most coals split into zones toward the basin axis, again suggesting strong tectonic accommodation controls, at least on regional coal thickness and preservation. Coal-bed thickness and quality also vary temporally in the eastern Kentucky coal field (Figures 13–16). Increasing sulfur and ash contents of Appalachian coals from the lower to upper Pennsylvanian have been interpreted to reflect changing paleoclimates, from ever wet in the early and middle Pennsylvanian to seasonally drier in the late Pennsylvanian (Cecil et al., 1985; Cecil, 1990; Donaldson and Eble, 1991). The sharp increase in sulfur content above the Princess 5 (Richardson) coal bed in the Princess Formation (Figure 16) corresponds to major changes in palynology and has been interpreted as marking a shift in paleoclimates. This time also corresponds to the apparent shift in coal depocenters as mentioned above. Another interesting change is the overall increase in ash content and ash variability above the Kendrick Shale (Figure 16). Changing paleoclimates might explain the overall temporal changes from the early to late Pennsylvanian, but this is not a gradual change, and the change is much older than the sudden change in sulfur content. Perhaps changes in sedimentation or alluvial influence followed regression of the Kendrick seas. Post-Kendrick deposition was influenced by less tectonic accommodation than previously, which may have allowed for relatively slower burial and a greater potential for sediment influences than occurred previously. Perhaps the change is tied to the extent of the post-Kendrick coastal plain, which was more widespread and less marine influenced than in the preceding sequence, and not bisected by a broad longitudinal drainage system as in the early Pennsylvanian. Whatever the causes, thicker, lower-ash, and lower-sulfur coal beds tend to occur toward the zone of minimum marine influence in each sequence.

SUMMARY AND CONCLUSIONS

Lower and middle Pennsylvanian coals of the central Appalachian basin were deposited on coastal plains building northwestward from the rising Appalachian Mountains. Two different styles of depositional sequences developed. The first style consisted of transverse depositional systems prograding northwestward from the Appalachians. The second was a longitudinal system developed on the western limb of the basin. Lower Pennsylvanian depositional sequences consist of updip quartzarenite belts (longitudinal systems) and downdip coal-measure facies (mostly transverse systems). At least four genetic sequences can be defined, based on inferred maximum flooding surfaces above each major quartzarenite belt. Because each belt contains at least a pair of vertically stacked sandstones, the quartzarenite belts may represent parts of as many as eight genetic sequences (Figure 18). Quartzarenite belts are juxtaposed and truncate lateral coal-measure facies, intertonguing at the tops of each of eight sandstones, two each in the four quartzarenite belts. Fourth-order coal-clastic cycles were truncated along their western and northwestern margins (structurally updip) during incision of third-order paleovalleys in the longitudinal quartzarenite belts. Longitudinal paleovalleys were converted to estuaries during base-level rise, and quartzarenites were reworked into lateral coal-measure deposystems at the top of each genetic sequence.

Middle Pennsylvanian third-order sequences prograded further to the northwest than their lower Pennsylvanian counterparts and were not bisected by a longitudinal drainage network. By the middle Pennsylvanian, the northeastern source was no longer supplying sediment to the eastern Kentucky part of the basin, coals became more widespread and less split into zones, and marine influence increased.

Lateral rates of third- and fourth-order sequence thickness change were strongly influenced by tectonic subsidence in the foreland basin, which changed from the lower into the middle Pennsylvanian. Fourth-order coal-clastic cycles increase in thickness and number in an axial direction to the southeast. Coals in coal-clastic cycles commonly occur as zones to the southeast (axial direction) and thin, pinch out, or are truncated to the

northwest on the basin margin. The greatest increases in stratal thickness occur between the Walbridge fault and Paintsville-Warfield anticline, which formed a hinge line on the northern margin of the basin. Spatial changes in tectonic influences decreased from the early to late middle Pennsylvanian, resulting in more uniform sedimentation patterns and more widespread coal beds in the middle than lower Pennsylvanian. Overall stratigraphic patterns illustrate how changing influences of tectonics, sedimentation, climate, and eustasy influence stratal stacking and coal distribution patterns in a foreland basin.

ACKNOWLEDGMENTS

The authors wish to thank Frank Ettensohn and Jack Beuthin for technical reviews. Their thoughtful suggestions greatly aided the manuscript. Thanks also to Meg Smath at the Kentucky Geological Survey for grammatical review.

REFERENCES CITED

Aitken, J. F., and S. S. Flint, 1994, High-frequency sequences and the nature of incised-valley fills in fluvial systems of the Breathitt Group (Pennsylvanian), Appalachian foreland basin, eastern Kentucky, in R. Dalrymple, R. Boyd, and B. Zaitlin, eds., Incised valley systems— Origin and sedimentary sequences: SEPM Special Publication 51, p. 353–368.

Aitken, J. F., and S. S. Flint, 1995, The application of high-resolution sequence stratigraphy to fluvial systems— A case study from the Upper Carboniferous Breathitt Group, eastern Kentucky, U.S.A.: Sedimentology, v. 42, p. 3–30.

Archer, A. W., and S. F. Greb, 1995, An Amazon-scale drainage system in the early Pennsylvanian of central North America: Journal of Geology, v. 103, p. 611–627.

Arkle, T., Jr., D. R. Beissel, R. E Larese, E. B. Nuhfer, D. G. Patchen, R. A. Smosna, W. H. Gillespie, R. Lund, C. W. Norton, and II. W. Pfefferkorn, 1979, The Mississippian and Pennsylvanian (Carboniferous) systems in the United States— West Virginia and Maryland: U.S. Geological Survey Professional Paper 110 D, p. D1 D35.

Barnhill, M. L., 1994, The sedimentology of the Corbin Sandstone Member, Lee Formation, eastern Kentucky, and a comparison to the age-equivalent rocks of the Illinois Basin, southwestern Indiana: Ph.D. dissertation, University of Cincinnati, Cincinnati, 271 p.

Bement, W. O., 1976, Sedimentological aspects of middle Carboniferous sandstones on the Cumberland overthrust sheet: Ph.D. dissertation, University of Cincinnati, Cincinnati, 182 p.

Bennington, J. B., 1992, Preliminary analysis of a marine

interval near the Howland Lookout Tower on Kentucky AA Highway, *in* C. L. Rice, R. L. Martino, and E. R. Slucher, eds., Regional aspects of Pottsville and Allegheny stratigraphy and depositional environments: U.S. Geological Survey Open-File Report 92-558, p. 34–40.

Bennington, J. B., 1996, Depositional and biofacies patterns in the middle Pennsylvanian Magoffin marine unit in the Appalachian basin, U.S.A.: International Journal of Coal Geology, v. 31, p. 169–194.

Beuthin, J. D., 1997, Paleopedological evidence for a eustatic Mississippian–Pennsylvanian (mid-Carboniferous) unconformity in southern West Virginia: Southeastern Geology, v. 37, p. 25–37.

Blake, B. M., A. F. Keiser, and C. L. Rice, 1994, Revised stratigraphy and nomenclature for the middle Pennsylvanian Kanawha Formation in southwestern West Virginia, *in* C. L. Rice, ed., Elements of Pennsylvanian stratigraphy, central Appalachian basin: Geological Society of America Special Paper 294, p. 41–53.

Blake, B. M. Jr., A. T. Cross, C. F. Eble, W. H. Gillespie, and H. W. Pfetferkorn, 2003, Selected plant megafossils from the Carboniferous of the Appalachian region, eastern United States— Geographic and stratigraphic distribution, *in* L. V. Hills, C. M. Henderson, and E. W. Bamber, eds., Carboniferous and Permian of the world, XIV International Congress of the Carboniferous and Permian (ICCP) Proceedings: Canadian Society of Petroleum Geologists Memoir 19, p. 259–335.

Brant, R. A., 1983a, Coal resources of the Princess District, Kentucky: University of Kentucky, Institute for Mining and Minerals Research, Energy Resource Series, 61 p.

Brant, R. A., 1983b, Coal resources of the Southwestern District, Kentucky: University of Kentucky, Institute for Mining and Minerals Research, Energy Resource Series, 89 p.

Brant, R. A., D. R. Chesnut Jr., W. T. Frankie, and E. R. Portig, 1983a, Coal resources of the Big Sandy District, Kentucky: University of Kentucky, Institute for Mining and Minerals Research, Energy Resource Series, 47 p.

Brant, R. A., D. R. Chesnut Jr., W. T. Frankie, and E. R. Portig, 1983b, Coal resources of the Hazard District, Kentucky: University of Kentucky, Institute for Mining and Minerals Research, Energy Resource Series, 49 p.

Brant, R. A., D. R. Chesnut Jr., W. T. Frankie, and E. R. Portig, 1983c, Coal resources of the Licking River District, Kentucky: University of Kentucky, Institute for Mining and Minerals Research, Energy Resource Series, 57 p.

Brant, R. A., D. R. Chesnut Jr., E. R. Portig, and R. A. Smath, 1983d, Coal resources of the upper Cumberland District, Kentucky: University of Kentucky, Institute for Mining and Minerals Research, Energy Resource Series, 41 p.

Brown, W. R., 1977, Geologic map of the Willard quadrangle, eastern Kentucky: U.S. Geological Survey, Geologic Quadrangle Map GQ-1387, scale 1:24,000, 1 sheet.

Busch, R. M., and H. B. Rollins, 1984, Correlation of Carboniferous strata using a hierarchy of transgressive-regressive units: Geology, v. 12, p. 471–474.

Carlson, J. E., 1971, Geologic map of the Webbville quadrangle, eastern Kentucky: U.S. Geological Survey Geologic Quadrangle Map GQ-927, scale 1:24,000, 1 sheet.

Cavaroc, V. V., Jr., and J. C. Ferm, 1968, Siliceous spiculites as shoreline indicators in deltaic sequences, *in* J. C. Ferm and J. C. Horne, eds., Carboniferous depositional environments in the Appalachian region: Columbia, University of South Carolina, Department of Geology, Carolina Coal Group, p. 183–192.

Cecil, C. B., 1990, Paleoclimate controls on stratigraphic repetition of chemical and siliciclastic rocks: Geology, v. 12, p. 533–536.

Cecil, C. B., and K. J. Englund, 1989, Origin of coal deposits and associated rocks in the Carboniferous of the Appalachian Basin, *in* Cecil, C. B., Cobb, J. C., Chestnut, D. R. Jr., Damberger, II., and Englund, K. J., (trip leaders), Carboniferous geology of the eastern United States: Washington, D. C., American Geophysical Union, 28th International Geological Congress, Field Trip Guidebook no. 143, p. 84–90.

Cecil, C. B., R. W. Stanton, S. G. Neuzil, F. T. Dulong, C. F. Ruppert, and B. S. Pierce, 1985, Paleoclimate controls on late Paleozoic sedimentation and peat formation in the central Appalachian basin (U.S.A.): International Journal of Coal Geology, v. 5, p. 195–230.

Chesnut, D. R., Jr., 1981, Marine zones of the Upper Carboniferous of eastern Kentucky, *in* J. C. Cobb, D. R. Chesnut Jr., N. C. Hester, and J. C. Hower, eds., Coal and coal-bearing rocks of eastern Kentucky: Geological Society of America Coal Geology Division field trip, November 5–8, 1981: Kentucky Geological Survey, series 11, p. 57–66.

Chesnut, D. R., Jr., 1991a, Paleontological survey of the Pennsylvanian rocks of the eastern Kentucky coal field: Kentucky Geological Survey, series 11, Information Circular 36, 71 p.

Chesnut, D. R., Jr., 1991b, Timing of Alleghanian tectonics determined by central Appalachian foreland basin analysis: Southeastern Geology, v. 31, p. 203–221.

Chesnut, D. R., Jr., 1992, Stratigraphic and structural framework of the Carboniferous rocks of the central Appalachian basin: Kentucky Geological Survey, series 11, Bulletin 3, 42 p.

Chesnut, D. R., Jr., 1994, Eustatic and tectonic control of deposition of the lower and middle Pennsylvanian strata of the central Appalachian basin, *in* J. M. Dennison and F. R. Ettensohn, eds., Tectonic and eustatic controls on sedimentary cycles: SEPM Concepts in Sedimentology and Paleontology, v. 4, p. 51–64.

Chesnut, D. R., Jr., 1996, Geologic framework for the coal-bearing rocks of the central Appalachian basin: International Journal of Coal Geology, v. 31 p. 55–66.

Churnet, H. G., 1996, Depositional environments of lower Pennsylvanian coal-bearing siliciclastics of southeastern Tennessee, northwestern Georgia, and northeastern

Alabama, U.S.A.: International Journal of Coal Geology, v. 31, p. 21–54.

Cobb, J. C., and D. R. Chesnut Jr., 1989, Resource perspectives of coal in eastern Kentucky, in C. B. Cecil, J. C. Cobb, D. R. Chesnut Jr., H. Damberger, and K. J. Englund, eds., Carboniferous geology of the eastern United States: 28th International Geological Congress Field Trip Guidebook T143, Washington D.C., American Geophysical Union, 28th International Geological Congress, Field Trip Guidebook no. 143, p. 64–66.

Davis, M. W., and R. Ehrlich, 1974, Late Paleozoic crustal composition and dynamics in the southeastern United States, in G. Briggs, ed., Carboniferous of the southeastern United States: Geological Society of America Special Paper 148, p. 171–186.

Donaldson, A. C., 1974, Pennsylvanian sedimentation of central Appalachians, in G. Briggs, ed., Carboniferous of the southeastern United States: Geological Society of America Special Paper 148, p. 47–78.

Donaldson, A. C., and C. F. Eble, 1991, Pennsylvanian coals of central and eastern United States, in H. J. Gluskoter, D. D. Rice, and R. B. Taylor, eds., Economic geology, United States: Geological Society of America, The geology of North America, v. P2, p. 523–515.

Donaldson, A. C., and R. C. Shumaker, 1981, Late Paleozoic molasses of central Appalachians, in A. D. Miall, ed., Sedimentation and tectonics in alluvial basins: Geological Association of Canada Special Paper 23, p. 99–124.

Donaldson, A. C., J. J. Renton, and M. W. Presley, 1985, Pennsylvanian deposystems and paleoclimates of the Appalachians: International Journal of Coal Geology, v. 5, p. 167–193.

Eble, C. F., 1996, Lower and lower middle Pennsylvanian palynofloras, southwestern Virginia: International Journal of Coal Geology, v. 31, p. 67–114.

Englund, K. J., 1974, Sandstone distribution patterns in the Pocahontas Formation of southwest Virginia and southern West Virginia, in G. Briggs, ed., Carboniferous of the southeastern United States: Geological Society of America Special Paper 148, p. 31–45.

Englund, K. J., 1976, Geologic map of the Grahn quadrangle, Carter County, Kentucky: U.S. Geological Survey, Geologic Quadrangle Map GQ-1262, scale 1:24,000, 1 sheet.

Englund, K. J., 1979, The Mississippian and Pennsylvanian (Carboniferous) systems in the United States—Virginia: U.S. Geological Survey Bulletin, v. 1110-C, 21 p.

Englund, K. J., and A. O. DeLaney, 1966a, Intertonguing relations of the Lee Formation in southwestern Virginia: U.S. Geological Survey Professional Paper 550-D, p. D47–D52.

Englund, K. J., and A. O. DeLaney, 1966b, Geologic map of the Isonville quadrangle, Pike County, Kentucky: U.S. Geological Survey, Geologic Quadrangle Map GQ-501, scale 1:24,000, 1 sheet.

Englund, K. J., and R. E. Thomas, 1990, Late Paleozoic depositional trends in the central Appalachian basin: U.S. Geological Survey Bulletin, v. 1839-F, p. F1–F119.

Englund, K. J., and R. E. Thomas, 1991, Coal resources of Tazewell County, Virginia, 1980: U.S. Geological Survey Bulletin, v. 1913, 17 p.

Ferm, J. C., 1974, Carboniferous environmental models in the eastern United States and their significance, in G. Briggs, ed., Carboniferous of the southeastern United States: Geological Society of America Special Paper 148, p. 79–96.

Ferm, J. C., 1979, Stratigraphic evidence for the position of Pocahontas basin source terranes, in J. C. Ferm and J. C. Horne, eds., Carboniferous depositional environments in the Appalachian region: Columbia, University of South Carolina, Department of Geology, Carolina Coal Group, p. 280–282.

Ferm, J. C., and V. Cavaroc, 1969, A field guide to Allegheny deltaic deposits in the upper Ohio Valley: Pittsburgh, Pennsylvania, Pittsburgh Geological Society, 21 p.

Ferm, J. C., and G. A. Weisenfluh, 1989, Evolution of some depositional models in Late Carboniferous rocks of the Appalachian coal fields: International Journal of Coal Geology, v. 12, p. 259–292.

Ferm, J. C., J. C. Horne, J. P. Swinchatt, and P. W. Whaley, 1971, Carboniferous depositional environments in northeastern Kentucky, Geological Society of Kentucky Guidebook for Annual Spring Conference, April 1971: Kentucky Geological Survey, series 10, 30 p.

Galloway, W. E., 1989, Genetic stratigraphic sequences in basin analysis: Architecture and genesis of flooding-surface bounded depositional units: AAPG Bulletin, v. 73, p. 125–142.

Greb, S. F., and D. R. Chesnut Jr., 1989, Geology of lower Pennsylvanian strata along the western outcrop belt of the eastern Kentucky coal field: Illinois Basin Consortium, Illinois Basin Studies, v. 1, 106 p.

Greb, S. F., and D. R. Chesnut Jr., 1992, Transgressive channel filling in the Breathitt Formation (Upper Carboniferous), eastern Kentucky coal field, U.S.A.: Sedimetary Geology, v. 75, p. 209–221.

Greb, S. F., and D. R. Chesnut Jr., 1994, Paleoecology of an estuarine sequence in the Breathitt Formation (Upper Carboniferous), eastern Kentucky coal field, U.S.A.: Sedimentary Geology, v. 75, p. 209–221.

Greb, S. F., and D. R. Chesnut Jr., 1996, Lower and lower middle Pennsylvanian fluvial to estuarine deposition, central Appalachian basin—Effects of eustasy, tectonics, and climate: Geological Society of America Bulletin, v. 108, p. 303–317.

Greb, S. F., and G. A. Weisenfluh, 1996, Paleoslumps in coal-bearing strata of the Breathitt Group (Pennsylvanian) in the eastern Kentucky coal field, U.S.A.: International Journal of Coal Geology, v. 31, p. 115–134.

Greb, S. F., D. R. Chesnut Jr., and K. A. Eriksson, 2001, Fluvial to estuarine transitions in lower Pennsylvanian strata of the Breathitt Group, eastern Kentucky coal field, central Appalachian basin, U.S.A., (abs.): University of Nebraska-Lincoln, Nebraska, 7th International Conference on Fluvial Sedimentology Program and Abstracts, p. 114.

Greb, S. F., C. F. Eble, and J. Hower, 1999, Depositional history of the Fire Clay coal bed (late Duckmantian), eastern Kentucky, U.S.A.: International Journal of Coal Geology, v. 40, p. 255–280.

Heckel, P. H., 1995, Glacial-eustatic base-level-climatic model for late middle to late Pennsylvanian coal-bed formation in the Appalachian basin: Journal of Sedimentary Research, v. B65, p. 348–356.

Henry, T. W., and M. Gordon, Jr., 1979, Late Devonian through Early Permian (?) faunas in proposed Pennsylvanian stratotype area, in K. J. Englund, H. H. Arndt, and T. W. Henry, eds., Proposed Pennsylvanian system stratotype Virginia and West Virginia: Washington, D.C., American Geological Institute Selected Guidebook Series 1, p. 97–103.

Horne, J. C., 1979, Sedimentary responses to contemporaneous tectonism, in J. C. Ferm and J. C. Horne, eds., Carboniferous depositional environments in the Appalachian region: Columbia, University of South Carolina, Department of Geology, Carolina Coal Group, p. 259–265.

Horne, J. C., J. P. Swinchatt, and J. C. Ferm, 1971, Lee-Newman barrier shoreline model, in J. C. Ferm, J. C. Horne, J. P. Swinchatt, and P. W. Whaley, eds., Carboniferous depositional environments in northeastern Kentucky, Geological Society of Kentucky Guidebook for Annual Spring Conference, April 1971: Kentucky Geological Survey, series 10, p. 5–9.

Horne, J. C., J. C. Ferm, F. T. Carrucio, and B. P. Baganz, 1978, Depositional models in coal exploration and mine planning in Appalachian region: AAPG Bulletin, v. 62, p. 2379–2411.

Houseknecht, D. W., 1980, Comparative anatomy of a Pottsville lithic arenite and quartz arenite of the Pocahontas Basin, southern West Virginia; petrogenetic, depositional, and stratigraphic implications: Journal of Sedimentary Petrology, v. 50, p. 3–20.

Hower, J. C., J. D. Pollock, and T. B. Griswold, 1991, Structural controls on the petrology and geochemistry of the Pond Creek coal seam, Pike and Martin Counties, eastern Kentucky, in D. C. Peters, ed., Geology in coal resource utilization: Fairfax, Virginia, Techbooks, p. 413–426.

Hower, J. C., E. J. Trinkle, J. D. Pollock, and C. T. Helfrich, 1992, Influence of penecontemporaneous tectonism on thickness and quality of Breathitt Formation coals, eastern Kentucky, in J. Platt, J. P. Price, M. Miller, and S. Suboleski, eds., 1.2— New perspectives on central Appalachian low-sulfur coal supplies: Fairfax, Virginia, Techbooks, Coal Decisions Forum Publication, p. 143–171.

Huddle, J. W., and K. J. Englund, 1962, Geology of the Varney quadrangle, Kentucky: U.S. Geological Survey, Geological Quadrangle Map, GQ-180, scale 1:24,000, 1 sheet.

Jenkins, E. C., 1966, Geologic map of the Millard quadrangle, Pike County, Kentucky: U.S. Geological Survey, Geologic Quadrangle Map GQ-659, scale 1:24,000, 1 sheet.

Jervey, M. T., 1988, Quantitative modeling of siliclastic rock sequences and their seismic expression, in C. K. Wilgus, B. S. Hastings, C. G. St. C. Kendall, H. W. Posamentier, C. A. Ross, and J. C. Van Wagoner, eds., Sea-level changes— An integrated approach: SEPM Special Publication 42, p. 47–70.

Klein, G. D., and D. A. Willard, 1989, Origin of the Pennsylvanian coal-bearing cyclothems of North America: Geology, v. 17, p. 152–155.

Korus, J. T., and K. A. Eriksson, 2002, Paleogeomorphology and facies architecture of Pennsylvanian trunk-tributary incised valley systems: New River Formation, southern West Virginia: 2002 Annual Meeting, AAPG, Abstracts with Programs, v. 11.

Lovett, J. A., W. W. Whitlock, W. S. Henika, and R. N. Diffenback, 1992, Geology of the Virginia portion of the Hurley, Panther, Wharncliffe, and Majestic quadrangles: Virginia Division of Mineral Resources Publication 121, 1 sheet.

Lyons, P. C., W. F. Outerbridge, D. M. Triplehorn, H. T. Evans Jr., R. D. Congdon, M. Capiro, J. C. Hess, and W. P. Nash, 1992, An Appalachian isochron— A kaolinized Carboniferous air-fall volcanic-ash deposit (tonstein): Geological Society of America Bulletin, v. 104, p. 1515–1527.

Martino, R. L., 1992, Marine rocks— Their recognition and role in coal correlation and quality in the Kanawha Formation (middle Pennsylvanian) of West Virginia, in J. Platt, J. Price, M. Miller, and S. Suboleski, eds., 1.2— New perspectives on central Appalachian low-sulfur coal supplies: Fairfax, Virginia, Techbooks, Coal Decisions Forum Publication, p. 125–142.

Martino, R. L., 1994, Facies analysis of middle Pennsylvanian marine units, southern West Virginia, in C. L. Rice, ed., Elements of Pennsylvanian stratigraphy, central Appalachian basin: Geological Society of America Special Publication 294, p. 69–86.

Martino, R. L., 1996, Stratigraphy and depositional environments of the Kanawha Formation (middle Pennsylvanian), southern West Virginia, U.S.A.: International Journal of Coal Geology, v. 31, p. 217–248.

Martino, R. L., and C. L. Rice, 1992a, Stop 5— Basal Pennsylvanian strata, Kentucky AA Highway, 0.5 km (0.3 mi) west of Kentucky Route 7, in C. L. Rice, R. L. Martino, and E. R. Slucher, eds., Regional aspects of Pottsville and Allegheny stratigraphy and depositional environments: U.S. Geological Survey Open-File Report 92-558, p. 28–30.

Martino, R. L., and C. L. Rice, 1992b, Stop 6— Basal Pennsylvanian strata, Kentucky AA Highway, 5.5 km (3.4 mi) west of Kentucky Route 7, in C. L. Rice, R. L. Martino, and E. R. Slucher, eds., Regional aspects of Pottsville and Allegheny stratigraphy and depositional environments: U.S. Geological Survey Open-File Report 92-558, p. 31–33.

Meissner, C. R., Jr., and R. L. Miller, 1981, Geologic map of the Honaker quadrangle, Russell, Tazewell, and Buchanan Counties, Virginia: U.S. Geological Survey, Geological Quadrangle Map GQ-1542, scale 1:24,000, 1 sheet.

Miller, M. S., 1974, Stratigraphy and coal beds of upper Mississippian and lower Pennsylvanian rocks in southwestern Virginia: Virginia Division of Mineral Resources Bulletin, v. 84, 211 p.

Morse, W. C., 1931, The Pennsylvanian invertebrate fauna, *in* Paleontology of Kentucky: Kentucky Geological Survey, series 6, v. 36, p. 293–348.

Nolde, J. E., 1989, Geology of the Keen Mountain quadrangle, Virginia: Virginia Division of Mineral Resources Publication 96, 1 sheet, scale 1:24,000.

Nolde, J. E., 1994, Devonian to Pennsylvanian stratigraphy and coal beds of the Appalachian Plateaus Province, *in* J. E. Nolde, W. W. Whitlock, J. A. Lovett, and W. S. Henika, eds., Geology and mineral resources of the southwest Virginia coalfield: Virginia Division of Mineral Resources Publication 131, p. 1–85.

Outerbridge, W. F., 1963, Geology of the Inez quadrangle, Kentucky: U.S. Geological Survey, Geologic Quadrangle Map GQ-226, scale 1:24,000, 1 sheet.

Outerbridge, W. F., 1976, The Magoffin Member of the Breathitt Formation: U.S. Geological Survey Bulletin, v. 1422-A, p. A64–A65.

Outerbridge, W. F., and R. Van Vloten, 1968, Geologic map of part of the Jamboree quadrangle, Pike County, Kentucky: U.S. Geological Survey, Geologic Quadrangle Map GQ-775, scale 1:24,000, 1 sheet.

Outerbridge, W. F., D. M. Triplehorn, and P. C. Lyons, 1990, The Princess No. 6 middle Pennsylvanian volcanic ash fall (tonstein), Kentucky and West Virginia, central Appalachian basin: Southeastern Geology, v. 31, p. 63–78.

Padgett, G., and R. Ehrlich, 1979, An analysis of two tectonically controlled integrated drainage nets of middle Carboniferous age in southern West Virginia, *in* J. C. Ferm and J. C. Horne, eds., Carboniferous depositional environments in the Appalachian region: Columbia, University of South Carolina, Department of Geology, Carolina Coal Group, p. 266–276.

Pillmore, C. L., and C. W. Connor, 1978, Geologic map of the Blaine quadrangle, Lawrence County, Kentucky: U.S. Geological Survey, Geological Quadrangle Map, GQ-1507, scale 1:24,000, 1 sheet.

Potter, P. E., and R. Siever, 1956, Sedimentary petrology, [Part 2] Sources of basal Pennsylvanian sediments in the Eastern Interior Basin: Journal of Geology, v. 64, p. 317–335.

Powell, L. R., 1979, Breathitt depositional systems in Martin County, Kentucky, *in* A. C. Donaldson, M. W. Presley, and J. J. Renton, eds., Carboniferous coal, short course and guidebook, AAPG Eastern Section: Morgantown, West Virginia University, Geology and Geography Department, p. 51–100b.

Quinlan, G. M., and C. Beaumont, 1984, Appalachian thrusting, lithospheric flexure, and the Paleozoic stratigraphy of the eastern interior of North America: Canadian Journal of Earth Sciences, v. 21, p. 973–996.

Rice, C. L., 1984, Sandstone units of the Lee Formation and related strata in eastern Kentucky: U.S. Geological Survey Professional Paper 1151-G, 53 p.

Rice, C. L., and J. K. Hiett, 1994, Revised correlation chart of coal beds, coal zones, and key stratigraphic units in the Pennsylvanian rocks of eastern Kentucky: U.S. Geological Survey Miscellaneous Field Studies Map, MF-2275, 1 sheet.

Rice, C. L., and J. F. Schwietering, 1988, Fluvial deposits in the central Appalachians during the early Pennsylvanian: U.S. Geological Survey Bulletin, v. 1839, p. 1–10.

Rice, C. L., R. G. Ping, and J. L. Barr, 1977, Geologic map of the Belfry quadrangle, Pike County, Kentucky: U.S. Geological Survey, Geologic Quadrangle Map GQ-1369, scale 1:24,000, 1 sheet.

Rice, C. L., E. G. Sable, G. R. Dever Jr., and T. M. Kehn, 1979, The Mississippian and Pennsylvanian (Carboniferous) systems in the United States— Kentucky: U.S. Geological Survey Professional Paper 110-F, p. F1–F32.

Rice, C. L., J. C. Currens, J. A. Henderson Jr., and J. E. Nolde, 1987, The Betsie Shale Member— A datum for exploration and stratigraphic analysis of the lower part of the Pennsylvanian in the central Appalachian basin: U.S. Geological Survey Bulletin, v. 1834, 17 p.

Rice, C. L., R. M. Kosanke, and T. W. Henry, 1994a, Revision of nomenclature and correlations of some middle Pennsylvanian units in the northwestern part of the Appalachian basin, Kentucky, Ohio, and West Virginia, *in* C. L. Rice, ed., Elements of Pennsylvanian stratigraphy, central Appalachian basin: Geological Society of America Special Paper 294, p. 41–53.

Rice, C. L., T. W. Henry, and D. R. Chesnut Jr., 1994b, The Crummies Member (new name) of the Pennsylvanian Breathitt Formation, eastern Kentucky; its distribution and biostratigraphy, *in* Sando, W. J., ed., Shorter contributions to paleontology and stratigraphy 1993: U.S. Geological Survey Bulletin, v. 2073, p. A1–A6.

Sanchez, J. D., D. C. Alvord, and P. T. Hayes, 1978, Geologic map of the Richardson quadrangle, Lawrence and Johnson Counties, Kentucky: U.S. Geological Survey, Geologic Quadrangle Map GQ-1460, scale 1:24,000, 1 sheet.

Sheppard, R. A., 1964, Geology of the Tygarts Valley quadrangle, Kentucky: U.S. Geological Survey, Geologic Quadrangle Map GQ-289, scale 1:24,000, 1 sheet.

Staub, J. R., 1991, Comparisons of central Appalachian Carboniferous coal beds by benches and a raised Holocene peat deposit: International Journal of Coal Geology, v. 18, p. 45–69.

Tankard, A. J., 1986, Depositional response to foreland deformation in the Carboniferous of eastern Kentucky: AAPG Bulletin, v. 70, p. 853–868.

Taylor, A. R., 1989, Geology of the Grundy quadrangle, Virginia: Virginia Division of Mineral Resources Publication 97, 1 sheet, scale 1:24,000.

Thomas, W. A., 1995, Diachronous thrust loading and fault partitioning of the Black Warrior foreland basin within the Alabama recess of the late Paleozoic Appalachian–Ouachita thrust belt, *in* S. L. Dorobek and G. M. Ross, eds., Stratigraphic evolution of foreland basins: SEPM Special Publication 52, p. 111–126.

Trent, V. A., 1965, Geology of the Matewan quadrangle in Kentucky: U.S. Geological Survey, Geologic Quadrangle Map GQ-373, scale 1:24,000, 1 sheet.

Wanless, H. R., 1975, The Appalachian region, *in* E. D. McKee and E. J. Crosby, eds., Paleotectonic investigations of the Pennsylvanian system in the United States: U.S. Geological Survey Professional Paper 853-C, 62 p.

Wanless, H. R., and F. P. Shepard, 1936, Sea level and climatic changes related to late Paleozoic cycles: Geological Society of America Bulletin, v. 47, p. 1177–1206.

Ward, D. E., 1978, Geologic map of the Adams quadrangle, Lawrence County, Kentucky: U.S. Geological Survey, Geologic Quadrangle Map GQ-1489, scale 1:24,000, 1 sheet.

Weisenfluh, G. A., and J. C. Ferm, 1991, Roof control in the Fire Clay coal group, southeastern Kentucky: Journal of Coal Quality, v. 10, p. 67–74.

Weller, J. M., 1930, Cyclic sedimentation of the Pennsylvanian period and its significance: Journal of Geology, v. 38, p. 97–135.

Weller, J. M., 1964, Development of the concept and interpretation of cyclic sedimentation, *in* D. F. Merriam, ed., Symposium on Cyclic Sedimentation: Kansas Geological Survey Bulletin, v. 169, p. 607–621.

Whitlock, W. W., 1994, Analysis of sandstone bodies associated with several major coal beds in the coalfield of southwest Virginia, *in* J. E. Nolde, W. W. Whitlock, J. A. Lovett, and W. S. Henika, eds., Geology and mineral resources of the southwest Virginia coalfield: Virginia Division of Mineral Resources Publication 131, p. 86–99.

Whittington, C. L., and J. C. Ferm, 1967, Geologic map of the Grayson quadrangle, Carter County, Kentucky: U.S. Geological Survey, Geologic Quadrangle Map GQ-640, scale 1:24,000, 1 sheet.

Wizevitch, M. C., 1992, Sedimentology of Pennsylvanian quartzose sandstones of the Lee Formation, central Appalachian basin—Fluvial interpretation based on lateral profile analysis: Sedimentary Geology, v. 78, p. 147.

Wolcott, D. E., and E. C. Jenkins, 1966, Geologic map of the Meta quadrangle, Pike County, Kentucky: U.S. Geological Survey, Geologic Quadrangle Map GQ-497, scale 1:24,000, 1 sheet.

Warwick, P. D., R. M. Flores, D. J. Nichols, and E. C. Murphy, 2004, Chronostratigraphic and depositional sequences of the Fort Union Formation (Paleocene), Williston Basin, North Dakota, South Dakota, and Montana, *in* J. C. Pashin and R. A. Gastaldo, eds., Sequence stratigraphy, paleoclimate, and tectonics of coal-bearing strata: AAPG Studies in Geology 51, p. 121–145.

6

Chronostratigraphic and Depositional Sequences of the Fort Union Formation (Paleocene), Williston Basin, North Dakota, South Dakota, and Montana

Peter D. Warwick

U.S. Geological Survey, Reston, Virginia, U.S.A.

Romeo M. Flores

U.S. Geological Survey, Denver, Colorado, U.S.A.

Douglas J. Nichols

U.S. Geological Survey, Denver, Colorado, U.S.A.

Edward C. Murphy

North Dakota Geological Survey, Bismarck, North Dakota, U.S.A.

ABSTRACT

The Fort Union Formation in the Williston Basin of North Dakota, South Dakota, and Montana comprises chronostratigraphic and depositional sequences of Paleocene age. Individual chronostratigraphic sequences are defined by palynostratigraphic (pollen and spore) biozones and radiometric (^{40}Ar/^{39}Ar) ages obtained from tonsteins or volcanic ash layers. Analyses of depositional sequences are based on lithofacies constrained by the radiometric ages and biozones.

The lower Paleocene (biozones P1–P3) contains three marine parasequences (landward stepping) in southwestern North Dakota that sequentially onlapped westward between 65 and 61 Ma (lower Ludlow and Cannonball Members). Maximum flooding (transgressive systems tract) occurred during an approximate 1-m.y. interval from 65 to 64 Ma, which regionally is correlated biostratigraphically to a tidally influenced, distributary-shoreface, and fluvial-channel complex in the Cave Hills, northwestern South Dakota, and to channel-dominated fluvial (lowstand incised paleovalley systems) and tidally influenced, flood-plain-deltaic transition facies in the Ekalaka area of southeastern Montana.

The progradational parasequences in the Cannonball Member consist of shoreface sandstone beds (with ravinement lag deposits) deposited by strand-plain barrier systems. Landward of the barrier systems, tidal-estuarine and mire deposits included thick but laterally discontinuous peat accumulations (e.g., Beta and Yule coal beds in the Ludlow Member, southwestern North Dakota). However, landward of the coastal deposits, the laterally equivalent T-Cross-Big Dirty coal zone (dated 64.78 Ma) in southeastern Montana formed as thick, laterally extensive peat accumulations in mires in a fluvial setting. In the flood-plain-deltaic, tidal transition zone near Ekalaka, Montana, the Ludlow Member consists of flood-plain facies, discontinuous coal beds, and rooted and burrowed horizons that contain the marine or brackish trace fossil *Skolithos*. The flood-plain-deltaic tidal transition zone facies are incised by a massive, agglomerated channel sandstone complex (paleovalley fill) that is exposed along the modern Snow Creek drainage south of Mill Iron, Montana. The flood-plain-tidal transition zone was reworked during the maximum sea level highstand during the early Paleocene. This event was followed by a fall of sea level and deposition of the paleovalley fill.

Sea level fall during the mid-Paleocene (biozones P3 and P4) produced a regressive shallow-marine and lower deltaic tidal system (seaward stepping) that deposited strata that thin toward the east. These strata are overlain by a widespread paleosol (Rhame bed) and, in turn, a lignite-bearing fluvial facies (Tongue River Member) containing the laterally persistent Harmon-Hanson coal zone (61.23 Ma). Upper Paleocene biozone P5 is represented by fluvial, coal-bearing strata that contain several economically minable coal beds (HT Butte, Hagel, and Beulah-Zap zones, Sentinel Butte Member).

The Fort Union Formation of the Williston Basin contains significant coal resources. These coal deposits are now being explored for their potential coal-bed gas resources. A better understanding of the depositional setting for these deposits can lead to improved exploration and exploitation practices and a better understanding of regional paleogeography and paleoclimate during the Paleocene.

INTRODUCTION

Background

Few stratigraphic studies of North American continental rocks have included an integrated analysis of lithofacies, biostratigraphy, and radiometric age determinations to define regional stratigraphic depositional sequences. Such an integrated approach is necessary because these components, used together, can provide a detailed analysis of basin fill and stratigraphic evolution. The objectives of this study are (1) to present new biostratigraphic and radiometric data based on outcrop and subsurface studies of the Fort Union Formation (Paleocene) in the Williston Basin of North Dakota, South Dakota, and Montana (Figures 1, 2), (2) to integrate these data into stratigraphic depositional sequences, and (3) to offer a detailed stratigraphic depositional sequence model for coal accumulation. In addition, the present study focuses on the Cannonball Member and adjacent Ludlow Member of the Fort Union Formation, which were influenced markedly by relative sea level changes. A detailed study of the lower Fort Union Formation is

warranted not only because of its economically important coal beds and potential coal-bed methane resources, but because the history of the sea level changes preserved in these rocks can be used to understand regional paleogeographic and paleoclimatic controls on the distribution and quality of energy resources. The new stratigraphic age correlations may contribute also to better understanding of the timing of the closing of the mid-continent seaway and associated paleoclimatic changes.

Previous Studies

The stratigraphy of the lower part of the Fort Union Formation in the southwestern part of the Williston Basin has been the topic of numerous studies over the past one-and-a-half centuries. Because of the occurrence of coal, most of the early studies focused on the economic geology and general stratigraphy of the region (Meek and Hayden, 1862; Leonard, 1908; Calvert, 1912; Thom and Dobbin, 1924; Hares, 1928). More recent studies have concentrated on the depositional setting and regional stratigraphy (Moore, 1976; Belt et al., 1984; Cherven and Jacob, 1985; Daly et al.,

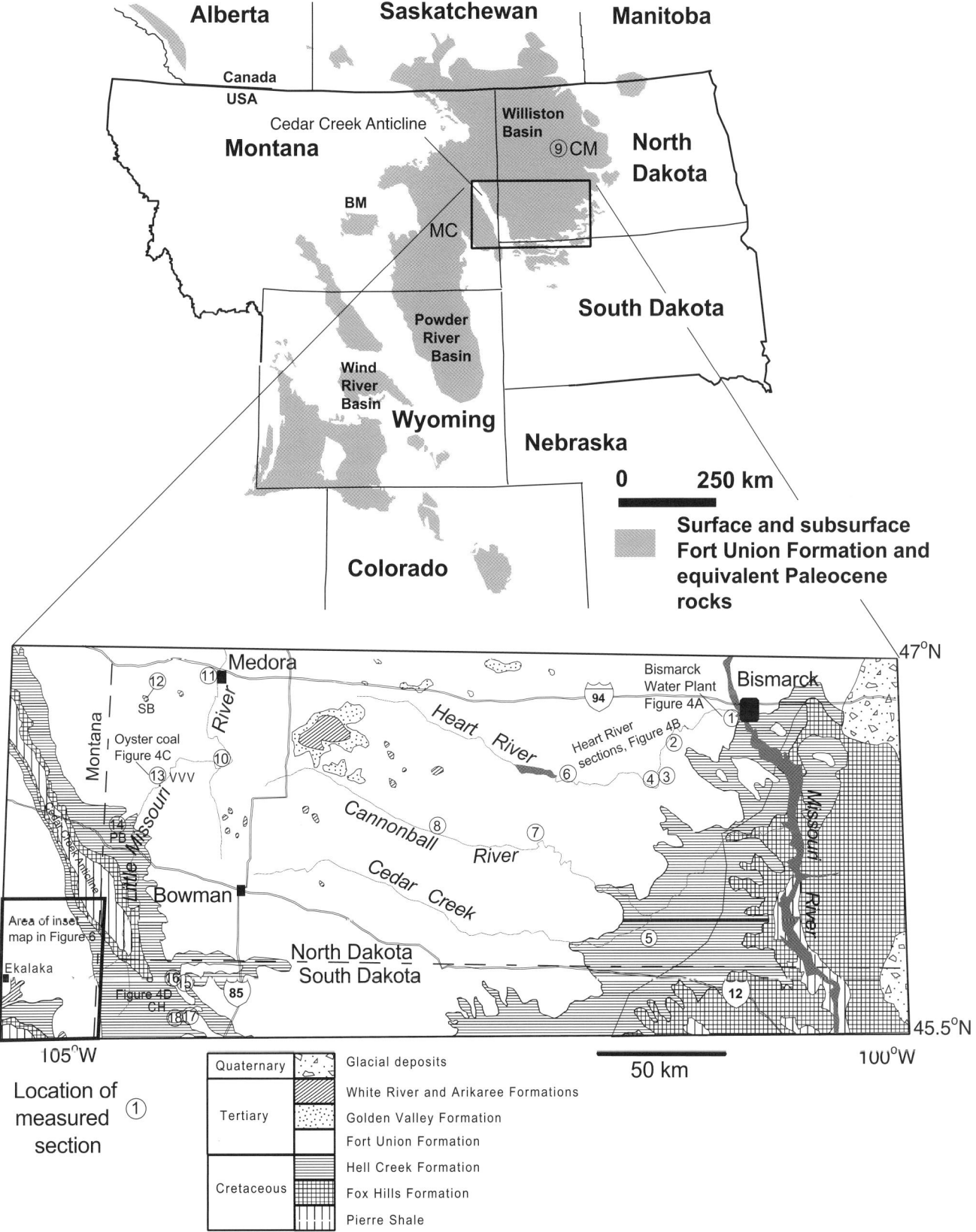

Figure 1. Location map of study area and composite sections (A–D) shown in Figure 4. Location details of the measured sections and sample localities are provided in Table 1. MC = Miles City, BM = Bull Mountains, PB = Pretty Butte, SB = Square Butte, VVV = Three V Crossing, CM = Coteau mine, CH = Cave Hills. Geology from Keefer (1974).

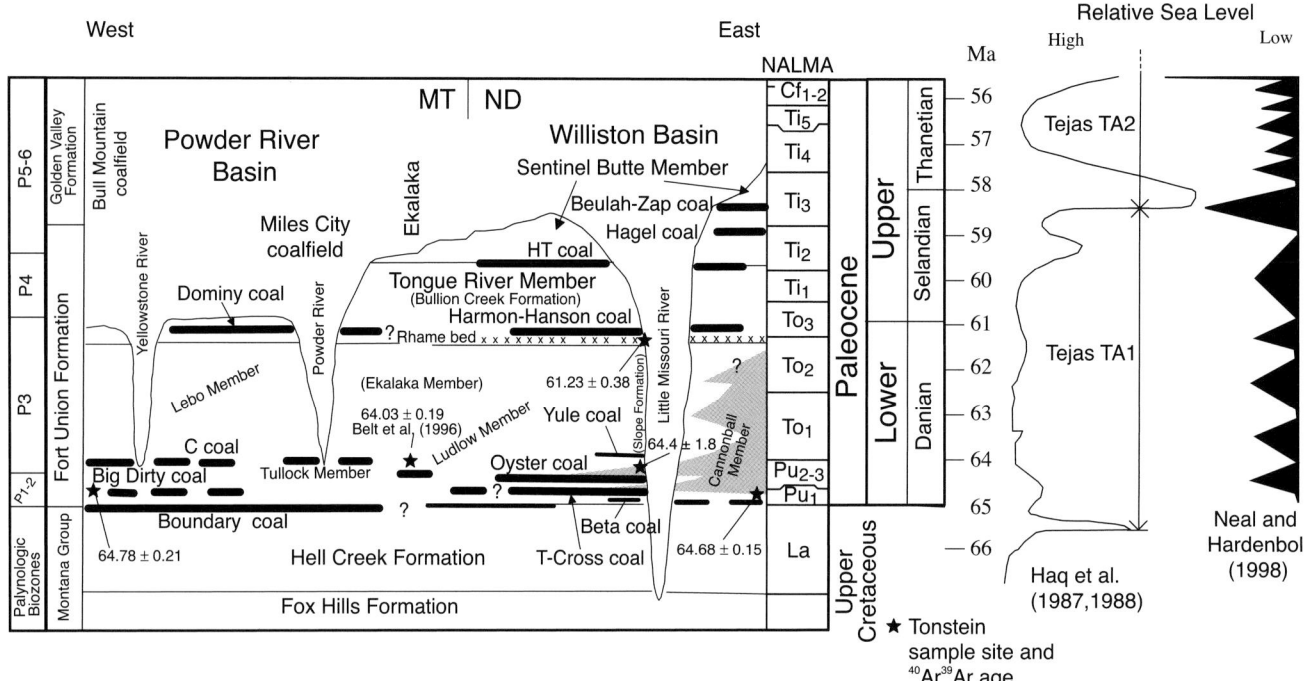

Figure 2. Regional chronostratigraphic correlations in the southwestern part of the Williston Basin. The North Dakota Geological Survey (NDGS) elevates Fort Union Formation member names to formation status, except for those NDGS formation names that are shown in parentheses (Clayton et al., 1977). The stratigraphic position of the Ekalaka Member (Belt et al., 2002) also is shown in parentheses. See Figure 1 for river and place locations. Relative sea level curves from Haq et al. (1987, 1988) and Neal and Hardenbol (1998); timescale in megaannum from Berggren et al. (1995). Correlation to North American Land Mammal Ages from Hartman and Kihm (1996), Kroeger and Hartman (1997), and Belt et al. (2002). La = Lancian; Pu = Puercan; To = Torrejonian; Ti = Tiffanian; Cf = Clarkforkian. Base diagram modified from Belt et al. (1984).

1985; Hartman and Kihm, 1991; Murphy et al., 1995). Cherven and Jacob (1985) provided a good summary of the stratigraphy and pertinent references to previous studies of the Paleocene rocks in the southwestern part of the Williston Basin. Archibald et al. (1982) reviewed floral and faunal extinctions and magnetostratigraphy at the Cretaceous–Tertiary boundary (see also Hartman et al., 2002). Swisher et al. (1993) provided radiometric and magnetostratigraphic ages for Upper Cretaceous and lower Paleogene mammal-bearing rocks in northeastern Montana. Brown (1962), Hickey (1984), and Johnson and Hickey (1990) provided reviews of the megaflora of the Fort Union Formation, whereas Hartman and Kihm (1991) and Cvancara and Hoganson (1993) provided an analysis of the vertebrate paleontology. Kroeger and Hartman (1997) described the paleoecological significance of Paleocene palynomorphs from this interval. Belt et al. (1992; 1993; 1996a, b; 1997a, b; 2002) described the Fort Union stratigraphy primarily in the Montana part of the Williston Basin (Figure 1). Belt et al. (1996b, 1997a, 2002) proposed the Ekalaka Member of the Fort Union Formation for an extension of Cannonball-equivalent marine and estuarine facies into the

southeastern part of Montana and introduced the possibility of two regional unconformities in the Ludlow and/or Tullock Members of the Fort Union Formation in southeastern Montana.

Perhaps the foremost studies of the marine Cannonball Member were performed by Cvancara (1965, 1966, 1972, 1976, 1980). In these studies, the depositional setting and the rich paleontological content of the interval were described. Cvancara (1965, 1966) interpreted the depositional setting as marine to marginal marine and described an abundant marine fauna that includes foraminiferids, corals, bryozoans, mollusks, ostracods, and vertebrates. Furthermore, Cvancara (1965, 1976, 1980) mapped the distribution of sandstone in southwestern North Dakota using subsurface data and outcrop observations and recognized that (1) the general thickness of the Cannonball Member increases eastward; (2) there are at least three widespread sandstone-rich intervals in the Cannonball Member; and (3) the percentage of sandstone in the member increases westward. Other reports focused on the Cannonball invertebrate fauna include a study of foraminiferids by Fenner (1976) and a study of bivalves by Lindholm (1984).

Other workers have contributed to an understanding of the depositional environments of the Cannonball Member, especially in the Cave Hills area of South Dakota (Figure 1). These include Pipiringos et al. (1965), Rich and Goodrum (1982), Goodrum (1983), and Best (1987). In these studies, the rocks of the Fort Union Formation, including the Ludlow, Cannonball, and Tongue River Members, were described as fluvial-deltaic-shoreline deposits. These workers recognized the westward pinch-out of the Cannonball sandstones and proposed lower and upper shoreface depositional environments for the sandstone-dominated interval in the Cave Hills area of northwestern South Dakota (Rich and Goodrum, 1982; Goodrum, 1983; Best, 1987).

Stratigraphy

The stratigraphic nomenclature of the exposed rocks in the southwestern Williston Basin has been the subject of considerable discussion in recent years. Two accepted sets of names currently exist for the lower Paleogene deposits in North Dakota. The North Dakota Geological Survey follows the nomenclature proposed by Clayton et al. (1977). Under this classification, the Fort Union Group (Paleocene), above the Hell Creek Formation (Cretaceous), consists of five formations (in stratigraphic order from bottom to top): the Ludlow, Cannonball, Slope, Bullion Creek (sometimes used interchangeably with Tongue River), and Sentinel Butte Formations. A white marker bed (Rhame bed) defines the transition zone (as much as 15 m [49 ft] thick) between the Slope and Bullion Creek Formations (Clayton et al., 1977; Wehrfritz, 1978; Belt et al., 1984).

In southeastern Montana, Belt et al. (1996b, 1997a, 2002) and Vuke et al. (2001) have proposed the name Ekalaka Member for the upper part of the Ludlow Member of the Fort Union Formation (Figure 2). The Ekalaka Member is defined as a sand-dominated interval that contains marine-brackish ichnotaxa and numerous slump blocks and megabreccia deposits that are interpreted as resulting from penecontemporaneous channel-wall failure. The Ekalaka Member is proposed to be a marine and estuarine equivalent to the Cannonball Member found in southeastern North Dakota and northeastern South Dakota.

The stratigraphic nomenclature used in the present study follows that of the U.S. Geological Survey as described by Swanson et al. (1981). Under this classification, the Fort Union Formation consists of four units, which starting from the bottom are the Ludlow, Cannonball, Tongue River, and the Sentinel Butte Members (Figure 2). The stratigraphic relations of the

lower Eocene	*Platycarya platycaryoides*
upper Paleocene	Zone P6
	Zone P5
	Zone P4
lower Paleocene	Zone P3
	Zone P2
	Zone P1
Upper Cretaceous (part)	*Wodehouseia spinata assemblage*

Figure 3. Diagram showing stratigraphic relations of uppermost Cretaceous, Paleocene, and lower Eocene palynostratigraphic biozones in the western interior of the United States.

uppermost Cretaceous, Paleocene, and lower Eocene palynostratigraphic biozones used in this study are shown in Figure 3. The Fort Union Formation may be as much as 590 m (1936 ft) thick (Bluemle et al., 1986) and overlies nonmarine carbonaceous shale and sandstone of the Hell Creek Formation (Cretaceous).

A detailed study of the Fort Union Formation has indicated that the nonmarine, lignite-bearing Ludlow Member interfingers toward the east with the brackish to marine Cannonball Member (Cvancara, 1966, 1976; Van Alstine, 1974; Moore, 1976; Belt et al., 1984; Cherven and Jacob, 1985; Hartman, 1989). Two marine tongues of the Cannonball Member have been identified in outcrops along the Little Missouri River in southwestern North Dakota (Moore, 1976) (Figures 1, 2). The lower tongue (Figure 2) is recognized by the presence of *Corbicula*, whereas the upper tongue is characterized by the presence of fossil oysters (Clayton et al., 1977; Belt et al., 1984). More recently, the *Corbicula* tongue has been called the Boyce tongue, and the Oyster tongue has been renamed the Three V tongue (Kroeger, 1995; Diemer et al., 1996; Belt et al.,

1997b). The nonmarine, lignite-bearing Tongue River Member (Bullion Creek Formation of Clayton et al., 1977) gradationally overlies the Ludlow Member along the Little Missouri River. Several authors have noted the presence of calcified and silicified paleosol horizons (Rhame bed), which mark the transition from the Ludlow to Tongue River Members (Hares, 1928; Moore, 1976; Clayton et al., 1977; Wehrfritz, 1978; Belt et al., 1984; Christensen, 1984). Diemer et al. (1996) and Belt et al. (1997b) have suggested that the Rhame bed is a paleosol that formed during non-deposition and represents a hiatus of about 1 m.y. (radiometric dates were not provided). The lignite beds of the Tongue River Member are more laterally persistent (as much as 250 km [155 mi]) and thicker (as much as 16 m [52 ft]) than lignite beds of the Ludlow Member (Rehbein, 1977). The Harmon coal bed, which is as much as 6 m (20 ft) thick, was mined in southwestern North Dakota (Flores et al., 1999a).

The Tongue River Member is overlain gradationally by similar lithologies of the Sentinel Butte Member. The two members are differentiated by the color of the sandstone and mudstone beds, which are tan, light gray to yellow, grayish white, and greenish white in the Tongue River. The overlying Sentinel Butte sandstone beds are dark-hued and somber gray, and the mudstone beds are steel bluish-gray. Sentinel Butte sandstone and mudstone beds are similar in color to those of the Ludlow Member of the Fort Union Formation and the Hell Creek Formation (Hares, 1928). The Hagel and Beulah-Zap coals, which are as much as 6.4 m (21 ft) thick and laterally extensive over a 5200-km^2 (2008-mi^2) area, occur in the Sentinel Butte Member and are mined in west-central North Dakota (Flores et al., 1999a).

The Fort Union Formation is overlain by the Golden Valley Formation (Paleocene–Eocene), the White River Group (Oligocene), and the Arikaree Formation (Oligocene–Miocene). These units are preserved in the southwestern Williston Basin almost exclusively as erosional remnants on high hills (Leonard, 1908; Holtzman, 1978; Bluemle et al., 1986) and consist of fluvial and lacustrine sediments (Bluemle et al., 1986; Murphy et al., 1993).

METHODS OF STUDY

Measured Sections

Because the stratigraphic transition between the Cannonball and Ludlow Members is poorly understood, we chose to focus our field investigations on this part of the Fort Union Formation. Four composite stratigraphic sections (Figure 4) were compiled from numerous measured stratigraphic sections of Cannonball and Ludlow rocks in southwestern North Dakota and northwestern South Dakota (Figure 1; Table 1). Where possible, the stratigraphic sections described by Cvancara (1965) were incorporated. In addition, we analyzed one core drilled by the North Dakota Geological Survey from a location near the Bismarck City Water Plant (Figures 1, 4A, 5E). The core contained parts of the Cannonball and Ludlow Members and the uppermost part of the Hell Creek Formation (Figure 2). Field investigations in southeastern Montana resulted in eight measured stratigraphic sections (Figures 1, 6).

Detailed descriptions of the measured stratigraphic sections and core included thickness of individual lithologic units, nature of contacts, grain size, color, sedimentary structures, and fossil content. Where possible, concealed intervals were exposed by excavation. The described rock types include sandstone, siltstone, mudstone, limestone, carbonaceous shale, and lignite. In addition, some sandstone and mudstone intervals in the Cannonball Member contain ironstone nodules and basal conglomeratic lag deposits.

Tonstein Samples

During the course of field study, 23 potential tonstein samples were collected from thin (1–3-cm [0.4–1.2-in.]) claystone beds found in mudstone or coal beds in the Ludlow and Tongue River Members of the Fort Union Formation (Table 2). Tonsteins are thin claystone bands of altered volcanic ash that commonly occur in coal beds (Bohor and Triplehorn, 1993). Many of the collected samples were determined not to be tonsteins or lacked sufficient quantities of sanidine crystals for radiometric dating. Apparent age determinations using the $^{40}Ar/^{39}Ar$ method on sanidine crystals were made using the interlaboratory standard Fish Canyon Tuff (age = 27.84 Ma) by geochronological laboratories at the U.S. Geological Survey and the New Mexico Bureau of Mines and Mineral Resources. In-depth description of the procedures used in this technique can be found in Steiger and Jager (1977), Dalrymple et al. (1981), Samson and Alexander (1987), and McDougall and Harrison (1988). Five Williston Basin tonstein ages are plotted on the stratigraphic diagram shown in Figure 2, and the stratigraphic positions and resulting ages are given in Table 2. Three samples are from this study. A fourth sample was collected from the Big Dirty coal bed in the Bull Mountains of Montana (Figure 1) by Carol W. Connor (U.S. Geological Survey, 1987–1988;

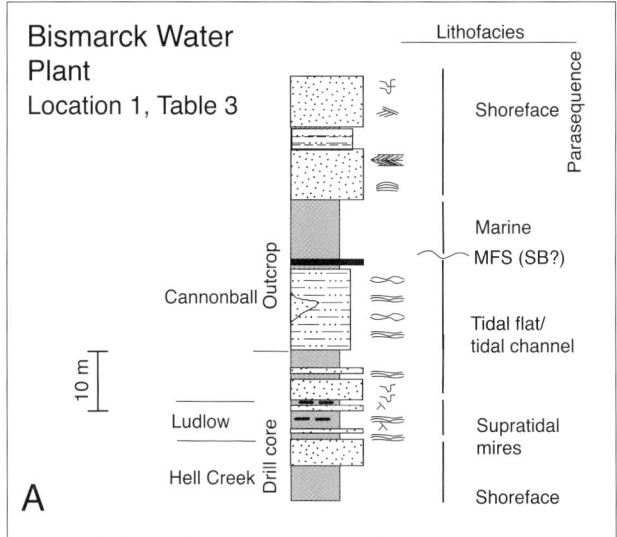

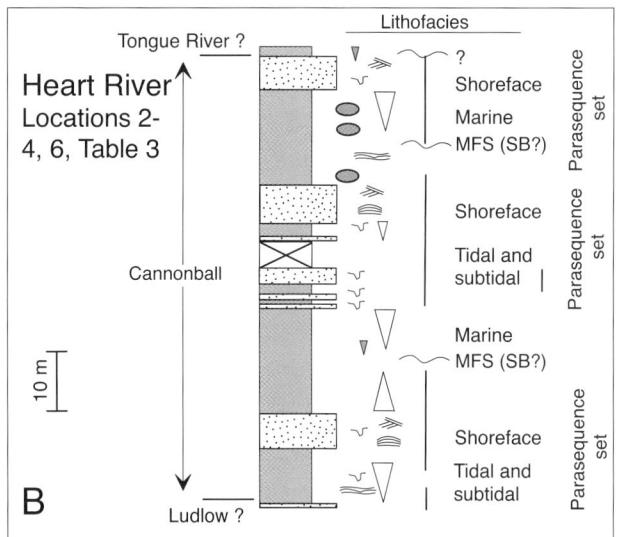

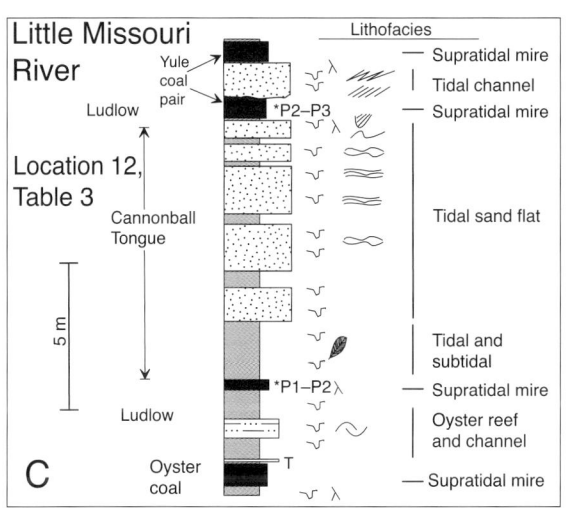

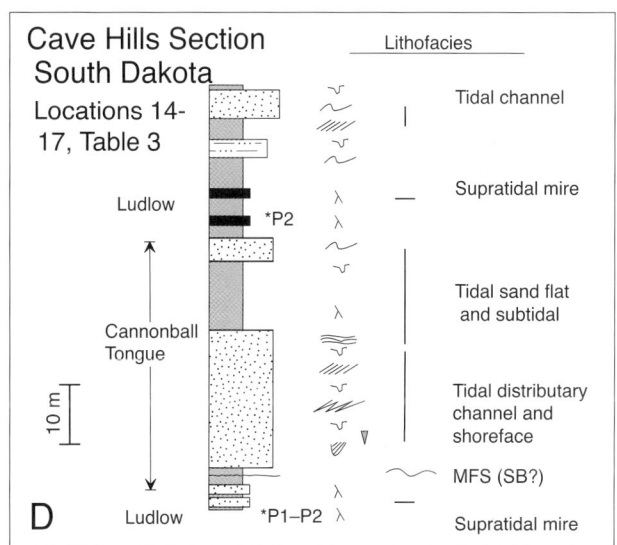

Legend

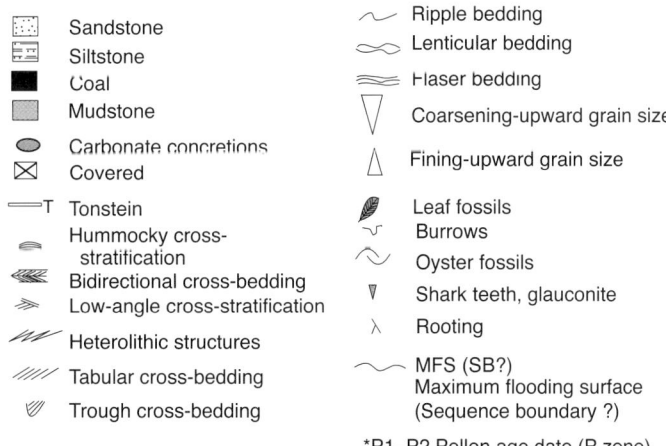

Sandstone
Siltstone
Coal
Mudstone
Carbonate concretions
Covered
Tonstein
Hummocky cross-stratification
Bidirectional cross-bedding
Low-angle cross-stratification
Heterolithic structures
Tabular cross-bedding
Trough cross-bedding

Ripple bedding
Lenticular bedding
Flaser bedding
Coarsening-upward grain size
Fining-upward grain size
Leaf fossils
Burrows
Oyster fossils
Shark teeth, glauconite
Rooting
MFS (SB?)
Maximum flooding surface (Sequence boundary ?)

*P1–P2 Pollen age date (P zone)

Figure 4. Composite stratigraphic sections (A–D) and lithofacies in the Hell Creek Formation and in the Ludlow and Cannonball Members of the Fort Union Formation in the Williston Basin (North Dakota and South Dakota). Section localities are shown in Figure 1 and are described in Table 1. Note that the vertical scale on the Little Missouri River section (C) is 2.5 times that of the other sections.

Table 1. Locations of sample sites and measured sections used in this study.

Location	State	Latitude	Longitude	Township/Range	Section
1) Bismarck Water Plant	North Dakota	46°48′47″N	100°49′14″W	139N/80W	31, NW/NE/SE
2) Grove Ranch	North Dakota	46°46′44″N	101°08′23″W	138N/83W	10, SE/SE/SE
3) Kopp Ranch	North Dakota	46°42′45″N	101°11′11″W	137N/83W	5, SE/SE/NW
4) Stark Ranch	North Dakota	46°42′39″N	101°12′55″W	137N/84W	6, SE/SW/NW
5) Section 18*	North Dakota	46°03′47.4″N	101°16′42.6″W	130N/85W	23, SE/SW/SW
6) Lindstrom Creek	North Dakota	46°35′43″N	101°44′18″W	136N/88W	16, SE/SE/NE
7) Stewart Ranch	North Dakota	46°13′57″N	101°29′36″W	132N/86W	30, NE/NE/NE
8) Elgin	North Dakota	42°21′28″N	101°48′48″W	133N/89W	2, SE/NW/SE
9) Coteau mine	North Dakota	47°24′11.4″N	101°50′55.2″W	145N/88W	1, SW/SW/NE
10a) Hanson Ranch (upper part)	North Dakota	46°37′18.6″N	103°32′10.8″W	136N/102W	6, NE/SW/NE
10b) Hanson Ranch (lower part)	North Dakota	46°37′20.4″N	103°32′39″W	136N/102W	6, NE
11) Medora	North Dakota	46°55′53.4″N	103°34′40.8″W	140N/102W	17, SW/SE/SE
12) Square Butte	North Dakota	46°51′42.2″N	103°42′36.6″W	139N/103W	8, SE and 17, NE
13a) Slope type sec. (lower part)	North Dakota	46°36′35.1″N	103°51′36.7″W	135N/105W	10, SW
13b) Slope type sec. (upper part)	North Dakota	46°30′48.6″N	103°51′39.6″W	135N/105W	15, NW
14) Pretty Butte	North Dakota	46°23′07.3″N	103°56′07.7″W	134N/106W	25, SW/SE/SW
15) North Cave Hills	South Dakota	45°50′29″N	103°28′34″W	22N/5E	26, NW
16) North Cave Hills (top of butte)	South Dakota	45°49′51″N	103°27′47″W	22N/5E	26
17) South Cave Hills	South Dakota	45°45′29″N	103°35′26″W	21N/4E	26, SE
18) South Cave Hills (road section)	South Dakota	45°45′42″N	103°35′33″W	21N/4E	23, SW
19) Jardee Ranch	Montana	45°53′09.5″N	104°15′53.1″W	2N/60E	33, SE
20) Snow Creek	Montana	45°46′32″N	104°12′52″W	1N/59E	5, SE,NE
21) Snow Creek 2	Montana	45°46′16.1″N	104°12′28.1″W	1N/59E	5, SE,SE
22) Beach Ranch	Montana	45°57′28.9″N	104°24′13.3″W	2N/59E	4, SW/NW
23) Chimney Rock	Montana	45°58′16.7″N	104°24′47.4″W	3N/59E	32, SE/SE
24) Highway 323	Montana	45°49′39.9″N	104°26′53″W	1N/59E	19, SW/NE
25) Medicine Rocks	Montana	46°03′36.8″N	104°28′57.8″W	4N/58E	35, SE
26) Ekalaka	Montana	45°53′42.9″N	104°32′27.3″W	2N/58E	28, SW/NW

Section 18, from Murphy et al. (1995).

the sample was analyzed by J. D. Obradovich, U.S. Geological Survey, 1996, personal communication); the age of this sample (64.78 ± 0.21 Ma) was reported by Warwick et al. (1997). A fifth sample was collected by Belt et al. (1996b) from the Ludlow Member near Ekalaka, Montana (Figure 6) and was dated at 64.03 ± 0.19 Ma.

Palynology Samples

A palynostratigraphic framework for the Paleocene rocks of the region was constructed based on reference samples collected from four measured sections in the southwestern part of the Williston Basin in North Dakota (Table 1). The Ludlow Member was sampled at Pretty Butte (location 14 in Table 1; 15 samples, including 3 from the upper part of the underlying Hell Creek Formation). The Ludlow and Cannonball Members were sampled at the type section of the Slope Formation of Clayton et al. (1977) along the Little Missouri River (location 13 in Table 1; 18 samples). The Tongue River Member was sampled along the Little Missouri River southeast of Bullion Butte, in a section on the Hanson Ranch (location 10 in Table 1; 22 samples), and near Medora, North Dakota (location 11 in Table 1; 7 samples). Samples from these measured sections were supplemented by

Figure 5. Representative photographs showing lithologies and sedimentary features for selected rocks in the North Dakota and South Dakota study areas. (A) Coarsening-upward mudstone and silty sandstone interval in the lower part of the Heart River section (Figure 4B). Arrow points to hardened hummocky cross-stratified sandstone in the upper part of the coarsening-upward sequence. (B) Ravinement lag consisting of mudstone rip-up clasts, fossil bone fragments, and shark teeth marked by the hammer pick (8 cm [3 in.] length). These units are found at the base of some of the major sandstones in the Cannonball Member. Photograph location is in the Heart River outcrop area. (C) Low-angle cross-stratification is a common feature in Cannonball coarsening-upward sandstone units. Hammer length is about 25 cm (10 in.). Photograph location is in the Heart River outcrop area. (D) *Ophiomorpha*-type burrows (4 cm [1.6 in.] diameter, 15 cm [6 in.] long) are common throughout the coarsening-upward sandstones of the Cannonball Member. Photograph location is the North Cave Hills, South Dakota (Figures 1, 4D). Scale in centimeters. (E) Interbedded silty sandstone and mudstone showing flaser and lenticular bedding and horizontal and vertical burrows in the Cannonball and Ludlow Members. A few lenticular beds show bipolar ripple laminations. Core is from the Bismarck City Water Plant location (Figure 1). Note coin (2 cm [0.8 in.] diameter) for scale; arrow indicates hammer scrape marks. (F) Large cannonball-like concretion in the Cannonball Member exposed along the Heart River valley wall (Figure 1). Scale in centimeters. (G) Oyster coal locality along the Little Missouri River valley wall in southwestern North Dakota (see Figures 1, 4D). Large arrow indicates the location of the tonstein layer sampled for radiometric age dating, which is overlain by oyster shells (white specks). The interval between the Oyster coal bed and the Yule coal pair is about 20 m (66 ft). (H) Arrow points to the top of a massive cliff-forming Cannonball sandstone in the Cave Hills area of South Dakota (Figures 1, 4D). The sandstone ranges from 20 to 25 m (66 to 82 ft) thick.

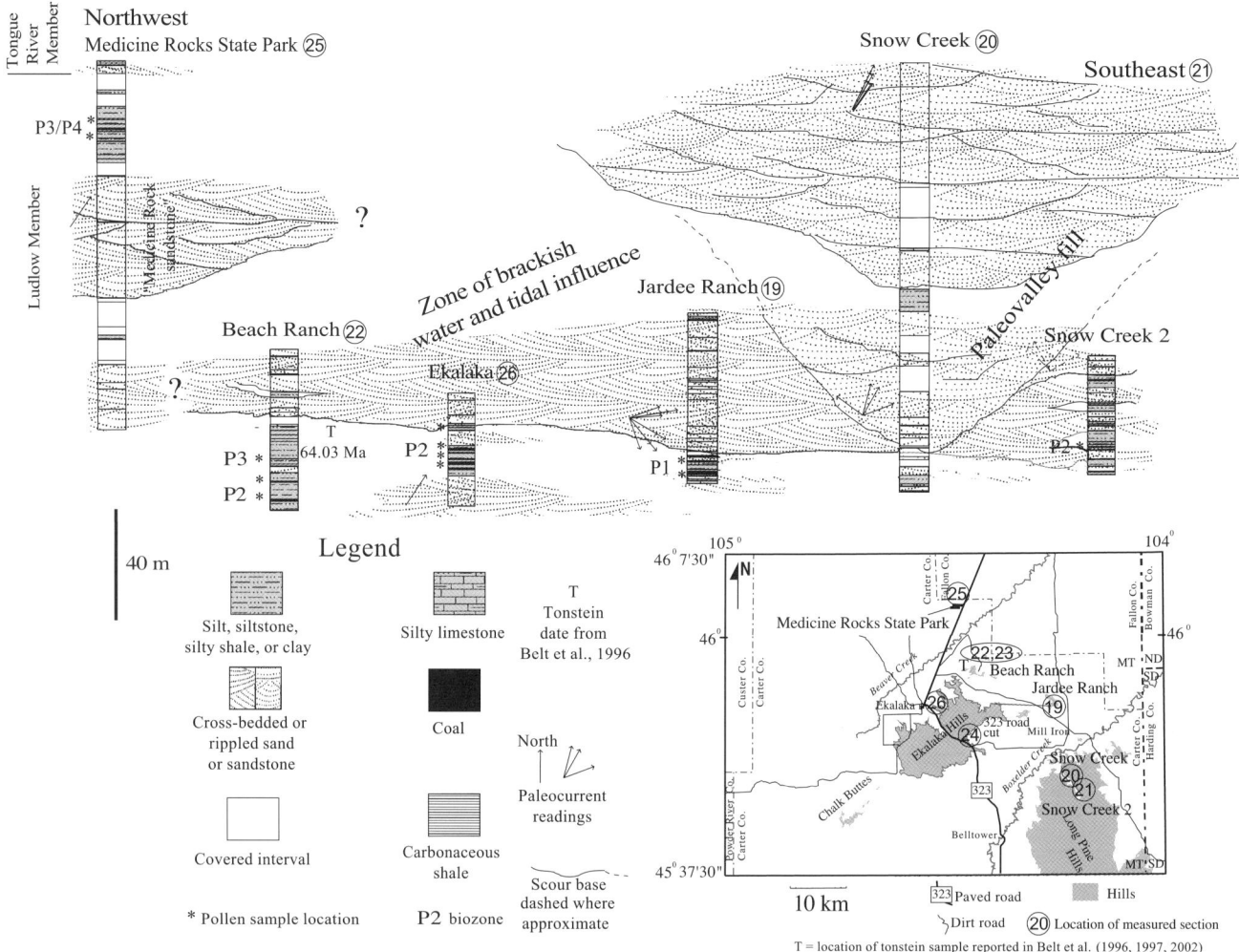

Figure 6. Northwest to southeast cross section showing the vertical and lateral lithostratigraphic variations of the Ludlow and Tongue River members of the Fort Union Formation in the Ekalaka area, Montana. Inset map shows the location of the measured sections used in the cross section. Refer to Table 1 for section location details.

one from the Sentinel Butte Member at Square Butte in Golden Valley County (Figure 1; location 12 in Table 1) and by five others from the Sentinel Butte Member collected in the Coteau Properties Freedom Mine near Beulah, Mercer County, North Dakota, in the eastern part of the basin (Figure 1; location 9 in Table 1). In this composite palynostratigraphic reference section, occurrences of 95 species of pollen and spores in these 68 samples were used to recognize five palynostratigraphic zones in the total interval, which comprises about 450 m (1476 ft) of the Fort Union Formation. Samples are listed in Table 3.

We also collected samples for palynological age determinations from sections measured at localities in the Williston Basin in Montana and South Dakota. The lower part of the Ludlow Member was sampled at five localities, and the upper part was sampled at one locality in eastern Montana (Figure 6; locations

19, 21, 22, 24, 25, and 26 in Table 1). In South Dakota, the Ludlow Member was sampled in the North Cave Hills area (Figure 1; location 16 in Table 1). Results of analyses of these samples are discussed later; the productive samples are listed in Table 3.

The palynostratigraphic zonation used is based on that originally developed in the Wind River Basin of central Wyoming (Nichols and Ott, 1978) and subsequently extended throughout Wyoming and adjacent Montana and North Dakota (Nichols, 1994, 1999, 2003). With modifications necessitated by minor differences in local stratigraphic ranges of some palynomorph species, Paleocene palynostratigraphic zones P1–P5 can be identified and correlated across the Williston Basin. The occurrence of the palynomorph taxa that define the biozones is independent of facies, as has been demonstrated in numerous studies (Jacobson and Nichols, 1982; Honey and Hettinger,

Table 2. Tonstein samples from the Fort Union Formation and associated $^{40}Ar/^{39}Ar$ ages.*

Coal bed/zone	Member	Location		Laboratory	Number of Analyses	Weighted Mean Age (Mean ± S.E.) (Ma)
		Lat.	Long.			
H coal	Tongue River	46°30′48″N	103°51′38.4″W	USGS	NR	61.23 ± 0.38
Unnamed	Ludlow	Near Ekalaka (see Figure 6)		USGS	NR	64.03 ± 0.19**
Oyster coal	Ludlow	46°30′54.14″N	103°52′07.44″W	USGS	1	64.40 ± 1.8
Unnamed	Ludlow	46°03′47.4″N	101°16′42.6W	NMBMMR	>30	64.68 ± 0.15
Big Dirty	Tullock	Bull Mountains (see Figure 1)†		USGS	NR	64.78 ± 0.21

*Lat. = latitude; Long. = longitude; USGS = U.S. Geological Survey; NMBMMR = New Mexico Bureau of Mines and Mineral Resources, NR = not reported, S.E. = standard error.
**From Belt et al. (1996a, b).
†For sample site description, see Connor (1984).

1989; Franczyk et al., 1990; Hettinger et al., 1991; Nichols and Flores, 1993; Nichols, 1994; Roberts and Stanton, 1994; Nichols, 1998; Roberts, 1998). Zones P1–P3 are early Paleocene, and zones P4 and P5 are late Paleocene in age (latest Paleocene zone P6 was not identified in the sampled interval, but may be present in the unsampled uppermost part of the Sentinel Butte Member or lowermost part of the overlying Golden Valley Formation). Because these biozones are based on nonmarine fossils (pollen), the terms "early Paleocene" and "late Paleocene" used herein may not correspond precisely with those terms as defined and used by Berggren et al. (1995). The use of palynostratigraphic biozones P1–P5 helped to subdivide the stratigraphic interval under study into time-correlative packages. The biozones are shown in Figure 3. Radiometric age data from the present study provide calibration points for parts of the palynostratigraphic zonation in the Williston Basin and, by implication, throughout the region.

RESULTS

Lithostratigraphy and Lithofacies

Sandstone bodies in the Cannonball and Ludlow Members are generally yellowish gray (5Y 7/2) and poorly consolidated. Grain size in a single body typically ranges from very fine sand to coarse silt. Sandstone bodies are laterally continuous across the outcrop, can be traced for several kilometers along river banks, and range from 20 to 45 m (66 to 148 ft) in thickness. Sandstone bodies generally have blocky outcrops and typically have a slight fining upward in grain size. The bases of these bodies are generally

sharp in contact with the underlying mudstone units, which coarsen upward (Figure 5), and these bodies contain ironstone rip-up clasts, glauconite, shark teeth, and vertebrate bone fragments in a sandstone matrix (Figure 5B).

Sedimentary structures in the sandstone bodies (Figures 4, 5C) consist of planar and trough cross-beds that commonly occur near the base of the unit. Hummocky and low-angle cross-beds, as well as bi-directional cross-bedding, are present in the lower part (Figures 4, 5A). Many of the thinner sandstone units exhibit flaser, lenticular, and ripple bedding. Most sandstone units are burrowed both horizontally and vertically, with burrows that are commonly 2–3 cm (0.8–1.2 in.) in diameter and 5–10 cm (2–4 in.) in length. *Ophiomorpha*-type burrowing (as much as 7.5 cm [3 in.] in diameter and >60 cm [>24 in.] long) is common in some of the sandstone-dominated intervals (Figure 5D).

Siltstone and mudstone are associated lithologies. Mudstone beds are as thick as 15 m (49 ft) and commonly occur as interbeds (a few centimeters thick) in the sandstone units. Individual mudstone units exhibit both coarsening- and fining-upward grain size, exhibit flaser bedding, lenticular bedding, and are almost always burrowed where they are coarsely interspersed with sandstone (Figures 4; 5A, E). In some intervals, especially near the transition between the Ludlow and Cannonball Members, mudstone units are rooted and rich in organic material that sometimes includes carbonaceous films and imprints of whole leaves. Limestone and ironstone concretions, as much as 1 m (3.3 ft) in diameter, can be found in layers in the upper part of the Cannonball Member

Table 3. Palynology samples from reference sections and study sections, with associated age determinations.*

Sample Number	Location	Startigraphic Unit	Stratigraphic Position	Age
		Reference sections		
D8157-A	Square Butte	Sentinel Butte Mbr, FUF	middle to upper part	zone P5
D8060-E	Coteau mine	Sentinel Butte Mbr, FUF	Schoolhouse rider	zone P5
D8060-D	do.	do.	1.5 m above Beulah-Zap coal	do.
D8060-C	do.	do.	Beulah-Zap coal	do.
D8060-B	do.	do.	do.	do.
D8060-A	do.	do.	do.	do.
D8104-G	Medora	Tongue River Mbr, FUF	13 m above D8104-F	zone P5
D8104-F	do.	do.	8.5 m above D8104-E	do.
D8104-E	do.	do.	coal bed 22 m above D8104-D	do.
D8104-D	do.	do.	8 m above D8104-C	do.
D8104-C	do.	do.	coal bed 22 m above D8104-B	do.
D8104-B	do.	do.	just above D8104-A	do.
D8104-A	do.	do.	coal, possibly Meyer bed	do.
D9122-D	Hanson Ranch	Tongue River Mbr, FUF	uppermost sample in section	do.
D9122-C	do.	do.	uppermost coal bed in section	do.
D8103-B	do.	do.	coal bed, unit 66 of Warwick	do.
D8103-A	do.	do.	"Medora Member"	do.
D9122-B	do.	do.	coal bed 5.5 m above D9122-A	do.
D9122-A	do.	do.	just above tonstein in D8102 coal	do.
D8102-A	do.	do.	just above D8100-B	do.
D8102-B	do.	do.	coal just above D8100-C	do.
D8102-C	do.	do.	parting 0.6 m above D8100-D	do.
D8102-D	do.	do.	coal bed 24 m above D8101-C	do.
D8101-C	do.	do.	11 m above D8101-B	do.
D8101-B	do.	do.	coal bed 13.5 m above D8101-A	do.
D8101-A	do.	do.	coal bed 8.5 m above D8100-I	do.
D8100-I	do.	do.	15.5 m above D8100-H	zone P4
D8100-H	do.	do.	just above D8100-G coal bed	do.
D8100-G	do.	do.	coal bed above Harmon	do.
D8100-F	do.	do.	top of Harmon coal	do.
D8100-E	do.	do.	base of Harmon coal	do.
D8100-D	do.	do.	parting in Hanson coal	zone P3
D8100-C	do.	do.	base of Hanson coal	do.
D8100-B	do.	do.	upper part of "H" coal	do.
D8100-A	do.	do.	lower part of "H" coal	do.
D8099-E	Slope type sec.	Ludlow Mbr, FUF	parting near top Yule coal	zone P3
D8099-D	do.	do.	parting in upper part no. 1 coal	do.
D8099-C	do.	do.	parting in lower part no. 1 coal	do.
D9123	do.	do.	just above "Oyster tonstein"	do.
D7681-F	do.	do.	1.5 m above Cannonball	do.
D8099-B	do.	do.	just above Cannonball	zone P2
D7681-E	do.	Cannonball Mbr, FUF	top of Cannonball Member	do.
D8099-A	do.	do.	middle of Cannonball	do.
D7681-D	do.	do.	middle of Cannonball	do.
D7681-C	do.	do.	lower part of Cannonball	do.
D7681-B	do.	do.	coal at base of Cannonball	do.
D7681-A	do.	Ludlow Mbr, FUF	just below Cannonball Member	do.
D8098-F	do.	do.	lower bed of Upper Coal Pair	do.

Table 3. Palynology samples from reference sections and study sections, with associated age determinations* (cont.).

Sample Number	Location	Startigraphic Unit	Stratigraphic Position	Age
D8098-E	do.	do.	upper bed of Lower Coal Pair	do.
D8098-D	do.	do.	lower bed of Lower Coal Pair	do.
D8098-C	do.	do.	near top of T-Cross coal	do.
D8098-B	do.	do.	parting in T-Cross coal	do.
D8098-A	do.	do.	parting in Beta coal	do.
D8045-E	Pretty Butte	do.	67.8 m above base	zone P1
D8045-C	do.	do.	67 m above base	do.
D8045-B	do.	do.	62.6 m above base	do.
D8045-A	do.	do.	51.7 m above base	do.
D7656-H	do.	do.	34.2 m above base	do.
D7656-F	do.	do.	30.4 m above base	do.
D7656-E	do.	do.	28.1 m above base	do.
D7656-C	do.	do.	21 m above base	do.
D7656-A	do.	do.	5 m above base	do.
D7788-BB	do.	do.	1.4 m above base	do.
D7788-CC	do.	do.	0.8 m above base	(Cretaceous)
D7788-DD	do.	Ludlow Mbr, FUF	0.3 m above base of FU Fm	do.
D9094-D	do.	Hell Creek Formation	0.1 m below top	do.
D7788-EE	do.	do.	23 m below top	do.
D7788-FF	Pretty Butte	do.	26 m below top of HC Fm	do.

Study sections

Sample Number	Location	Startigraphic Unit	Stratigraphic Position	Age
D9115-B	Medicine Rocks	Ludlow Mbr, FUF	33 m below Tongue River Mbr	zone P3
D9116	do.	do.	11.6 m below D9115	do.
D9111-A	Beach Ranch	Ludlow Mbr, FUF	coaly interval 8 m above D9110	zone P3
D9110-B	do.	do.	coaly interval 6 m above D9110	do.
D9109-C	do.	do.	middle of lowest coal bed	zone P2
D9109-B	do.	do.	base of lowest coal bed	do.
D9120-D	North Cave Hills	Ludlow Mbr, FUF	0.2 m above D9120-C	zone P3
D9120-C	do.	do.	1 m above D9120-B	zone P2
D9120-B	do.	do.	near base of measured section	do.
D9113-I	Ekalaka	Ludlow Mbr, FUF	7 m above D9113-H	zone P2
D9113-H	do.	do.	3.5 m above fourth coal bed	do.
D9113-G	do.	do.	fourth coal bed	do.
D9113-F	do.	do.	third coal bed	do.
D9113-D	do.	do.	second coal bed	do.
D9113-B	do.	do.	lowermost coal bed	do.
D9113-A	do.	do.	base of section	do.
D9119-D	Snow Creek	Ludlow Mbr, FUF	top of second coal bed	zone P2
D9119-C	do.	do.	parting in second coal bed	do.
D9119-B	do.	do.	base of second coal bed	do.
D9117-C	Highway 323	Ludlow Mbr, FUF	top of coal bed in road cut	zone P2
D9117-B	do.	do.	middle of coal bed in road cut	do.
D9117-A	do.	do.	base of coal bed in road cut	do.
D9121-B	Jardee Ranch	Ludlow Mbr, FUF	coal bed 3.6 m above D9121-A	zone P1
D9121-A	do.	do.	coal bed near base of section	do.

*Most ages given with reference to zones P1–P3 (early Paleocene) or zones P4–P5 (late Paleocene). Detailed sample location information provided in Table 1. FUF = Fort Union Formation; Mbr = Member.

(Figure 5F). Cvancara (1976) reported that some of the concretions may be fossiliferous.

Lignite and carbonaceous shale beds are not found in the Cannonball Member, but are present in the Ludlow, Tongue River, and Sentinel Butte Members (Figures 2, 4, 6). The uppermost stratigraphic occurrence of coal or carbonaceous beds traditionally has been used to mark the transition between the terrestrial Ludlow and the marine Cannonball Members (Moore, 1976). The coal beds of the Ludlow commonly are less developed and not as widespread as those in the Tongue River and Sentinel Butte Members (Hares, 1928). A few beds in the Ludlow, such as the T-Cross and Yule coal beds that outcrop along the Little Missouri River (Figure 2), reach a thickness of several meters, are locally widespread, and are laterally equivalent to the Cannonball Member (Belt et al., 1984). In the subsurface, the T-Cross can be as much as 6 m (20 ft) thick (Flores et al., 1999b). The coal beds are generally rich in woody material, and bedded lignitic logs are common. Claystone partings are common, and some contain volcanic ash layers or tonsteins (see below).

Bismarck City Water Plant Locality

The composite section from the Bismarck City Water Plant (Figure 4A) is derived from core and outcrop observations that include the uppermost part of the Hell Creek Formation, a thin interval of the Ludlow Member (about 5 m [16 ft]), and the lowermost part of the Cannonball Member. Somber gray, clay-rich, siltstone, sandstone, and mudstone characterize the uppermost part of the Hell Creek Formation. The transition between the Hell Creek Formation and the Ludlow Member is marked primarily by a color change from somber gray (5Y 5/2–5Y 6/4) to the lighter gray (5Y 7/2) color in the Ludlow. The mudstone beds of the upper Hell Creek also are characterized by flaser-bedded siltstones. The Ludlow is dominated by interbedded mudstone, siltstone, and very fine-grained sandstone intervals that show flaser beds and vertical burrows (Figure 5E). Fossil roots are common, and thin lignite and carbonaceous horizons (<5 cm [<2 in.] thick) are present just below the Cannonball Member. The Cannonball Member in outcrop is dominated by burrowed mudstone, siltstone, and very fine-grained sandstone intervals that have lenticular beds with bipolar ripple lamination (Figure 5E). Two sandstone units, each about 12 m (39 ft) thick, dominate the upper part of this outcrop. The lower part of the sandstone units exhibit hummocky and bidirectional cross-beds and low-angle tabular cross-beds; vertical burrows are common in the upper part of the sandstone units.

The lithofacies in the Hell Creek–Ludlow interval contain flaser beds similar to those found in the part of the section composed of the Cannonball Member lithofacies. The similarity of sedimentary structures in the lower and upper part of this section indicates that the lithofacies for these three stratigraphic units (uppermost part of the Hell Creek, Ludlow, and Cannonball) were deposited in tidally influenced environments, such as in supratidal mires, tidal flats, and possibly upper shoreface settings. The Cannonball Member contains a lower coarsening-upward, siltstone-dominated interval and two overlying coarsening-upward, sandstone-dominated intervals. The intervening thick mudstone interval probably indicates marine mud deposition with a marine flooding surface located just above the lower burrowed siltstone interval (Figure 4A). The marine flooding surface is marked by a thin (1–2 cm [0.4–0.8 in.] thick) layer containing fossil shark teeth. According to Walker and Plint (1992), mudstone and clay layers containing shark teeth and glauconite are indicative of flooding surfaces. Similar to Cvancara (1980), the coarsening-upward units most likely represent lower and upper shoreface deposits of landward-stepping parasequences. Dalrymple (1992) and Reinson (1992) have described other shoreface deposits that are similar to the Cannonball deposits.

Heart River

The Heart River composite section is based on four detailed, measured sections from the Cannonball Member exposed along the Heart River valley walls in southwestern North Dakota (Figures 1, 4B). We also incorporated data from the sections that Cvancara (1965) described along the Heart and Cannonball Rivers (Figure 1). Although the lower contact with the Ludlow Member is not always present, the exposed section of the Cannonball Member commonly starts just above the transition zone. The lower part of the exposed Cannonball section along the Heart River contains a coarsening-upward, mudstone and sandstone (very fine- to fine-grained) interval that contains flaser beds and burrows in siltstone (Figure 5A) similar to that found at Bismarck (Figure 4A). This interval is overlain by three prominent coarsening-upward sandstone lithofacies that are traceable laterally for several kilometers along the Heart River valley walls. In many places, these sandstone lithofacies contain a basal lag with rip-up clasts and vertebrate bone fragments. The sandstone beds contain hummocky and low-angle cross-beds (Figure 5A, B, C), and vertical burrows are present throughout the coarsening-upward sandstone units. The upper and lower coarsening-upward sandstone lithofacies are about 8 m (26 ft) thick, whereas the middle coarsening-upward lithofacies contains multiple bedded sandstone layers a few meters thick that are overlain by a thicker (about 8 m [26 ft]) sandstone unit. As at the Bismarck section, the coarsening-upward sandstone units are separated by mudstone intervals that contain thin layers of claystone with shark teeth and glauconite, which probably represent marine flooding surfaces (Figure 4B). The mudstone units also contain large carbonate concretions (Figures 4B, 5F). The three sandstone units of the Cannonball Member described along the Heart River valley walls probably represent the laterally extensive shoreface sandstone deposits identified by Cvancara (1965, 1980). These coarsening-upward sandstone units were deposited as landward-stepping, progradational parasequences. Sea level rise is indicated by multiple flooding surfaces and associated ravinement lags in the marine mudstone.

Little Missouri River

The stratigraphic section containing the Oyster tongue of the Cannonball Member is exposed along the Little Missouri River valley wall (Figures 4C, 5G). These exposures previously have been described by Hares (1928), Clayton et al. (1977), Belt et al. (1984), and Lindholm (1984). The interval examined is about 20 m (66 ft) thick and is stratigraphically between the Oyster and Yule coal beds (Figures 4C, 5G). This interval is burrowed and contains multiple stacked sandstone units a few meters thick. These sandstone beds contain flaser beds and tidal bundles, indicating that the interval was deposited in tidal-flat or tidally influenced environments. This tidal lithofacies is thicker than described by previous workers and contains burrowed mudstone units that probably were deposited in subtidal environments. The Oyster coal bed is overlain by a sharp-based, silty sandstone bed (1 m [3.3 ft] thick) with oyster shells that, in some places, occur as coquina-like masses, which probably accumulated as small reefs in a tidal creek. These sandstone-mudstone deposits are similar to Pennsylvanian and Cretaceous tidal and intertidal sequences described by Reinson (1992, figure 13).

A thin clay layer about 1 cm (0.4 in.) thick was sampled from a carbonaceous shale horizon about 2 m (7 ft) above the Oyster coal bed (Figures 4C, 5G;

Tables 1, 2). The clay layer is an altered volcanic ash and was dated by $^{40}Ar/^{39}Ar$ methods (64.4 ± 1.8 Ma).

Results of palynostratigraphic analysis of Ludlow and Cannonball Member samples show that the upper part of zone P1, all of zone P2, and the lower part of zone P3 are present. The sampled interval extends from the Beta coal to the Yule coal and includes the T-Cross coal and the Oyster tongue of the Cannonball Member (Figure 2). The boundary of zones P1–P2 is just below the Oyster tongue, which is entirely in zone P2. The boundary of zones P2–P3 is just below the lower bed of the Yule coal pair (Figure 4C), which marks the contact with the overlying upper Ludlow Member.

Cave Hills, Northwestern South Dakota

Four measured stratigraphic sections of rocks exposed in the Cave Hills area of South Dakota (Figures 1, 5H) are represented by a composite section shown in Figure 4D. In addition to our measured sections, the exposures were described previously by Pipiringos et al. (1965), Monnens (1980), Rich and Goodrum (1982), Goodrum (1983), Best (1987), and Belt et al. (1997a, 2002). The Cannonball Member is represented by a prominent sandstone interval that is bound on the bottom and top by coal beds of the Ludlow Member. The lower coal beds are in either palynostratigraphic zones P1 or P2, whereas the upper coal beds are in zone P2 (Figure 4D). The lower coal beds are overlain by a sandstone interval (20–25 m [66–82 ft] thick) that forms a major cliff face in the outcrop area (Figure 5H). The lower half of this coarse- to very fine-grained sandstone lithofacies contains heterolithic bedding, trough and low-angle (bidirectional) cross-bedding, and abundant large *Ophiomorpha* burrows (as much as 5 cm [2 in.] in diameter and 30 cm [12 in.] long) (Figure 5D). Fossil shark teeth are common clasts in basal lags of the trough cross-bedded sandstone. The lower part of this lithofacies is interpreted to be a shoreface-tidal deposit, whereas the upper part of this unit, which preserves heterolithic cross-beds and small *Ophiomorpha* burrows in scour-based, thin (3 m [10 ft] thick) sandstone bodies, reflects shallow-marine and tidal-channel depositional environments. This tidal-channel lithofacies is coeval to a thick (10–15 m [33–49 ft]), scour-based, trough cross-bedded sandstone unit that is interpreted to be a distributary tidal-channel deposit. The entire coarsening-upward sandstone probably represents a landward-stepping progradational parasequence. The coarsening-upward sandstone is overlain by a mudstone and sandstone interval with coal beds of the Ludlow Member. This mudstone-sandstone interval contains lenticular beds that are burrowed extensively. The sandstone beds contain tabular cross-beds. This mudstone-sandstone lithofacies interval probably was deposited in tidal sand flat, tidal-channel, and supratidal mire environments. Rich and Goodrum (1982) have proposed similar tidal and shoreface interpretations for the Cave Hills stratigraphic section.

Ekalaka Area, Southeastern Montana

In the Ekalaka area of southeastern Montana (Figures 1, 6), the Ludlow Member is characterized by flood-plain facies, discontinuous coal beds, and rooted horizons that interfinger laterally with thin to thick (3–10 m [10–33 ft]), scour-based sandstone units that are incised by a very thick (120 m [394 ft]), scour-based amalgamated sandstone complex. The thin-bedded sandstone beds are trough cross-bedded and show east to southeast paleoflow directions (Figure 6). The sandstone probably represents distributary channel deposits. The uppermost part of the sandstone lithofacies (e.g., the Jardee Ranch section; Figure 6) preserves Skolithos-like burrows (Belt et al., 1997a) and other burrowed horizons that are interbedded with intensely rooted zones (Figure 7B, C). The axis of the amalgamated sandstone complex is exposed along the modern Snow Creek drainage, south of Mill Iron, Montana (Figures 6, 7). Cross-beds show a generally northeast paleoflow direction (Figure 6). The sandstone complex (Figure 7A) was deposited by rivers that flowed eastward toward the Cave Hills (South Dakota; Figure 1) tidal, distributary-shoreface, and fluvial-channel complex deposits. These burrowed horizons probably represent deposits of landward extensions of brackish waters along channel basins that probably are correlative to the tidal-channel/shoreface and distributary channel deposits at the Cave Hills area of South Dakota.

North of Medicine Rocks State Park, Montana (Figure 6), scour-based, trough cross-bedded sandstone bodies (30–40 m [98–131 ft] thick) equivalent to the bioturbated channels exposed in the Snow Creek sections are multiscoured, multilateral, and are not burrowed. These sandstone beds, which occur below coal beds of zone P3 age (Figure 6), were deposited by northeastward-flowing rivers. This suggests that the Cedar Creek anticline was not active or emergent at this time (early to mid-Paleocene), as suggested by Clement (1986).

Results of palynologic analyses from the measured sections near Ekalaka (Figure 6) indicate that the Ludlow Member beds exposed in the butte just northeast of the town (Ekalaka Town section, Figure 6) are in lower Paleocene zone P2. Results from the Beach Ranch section in Carter County, Montana (Figure 6), indicate that the beds also are in zone P2 or zone P3. Sample results from measured section 2 in the Snow Creek area in Carter County indicate that the Ludlow beds near the base of the section are in zone P2. Similar results were obtained from samples collected from the Highway 323 road-cut locality in Carter County. Most samples from the measured section at the Jardee Ranch in Carter County lacked palynostratigraphically useful species, but sparse assemblages from near the base of the section indicate that the Ludlow at this locality most likely is in zone P1. Samples from the measured section near Medicine Rocks State Park in Fallon County, Montana, yielded species characteristic of both the upper part of zone P3 and the lower part of zone P4. It is evident that this interval, which is in the upper part of the Ludlow Member, is near the boundary of zone P3–P4.

CHRONOSTRATIGRAPHIC SEQUENCES

Palynostratigraphic Zonation

The palynostratigraphic zonation of the Paleocene in the Williston Basin in western North and South Dakota and eastern Montana differs in detail from the Paleocene zonations in the Wind River Basin (Nichols and Ott, 1978) and Powder River Basin (Nichols, 1999) in Wyoming. The designations P1–P5 are used in all three basins, but because local ranges of some species of fossil pollen differ in these areas (presumably with paleolatitude), the species used to recognize some of the zones differ. In some respects, the species content of some of the P zones in the Williston Basin resembles that of the Paleocene zones in

Figure 7. Representative photographs showing lithologies and sedimentary features for selected rocks in the Ekalaka area, Montana study area. (A) Amalgamated fluvial-channel sandstone complex of the paleovalley fill exposed along the valley walls of Snow Creek, south of Mill Iron, Montana (Figure 6). Height of person for scale. (B) Burrowed and rooted zones in sandstones in the Jardee Ranch section, north of Mill Iron (Figure 6). Height of person for scale. (C) Skolithos-like burrows in sandstones in the Jardee Ranch section, north of Mill Iron (Figure 6).

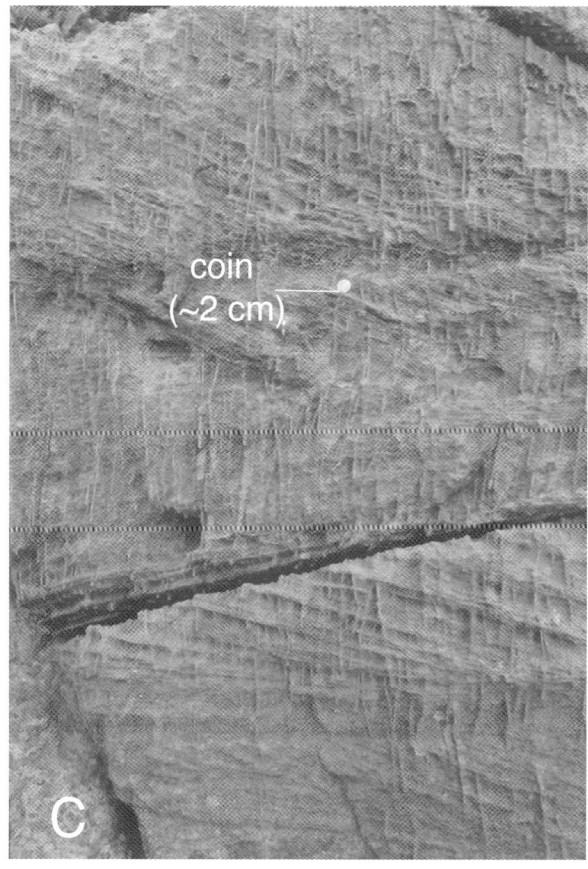

Alberta described by Demchuk (1990), but in others, they differ from those as well (Nichols, 2003).

Basal Paleocene zone P1 in the Williston Basin can be subdivided into lower and upper parts. The lower part of zone P1 is characterized by *Discoidites parvistriatus*, *Kurtzipites trispissatus*, and *Kurtzipites circularis* and may include uncommon occurrences of *Wodehouseia spinata* and *Wodehouseia fimbriata*; species of the genus *Momipites* are absent. Zone P2 in the Williston Basin is characterized by *Momipites tenuipolus*, *K. trispissatus*, *K. circularis*, and *Retitrescolpites anguloluminosus*; *D. parvistriatus* is absent. The upper part of zone P2 also includes *Momipites leffingwellii* and *Momipites waltmanensis*. A radiometric date from above the Cannonball Member in southwestern North Dakota (Table 2) indicates that the age of the boundary of zone P2–P3 is approximately 64.4 Ma. Zone P3 in the Williston Basin is characterized by other species of the genus *Momipites*, including *Momipites actinus*, *Momipites anellus*, *Momipites leboensis*, *Momipites triorbicularis*, and *Momipites ventifluminis*, and by the distinctive mid-Paleocene species of an otherwise more typically Upper Cretaceous genus, *Aquilapollenites spinulosus*. Also occurring in zone P3 in the Williston Basin is *Insulapollenites rugulatus*, which does not appear below zone P5 in the Wyoming basins. A radiometric date from the Hanson Ranch reference section indicates that the age of the boundary of zone P3–P4 is about 61.2 Ma (Table 2). Zone P4 in the Williston Basin is characterized by *Caryapollenites wodehousei*; *M. ventifluminis* and *Momipites wyomingensis* are common as well. Zone P5 in the Williston Basin is characterized by *Caryapollenites veripites* and *Pistillipollenites mcgregorii*; species of *Momipites* are present also, except for *M. actinus*, *M. leffingwellii*, and *M. waltmanensis*.

Radiometric Age Dating

The stratigraphically lowest, oldest age-dated sample (64.78 ± 0.21 Ma, Table 2) was collected from the Big Dirty coal bed in the Bull Mountains of Montana. Connor (1984, 1985) has described the location and stratigraphy of the Big Dirty coal. The next younger sample (64.68 ± 0.15 Ma) comes from about 60 cm (24 in.) above the Ludlow and Hell Creek contact in Sioux County, North Dakota (Table 2). The section has been described previously by Murphy et al. (1995, section 18). The next younger age-dated sample (64.4 ± 1.8) was collected from the outcrop of the Oyster tongue of the Cannonball Member that is exposed near the Little Missouri River (Figures 2, 4C, 5G). The sample collected by Belt et al. (1996b) was analyzed using $^{40}Ar/^{39}Ar$ methods and yielded an age

of 64.03 ± 0.19 Ma (Table 2). The stratigraphically youngest sample obtained (61.23 ± 0.38 Ma; Table 1) comes from the H coal bed, which is part of the Harmon-Hanson coal group exposed along the Little Missouri River valley wall near the base of the Hanson Ranch section (Figure 1). This section previously was described as section 95 by Warwick (1982) and Warwick and Luck (1995), is 14 m (46 ft) stratigraphically above the Rhame bed, and forms the base of the exposed section (Figure 2). Similar $^{40}Ar/^{39}Ar$ ages for lower Paleogene tonsteins in northeastern Montana have been reported by Swisher et al. (1993), but the exact stratigraphic correlation to equivalent age rocks in the southwestern part of the Williston Basin is unclear. The probable source for the volcanic ashes is the Cretaceous through Eocene igneous activity in western and central Montana that has been described by Marvin et al. (1980).

DISCUSSION OF DEPOSITIONAL SEQUENCES AND THEIR TIMING

Three landward-prograding marine parasequence sets (each 20–45 m [66–148 ft] thick) of the Cannonball Member of the Fort Union Formation interfinger with coastal plain deposits of the Ludlow Member in the southwestern part of the Williston Basin (Figure 8). These parasequence sets are composed, from bottom to top, of (1) lower shoreface lithofacies of burrowed mudstones (15–30 m [49–98 ft] thick) with elongate limestone nodules (as much as 20×50 cm [8×20 in.]); (2) coarsening-upward middle and upper shoreface lithofacies (5–10 m [16–33 ft] thick) of burrowed mudstone interbedded with hummocky cross-stratified, very fine-grained sandstone; (3) upper shoreface sandstone lithofacies (5–10 m [16–33 ft] thick) that are fine grained, burrowed, and trough and tabular cross-stratified; and (4) maximum flooding surfaces marked by a ravinement lag lithofacies (0.5 m [1.6 ft] thick) composed of claystone and ironstone rip-up clasts, glauconite, shark teeth, and bone fragments in a sandstone matrix (Figures 4, 5). The landward extent of these Cannonball parasequence sets along the Heart River may be represented by the two marine tongues (*Corbicula* and Oyster) that are exposed along the Little Missouri River (Figure 1). These brackish-marine tongues are composed of tidal lithofacies (5–25 m [16–82 ft] thick) that consist of burrowed mudstone and silty sandstone with flaser and lenticular beds marked by tidal bundles. The tidal lithofacies grade west-southwestward into the coal-bearing, fluvial-deltaic Ludlow Member.

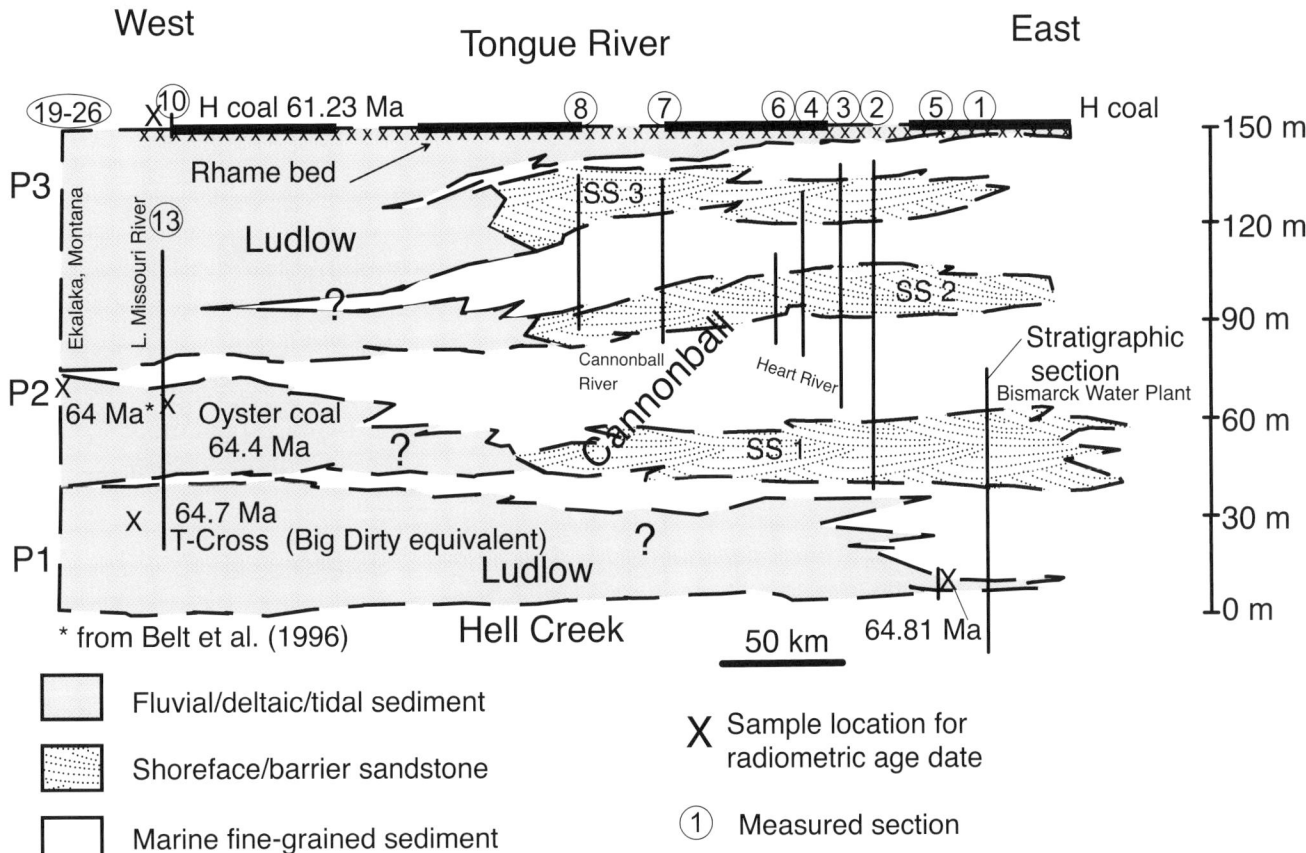

Figure 8. Generalized east-west cross section through the Ludlow and Cannonball Members of the Fort Union Formation, southwestern part of the Williston Basin. Vertical lines indicate the measured stratigraphic sections. The stratigraphic sections are positioned with reference to their elevation and distance below the laterally extensive Rhame bed. Refer to Table 1 for section location details.

The uppermost brackish-marine tongue of the Cannonball Member (the Oyster tongue) found along the Little Missouri River in southwest North Dakota may have encroached into the Cave Hills area (northwest South Dakota) and the Ekalaka area (southeast Montana). This regional correlation marking a maximum highstand of the Cannonball Sea is supported by the biozone P2–P3 ages for the rocks in the marine- and brackish-water-influenced lithofacies. Radiometric ages (Figure 2; Table 2) indicate that the maximum highstand of the Cannonball Sea probably occurred in about a 1-m.y. interval between 65 and 64(?) Ma. The youngest radiometric date from below the zone of brackish-water influence in southeastern Montana (Figure 6) is 64.03 ± 0.19 Ma (Belt et al., 1996b). The Cannonball maximum flooding event corresponds to the lowermost third-order sea level cycle of the Tejas A Super Cycle of the lower Tertiary described by Haq et al. (1987, 1988) and closely follows the European Paleocene sea level curves of Neal and Hardenbol (1998) (Figure 2).

Temporal Depositional Modeling

The lower Paleocene marine parasequence sets can be correlated across southwestern North Dakota by biostratigraphic zonations and by their stratigraphic and elevational position relative to the siliceous Rhame bed paleosol, which occurs between the Ludlow Member and the overlying Tongue River Member (Figure 9). The lower parasequence sets (SS 1 and SS 2; Figure 8) were correlated biostratigraphically and radiometrically with tidal, distributary, and fluvial channel complexes in the Cave Hills, northwestern South Dakota, and the Ekalaka area of southeastern Montana. This complex is interpreted as reworked sediments of the Late Cretaceous "Sheridan delta of the Hell Creek Formation" (Cherven and Jacob, 1985), resulting from wave and tidal processes related to Cretaceous–Paleocene transgressive paleoshoreline turnaround and regression (Figure 9). The stacked parasequence sets consist of shoreface sandstones (with ravinement lag deposits) deposited by barrier-strand-plain systems. Landward of the barrier

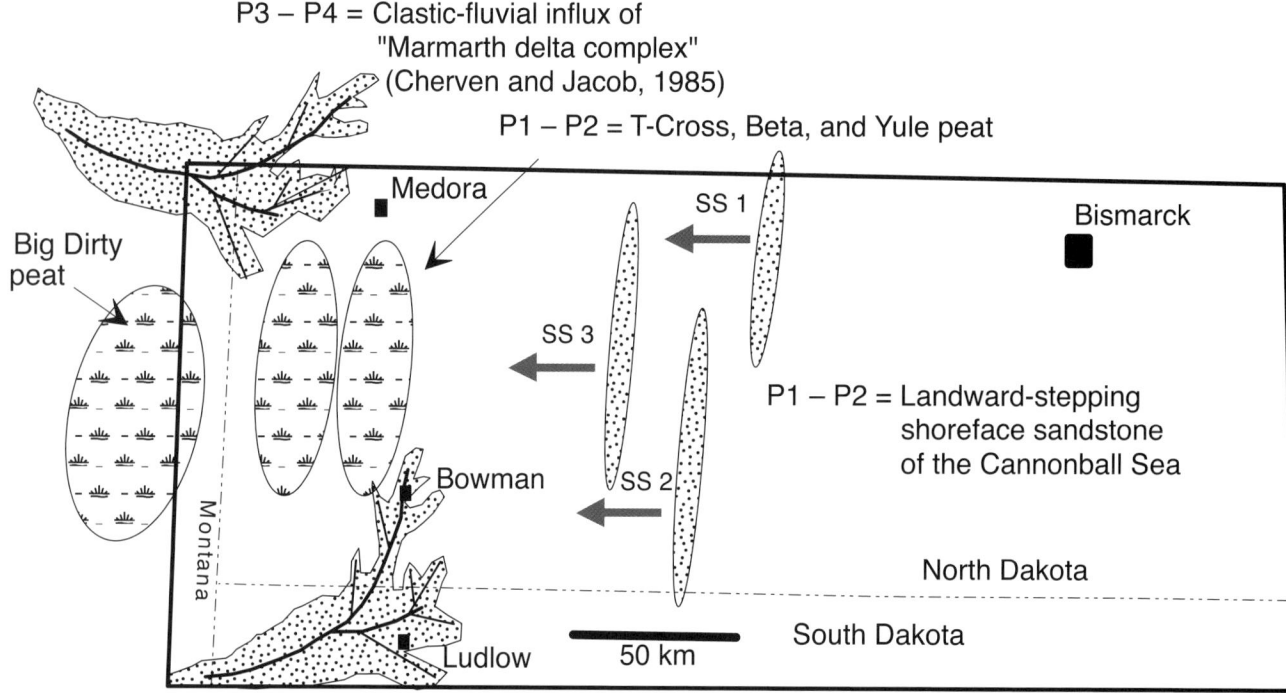

Figure 9. Generalized paleogeographic map of the study area showing zone P1–P4 (Figure 2) time slices based on palynostratigraphic biozonation and radiometric age determinations. Sediments of the landward-stepping shoreface deposits (SS 1, SS 2, and SS 3) of the Cannonball Sea were provided by reworking of the "Sheridan delta." This led to stacking of coarsening-upward sandstones along the coastal strandline during transgression maximum (P1–P2) and primary accumulation of economic coal beds (Big Dirty, T-Cross, etc.) in mires during periods of stagnation and slowly shifting shorelines.

systems, tidal-estuarine and mire deposits accumulated thin to thick laterally discontinuous peat beds during periods of stagnation and slow shoreline shifts (e.g., Beta and Yule coal beds in southwestern North Dakota; Figure 2). Landward in the coastal plain, the T-Cross coal bed and the laterally equivalent Big Dirty coal zone (dated at 64.78 ± 0.21 Ma) in southeastern Montana formed as thick, laterally extensive peat in fluvial mires. In addition, a 64.03 ± 0.19-Ma age for coal beds in the Ludlow Member near Ekalaka, Montana (Belt et al., 1996b), helps to constrain the relative ages of marine-influenced correlative units in the western part of the Williston Basin (Figure 2).

^{40}Ar/^{39}Ar ages obtained from two tonsteins, one layer located 2 m (7 ft) above the Oyster coal bed (middle of the Ludlow–Cannonball interval; 64.4 ± 1.8 Ma) and the other layer in the H coal bed (overlying the Rhame bed; 61.23 ± 0.38 Ma), indicate that the upper Cannonball parasequence sets accumulated during a period that lasted approximately 3 m.y., roughly between 64 and 61 Ma (Figure 2). These data

indicate that the last remnants of the once widespread intercratonic Cretaceous seaway underwent frequent base-level fluctuations prior to its final departure from the northern Rocky Mountain–Great Plains foreland during the early Paleocene (biozone P2 or about 64 Ma). This lowstand of sea level was marked by incised paleovalley infill by northeast-flowing rivers during the early to mid-Paleocene (biozone P3 or 64–61 Ma).

Mid-Paleocene (biozone P3–P4) strata consist of a regressive shallow-marine and lower delta-tidal system that thins toward the east ("Marmarth delta" of Cherven and Jacob, 1985) and an overlying lignite-bearing fluvial facies (Tongue River Member, Fort Union Formation) containing a laterally persistent paleosol (Rhame bed) and overlying Harmon-Hanson coal zone (61.23 ± 0.38 Ma). Biozone P5 is represented by fluvial-deltaic coal-bearing strata that contain several laterally persistent coal beds (HT Butte, Hagel, and Beulah-Zap coal zones, Sentinel Butte Member, Fort Union Formation) (Flores and Keighin, 1999).

CONCLUSIONS

1) Time-stratigraphic slices through the stratigraphic section of the Fort Union Formation (Paleocene) of the Williston Basin have been established using palynostratigraphic and radiometric dates.

2) At least three landward-stepping progradational parasequence sets can be identified in the Cannonball Member of the Fort Union Formation. These coarsening-upward sandstones probably were derived from reworked Cretaceous–lower Tertiary deltaic deposits of the "Sheridan delta" (Cherven and Jacob, 1985) (Figure 9).

3) Economically minable coal deposits (T-Cross and Big Dirty coal zones) in the Ludlow Member (North and South Dakota) and the equivalent Tullock Member of the Fort Union Formation in the adjacent Powder River Basin were deposited landward of the strand-plain/barrier systems.

4) Radiometric dating pinpoints the Cannonball Sea's maximum highstand to be limited to a 1-m.y. period from 65 to 64 Ma (Figure 2).

5) Limited evidence of brackish-marine influence in the Ludlow Member of the Fort Union Formation was found in the Ekalaka area of southeast Montana. These facies are characterized by an amalgamated fluvial-deltaic sandstone complex with laterally adjacent tidally influenced flood-plain and mire deposits. These deposits represent the westernmost extent (maximum flooding) of the Cannonball Sea prior to its final retreat.

6) The retreat of the Cannonball Sea is marked by a lowstand paleovalley incision in the Ekalaka area, which was infilled by northeast-flowing rivers that originated from the Powder River Basin (Flores, 1983; Flores and Bader, 1999). Regional correlations based on the Rhame bed as a datum indicate that the Cannonball Sea retreated from the western part of the Williston Basin within a 3-m.y. period by 61 Ma (Figures 2, 5).

7) Mid-Paleocene (biozones P3–P4) strata consist of regressive shallow-marine and lower delta-tidal systems, which thin toward the east ("Marmarth delta" of Cherven and Jacob, 1985) and an overlying lignite-bearing fluvial facies (Tongue River Member of the Fort Union Formation) containing the economically minable Harmon-Hanson coal zone (61.23 ± 0.38 Ma) (Figure 2) (Flores et al., 1999a, b).

8) Biozone P5 is represented in fluvial coal-bearing strata of the upper Tongue River and Sentinel Butte Members of the Fort Union Formation.

These strata contain several economically minable coal beds (HT Butte, Hagel, and Beulah-Zap zones; Flores et al., 1999a, b).

ACKNOWLEDGMENTS

We are indebted to the many property owners in the study area for granting us access during this study. We thank C. W. Keighin, A. M. Ochs, and J. D. Obradovich of the U.S. Geological Survey for their help in sample collection, preparation, and analysis. We also thank Lisa Peters and Matthew Heizler of the New Mexico Bureau of Geology & Mineral Resources, New Mexico Geochronological Research Laboratory, for their help in tonstein preparation and analysis. We appreciate the thoughtful comments of R. A. Gastaldo and two anonymous reviewers.

REFERENCES CITED

Archibald, J. D., P. D. Gingerich, E. H. Lindsay, W. A. Clemens, D. W. Krause, and K. D. Rose, 1982, First North American land mammal ages of the Cenozoic era, in M. O. Woodburne, ed., Cenozoic mammals of North America: Berkeley, California, University of California Press, p. 24–76.

Belt, E. S., R. M. Flores, P. D. Warwick, K. M. Conway, K. R. Johnson, and R. S. Waskowitz, 1984, Relationship of fluvio-deltaic facies to coal deposition in the lower Fort Union Formation (Paleocene), southwestern North Dakota, in R. A. Rahmani and R. M. Flores, eds., Sedimentology of coal and coal-bearing sequences: International Association of Sedimentologists Special Publication 7, p. 177–198.

Belt, E. S., S. E. H. Sakimoto, and B. W. Rockwell, 1992, A drainage-diversion hypothesis for the origin of widespread coal beds in the Williston Basin: Examples from Paleocene strata, eastern Montana, in M. A. Sholes, ed., Coal geology of Montana: Montana Bureau of Mines and Geology Special Publication 102, p. 21–60.

Belt, E. S., E. C. Beutner, and W. Coppinger, 1993, Slump structures below a major unconformity within the Paleocene Tongue River Member, and the Miocene unconformity above the Paleocene, southeastern Montana (abs.): The Sixth Keck Research Symposium in Geology, Abstract Volume, Whitman College, Walla Walla, Washington, p. 84–86.

Belt, E. S., J. A. Diemer, and E. C. Beutner, 1996a, Newly discovered estuarine facies in the Fort Union Formation (Paleocene), southeastern Montana: Geological Society of America Abstracts with Programs, v. 28, no. 4, p. 2.

Belt, E. S., S. M. Vuke, J. A. Diemer, and E. C. Beutner, 1996b, Ekalaka Member (new name) of the Fort Union Formation (Paleocene) southeastern Montana: Geological Society of America Abstracts with Programs, v. 28, no. 4, p. 2.

Belt, E. S., J. A. Diemer, and E. C. Beutner, 1997a, Marine ichnogenera within Torrejonian (Paleocene) of the Fort Union Formation, southeastern Montana, in J. H. Hartman, ed., Paleontology and geology in the northern Great Plains: University of Wyoming Contributions to Geology, v. 32, no. 1, p. 3–18.

Belt, E. S., J. F. Hicks, and D. A. Murphy, 1997b, A pre-Lancian regional unconformity and its relationship to Hell Creek paleogeography in southeastern Montana: University of Wyoming Contributions to Geology, v. 31, no. 2, p. 1–26.

Belt, E. S., J. A. Diemer, S. M. Vuke, E. C. Beutnerand, and B. S. Cole, 2002, The Ekalaka Member of the Fort Union Formation, southeastern Montana: Designating a new member and making a case for estuarine deposition and bounding unconformities: Montana Bureau of Mines and Geology Open-file Report 461, 56 p.

Berggren, W. A., D. V. Kent, C. C. Swisher III, and M.-P. Aubry, 1995, A revised Cenozoic geochronology and chronostratigraphy, in W. A. Berggren, D. V. Kent, M.-P. Aubry, and Jan Hardenbol, eds., Geochronology time scales and global stratigraphic correlation: SEPM Special Publication 54, p. 129–212.

Best, W. A., 1987, A sedimentologic and stratigraphic study of the Paleocene Fort Union Formation in the South Cave Hills of Harding County, South Dakota: M.S. thesis, South Dakota School of Mines and Technology, Rapid City, South Dakota, 124 p.

Bluemle, J. P., S. B. Anderson, J. A. Andrew, D. W. Fischer, and J. A. LeFever, 1986, North Dakota stratigraphic column: North Dakota Geological Survey, Miscellaneous Series no. 66, Plate 1, no scale.

Bohor, B. F., and D. M. Triplehorn, 1993, Tonsteins: Altered volcanic-ash layers in coal-bearing sequences: Geological Society of America Special Paper 285, 44 p.

Brown, R. W., 1962, Paleocene flora of the Rocky Mountains and Great Plains: U.S. Geological Survey Professional Paper 375, 119 p.

Calvert, W. R., 1912, Geology of certain lignite fields in eastern Montana: U.S. Geological Survey Bulletin, v. 471, p. 187–201.

Cherven, V. B., and A. F. Jacob, 1985, Evolution of Paleogene depositional systems, Williston Basin, in response to global sea level changes, in R. M. Flores and S. S. Kaplan, eds., Cenozoic paleography of the west-central United States: Denver, Colorado, Rocky Mountain Section SEPM, p. 127–170.

Christensen, K. C., 1984, The stratigraphy and petrography of a light-colored siliceous horizon within the Fort Union Formation (Paleocene), southeastern Montana: M.S. thesis, Montana College of Mineral Science and Technology, Butte, Montana, 183 p.

Clayton, L., C. G. Carlson, W. L. Moore, G. Groenewold, F. D. Holland Jr., and S. R. Moran, 1977, The Slope (Paleocene) and Bullion Creek (Paleocene) Formations of North Dakota: North Dakota Geological Survey Report of Investigation 59, 14 p.

Clement, J. H., 1986, Cedar Creek: A significant paleotectonic feature of the Williston Basin: AAPG Memoir 41, p. 213–240.

Connor, C. W., 1984, Ash-fall sequences in a Paleocene coal-potential indicator of synchroneity between Montana and Wyoming basins, in R. L. Houghton and E. N. Clausen, eds., 1984 Symposium on the Geology of Rocky Mountain Coal: North Dakota Geological Society Publication 84-1, p. 137–151.

Connor, C. W., 1985, Sixty-five volcanic events recorded in a single coal bed: AAPG Bulletin, v. 69, p. 246.

Cvancara, A. M., 1965, Bivalves and biostratigraphy of the Cannonball Formation (Paleocene) in North Dakota: Ph.D. dissertation, University of Michigan, Ann Arbor, Michigan, 470 p.

Cvancara, A. M., 1966, Revision of the fauna of the Cannonball Formation (Paleocene) of North and South Dakota: Part 1. Bivalvia: Contributions from the Museum of Paleontology, University of Michigan, v. 20, no. 10, p. 277–374.

Cvancara, A. M., 1972, Summary of the Cannonball Formation (Paleocene) in North Dakota: North Dakota Geological Survey Miscellaneous Series, v. 50, no. 3, p. 69–75.

Cvancara, A. M., 1976, Geology of the Cannonball Formation (Paleocene) in the Williston Basin, with reference to uranium potential: North Dakota Geological Survey Report of Investigation 57, 22 p.

Cvancara, A. M., 1980, Bench-forming sandstone as a correlation tool; Cannonball Formation (Paleocene), North Dakota, in A. W. Johnson, ed., Proceedings of the 72nd Annual Meeting of the North Dakota Academy of Science, Fargo, North Dakota, April 25–26, 1980: Proceedings of the North Dakota Academy of Science, v. 34, p. 34.

Cvancara, A. M., and J. W. Hoganson, 1993, Vertebrates of the Cannonball Formation (Paleocene) in North and South Dakota: Journal of Vertebrate Paleontology, v. 13, p. 1–23.

Dalrymple, R. W., 1992, Tidal depositional systems, in R. G. Walker and N. P. James, eds., Facies models: Response to sea level change: Geological Association of Canada, p. 195–218.

Dalrymple, G. B., E. C. Alexander Jr., M. A. Lanphere, and G. P Kraker, 1981, Irradiation of samples for $^{40}Ar/^{39}Ar$ dating using the Geological Survey TRIGA reactor: U.S. Geological Survey Professional Paper 1176, p. 55.

Daly, D. J., G. H. Groenewold, and C. R. Schmit, 1985, Paleoenvironments of the Paleocene Sentinel Butte Formation, Knife River area, west-central North Dakota, in R. M. Flores and S. S. Kaplan, eds., Cenozoic paleogeography of the west-central United States: Rocky Mountain Section SEPM, Rocky Mountain Paleogeography Symposium, v. 3, p. 171–185.

Demchuk, T. D., 1990, Palynostratigraphic zonation of Paleocene strata in the central and south-central Alberta plains: Canadian Journal of Earth Sciences, v. 27, p. 1263–1269.

Diemer, J. A., E. S. Belt, and J. H. Hartman, 1996, Base level changes as a consequence of tectonic, eustatic and autogenic processes in Late Cretaceous and Paleocene strata, western Williston Basin: Geological Society of America Abstracts with Programs, v. 28, no. 7, p. 141.

Fenner, W. E., 1976, Foraminiferids of the Cannonball Formation (Paleocene, Danian) in western North Dakota: Ph.D. dissertation, University of North Dakota, Grand Forks, North Dakota, 216 p.

Flores, R. M., 1983, Basin facies analysis of coal-rich Tertiary fluvial deposits in the northern Powder River Basin, Montana and Wyoming, in J. D. Collinson and J. Lewin, eds., Modern and ancient fluvial systems: International Association of Sedimentologists Special Publication 6, p. 501–515.

Flores, R. M., and L. R. Bader, 1999, Fort Union coal in the Powder River Basin, Wyoming and Montana: A synthesis, in Fort Union Coal Assessment Team, 1999 Resource assessment of selected Tertiary coal beds and zones in the northern Rocky Mountains and Great Plains region: U.S. Geological Survey Professional Paper 1625-A, chapter PS, disc 1, v. 1.0, p. PS-1–PS-71.

Flores, R. M., and C. W. Keighin, 1999, Fort Union in the Williston Basin, North Dakota: A synthesis, in Fort Union Coal Assessment Team, 1999 Resource assessment of selected Tertiary coal beds and zones in the northern Rocky Mountains and Great Plains region: U.S. Geological Survey Professional Paper 1625-A, chapter WS, disc 1, v. 1.0, p. WS-1–WS-41.

Flores, R. M., C. W. Keighin, A. M. Ochs, P. D. Warwick, L. R. Bader, and E. C. Murphy, 1999a, Framework geology of the Fort Union coal in the Williston Basin, North Dakota, in Fort Union Coal Assessment Team, 1999 Resource assessment of selected Tertiary coal beds and zones in the northern Rocky Mountains and Great Plains region: U.S. Geological Survey Professional Paper 1625-A, chapter WF, disc 1, v. 1.0, p. WF-1–WF-64.

Flores, R. M., A. M. Ochs, G. D., Stricker, S. B. Roberts, M. E. Ellis, C. W. Keighin, E. C. Murphy, V. V. Cavaroc Jr., R. C. Johnson, and E. M. Wilde, 1999b, National coal resource assessment non-proprietary data: Location, stratigraphy, and coal quality of selected Tertiary coals in the northern Rocky Mountains and Great Plains region: U.S. Geological Survey Open-file Report 99-376, 12 p., 6 figures, 23 spreadsheets.

Franczyk, K. J., J. K. Pitman, and D. J. Nichols, 1990, Sedimentology, mineralogy, palynology, and depositional history of some uppermost Cretaceous and lowermost Tertiary rocks along the Utah Book and Roan Cliffs east of the Green River: U.S. Geological Survey Bulletin, v. 1787-N, 27 p.

Goodrum, C. K., 1983, A paleoenvironmental and stratigraphic study of the Paleocene Fort Union Formation in the Cave Hills area of Harding County, South Dakota: M.S. thesis, South Dakota School of Mines and Technology, Rapid City, South Dakota, 142 p.

Haq, B. U., J. Hardenbol, and P. R Vail, 1987, The new chronostratigraphic basis of Cenozoic and Mesozoic sea level cycles, in C. A. Ross and D. Haman, eds., Timing and depositional history of eustatic sequences; constraints on seismic stratigraphy: Cushman Foundation Special Publication 24, p. 7–13.

Haq, B. U., J. Hardenbol, and P. R Vail, 1988, Mesozoic and Cenozoic chronostratigraphy and cycles of relative sea level change, in C. K. Wilgus, B. S. Hastings, G. C. St. C., Kendall, H. W. Posamentier, C. A. Ross, and J. C. Van Wagoner, eds., Sea-level changes: An integrated approach: SEPM Special Publication 42, p. 71–108.

Hares, C. J., 1928, Geology and lignite resources of the Marmarth field, southwestern North Dakota: U.S. Geological Bulletin, v. 775, 110 p.

Hartman, J. H., 1989, The T-Cross coal bed (Paleocene, North Dakota): The importance of reevaluating historic data in geologic research: Proceedings of the North Dakota Academy of Science, v. 43, p. 49.

Hartman, J. H., and A. J. Kihm, 1991, Stratigraphic distribution of the Titanoides (Mammalia: Pantodonta) in the Fort Union Group (Paleocene) of North Dakota, in J. E. Christopher and F. M. Haidl, eds., Proceedings, Sixth International Williston Basin Symposium: Saskatchewan Geological Society Special Publication 11, p. 207–215.

Hartman, J. H., and A. J. Kihm, 1996, Bio- and magnetostratigraphy of the uppermost Cretaceous and lower Tertiary strata of North Dakota, in Sixth North American Paleontological Convention Abstracts of Papers: The Paleontological Society Special Publication 8, p. 163.

Hartman, J. H., K. R. Johnson, and D. J. Nichols, eds., 2002, The Hell Creek Formation and the Cretaceous–Tertiary boundary in the northern Great Plains— An integrated continental record of the end of the Cretaceous: Geological Society of America Special Paper 361, 520 p.

Hettinger, R. D., J. G. Honey, and D. J. Nichols, 1991, Chart showing correlations of Upper Cretaceous Fox Hills Sandstone and Lance Formation, and lower Tertiary Fort Union, Wasatch, and Green River Formations, from the eastern flank of the Washakie Basin to the southeastern part of the Great Divide Basin, Wyoming: U.S. Geological Survey Miscellaneous Field Investigations Map I-2151, 1 sheet, no scale.

Hickey, L. O., 1984, Changes in the angiosperm flora across the Cretaceous–Tertiary boundary, in W. A. Berggren and J. A. Van Couvering, eds., Catastrophes and Earth history: Princeton, New Jersey, Princeton University Press, p. 279–314.

Holtzman, R. C., 1978, Late Paleocene mammals of the Tongue River Formation, western North Dakota: North Dakota Geological Survey Report of Investigation 65, 88 p.

Honey, J. G., and R. D. Hettinger, 1989, Cross section showing correlations of Upper Cretaceous Fox Hills Sandstone and Lance Formation, and lower Tertiary Fort Union and Wasatch Formations, southeastern

Washakie Basin, Wyoming, and eastern Sand Wash Basin, Colorado: U.S. Geological Survey Miscellaneous Investigations Series Map I-1964, 1 sheet, no scale.

Jacobson, S. R., and D. J. Nichols, 1982, Palynological dating of syntectonic units in the Utah–Wyoming thrust belt: The Evanston Formation, Echo Canyon conglomerate, and Little Muddy Creek conglomerate, *in* R. B. Powers, ed., Geologic studies of the Cordilleran thrust belt: Denver, Colorado, Rocky Mountain Association of Geologists, v. 2, p. 735–750.

Johnson, K. R., and L. J. Hickey, 1990, Megafloral change across the Cretaceous/Tertiary boundary in the northern Great Plains and Rocky Mountains, U.S.A., *in* V. L. Sharpton and P. D. Ward, eds., Global catastrophes in Earth history: An interdisciplinary conference on impacts, volcanism, and mass mortality: Geological Society of America Paper 2476, p. 433–444.

Keefer, W. R., 1974, Regional topography, physiography, and geology of the northern Great Plains/prepared for the Northern Great Plains Resources Program: U.S. Geological Survey Open-file Report 74-50, 20 p.

Kroeger, T. J., 1995, The paleoecologic significance of Paleocene palynomorph assemblages from the Ludlow, Slope, and Cannonball Formations, southwestern North Dakota: Ph.D. dissertation, University of North Dakota, Grand Forks, North Dakota, 389 p.

Kroeger, T. J., and J. H. Hartman, 1997, Paleoenvironmental distribution of Paleocene palynomorph assemblages from brackish water deposits in the Ludlow, Slope, and Cannonball Formations, southwestern North Dakota, *in* J. H. Hartman, ed., Paleontology and geology in the northern Great Plains: University of Wyoming Contributions to Geology, v. 32, no. 1, p. 115–129.

Leonard, A. G., 1908, The geology of southwestern North Dakota with special reference to the coal: North Dakota Geological Survey 5th Biennial Report, p. 27–114.

Lindholm, R. M., 1984, Bivalve associations of the Cannonball Formation (Paleocene, Danian) of North Dakota: M.S. thesis, University of North Dakota, Grand Forks, North Dakota, 184 p.

Marvin, R. F., B. C. Hearn Jr., H. H. Mehnert, C. W. Naeser, R. E. Zartman, and D. A. Lindsey, 1980, Late Cretaceous–Paleocene–Eocene igneous activity in north-central Montana: Isochron/West, no. 29, p. 5–25.

McDougall, I., and T. M. Harrison, 1988, Geochronology and thermochronology by the ^{40}Ar-^{39}Ar method: Oxford Monographs on Geology and Geophysics, v. 9, 212 p.

Meek, F. B., and F. V. Hayden, 1962, Descriptions of new lower Silurian (primordial), Jurassic, Cretaceous, and Tertiary fossils collected in Nebraska territory, with some remarks on the rocks from which they were obtained: Philadelphia Academy of Natural Sciences 1861, v. 13, p. 415–435.

Monnens, L. E., 1980, The stratigraphy of the Cenozoic deposits in the Cave Hills, northwestern South Dakota: M.S. thesis, Iowa State University, Ames, Iowa, 102 p.

Moore, W. L., 1976, The stratigraphy and environments of deposition of the Cretaceous Hell Creek Formation (reconnaissance) and the Paleocene Ludlow Formation (detailed), southwestern North Dakota: North Dakota Geological Survey Report of Investigation 56, 40 p.

Murphy, E. C., J. W. Hoganson, and N. F. Forsman, 1993, The Chadron, Brule, and Arikaree Formations in North Dakota— The Buttes of southwestern North Dakota: North Dakota Geological Survey Report of Investigation 96, 144 p.

Murphy, E. C., D. J. Nichols, J. W. Hoganson, and N. F. Forsman, 1995, The Cretaceous/Tertiary boundary in south-central North Dakota: North Dakota Geological Survey Report of Investigation 98, 74 p.

Neal, J. E., and J. Hardenbol, 1998, Introduction to the Paleocene, *in* P.-C. de Graciansky, J. Hardenbol, T. Jacquin, and P. R Vail, eds., Mesozoic and Cenozoic sequence stratigraphy of European basins: SEPM Special Publication 60, p. 87–90.

Nichols, D. J., 1994, Palynostratigraphic correlation of Paleocene rocks in the Wind River, Bighorn, and Powder River Basins, Wyoming, *in* R. M. Flores, K. T. Mehring, R. W. Jones, and T. L. Beck, eds., Organics and the Rockies field guide: Wyoming State Geological Survey Public Information Circular 33, p. 17–29.

Nichols, D. J., 1998, Palynological age determinations of selected outcrop samples from the Lance and Fort Union Formations in the Bighorn Basin, Montana and Wyoming, *in* W. R. Keefer and J. E. Goolsby, eds., Cretaceous and lower Tertiary rocks of the Bighorn Basin, Wyoming and Montana: Wyoming Geological Association, Forty-Ninth Guidebook, p. 117–129.

Nichols, D. J., 1999, Stratigraphic palynology of the Fort Union Formation (Paleocene) in the Powder River Basin, Montana and Wyoming— A guide to correlation of methane-producing coal zones, *in* W. R. Miller, ed., Coalbed methane and the Tertiary geology of the Powder River Basin, Wyoming and Montana: Wyoming Geological Association Fifteenth Field Conference Guidebook 1999, p. 25–41.

Nichols, D. J., 2003, Palynostratigraphic framework for age determination and correlation of the nonmarine lower Cenozoic of the Rocky Mountains and Great Plains region, *in* R. G. Raynolds and R. M. Flores, eds., Cenozoic Systems of the Rocky Mountain Region: Denver, Colorado, Rocky Mountain Section of the Society for Sedimentary Geology (SEPM), p. 107–134.

Nichols, D. J., and H. L. Ott, 1978, Biostratigraphy and evolution of the *Momipites-Caryapollenites* lineage in the early Tertiary in the Wind River Basin, Wyoming: Palynology, v. 2, p. 93–112.

Nichols, D. J., and R. M. Flores, 1993, Palynostratigraphic correlation of the Fort Union Formation (Paleocene) in the Wind River Reservation and Waltman area, Wind River Basin, Wyoming, *in* W. R. Keefer, W. J. Metzger, and L. H. Godwin, eds., Oil and gas resources of the Wind River Basin, Wyoming: Wyoming Geological Association Special Symposium, p. 175–189.

Pipiringos, G. N., W. A. Chisholm, and R. C. Kepferle, 1965, Geology and uranium deposits in the Cave Hills area, Harding County, South Dakota: U.S. Geological Survey Professional paper 476-A, 64 p.

Rehbein, E. A., 1977, Preliminary report on stratigraphy, depositional environments, and lignite resources in the Fort Union Formation, west-central North Dakota: U.S. Geological Survey Open-file Report 77-69, 23 p.

Reinson, G. E., 1992, Transgressive barrier island and estuarine systems, *in* R. G. Walker and N. P. James, eds., Facies models: Response to sea level change: Geological Association of Canada, p. 179–194.

Rich, F. J., and C. K. Goodrum, 1982, Paleoecology and sedimentology of the Fort Union Formation, Harding County, South Dakota, *in* Proceedings, Fifth Symposium on the Geology of Rocky Mountain Coal: Utah Geological and Mineral Survey Bulletin, v. 118, p. 158–162.

Roberts, S. B., 1998, An overview of the stratigraphic and sedimentologic characteristics of the Paleocene Fort Union Formation, southern Bighorn Basin, Wyoming, *in* W. R. Keefer and J. E. Goolsby, eds., Cretaceous and lower Tertiary rocks of the Bighorn Basin, Wyoming and Montana: Wyoming Geological Association, Forty-Ninth Guidebook, p. 91–116.

Roberts, S. B., and R. W. Stanton, 1994, Stratigraphy and depositional setting of thick coal beds in the Grass Creek coal mine, southwestern Bighorn Basin, Wyoming, *in* R. M. Flores, K. T. Mehring, R. W. Jones, and T. L. Beck, eds., Organics and the Rockies field guide: Wyoming State Geological Survey Public Information Circular 33, p. 125–138.

Samson, S. D., and E. C. Alexander Jr., 1987, Calibration of the interlaboratory $^{40}Ar/^{39}Ar$ dating standard, MMhb-1: Chemical geology; Isotope Geoscience Section, v. 66, nos. 1–2, p. 27–34.

Steiger, R. H., and E. Jager, 1977, Subcommission on geochronology: Convention on the use of decay constants in geo- and cosmochronology: Earth and Planetary Science Letters, v. 36, p. 359–362.

Swanson, R. W., M. L. Hurburt, G. W. Luttrell, and V. M. Jussen, 1981, Geologic names of the United States

through 1975: U.S. Geological Survey Bulletin, v. 1535, 643 p.

Swisher, C. C., L. Dingus, and R. F. Butler, 1993, $^{40}Ar/^{39}Ar$ dating and magnetostratigraphic correlation of the terrestrial Cretaceous–Paleogene boundary and Puercan Mammal age, Hell Creek–Tullock Formations, eastern Montana: Canadian Journal of Earth Sciences, v. 30, p. 1981–1996.

Thom, W. T., Jr., and C. E. Dobbin, 1924, Stratigraphy of Cretaceous–Eocene transition beds in eastern Montana and the Dakotas: Geological Society of America Bulletin, v. 35, p. 481–505.

Van Alstine, J. B., 1974, Paleontology of brackish-water faunas in two tongues of the Cannonball Formation (Paleocene, Danian), Slope and Golden Valley Counties, southwestern North Dakota: M.S. thesis, University of North Dakota, Grand Forks, North Dakota, 101 p.

Vuke, S. M., E. M. Wilde, R. N. Bergantino, and R. B. Colton, 2001, Geologic map of the Ekalaka 30′ × 60′ quadrangle, eastern Montana and adjacent North and South Dakota: Montana Bureau of Mines and Geology Open-file Report, MBMG 430, scale 1:100,000, 1 sheet.

Walker, R. G., and A. G. Plint, 1992, Wave- and storm-dominated shallow marine systems, *in* R. G. Walker and N. P. James, eds., Facies models: Response to sea level change: Geological Association of Canada, p. 219–238.

Warwick, P. D., 1982, The geology of some lignite-bearing fluvial deposits (Paleocene), south-western North Dakota: M.S. thesis, North Carolina State University, Raleigh, North Carolina, 116 p.

Warwick, P. D., and K. R. Luck, 1995, Stratigraphic sections of the lignite-bearing Tongue River Member, Fort Union Formation (Paleocene), southwestern North Dakota: U.S. Geological Survey Open-file Report 95-676, 39 p.

Warwick, P. D., R. M. Flores, D. J. Nichols, E. C. Murphy, and J. D. Obradovich, 1997, Fort Union chronostratigraphic and depositional sequences, Williston Basin, North Dakota, South Dakota, and Montana: Geological Society of America Abstracts with Programs, v. 29, no. 6, p. A-204.

Wehrfritz, B. D., 1978, The Rhame bed (Slope Formation, Paleocene), a silcrete and deep-weathering profile, in southwestern North Dakota: M.S. thesis, University of North Dakota, Grand Forks, North Dakota, 158 p.

Holz, M., and W. Kalkreuth, 2004, Sequence stratigraphy and coal
petrology applied to the Early Permian coal-bearing Rio Bonito Formation,
Paraná Basin, Brazil, *in* J. C. Pashin and R. A. Gastaldo, eds., Sequence
stratigraphy, paleoclimate, and tectonics of coal-bearing strata: AAPG
Studies in Geology 51, p. 147–167.

7

Sequence Stratigraphy and Coal Petrology Applied to the Early Permian Coal-bearing Rio Bonito Formation, Paraná Basin, Brazil

Michael Holz
Instituto de Geociências, Universidade Federal do Rio Grande do Sul, Porto Alegre, Brazil

Wolfgang Kalkreuth
Instituto de Geociências, Universidade Federal do Rio Grande do Sul, Porto Alegre, Brazil

ABSTRACT

The coal-bearing Early Permian succession of the Paraná Basin in southernmost Brazil is linked to a third-order depositional sequence, where the most important coals occur in the initial transgressive systems tract. In the Candiota area, the main coal zone consists of 17 seams, which were analyzed for petrographic properties (macerals, gelification index, tissue-preservation index, vitrinite reflectance). These results are compared to the high-resolution sequence-stratigraphic framework to enhance our understanding of the stratigraphic controls on coal formation, coal distribution, and coal quality, providing guidelines for optimal exploitation.

The results show that local changes in accommodation trends and high sediment influx practically preclude coal formation in the lowstand and highstand systems tracts, whereas major coal development occurred in the transgressive systems tract. Seam distribution and thickness are controlled directly by flooding events, as depicted by the parasequences mapped in the study area. The main variations in thickness and extent occur at, or close to, the parasequence bounding surfaces. The most important coals, which are as much as 2.50 m in thickness, occur in the initial transgressive systems tract.

Coal petrographic parameters suggest an overall drying-upward trend in the coal seams developed in the upper part of third-order sequence 2, with significant differences of coal properties relative to their stratigraphic position between and within the parasequences. Detailed petrographic analysis of seam subsections indicates a transgressive nature for the thick coal seams occurring in parasequence 4 (CCI and CCS seams). These are characterized by decreased vitrinite reflectance at the base and top of the coal seams. The high inertinite content of the overlying BL seam at the top of PS 4 suggests accumulation of the precursor peat in a regressive phase of the parasequence.

INTRODUCTION

The southern region of Brazil (Figure 1), comprising the Paraná, Santa Catarina, and Rio Grande do Sul states, has been known for its abundant and economically important coal seams since the beginning of the 20th century (White, 1906). The southernmost state of Rio Grande do Sul contains the majority of the coal reserves, whereas Santa Catarina has the greatest annual coal production.

These coal reserves are assigned to the Rio Bonito Formation (Figure 2), a fluvial to marine sandstone- and shale-prone lithostratigraphic unit of Early Permian age (Artinskian–Kungurian). The coal beds have characteristics that are indicative of an origin in limnotelmatic mires, where arborescent and herbaceous plant material accumulated after some transport, promoting parautochthonous coal seams rich in inertinite (Correa da Silva, 1991). Coals in Rio Grande do Sul were deposited in a back-barrier depositional setting, an interpretation based on regional sequence-stratigraphic analysis (Holz, 1998) and on tissue-preservation and gelification indices derived from maceral analysis (Alves and Ade, 1996).

This chapter completes the preliminary results presented by Holz et al. (2002), focusing on conditions of coal formation in the Early Permian, and investigates petrographical and geochemical characters of

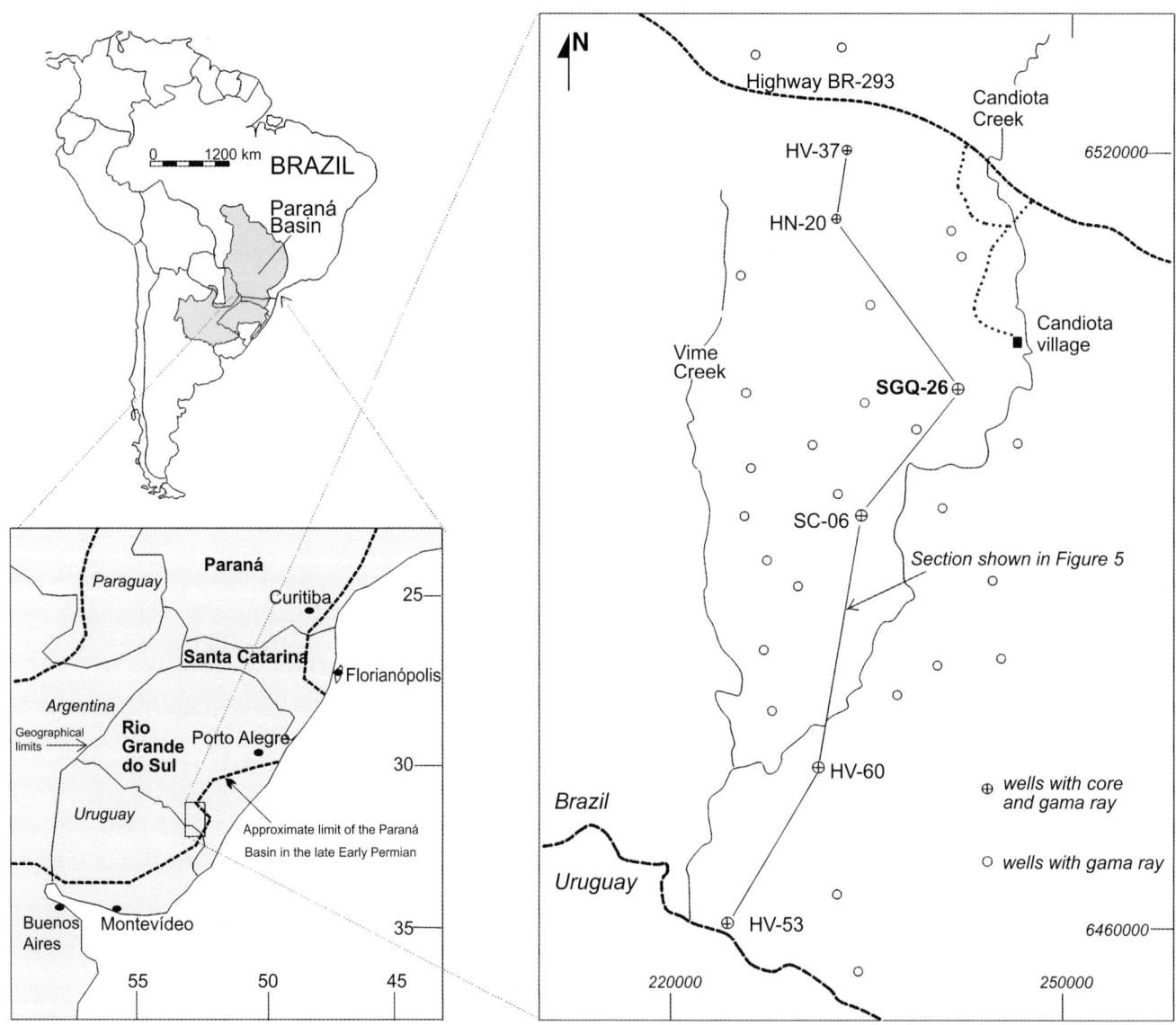

Figure 1. Map showing the location of the Paraná Basin, simplified geological setting, and distribution of boreholes investigated in this study. Note the location of borehole SGQ-26, for which sequence stratigraphy, coal petrological data, and chemical characteristics are detailed in Figures 9–11.

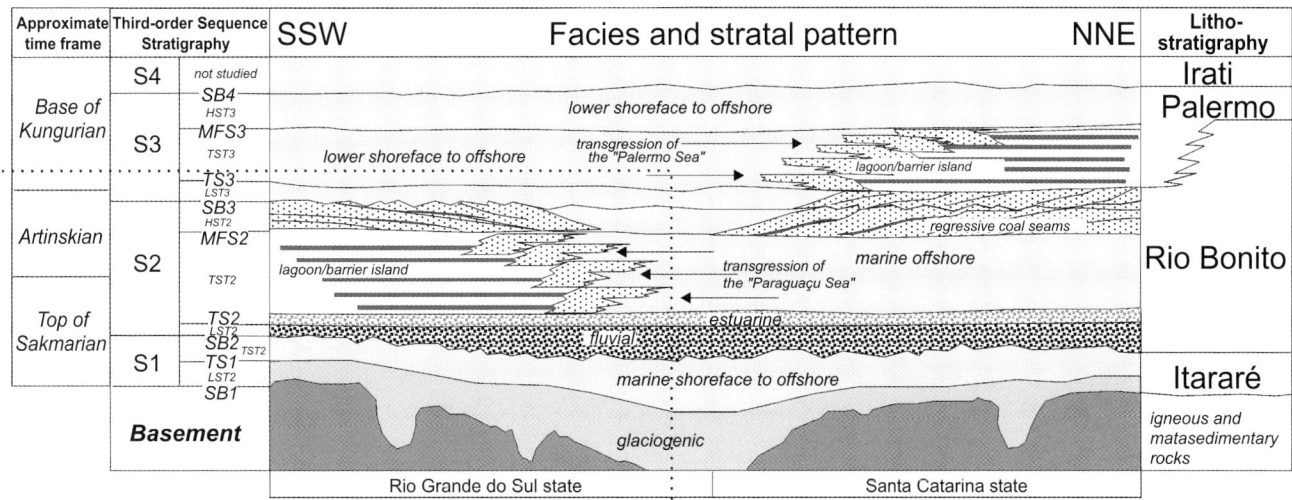

Figure 2. Stratigraphic overview of the Early Permian succession in southernmost Brazil, showing a north-northeast-south-southwest–oriented composite cross section representing the stratigraphic relationship between the coal-bearing strata in the Santa Catarina and the Rio Grande do Sul states (dotted rectangle indicates studied interval and study area). The interval records a second-order transgression with most coals developed in third-order sequences. The first barrier-lagoonal system developed along the paleoshoreline in Rio Grande do Sul state, and at that time, the Santa Catarina area was topographically low and characterized by a marine embayment. After a tectonic movement generating sequence boundary SB 3, Rio Grande do Sul state was flooded, whereas in Santa Catarina, a coal-forming coastal system developed (modified from Holz et al., 2000).

coal seams formed in transgressive and regressive depositional settings. This is accomplished by comparison between a fourth-order sequence-stratigraphic framework and the vertical variation of petrographic parameters, such as vitrinite content, gelification index, and tissue-preservation index. In addition, geochemical parameters, such as boron and sulfur content, have been used.

GEOGRAPHICAL AND GEOLOGICAL CHARACTERIZATION OF THE STUDY AREA

The study area is part of a tectonic unit in southwestern Gondwana known as the Paraná Basin, a large intracratonic basin (Milani et al., 1994). This basin is located at the east-central part of the South American continental platform and covers a surface area about 1,700,000 km² (656,374 mi²). The basin has a northeast-southwest–elongate shape and is approximately 1750 km (1087 mi) long and 900 km (559 mi) wide.

Sedimentation began after the Brasiliano cycle (the Brazilian tectonic equivalent of the formation of Gondwana during the Proterozoic), with plutonic events (610–580 Ma) accompanied by subsequent cooling and thermal subsidence in the Cambrian–Ordovician (500–450 Ma). The sedimentary fill of the basin was controlled from Ordovician to Cretaceous by Paleozoic orogenic movements (accretion of ter-

rains on the western margin of Gondwana) and climatic events (Early Permian deglaciation) and by the Mesozoic opening of the South Atlantic (e.g., Zálan et al., 1990; Milani et al., 1994). The prevalence of eustatic-tectonic cycles that controlled sedimentation in the Paraná Basin has generated a rock record that is marked by numerous stratigraphic gaps. Milani et al. (1994) subdivided the basin fill into six second-order depositional sequences (Ordovician–Silurian to Late Cretaceous). The coal-bearing Rio Bonito Formation (Figure 2), which is the focus of this study, is located at the base of the third sequence of Milani et al. (1994), namely, the Carboniferous–Lower Triassic sequence, which forms the thickest sedimentary sequence of the basin (as much as 2800 m [9186 ft]). The base of the Carboniferous sequence occurs only in the depocenter, specifically in Santa Catarina and Paraná states. During the Late Carboniferous and Early Permian, strata lapped onto the basin margins, as in Rio Grande do Sul, where the oldest rocks of this depositional sequence have a Sakmarian to Artinskian age.

During the Early Permian, the study area was located approximately 41° south (Smith et al., 1981). During summer in the southern hemisphere, a low-pressure system over central Africa and a contrasting high-pressure system over the Panthalassa Ocean created an atmospheric gradient that was responsible for westerly summer winds. These winds brought humidity to the Paraná Basin, as well as to the Kalahari and Karoo Basins of Africa (Holz, 1998). This seasonally humid climate sustained an abundant peat-forming vegetation that accumulated in the Early Permian of meridional Gondwana (e.g., the Rio Bonito Formation, Paraná Basin, and the Vryheid Formation, Karoo Basin).

The study area is in the southwestern region of the Rio Grande do Sul state (Figure 1) and covers about 2000 km^2 (772 mi^2). This area includes Brazil's most important coalfield, the Candiota coalfield, discovered in the 1970s by the Brazilian agency Companhia de Pesquisas de Recursos Minerais. This agency carried out an extensive regional research program to identify coal reserves and drilled more than 90 boreholes to estimate regional coal potential. The regional company Companhia Riograndense de Mineração, which operates the Candiota mine, provided access to new drilling data and cores. The well logs and cores from this exploration program and data obtained from outcrop locations constitute the basis for the stratigraphic and petrographic analysis of the current study.

FACIES AND DEPOSITIONAL SYSTEMS OF THE COAL-BEARING SUCCESSION

An overview of the general stratigraphy of the coal-bearing succession shows the entire Lower Permian interval in southernmost Brazil (Sakmarian to Kungurian–Ufimian) (Figure 2). This interval comprises the Itararé, Rio Bonito, Palermo, and Irati (base) lithostratigraphic units; it records a second-order transgressive cycle that began at the time of deposition of the topmost Itararé unit and has its maximum flooding surface in the Palermo Formation (e.g., Milani et al., 1994; Holz, 1999). This second-order cycle is punctuated by important third-order base-level falls, which are interpreted to have generated several third-order depositional sequences. The coal-bearing Rio Bonito Formation is linked to sequence 2 and the base of sequence 3 (Figure 2), and in the Rio Grande do Sul state, most of the coals occur in the transgressive systems tract of sequence 2 (Holz, 1998; Holz et al., 2000).

To understand the sequence stratigraphy of the coal-bearing interval and to correlate the coal parameters, a detailed stratigraphic framework was established. Fifteen lithofacies were recognized; the description and interpretation are provided in Table 1. The spatial relationship of these lithofacies permits the recognition of four main depositional systems—alluvial fan, delta, lagoonal estuary, and barrier shore-face—with several subsystems (Table 2). The coals are linked to swamps and marshes in the lagoonal estuary setting. A typical well log is shown in Figure 3 for facies illustration.

GENERAL SEQUENCE STRATIGRAPHY OF THE STUDY AREA

Concepts of sequence stratigraphy (Wilgus et al., 1988) focus on the interpretation of chronostratigraphic surfaces that represent events of rise or fall of base level. These surfaces serve as boundaries for system tracts, which are associations of genetically and spatially related depositional systems, and for depositional sequences, which are major stratigraphic units bounded by unconformities. Every systems tract has a well-defined stratigraphic position in the depositional sequence and is the result of a particular sedimentary regime, dictated by the combined influence of sea level fluctuation, or eustasy, and basin tectonics, or subsidence. Accordingly, stratigraphers primarily recognize systems tracts that develop during three distinct phases of relative sea level. The lowstand systems tract is characterized mainly by aggradation of sediment, although some progradation may occur. The transgressive systems tract is dominated by retrogradation, whereas the highstand systems tract is dominated by progradation (also, see discussions of regressive systems tracts, e.g., Hunt and Tucker, 1992; Helland-Hansen and Gjelberg, 1994; Plint, 1996).

During lowstand, progradation and aggradation of fluvial-deltaic and shoreface sediment is characteristic in nearshore and transitional settings on the continental shelves, whereas during transgression, there is overall retrogradation. Regrogradation is punctuated by a maximum flooding event in which the sea reaches its maximum extent, and almost all clastic sediment is trapped near the coastline. At this time, only fine-grained sediment is deposited over much of the basin floor, forming a thin layer of muddy sediment called a condensed section. The phase of maximum flooding is followed by times of stationary and regressive shorelines caused by a progradational regime during deposition of the highstand systems tract. Therefore, the events of rise and fall of base level and the subsequent conditions of sedimentation (aggradation, progradation, or retrogradation) are mapped and combined into a chronostratigraphic framework, which is the essence of sequence-stratigraphic analysis.

Sequence-stratigraphic analysis is based on the concept and recognition of parasequences, which are

Table 1. Lithofacies of the coal-bearing succession.

Identification	Lithology/Texture	Sedimentary Structures	Interpretation
Gm	matrix-supported gravel	none	debris flow
Gc	clast-supported gravel	normal grading, trough cross-bedding	migration of subaqueous dunes under very high flow energy
Gi	intraclastic conglomerate	none	reworking of exposed sediments with some lithification
Sb	very fine sandstone	bioturbation (*Diplocraterion*–?)	migration of subaqueous ripples and dunes under low flow energy, low sedimentation rate
St	fine to coarse sandstone	trough and planar cross-bedding	migration of subaqueous ripples and dunes under low to high flow energy
Sf	fine to medium sandstone, sometimes coarse-grained	flaser bedding, sometimes with double mud drapes	alternation of ripple migration and settling of suspended sediments
Sw	fine to medium sandstone	combined oscillation and current wavy bedding (including small-scale hummocky cross-stratification), some planolites and skolithos	alternating flow energy under wave influence
Ss	fine to medium sandstone	swash cross-stratification	deposition under high-energy condition (upper flow regime)
Shcs	fine sandstone	hummocky and swaley cross-bedding, sometimes heavily sideritized	settling of suspended sediments under high-energy wave influence (storm waves), i.e., proximal tempestite deposition
SMhcs	fine sandstone interlayered with silty mudstone	hummocky cross-bedding	settling of suspended sediments under low-energy wave influence, i.e., distal tempestite deposition
R	rhythmic silt/clay couplets	millimetric parallel bedding, irregular bedding planes, frequent slumping	low-energy underflows alternating with settling of suspended sediments
MS	mudstone and fine sandstone	lenticular bedding, some climbing ripple lamination, frequent slumping	ripple migration under low-energy conditions combined with settling of suspended sediment
Mb	mudstone and very fine sandstone	bioturbation (*Cruziana-Thallasinoides* and *Teichnicus*), some lenticular bedding	settling of suspended sediments
Ml	mudstone with some very fine- to fine-grained sandstone	lenticular bedding	settling of suspended sediments alternating with some ripple migration under low-energy conditions
CC	coals and coaly shales and mudstones	root marks	aggradational accumulation of plant material and fine-grained sediments

defined as comfortable successions of genetically related beds or bedsets bounded by marine flooding surfaces (Van Wagoner et al., 1990). The stacking pattern of parasequences is an important criterion for delimiting systems tracts. As the record of carbonaceous facies is interdependent with sea level change, one may try to compare sequence stratigraphy and coal-seam characteristics to investigate controls on sedimentation and to depict the environmental controls on peat formation. Perhaps the most explicit

Table 2. Depositional systems of the coal-bearing succession.

Identification	Depositional System	Subsystem	Main Lithofacies	Approximate Thickness Range (in meters)
0	alluvial fan	not detailed	Gm, Gc	2.00–10.00
1	delta	1A: distributary channel	fine to medium St	3.00–5.00
		1B: interdistributary bay	MS; CC,SB	1.00–2.00
		1C: proximal distributary mouth bar	coarse to medium St; Gc	8.00–10.00
		1D: distal distributary mouth bar	medium St	5.00–10.00
		1E: very distal mouth bar and prodelta	R	1.00–3.00
2	lagoonal estuary	2A: proximal bay-head delta	coarse-grained St, Gc	5.00–10.00
		2B: distal bay-head delta	fine-grained St; ML	10.00–15.00
		2C: tidal sand bars/tidal delta	Sf, St	2.00–8.00
		2D: muddy and sandy tidal flats	MS, Ml, Sf, Sw	1.00–3.00
		2E: marshes and swamps	CC	1.00–15.00
		2F: washover fans	Gi, Gm, St	0.05–0.30
3	barrier	3A: foreshore	Ss	1.00–10.00
		3B: upper shoreface	St	5.00–10.00
		3C: middle shoreface	Shcs; SMhcs; Sw	5.00–10.00
		3D: lower shoreface to offshore	Mb, some SMhcs	10.00–12.00

model of coal formation from a sequence-stratigraphic viewpoint is that of Bohacs and Suter (1997). These authors discussed the controls of carbonaceous rock formation, emphasizing that the fundamental control on peat accumulation and preservation is the accommodation rate in relation to peat production. As already pointed out by Gastaldo et al. (1993), Aitken and Flint (1995), and others, Bohacs and Suter (1997) showed that most thick, widespread peat accumulates when accommodation rate increases and clastic influx is moderate, as in the late lowstand and early transgressive systems tracts. The authors predict symmetrical pairs of thickness-geometry attributes throughout the cycle of sea level change, as mires should respond mainly to the rate of change of base level and not to the direction of that change.

To establish the sequence-stratigraphic framework of the coal-bearing interval in the Candiota area, a data set acquired from 56 log analyses (gamma-ray and resistivity logs), core descriptions from 14 boreholes, and 6 additional outcrop sections was used. The location of most of the boreholes, especially those used to compose the correlation section of Figure 5, are shown in Figure 1. The regional correlation of lithofacies in the different depositional systems was interpreted in a high-resolution, third-order sequence-stratigraphic framework. The sequence boundaries, parasequence boundaries, systems tracts, and major flooding surfaces for the third-order sequences S2 and S3 are shown in Figure 4.

Sequence boundaries SB 1 (between crystalline basement and Permian strata) and SB 2 (fluvial sediment overlying marine shale and sandstone) are easily recognized throughout the study area and enclose sequence 1, where no coal occurs (see Figure 2). Sequence boundary 3 has a different signature, reflecting differential basinal subsidence; some areas clearly underwent temporary regression and basinward shift of facies. In other areas, the transgression rapidly reworked the regressive sediments and left only a thin veneer of pebbly sandstone, the typical signature of a transgressive surface coinciding with a sequence boundary. The coastal encroachment during regional transgression reached about 70 km (43 mi) landward (Holz, 1998).

Seven parasequences (PS) are recognized in third-order depositional sequence 2 (Figure 4). Two parasequences represent the lowstand systems tract (PS 0 and PS 1), four represent the transgressive systems tract (PS 2–5), and one represents the highstand

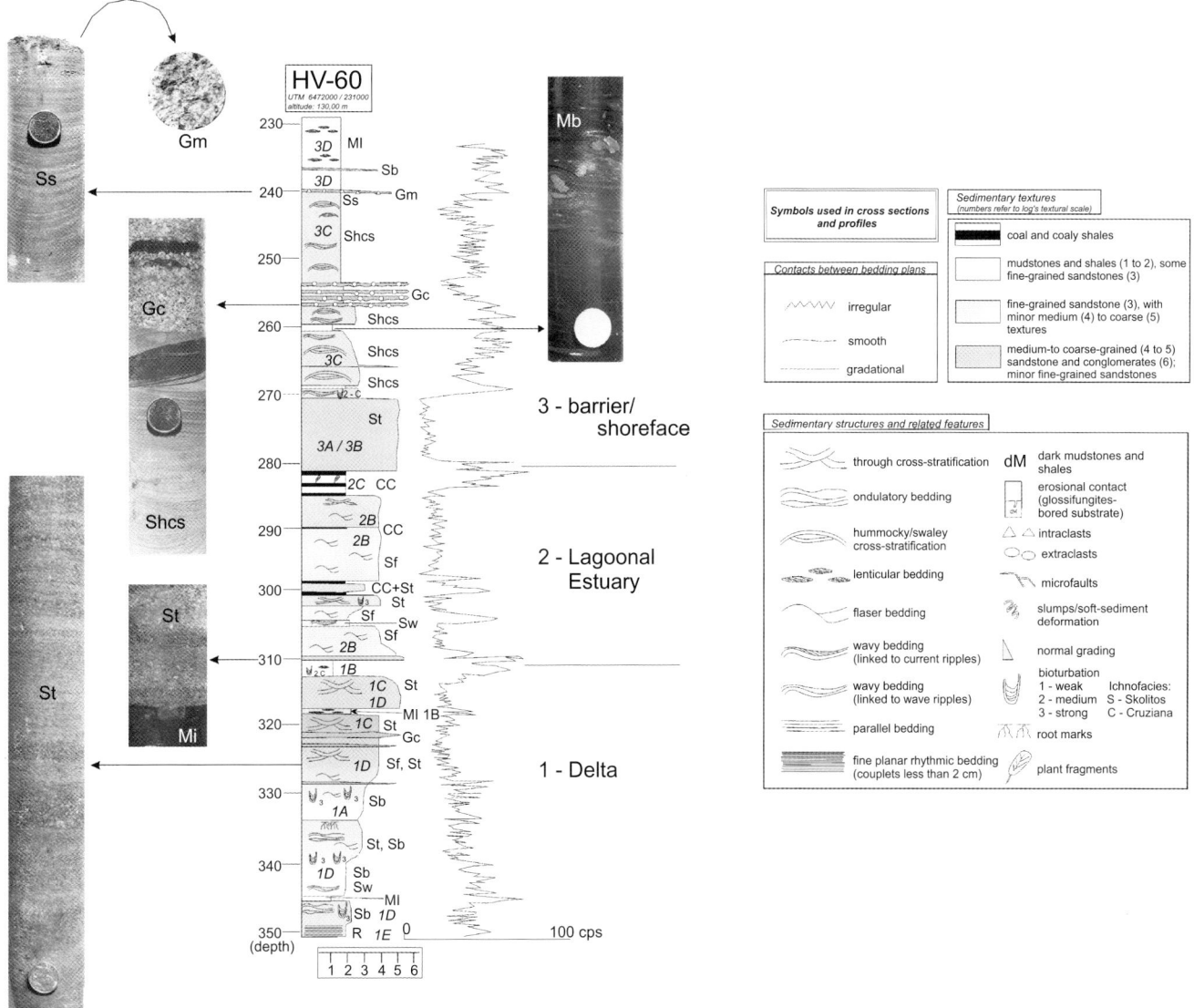

Figure 3. Representative well log showing typical lithofacies of the coal-bearing interval. The code for the facies and depositional systems are provided in Tables 1 and 2. See legend for other symbols. The delta depositional system is linked to the lowstand systems tract with little coal development, whereas the lagoonal-estuary and barrier-shoreface systems are mostly linked to a transgressive systems tract where thick coal seams are developed. Note the erosive contact between the delta and lagoonal-estuary systems, which is characterized by the occurrence of *Glossifungites* ichnofacies.

systems tract (PS 6). As shown by the correlation in Figure 5, the transgressive systems tract is the most extensive and thickest of sequence 2.

Third-order depositional sequence S3 is topped by boundary SB 4 (see Figure 2). As this sequence has only a few coal seams at its base (lowstand systems tract), the correlation section (Figure 5) only shows the basal portion of this sequence. Depositional sequence 3 has coal only in the northern part of the study area, because the ongoing transgression sustained marine conditions in most of the region. Two parasequences, PS 7 and PS 8, are delimited for the

lowstand systems tract of sequence 3. Each parasequence begins with facies indicative of a flooding event overlain by progradational sediment. Therefore, during initial times of parasequence development, the associated peat-forming environments are strongly transgressive. Toward the top of each parasequence, coal formed in a more regressive depositional environment, because sediment was prograding toward the basin.

The parasequences mapped in the study area have a variable thickness (2–12 m [7–39 ft]), and their bounding surfaces are marked by fine-grained sandstone

Figure 4. Schematic stratigraphic framework of the study area detailing the parasequences for each systems tract of depositional sequences S2 and S3. The erosional truncations of parasequences by sequence boundary and transgressive surfaces are based on regional correlation as shown in Figure 5. SB = sequence boundaries, eTS = transgressive surface of erosion, MFS = maximum flooding surface, LST = lowstand systems tract, TST = transgressive systems tract, HST = highstand systems tract, PS = parasequence.

Third-order depositional sequences	Systems tracts	Sequence boundaries and parasequence limits	Lithostratigraphy
S 3 (min. 20 m)	eTS2 / LST 3 / HST 2 MFS	*interval not analyzed (no coals)* / PS8 / PS 7 / PS 6 — SB 3 — / PS 5	Palermo
S 2 (~120 m)	TST 2	PS 4 / PS 3 / PS 2 / PS 1 / eTS1 — PS 0	Rio Bonito
S 1 (~60 m)	LST 2	— SB 2 — / *sequence stratigraphy here not analyzed (no coals) - for detailed stratigraphy, see Holz, 1999* / — SB 1 —	Itararé
		basement (Late Proterozoic to Cambrian–Ordovician igneous and metasedimentary rocks)	Cambaí / Guaritas

(60 m = maximum thickness)

with a wave-dominated or wave-influenced origin (hummocky cross-bedding or wavy and lenticular bedding). Overlying these facies are more current-dominated facies (fine- to coarse-grained sandstone with trough and planar cross-bedding) capped by coal seams. In our stratigraphic framework (Figure 4), parasequences PS 2 and PS 8 have sharp bases constituting transgressive surfaces of erosion. The base of parasequence PS 2 is marked by the occurrence of *Glossifungites* ichnofossils, which indicate erosion of the distal shoreface facies and recolonization of the substrate by firmground-boring organisms (Figure 3). Parasequence PS 8 has an intraclastic veneer composed of nodules (chert?), shell fragments, and muddy rip-up clasts at

its base (see facies Gm in Figure 3). This veneer is interpreted as a product of transgressive reworking of nearshore facies.

SEQUENCE STRATIGRAPHY AND COAL DISTRIBUTION

Coal seams occur in all parasequences, but the distribution of coal can vary in terms of the number of seams, seam thickness, and lateral continuity (Table 3). To illustrate and permit accurate analysis of the pattern of coal thickness, cumulative coal isopach maps for each parasequence were constructed (Figure 6).

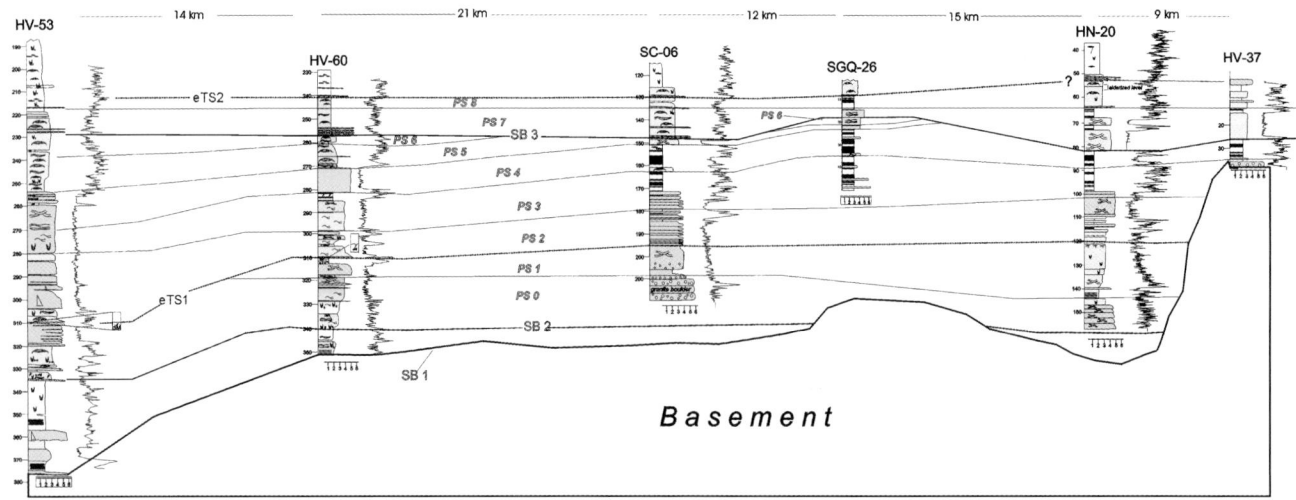

Figure 5. A depositional-dip-oriented correlation section showing the regional distribution of the parasequence bounding surfaces and coal distribution in the study area. Note truncation of PS 5 and PS 6 by overlying sequence boundary, SB 3. See Figure 1 for location of section (modified from Holz et al., 2000).

Table 3. Coal distribution in the sequence-stratigraphic framework.

Systems Tract	Parasequence	Depositional Setting	Maximum Number of Coal Seams	Average Number of Coal Seams	Thickness Range (m)	Average Cumulative Thickness (m)	Overall Coal Distribution Trend	Comparison with Model of Bohacs and Suter (1997)
Lowstand	PS 8	delta progradation	4	2.6	1.0–3.5	1.9	few, discontinuous and thin coal seams	moderately thick (~1–3 m [3.3–10 ft])
	PS 7	delta progradation	6	3.0	0.1–4.0	1.65		isolated seams
Highstand	PS 6	lagoonal estuary	no data on coal seams because this parasequence is mostly eroded by the overlying sequence boundary					
Final transgressive	PS 5	lagoonal estuary	3	1.6	0.30–0.95	0.50	few and thin coal seams	thin coal seams (~1 m [3.3 ft])
	PS 4	lagoonal estuary	12	4.3	0.20–11.90	5.50	many thick and continuous coal seams	very thick (~3 m [10 ft])
Initial transgressive	PS 3	lagoonal estuary	10	4.6	0.20–7.80	2.60	increasing thickness toward top of initial transgressive systems tract	continuous seams
Lowstand	PS 2	lagoonal estuary	5	2.0	0.30–1.0	0.60		
	PS 1	delta progradation	3	2.0	0.20–4.10	1.20	few, discontinuous and thin coal seams	moderately thick (~1–3 m [3.3–10 ft])
	PS 0	delta progradation	1	1.0	0–0.30	0.30		isolated seams

Coals are uncommon in the two parasequences of the lowstand systems tracts of sequence 2 (Figure 6A, B). Here, Bohacs and Suter's (1997) predicted relationship is not observed between systems tracts and coal geometry and thickness for moderate rates of accommodation and clastic influx. This is probably the result of a variation in accommodation rate in the study area. According to this model, low accommodation rate creates space that is promptly filled during lowstand. Next, the mire extends horizontally, forming a continuous body with resultant coal seams characterized by dulling-upward trends. During late lowstand, the increasing accommodation rate permits the peat to accumulate to its full capacity in place; hence, the mire does not need (or may not be able) to extend laterally, and thick but relatively isolated, laterally discontinuous coal seams are formed (Bohacs and Suter, 1997).

In the present study, the lowstand systems tract of sequence 2 is strongly progradational in the beginning, because of tectonic reactivation of source areas. Reactivation is observed not only regionally, but basinwide (e.g., Milani et al., 1994). Few coals were formed in the deltaic environments of sequence 2 because of a low rate of accommodation combined with high clastic input. The presence of a transgressive surface of erosion indicates that the late lowstand systems tract and its thicker coals might not be preserved. Hence, as much as 20 m (66 ft) of sediment is missing because of transgressive erosion.

The transgressive systems tract of sequence 2 is exemplary of a good geometric relationship between systems tracts and coal beds (Table 3), as predicted by the Bohacs and Suter (1997) model. The most prevalent coal seams occur in this transgressive systems tract and have a cumulative thickness of as much as 12 m (39 ft). The initial transgressive systems tract (parasequences PS 2–4) has thick and relatively continuous coal seams (Figure 6C–E), including the most important coal seams of the Candiota mining area (seams CCI and CCS). In the late transgressive systems tract (PS 5), the coals are thinner and dispersed (Figure 6F). The difference between coal thickness and continuity in the initial and late transgressive systems tracts is explained by the fact that in the late transgressive systems tract, the high accommodation rate precludes mires to accumulate until fulfilling the space available. Thus, only thin, dispersed coal beds are formed. In the early

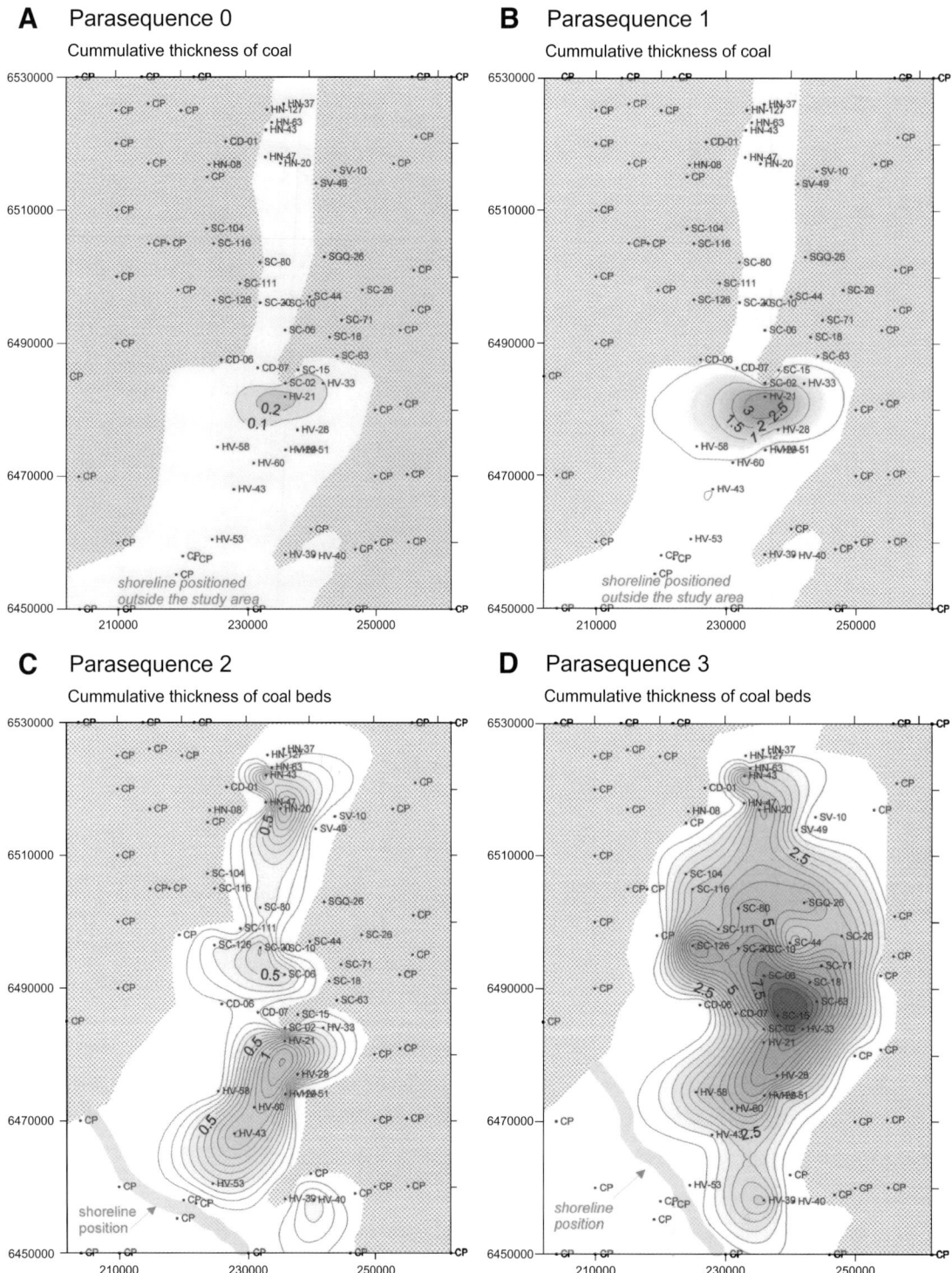

Figure 6. Isopach maps of coal seams, variation in basin configuration, and shoreline position in the Candiota area during the time of peat accumulation. (A, B) The lowstand systems tract of sequence 2, mainly fluvial-deltaic, has few, thin, and discontinuous coal seams. By that time, the shoreline is located southward, outside the study area. (C, D, E) Coal distribution in the parasequences of the initial transgressive systems tract, which is more aggradational than retrogradational, as shown by the vertically stacked coal depocenters and the almost stationary shoreline. (F) Late transgressive systems tract with few and thin coal seams. Note the northward-shifted coal depocenter and the increased coastal encroachment, as indicated by the new position of the shoreline compared to its position during the initial transgressive systems tract. (G) The highstand systems tract is mostly eroded by overlying sequence boundary, so the coal-bearing back-barrier setting is not preserved. (H, I) The mostly deltaic parasequences of the lowstand systems tract of sequence 3 have thin to moderately thick coal seams in the northeastern part of the study area. CP = control points. Shaded area = basement. HV-60 = boreholes. Locations in Universal Transverse Mercator (UTM) coordinates.

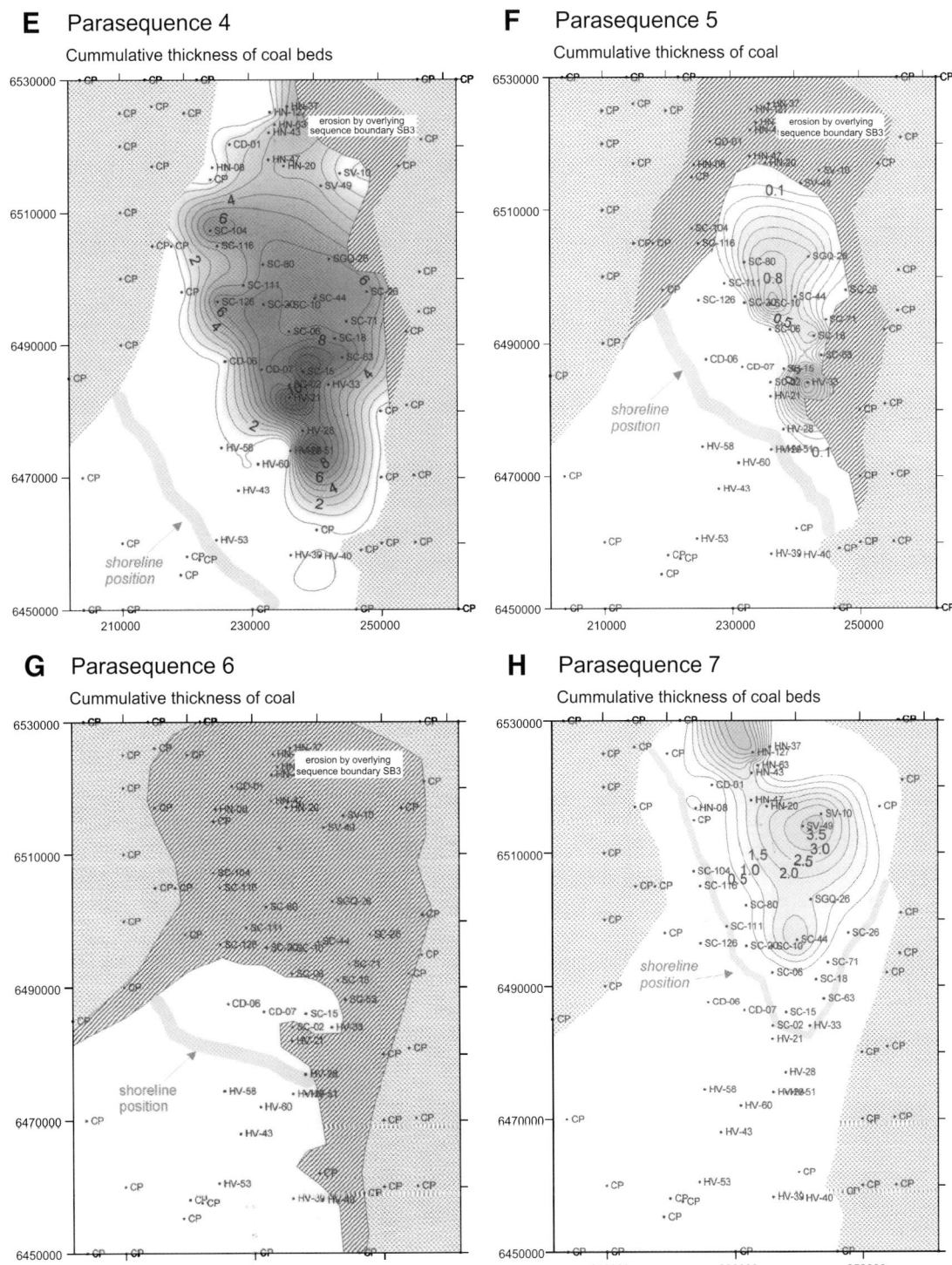

Figure 6. (cont.).

transgressive systems tract, by comparison, the accommodation rate permits the formation of thick and laterally continuous coal beds, because peat production can maintain pace with increasing accommodation.

The highstand systems tract of sequence 2, which is formed by PS 6, is mostly eroded by the overlying sequence boundary (Figure 6G). Therefore, no data on coal distribution are available. The overall trend of coal distribution in depositional sequence S2 is very similar to the pattern of coal distribution in the Cretaceous Ferron Sandstone Member of the Mancos Shale in Utah, United States (Ryer, 1981; Bohacs and Suter, 1997). In that section, the basal progradational parasequence has almost no coal, whereas the most

Parasequence 8

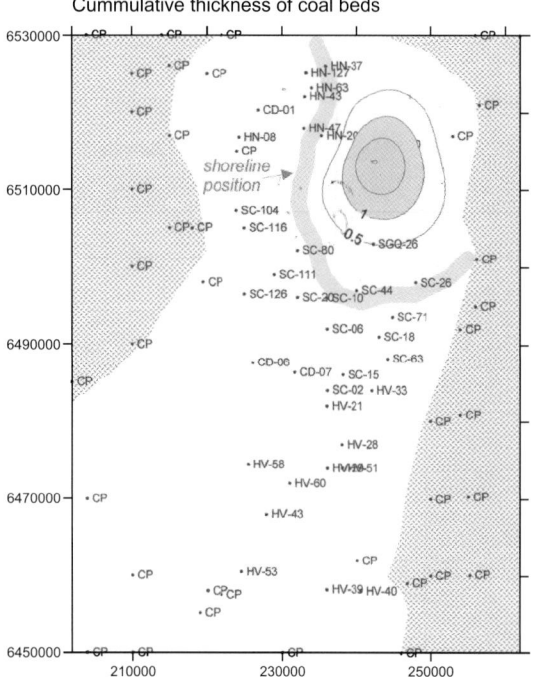

Figure 6. (cont.).

extensive coal beds are associated with aggradational to slightly retrogradational parasequence sets.

The lowstand systems tract of the overlying sequence 3 (LST3) is formed by parasequences PS 7 and PS 8. These are marine in most of the study area, as indicated by a peculiar sequence boundary (Figure 7). In the extreme northern part of the study area, near the margin of the basin, fluvial-deltaic and upper shoreface facies are recognized as the record of a forced regression indicative of a sequence boundary. By contrast, the regressive facies is represented by a veneer of coarse-grained sand and pebbles embedded in hummocky-statified facies in the central and southern (distal) parts of the study area. This indicates that base-level fall, which caused a basinward facies shift, was followed by a rapid base-level rise, causing erosion and reworking of the regressive facies in a large part of the study area. The coals of this systems tract are linked to a delta system that developed in the northern part of the study area (Figure 6H, I). Coal thickness and distribution is similar to that of the lowstand systems tract of the previous sequence, showing moderately thick (1.6–1.9 m [5–6 ft]) coal seams formed in a deltaic setting.

COAL PETROLOGY AND SEQUENCE STRATIGRAPHY

Petrologic and geochemical signatures of coal seams that formed in transgressive and regressive depositional settings have been studied by several authors (Diessel, 1992; Banerjee et al., 1996; Holz et al., 2002; Banerjee and Kalkreuth, 2002; Wadsworth et al., 2002). According to these studies, petrographic parameters, including vitrinite content, vitrinite reflectance, fluorescence properties, tissue preservation, and gelification indices commonly show significant variation from the base to the top of the seam and can be related to the depositional regime (transgressive vs. regressive) under which the precursor peat accumulated. The transgressive-regressive nature of coal seams also is reflected by chemical signatures, such as hydrogen

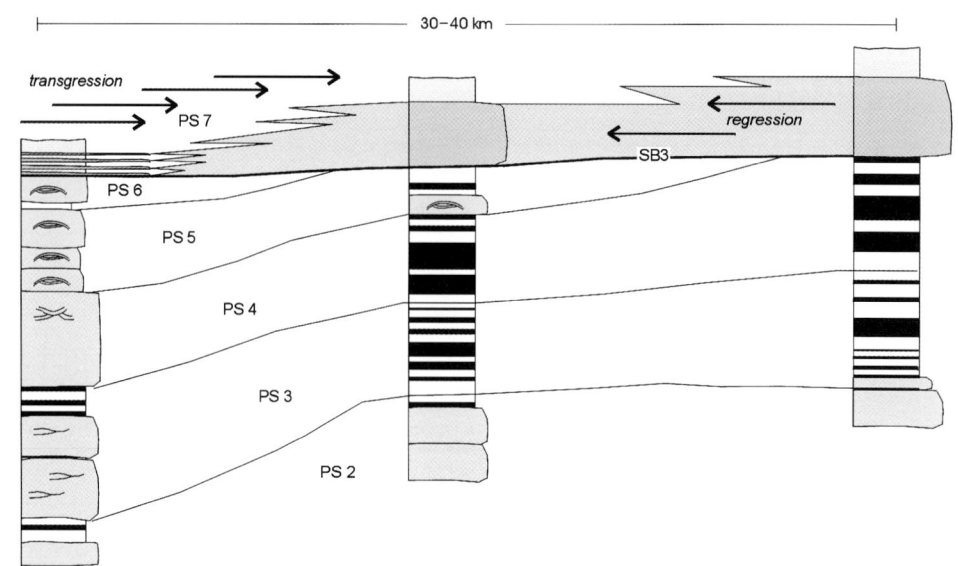

Figure 7. Formation of sequence boundary SB 3 in the study area in the proximal region of the estuary setting occurred by temporary regression and a basinward shift of facies, recorded by coarse-grained fluvial-deltaic sandstone overlying marine and estuarine parasequences. In the distal part of the transgression, rapidly reworked regressive sediments left only a thin veneer of pebbly sandstone, the typical signature of a transgressive surface coinciding with a sequence boundary. This pattern is thought to reflect differential basin subsidence.

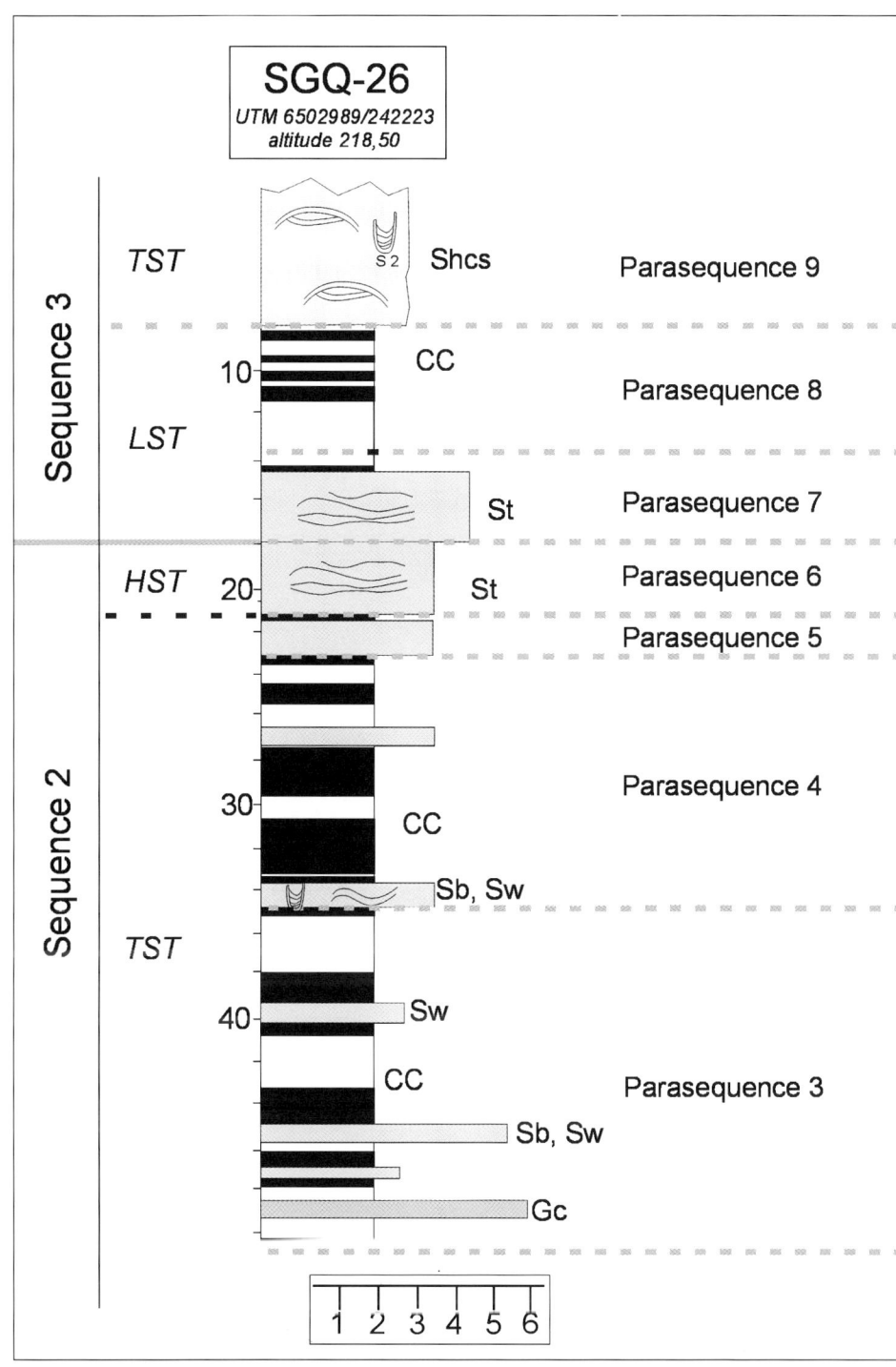

Figure 8. Lithologic log of borehole SGQ-26 used for the petrographic analysis of the coal seams. The log shows that at the base of intercalated estuarine and shallow-marine sandstone beds, the coal-bearing interval can be subdivided into several parasequences. Symbols are explained in the legend of Figure 3. See Table 1 for facies codes.

samples of seam subsections, or benches (most having a thickness of about 30 cm [12 in.]), to study in-seam variation.

The petrographic composition of the samples was determined by maceral analysis following standard procedures (Bustin et al., 1989). Maceral groups (mineral-matter-free base) and mineral matter are reported in volume %. Maceral classification and nomenclature is that of the International Committee for Coal and Organic Petrology (ICCP, 1963, 1994). Gelification (GI) and tissue-preservation (TPI) indices were calculated from the maceral data using the concept of Diessel (1986). Mean vitrinite reflectance was determined randomly (Bustin et al., 1989) based on 50 readings for the full-seam channel samples and 25 readings for the seam subsamples.

and sulfur content (Diessel, 1992), as well as by variations in palynomorph assemblages (Banerjee et al., 1996).

Sampling and Analytical Procedures

For the present study, 17 coal seams were analyzed from borehole SGQ-26, which spans the complete coal-bearing interval of the Rio Bonito Formation (parasequences PS 3–8; Figure 8). The seams were sampled as full-seam channel samples and channel

Petrologic Characteristics of Full-seam Channel Samples

Coal distribution in borehole SGQ-26, along with petrographic characteristics and sequence-stratigraphic interpretations (bounding surfaces of parasequences, third-order sequence boundaries, and system tracts), are shown in Figure 9. According to the present sequence-stratigraphic interpretation, the top seams (seams S3–S7) form part of third-order sequence 3.

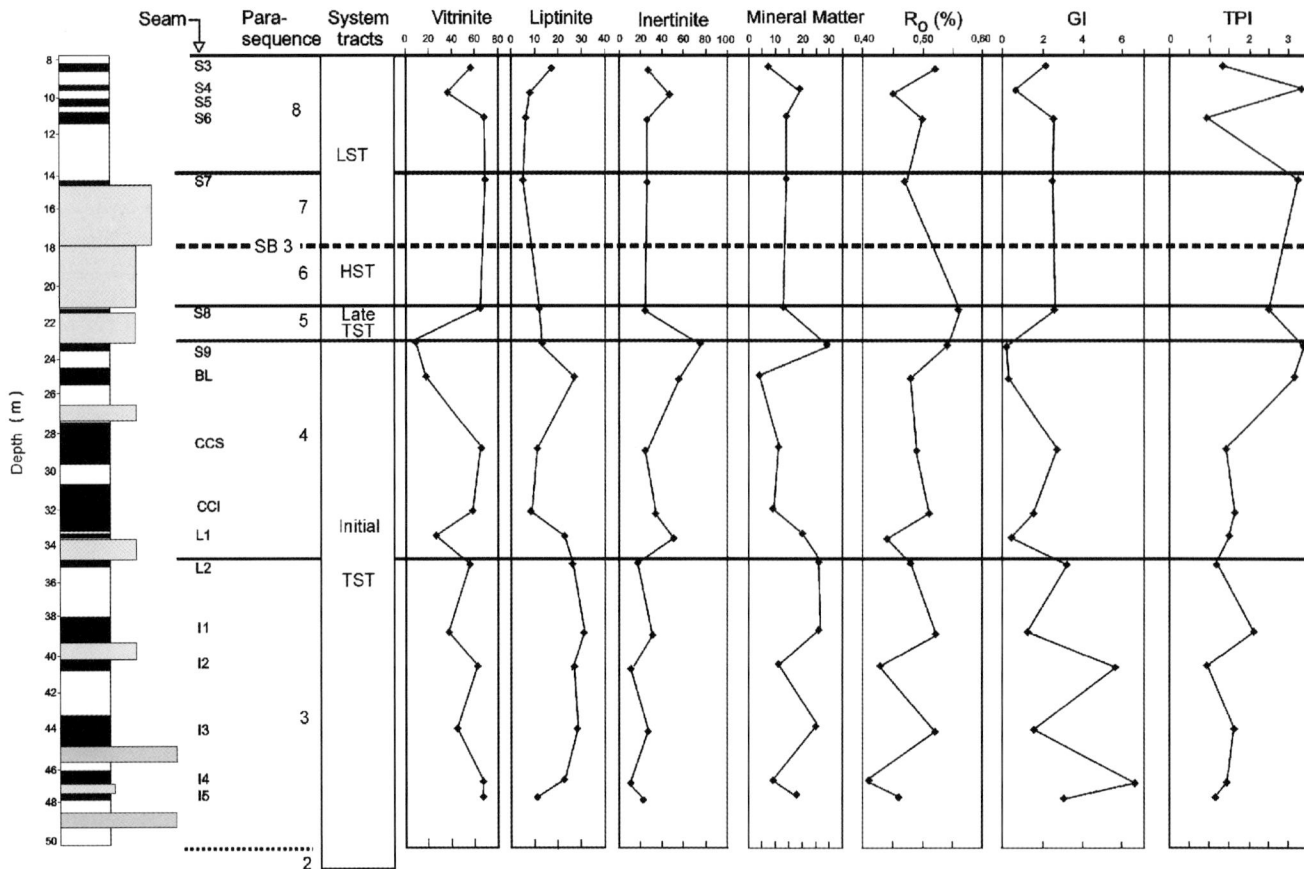

Figure 9. Sequence-stratigraphic interpretation of the coal-bearing strata in borehole SGQ-26, Candiota coalfield, and coal petrographic characteristics of enclosed coal seams. Maceral groups (mineral-matter-free basis) and mineral matter are reported in percent by volume. GI = gelification index, TPI = tissue-preservation index (for explanation, see text), SB 3 = lower boundary of third-order sequence 3. For textural scale, see legend in Figure 3.

Thin coals are developed at the top of third-order sequence 2 (PS 5, seam S8) and also in PS 3 (seams L2, I1–I5) at the base of the coal-bearing interval. Maximum coal development occurs in PS 4 (seams S9, BL, CCS, CCI, and L1; Figure 9).

The petrographic composition of the coal seams is shown in terms of maceral groups (vitrinite, liptinite, and inertinite) and mineral matter content (Figure 9). There is an overall trend of decreasing vitrinite content from the base to the top of the coal-bearing interval of third-order sequence 2 (PS 3–4), which is paralleled by an overall increase in inertinite content. The sharp increase in inertinite macerals at the top of PS 4 is caused by a high contribution of fusinite in seam BL (47%) and by a combination of high fusinite, semifusinite, and inertodetrinite contents in seam S9 (79% total). This trend is reversed in PS 5, where seam S8 shows higher vitrinite content and a relatively low inertinite content. Coal seams above PS 5 continue to have relatively high vitrinite and lower inertinite contents (Figure 9), with the exception of seam S4.

Mineral-matter content is highly variable, ranging from 4 to 30% (Figure 9). In parasequences PS 3 and PS 4, the uppermost coal seams are enriched in mineral matter compared to the underlying seams. This is most likely related to a larger influx of clastic sediment during the most regressive phase of these parasequences.

Vitrinite reflectance values follow roughly the trend shown by the inertinite contribution (parasequences PS 3–5), indicating that slightly increased reflectance values are associated with higher inertinite content and vice versa. In the upper part of the study interval (parasequences PS 7–8), the relationships among vitrinite reflectance, coal composition, and stratigraphic position are not well defined.

The gelification index (GI) and tissue-preservation index (TPI) have been used widely in coal petrographic studies to assess depositional environment and coal facies, since Diessel introduced these parameters in 1986. The basic concept is that the GI contrasts macerals of vitrinite and inertinite groups that have undergone gelification with those that have not (GI = [vitrinite + macrinite]/[fusinite + semifusinite + inertodetrinite]). As such, the GI is considered a measure of relative humidity during early peat formation, with high values indicating relative high water tables. The TPI contrasts macerals of the vitrinite and inertinite groups exhibiting original botanical cell structure with those where no botanical cell structure is visible (TPI = [telinite + collotelinite + fusinite + semifusinite]/[collodetrinite + vitrodetrinite + macrinite + inertodetrinite]). As such, the TPI is considered to reflect the precursor plant material (arborescent, or woody, vs. herbaceous) and also the degree of degradation.

When applying the GI-TPI concept to Candiota coals, it is apparent that the GI values roughly parallel the vitrinite contents determined in the samples, suggesting successively drier conditions during peat accumulation (from basal coal seams in PS 3 to top of PS 4). There is also a trend toward higher TPI values in the same interval, suggesting higher input of arborescent material and better overall preservation. This is particularly true for the fusinite-rich seams BL and S9 of PS 4. The trend toward relatively high TPI values actually continues into PS 7 (Figure 9), whereas greater amounts of structureless collodetrinite in seams S6, S5, and S3 account for lower TPI values in PS 8.

Recent studies on the distribution of palynomorph assemblages in the Candiota coal seams (Meyer, 1999; Meyer and Marques-Toigo, 2000; Cazzulo-Klepzig, 2001) suggest the predominance of a pteridophytic flora (ferns and fernlike foliage) throughout the coal-bearing interval investigated (seam BL to I4; for stratigraphic position, see Figure 9). Palynomorphs representing lycopods (*Lundbladispora* sp., *Cristatisporitis* sp., *Punctatisporites* sp., and others) and ferns were found to be most abundant, with minor contributions from the Sphenophyta (horsetails). Lycopods generally have been considered to represent a herbaceous flora, but there is evidence (Cazzulo-Klepzig, 2001) that some forms, such as *Lundbladispora braziliensis* and others, are arborescent. This explains the commonly well-preserved botanical structures in the vitrinite group (collotelinite) of the Candiota coal seams.

Palynomorphs representing gymnosperms (Coniferales, Cordaitales) and pteridosperms, including *Glossopteris*, were observed in all coal seams investigated (Meyer, 1999). Quantitatively, these plants constitute only a minor proportion of the palynomorph assemblage. The palynological data (Cazzulo-Klepzig, 2001) indicate that herbaceous and arborescent lycopods dominated the coastal mires, whereas *Glossopteris*, cordaitales, and coniferales occupied the more distant and well-drained areas.

The high fusinite content in seams BL and S9 (44 and 47%, respectively) is remarkable. Although Taylor et al. (1998) cited four possible pathways for the incorporation of this maceral into coal (pyrofusinite, degradofusinite, rank fusinite, and primary fusinite), studies on the origin of fusain and fusain-derived fusinite (Scott, 1989; Jones et al., 1993; Scott and Jones, 1994) suggest that most, if not all, fusinite is derived from wildfires. This interpretation is based on the similarity of fusinite obtained from recent charcoal experiments with those observed in Jurassic and Carboniferous coals.

Fusinite commonly shows excellent preservation of the original botanical structures that allows for the identification of the precursor material (Scott, 1989). In the BL and S9 seams, paleobotanical examination of several well-preserved fusinite tissues suggest that most fusinite is derived from gymnospermous wood (M. Guerra-Sommer, 2003, personal communication).

Petrologic Characteristics of Seam Subsections

The petrological characteristics of seam subsections are discussed for seams developed in PS 4 (Figure 10), which comprise seams L1, CCI, CCS, BL, and S9. The first coal to develop in PS 4 is a 30-cm (12-in.)-thick seam (L1), which is characterized by low vitrinite content (Figure 10) that increases slightly upward in section (14 and 18%). Liptinite volume, mainly in the form of sporinite, is 15 and 19%, respectively. The remainder is made up of inertinite macerals, mainly fusinite and inertodetrinite (26 and 33%). The overlying thick coals (CCI and CCS) show a significant increase in vitrinite macerals and display a similar pattern in terms of in-seam maceral distribution. That is, the highest vitrinite content is at the base of the seam, vitrinite content is reduced toward the center of the seam, and vitrinite content increases toward the top of the seam (except in the uppermost sample).

A return to low vitrinite content is observed in coal in the upper part of PS 4 (BL, S9). The three subsection samples from seam BL indicate successively lower vitrinite contents from seam base to seam top, accompanied by relatively high liptinite content (17–27%), mainly in the form of sporinite. The remainder are inertinite macerals, predominantly in the form of fusinite (22–53%). The trend of low vitrinite and high inertinite content continues toward the top of PS 4, where seam S9 is characterized by very low vitrinite

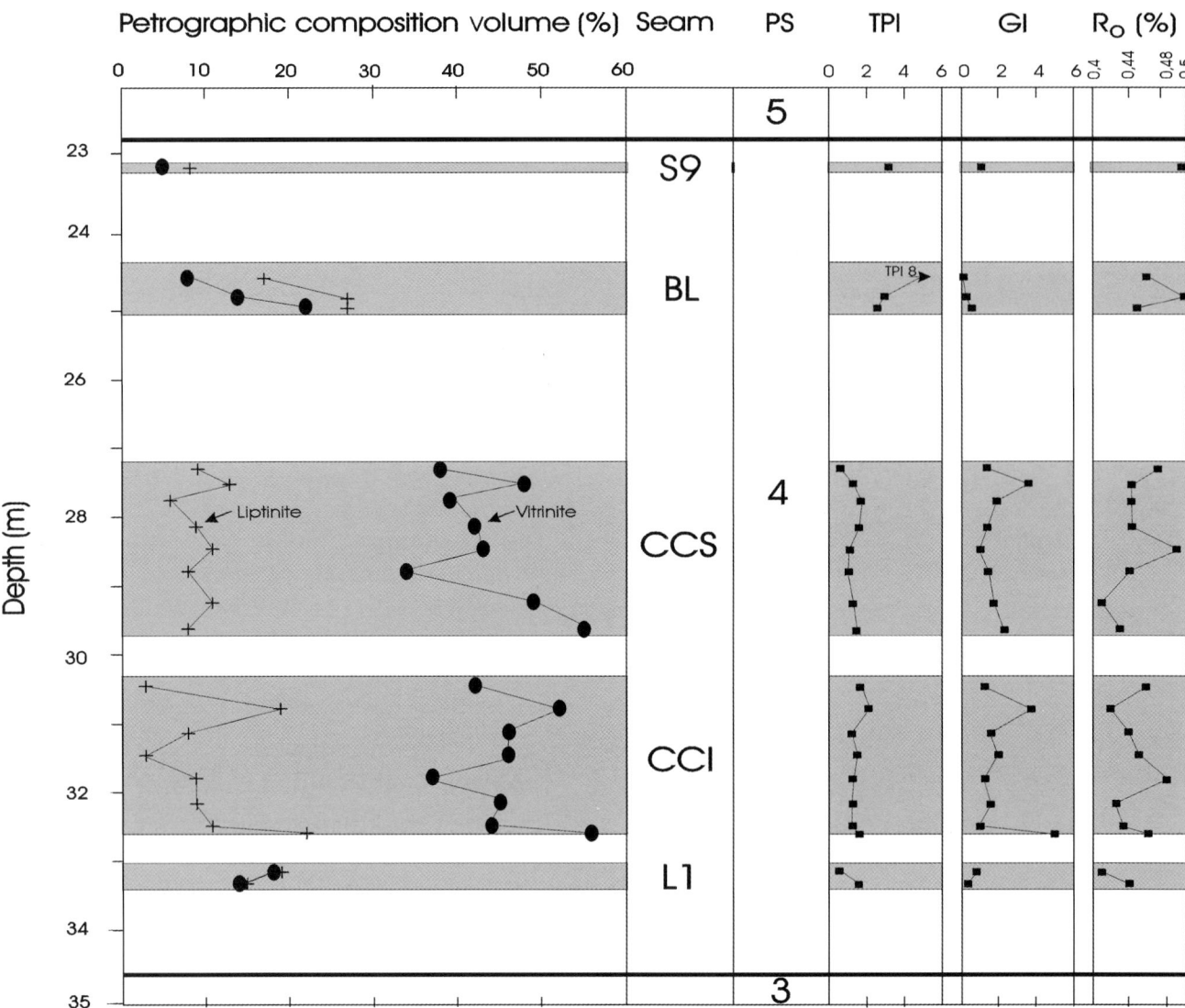

Figure 10. Petrographic characteristics of coal seam subsections in PS 4, borehole SGQ-26, Candiota coalfield. Maceral groups and mineral matter in volume percent. GI = gelification index, TPI = tissue-preservation index, PS = parasequence.

content (6%) and high inertinite content (55%). Fusinite comprises 31%, semifusinite comprises 14%, and inertodetrinite comprises 10% of seam S9.

Vitrinite reflectance shows a distinctive pattern in which values tend to increase toward the central parts of the three major seams in PS 4 (BL, CCS, CCI). TPI values in seams CCI and CCS show a similar trend, with little variation in the basal part of the seam (Figure 10). Values approach a maximum in the upper part of the seams and decrease at the very top. In the overlying BL seam, very high fusinite content accounts for high TPI values that reach 8.0 in the upper part of the seam (Figure 10). The GI values in seams CCI and CCS essentially parallel the trend shown for vitrinite content, with peak values at the seam base, followed by reduced values in the central

parts of the seams and a return to higher values in the top part, except for the uppermost seam subsection. In seam BL, the GI values are extremely low as a result of high inertinite content.

CHEMICAL INDICATORS FOR PALEOENVIRONMENTAL CONDITIONS OF CANDIOTA COALS

Boron and sulfur distributions in coal seams have been widely used to interpret paleodepositional environments of peat accumulation with respect to marine, brackish, and freshwater influence (Swaine, 1971, 1983; Boher and Gluscoter, 1973; Casagrande et al., 1977; Cohen et al., 1984; Goodarzi, 1987; Diessel, 1992; Banerjee and Kalkreuth, 2002). Finkelman

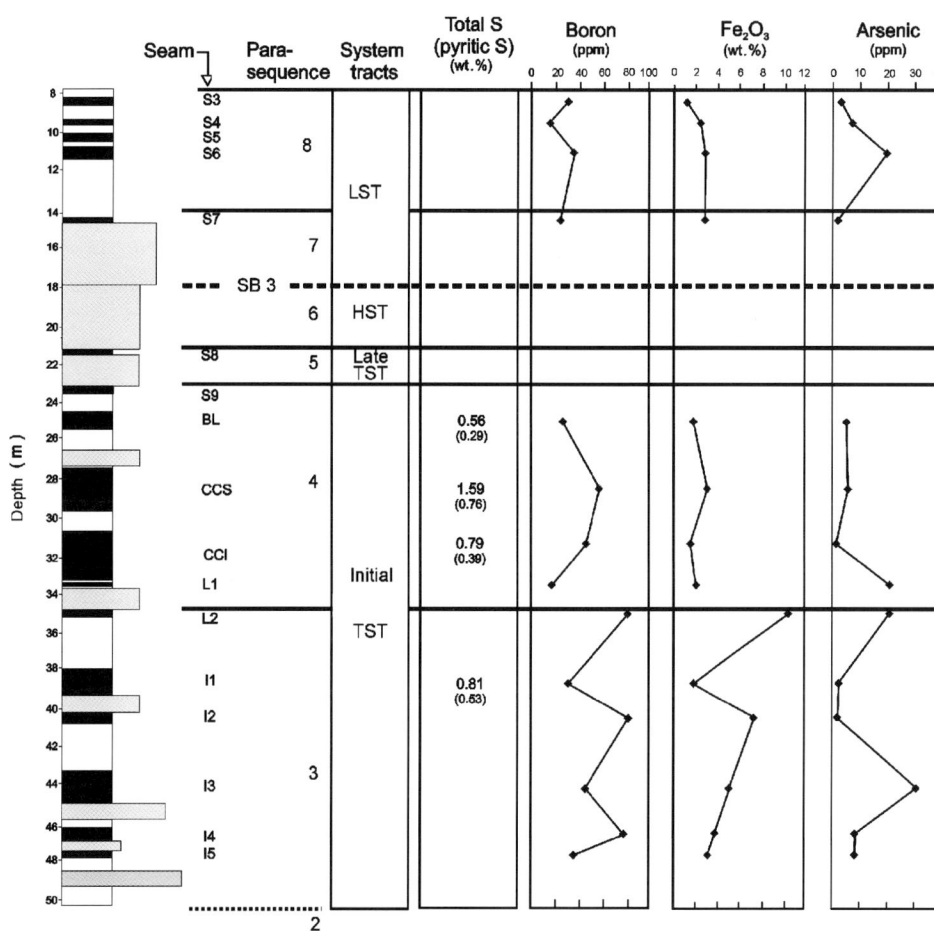

Figure 11. Sequence-stratigraphic interpretation of the coal-bearing strata in borehole SGQ-26, Candiota coalfield, and chemical characteristics of coal.

et al. (2000) and Kalkreuth et al. (2001) presented data on major and minor element distribution in Brazilian coals, including results for the majority of coal seams investigated in the present study. The data discussed below are based on these studies.

Boron

Boron in coal is considered to be associated with illite (Boher and Gluscoter, 1973) occurring with a world average of 55 ppm in coal seams (Bertine and Goldberg, 1971). In a study of Permian Bowen Basin coals of Australia, Swaine (1971) derived an average value of 18 ppm for freshwater-related coal seams and an average value of 164 ppm for marine-influenced coal. In a study of coal from the Illinois basin, Boher and Gluskoter (1973) considered boron values less than 125 ppm to be associated with freshwater deposition, whereas boron values greater than 125 ppm suggest marine- or brackish-water influence.

Boron values in the Paraná Basin coals at Candiota range from 15 ppm in S4 seam to 80 ppm in I2 seam (Figure 11). The relatively low values (<80 ppm) suggest peat accumulation in a predominantly fresh-

water regime. It is noteworthy, however, that boron concentrations appear to vary in respect to stratigraphic position and roof-rock lithology. The highest boron concentrations in PS 3 occur in seams I4, I2, and L2 (Figure 11), all of which are overlain by estuarine-lagoonal strata. The highest boron concentrations in PS 4 occur in the CCI and CCS seams (46 and 57 ppm, respectively), whereas values associated with the lowstand systems tract in PS 7 and PS 8 are generally lower. The presence of relatively high boron values in PS 3 and PS 4 are consistent with palynological observations (Meyer, 1999; Meyer and Marques-Toigo, 2000) documenting uncommon acritarchs associated with coal seams I3 and CCI, suggesting occasional inundation of the mire by brackish or marine water.

Sulfur

Sulfur content in coal is widely regarded to reflect the paleoenvironmental conditions under which the precursor peats accumulated (Diessel, 1992). Marine-influenced peat is characterized by relatively high sulfur content, mainly in the form of pyritic sulfur, whereas freshwater peat contains lower concentrations of sulfur with organically bound sulfur dominating.

In the coal seams investigated, only few sulfur values are available (Figure 11), ranging from 0.56% by weight in BL seam to 1.59% in seam CCS. Pyritic sulfur accounts for about half of the total sulfur. It is interesting to note that the sulfur values for seams BL, CCS, and CCI correspond to the boron values recorded for these seams, with the highest boron value corresponding to the highest sulfur value.

Ferric Oxide and Arsenic

From the data set on major and minor elements in the Candiota coals (Finkelmann et al., 2000), two elements were chosen to evaluate indirectly the relative amount of pyrite in the coal-bearing interval. The first is iron, which is determined in coal ash as ferric oxide (Fe_2O_3). In coal lacking carbonate, ferric oxide is considered to be associated mainly with pyritic sulfur. Arsenic is the second element for which a strong affinity with pyrite has been postulated (Ward et al., 1999; Finkelmann and Gross, 1999).

The ferric oxide values range from 1.28% by weight in seam S3 at the top of the coal-bearing sequence to 7.29% in seam L2 at top of PS 3 (Figure 11). The ferric oxide distribution follows roughly the trend outlined by the boron distribution (Figure 11). Again, the highest values in the seams are associated with marine-influenced roof rocks.

The majority of arsenic in coal is considered to be associated with the sulfide-rich fraction (pyrite, marcasite), with minor amounts of arsenic also contained in the organic matter (Valcovic, 1986). The arsenic values in the Candiota coals range from 2 to 32 ppm (Figure 11) and are similar to arsenic concentrations reported elsewhere for Permian Gondwana coals (Ward et al., 1999). With respect to stratigraphic position, and when compared to boron and Fe_2O_3 values, it is apparent that the arsenic values partially follow a similar trend. This suggests that arsenic distribution in these coal seams is probably more complex than its relationship just to sulfur and pyrite content as described by Ward et al. (1999).

DISCUSSION

Petrology of Full Seams and Sequence-stratigraphic Framework

Comparison of petrological coal characteristics with the sequence-stratigraphic framework of the enclosing strata shows that coal characteristics are controlled mostly by depositional setting. During the initial phase of transgressive systems tract, peat accumulation was associated with relatively high water tables, favoring the preservation of organic matter in the form of vitrinite (seams I4 and I5). A trend toward successively drier conditions exists higher in the section (Figure 9), as depicted by decreasing vitrinite and increasing inertinite content (seams I3 to L2). These trends are interpreted to reflect the increasing regressive conditions toward the top of PS 3.

The later stage of the initial transgressive systems tract (PS 4) shows a similar stratigraphic control on coal properties. The overall petrographic features suggest relatively stable conditions during peat formation for the CCI and CCS seams, in which plant growth and preservation were in equilibrium with basin subsidence. The petrographic features of seams BL and S9, which are developed toward the top of the parasequence, suggest a regressive nature of the seams (Figure 9), as indicated by the high inertinite content (fusinite and semifusinite account for 44–49% by volume). It has been suggested (Diessel, 1992) that coal of this type may have formed as back-barrier peat in a regressive phase of an overall transgressive period. The high amounts of fusinite apparently have an origin in forest fires at or near the mire margins, followed by transportation into the mire by wind or water (parautochthonous to allochthonous origin).

Botanical structures identified in many well-preserved fusinite tissues in samples from seams BL and S9 suggest gymnosperms as precursor plants. Their presence suggests a parautochthonous to allochthonous origin, because palynological studies for the same seam interval in nearby boreholes (Meyer, 1999) have shown a predominance of a pteridophytic flora with very little contribution by gymnosperms.

Although the petrographic composition of the single coal seam developed in PS 5 (seam S8) indicates a return to more moist conditions (Figure 9), the variations observed in vitrinite and inertinite contents and GI values for seams in PS 3–4 are indicative of a general drying-upward trend. Coal seams developed in the lowstand systems tract of PS 7 and PS 8 (Figure 9) show relatively high vitrinite contents at the top of PS 7 and at the base and top of the coal-bearing interval of PS 8. Somewhat drier conditions are interpreted for seam S4 (39% fusinite).

Petrology of Seam Subsections and Sequence-stratigraphic Framework

The in-seam characteristics for the CCI, CCS, and BL seams (Figure 10) show petrological characteristics similar to those reported elsewhere for transgressive-regressive coal seams (Diessel, 1992; Banerjee et al., 1995; Banerjee and Kalkreuth, 2002). The CCI and CCS seams of the Candiota area have strikingly similar petrographic characteristics, as documented for the marine-influenced Greta seam of the Sydney basin from Australia (Diessel, 1992). These include highest vitrinite content at the base of the seam (Figure 10), decreasing vitrinite content toward the middle, and a return to higher values near the top. The very top of the seam is characterized by a significant decrease in vitrinite content. The similarity of petrographic characteristics in CCI and CCS seams suggest that the precursor mires underwent similar wet-to-dry cycles during active peat accumulation.

The influence of brackish or marine conditions during early and late peat formation is reflected by vitrinite reflectance, GI, and TPI values (Figure 10). In transgressive seams, vitrinite reflectance is typically highest in the center of the seam and lower toward seam base and top. The lower vitrinite reflectances in those parts of the seam influenced by marine or brackish water have been explained by the incorporation of degraded algal material into the vitrinite and increased bacterial degradation (Diessel, 1992), thereby causing a suppression in reflectance of the associated vitrinite. The GI and TPI values also suggest a transgressive nature of the two seams (Figure 10), with GI values highest at the base and top of the coal (except for the uppermost sample), and drier conditions during accumulation of the central part of the peat body. The changes observed for TPI values are not that striking, although both seams show a slight trend toward lower TPI values at the top of the seam. This trend is related to accumulation of increasing amounts of detrital organic matter being deposited prior to drowning of the mire.

Results from the three subsections of the BL seam (PS 4) suggest successively drier conditions from the base to the top of the seam (Figure 10). This is indicated by very high inertinite content, which is dominated by fusinite and semifusinite (36–58%). The high percentage of structured inertinite is reflected in the high TPI values (2.4–8.0) and very low GI values (0.11–0.58). The overall petrologic properties of the seam reflect the regressive phase of that parasequence. The depositional model for accumulation of the precursor peat is that of a back-barrier mire, where the organic matter occasionally was exposed to oxidation processes during periods of low water tables, including the possibility of peat fires developing at the peat surface. In recent time, major fires in peat-forming environments, such as the Okefenokee swamp and East Kalimantan, Borneo, have been recorded (Scott and Jones, 1994). Alternatively, the fusinite may have formed during forest fires at mire margins or nearby areas, an origin favored for the Permian coal seams at Candiota. Supporting evidence is the identification of gymnosperm wood as the source for many of the fusinite tissues (M. Guerra-Sommer, 2003, personal communication), suggesting a parautochthonous to allochthonous origin of this material, which accumulated in a mire dominated by a pteridophyte flora (Meyer, 1999).

CONCLUSIONS

Sequence stratigraphy may be used to understand and explain variations of coal parameters, including coal petrography and coal chemistry. Additionally, high-resolution analysis of coal and acquisition of chemical characteristics can be a helpful tool to the sequence stratigrapher. We have shown conclusively that the thickest and most continuous coals of the study area in the Paraná Basin were formed in the early transgressive systems tract, demonstrating a clear stratigraphic control on the formation of the precursor peat, as has been postulated already by previous workers.

The coal parameters in each parasequence also are controlled stratigraphically. The shallowing-upward trend of the parasequences is reflected by the variation of coal parameters, such as vitrinite content, inertinite content, and related maceral ratios. The chemical signatures show similar trends, as indicated by the elevated boron and ferric oxide contents in seams overlain by parasequence bounding surfaces. With a high-resolution sequence-stratigraphic framework of a coal basin followed by detailed petrographic analyses of the coal seams, one may predict patterns of coal distribution and quality and provide guidelines for optimal exploitation.

ACKNOWLEDGMENTS

The study was supported by the Brazilian agency Conselho Nacional de Desenvolvimento Cientifico e Tecnológico (CNPq) (Research Grant 520332/96-2) and Fundação de Apoio à Pesquisa do Estado do Rio Grande do Sul (FAPERGS) (Research Grant 520271/97). The authors acknowledge Companhia Riograndense de Mineração and Companhia de Pesquisas de Recursos Minerais for supplying the well logs, cores, and other material essential for this research. CNPq is acknowledged for personal research and study grants to the authors (352887/96-6 and 300971/97-4, respectively). The manuscript benefited from helpful comments by Kevin Bohacs and Cortland Eble.

REFERENCES CITED

Aitken, J. F., and S. S. Flint, 1995, The application of high-resolution sequence stratigraphy to fluvial systems: A case study from the Upper Carboniferous Breathitt Group, eastern Kentucky, U.S.A.: Sedimentology, v. 42, p. 3–30.

Alves, R. G., and M. V. B. Ade, 1996, Sequence stratigraphy

and organic petrography applied to the study of Candiota coalfield, RS, South Brazil: International Journal of Coal Geology, v. 30, p. 231–248.

Banerjee, I., and W. Kalkreuth, 2002, Sedimentology, sequence stratigraphy, palynology, organic petrology and geochemistry of Mannville coals in south-central Alberta: Geological Survey of Canada Bulletin, v. 571, 57 p.

Banerjee, I., W. Kalkreuth, and E. Davies, 1995, Sequence stratigraphy of coal with examples from the Mannville Group in Central Alberta, in Proceedings of the Oil and Gas Forum´ 95, Energy from Sediments: Geological Survey of Canada Open-file Report 3058, p. 151–157.

Banerjee, J., W. Kalkreuth, and E. Davies, 1996, Coal seam splits and transgressive-regressive coal couplets: A key to stratigraphy of high-frequency sequences: Geology, v. 24, p. 1001–1004.

Bertine, K., and E. Goldberg, 1971, Fossil fuel combustion and the major sedimentary cycle: Science, v. 173, p. 223.

Bohacs, K., and J. Suter, 1997, Sequence stratigraphic distribution of coaly rocks: Fundamental controls and paralic examples: AAPG Bulletin, v. 81, p. 1612–1639.

Boher, B., and H. Gluskoter, 1973, Boron in illite as an indicator of paleosalinity in Illinois coals: Journal of Sedimentary Petrology, v. 43, p. 945.

Bustin, R., A., Cameron, D. Grieve, and W. Kalkreuth, 1989: Coal petrology— Its principles, methods and applications: Geological Association of Canada Short Course Notes 3, 3d ed., 276 p.

Casagrande, D., K., Siefert, C. Berschinski, and N. Sutton, 1977, Sulphur in peat-forming systems of Okefenokeee swamp and Florida Everglades: Origin of sulphur in coal: Geochimica et Cosmochimica Acta, v. 41, p. 161–167.

Cazzulo-Klepzig, M., 2001, Palinologia aplicada á reconstituição das unidades de paisagem e dinámica das turfeiras formadoras dos carvões permianos do Rio Grande do Sul: Ph.D. thesis, Curso de Pós-Graduação em Geociências, Universidade Federal do Rio Grande do Sul, Porto Alegre, RS, Brazil, 350 p.

Cohen, A., W. Spackman, and P. Delsen, 1984, Occurrence and distribution of sulphur in peat-forming environment of southern Florida: International Journal of Coal Geology, v. 4, p. 73–96.

Finkelmann, R. B., and P. Gross, 1999, The types of data needed for assessing the environmental and human health impacts of coal: International Journal of Coal Geology, v. 40, p. 91–101.

Correa da Silva, Z. C., 1991, The formation of coal deposits in South Brazil, in Gondwana 7 Proceedings: São Paulo, Universidade de São Paulo, p. 233–252.

Diessel, C. F. K., 1986, The correlation between coal facies and depositional environments, in Proceeding of the 20th Symposium of the Advances in the Study of the Sydney Basin: Australia, University of Newcastle, p. 19–22.

Diessel, C. F. K., 1992, Coal-bearing depositional systems: Berlin, Springer-Verlag, 721 p.

Finkelmann, R. B., W. Kalkreuth, and J. Willett, 2000, Characterization of coals from the Candiota, Butiá-Leão and Santa Terezinha coal deposits, Rio Grande do Sul, Brazil (abs.): Rio de Janeiro, Brazil, 31st International Geological Congress, Abstracts Volume, unpaginated CD-ROM.

Gastaldo, R. A, T. M. Demko, and Y. Liu, 1993, Application of sequence and genetic stratigraphic concepts to carboniferous coal-bearing strata: An example from the Black Warrior basin, U.S.A.: Geologische Rundschau, v. 82, p. 212–226.

Goodarzi, F., 1987, Concentration of elements in lacustrine coals from zone A Hat Creek deposit no. 1, British Columbia, Canada: International Journal of Coal Geology, v. 8, p. 247–268.

Helland-Hansen, W., and J. G. Gjelberg, 1994, Conceptual basis and variability in sequence stratigraphy: A different perspective: Sedimentary Geology, v. 92, p. 31–52.

Holz, M., 1998, The Eo-Permian coal seams of the Paraná Basin in southernmost Brazil: An analysis of the depositional conditions using sequence stratigraphic concepts: International Journal of Coal Geology, v. 36, p. 141–163.

Holz, M., 1999, Early Permian sequence stratigraphy and the palaeophysiographic evolution of the Paraná Basin in southernmost Brazil: Journal of African Earth Science, v. 29, p. 51–61.

Holz, M., P. E. Vieira, and W. Kalkreuth, 2000, The Early Permian coal-bearing succession of the Paraná Basin in southernmost Brazil: Depositional model and sequence stratigraphy: Revista Brasileira de Geociências, v. 30, p. 420–422.

Holz, M., W. Kalkreuth, and I. Banerjee, 2002, Sequence stratigraphy of paralic coal-bearing strata: An overview: International Journal of Coal Geology, v. 48, p. 147–179.

Hunt, D., and M. E. Tucker, 1992, Stranded parasequences and the forced regressive wedge systems tract: Deposition during base-level fall: Sedimentary Geology, v. 81, p. 1–9.

International Committee for Coal and Organic Petrology (ICCP), 1963, International handbook of coal petrography, 2d ed.: Paris, Centre National de la Recherche, 23 p.

International Committee for Coal and Organic Petrology (ICCP), 1994, Vitrinite classification, ICCP system: Paris, Centre National de la Recherche, 24 p.

Jones, T., A. C. Scott, and D. Mattey, 1993, Investigations of "fusain transition fossils" from the Lower Carboniferous: Comparison with modern partially charred wood: International Journal of Coal Geology, v. 22, p. 37–59.

Kalkreuth, W., M. Holz, M. Kern, M. Silva, R. B. Finkelmann, and J. Willett, 2001, Permian coal-measures in Rio Grande do Sul, Brazil— Geological setting, coal petrography and chemical characteristics: Montevideo, Uruguay, XI Congreso Latinoamericano de Geologia, unpaginated CD-ROM.

Meyer, K., 1999, Caracterização palinológica das camadas

de carvão da Malha IV na Mina de Candiota, RS, permiano da Bacia do Paraná: M.S. thesis, Curso de Pós-Graduação em Geociências, Universidade Federal do Rio Grande do Sul, Porto Alegre, RS, Brazil, 120 p.

Meyer, K., and M. Marques-Toigo, 2000, O significado paleoambiental da microflora dos carvões da Malha IV, Mina da Candiota, RS permiano da Bacia do Paraná, Brasil: Revista Universidade Guarulhos, Geociências V (numero especial), p. 17–20.

Milani, E. J., A. B. França, and R. L. Schneider, 1994, Bacia do Paraná, in F. J. Feijó, ed., Cartas estratigráficas das bacias sedimentares brasileiras: Rio de Janeiro, Boletim de Geociências da Petrobrás, v. 8, p. 69–82.

Plint, A. G., 1996, Recognition of marine and non-marine systems tracts in fourth-order sequences in the mid-Cenomanian Dunvegan Formation, northeastern British Columbia, Canada, in J. D. Aitken and J. Howell, eds., High resolution sequence stratigraphy: Innovations and applications: Geological Society (London) Special Paper 104, p. 159–191.

Ryer, T. A., 1981, Deltaic coals of Ferron Sandstone Member of Mancos Shale: Predictive model for Cretaceous coal-bearing strata of western interior: AAPG Bulletin, v. 65, p. 2323–2340.

Scott, A. C., 1989, Observations on the nature and origin of fusain: International Journal of Coal Geology, v. 12, p. 443–475.

Scott, A. C., and T. Jones, 1994, The nature and influence of fire in Carboniferous ecosystems: Palaeogeography, Palaeoclimatology, Palaeoecology, v. 106, p. 91–112.

Smith, A. G, A. M. Hurley, and J. C. Briden, 1981, Phanerozoic paleocontinental world maps: Cambridge, Cambridge University Press, 102 p.

Swaine, D., 1971, Boron in coals of the Bowen Basin as an environmental indicator: Geological Survey of Queensland Report 62, 32 p.

Swaine, D., 1983, Geological aspects of trace elements in coal, in S. Augustithis, ed., The significance of trace elements in solving petrogenetic problems and controversies: Athens, Greece, Theophrastus Publication, p. 521–532.

Taylor, J., M. Teichmüller, A. Davis, C. Diessel, R. Littke, and P. Robert, 1998, Organic petrology: Berlin, Gebrüder Borntraeger, 704 p.

Van Wagoner, J. C., R. M. Mitchum, K. M. Campion, and V. D. Rahmanian, 1990, Siliciclastic sequence stratigraphy in well logs, cores, and outcrops: Concepts for high-resolution correlation of time and facies: AAPG Methods in Exploration, v. 7, 55 p.

Valkovic, V., 1986, Trace elements in coal: Boca Raton, Florida, CRC Press, v. 1, 210 p.

Wadsworth, J., R. Boyd, C. Diessel, D. Leckie, and B. Zaitlin, 2002, Stratigraphic style of coal and non-marine strata in a tectonically influenced intermediate setting: The Mannville Group of the western Canadian sedimentary basin: Bulletin of Canadian Petroleum Geology, v. 50, p. 507–541.

Ward, C. R., D. A. Spears, C. Booth, I. Staton, and L. Gurba, 1999, Mineral matter and trace elements in coals of the Gunnedah basin, New South Wales, Australia: International Journal of Coal Geology, v. 40, p. 281–308.

White, I. C., 1906, Relatório final da comissão de estudos das minas de carvão de pedra do Brazil, 1 de julho de 1904 a 31 de maio de 1906. Edição fac-similar: São Paulo, Departamento Nacional da Produção Mineral, Seventh Gondwana Symposium, 617 p.

Wilgus, C. K., B. S. Hastings, C. G. St. C. Kendall, H. W. Posamentier, C. A. Ross, and J. C. Van Wagoner, eds., 1988, Sea-level changes— An integrated approach: SEPM Special Publication 42, 407 p.

Zálan, P. V., S. Wolff, J. C. Conceição, A. Marques, M. A. M. Astolfi, I. S. Vieira, C. J. Appi, and O. A. Zanotto, 1990, Bacia do Paraná, in G. P. Raja Gabaglia and E. J. Milani, cords., Origem e evolução de bacias sedimentares: Rio de Janeiro, Petrobrás, p. 135–168.

Gibling, M. R., K. I. Saunders, N. E. Tibert, and J. A. White, 2004, Sequence sets, high-accommodation events, and the coal window in the Carboniferous Sydney coalfield, Atlantic Canada, in J. C. Pashin and R. A. Gastaldo, eds., Sequence stratigraphy, paleoclimate, and tectonics of coal-bearing strata: AAPG Studies in Geology 51, p. 169–197.

8

Sequence Sets, High-accommodation Events, and the Coal Window in the Carboniferous Sydney Coalfield, Atlantic Canada

Martin R. Gibling
Department of Earth Sciences, Dalhousie University, Halifax, Nova Scotia, Canada

K. I. Saunders
Petro-Canada, Calgary, Alberta, Canada

N. E. Tibert
Department of Environmental Science and Geology, Mary Washington College, Fredericksburg, Virginia, U.S.A.

J. A. White
Department of Earth Sciences, Dalhousie University, Halifax, Nova Scotia, Canada

ABSTRACT

Economic coals of the Sydney Basin lie in high-frequency sequences mostly bounded by calcretes, indicative of base-level lowering on the coastal plain. Most coals represent blanket coastal peats that accumulated just prior to maximum transgression, which is commonly marked by dark limestone and shale with a restricted marine fauna. Thin coals are also present in highstand deposits. Coal and sequence thickness show a general correlation, confirming a link between potential accommodation and peat accumulation. Composite sequences commence with sustained coastal progradation and base-level lowering, and coals are especially prominent in a transgressive sequence set with a slightly retrogradational to aggradational style. Sequence architecture was controlled by high-accommodation events of relative sea level rise followed by relative falls, the expression of glacioeustatic events in a cratonic basin with moderate subsidence rate. Although thin (average 19 m [62 ft]) and thick (average 55 m [180 ft]) sequences show apparently similar architecture, thick sequences contain thick alluvium with cryptic transgressive units, and their detailed architecture probably reflects channel switching, climatic, and/or tectonic effects during prolonged, low-accommodation periods. Thick sequences represent the most landward transgressions and probably pass basinward into composite sequences.

The coal-bearing interval or "coal window" in the basin fill is about 1.5 km (0.9 mi) thick and reflects long-term accommodation driven by subsidence, as well as climatic control. Economic coals formed while the outcrop belt lay within range of high-accommodation, relative sea level events, and they show little petrographic change through the coal window. Upward loss of coals reflects regional progradation of the alluvial plain, coupled with climatic change as Pangea drifted northward.

INTRODUCTION

Coal distribution within stratigraphic successions is commonly predictable because accumulation of peat (the precursor to coal) reflects the dependence of plant growth on water availability, which in turn may be linked to accommodation change (Ryer, 1983; Cross, 1988; Bohacs and Suter, 1997; Diessel et al., 2000). The base level of deposition for peat accumulation is a surface or water level above which organic decay (humification) exceeds organic supply (Flint et al., 1995). This surface approximates the ground-water table for flow-fed (rheotrophic) peat, but may be considerably elevated for rain-fed (ombrotrophic) peat.

In Quaternary coastal settings, sea level rise over periods of thousands of years can cause ground water to pond landward of the transgressive limit, promoting the formation of extensive freshwater peat (Kosters and Suter, 1993; Supardi et al., 1993), and such effects can extend many tens of kilometers inland (Bohacs and Suter, 1997). The influence of relative sea level rise on peat accumulation and alluvial architecture also can be documented in the ancient record where suitable outcrop belts exist (Shanley and McCabe, 1993). Coals featured prominently in early interpretations of the transgressive-regressive patterns of Pennsylvanian cyclothems (Udden, 1912; Weller, 1930), and numerous authors have linked coal to potential accommodation created during the transgressive to highstand part of base-level transit cycles, typically in the Milankovitch band (Arditto, 1991; Gastaldo et al., 1993; Hartley, 1993; Shanley and McCabe, 1993; Gibling and Bird, 1994; Flint et al., 1995; O'Mara and Turner, 1999; among many others). In a comprehensive overview, Bohacs and Suter (1997) noted that coals are especially prominent in the late lowstand-early transgressive and late transgressive-early highstand systems tracts of high-frequency sequences. In inland settings, peat accumulation may reflect lake level, the ground-water table in river valleys, and recharge zones that border uplands (Courel et al., 1986; Rust and Gibling, 1990; Diessel, 1992; Calder, 1994).

Peat accumulation also reflects factors other than short-term accommodation creation, some of which are operative over long periods. In tectonically active areas, differential subsidence linked to faults and inherited basement topography can generate local accommodation and promote peat accumulation over a range of timescales (Weisenfluh and Ferm, 1984; Fielding, 1987; Titheridge, 1993; Hartley, 1993; Greb and Chesnut, 1996; Pashin, 1998). Temporary drainage diversions also may allow peat to accumulate where sediment supply is reduced (Wise et al., 1991). On a basinal scale and over millions of years, subsidence rate provides a fundamental control on accommodation, with a rate of change much slower than that of the ground-water table (Bohacs and Suter, 1997), and the geographic distribution of peat and coal is linked closely to climatic factors, especially the precipitation/evaporation ratio and the degree of seasonality (Lottes and Ziegler, 1994).

Because the relative importance of controls on coal distribution is related to the timescale under consideration, we examine coals in their sequence-stratigraphic context through a thick interval in the Carboniferous Sydney Basin of Atlantic Canada. In particular, we present a previously unpublished 130-m (427-ft) section to illustrate lines of evidence used in inferring key bounding surfaces (sequence boundaries and flooding surfaces). We then present models for peat accumulation on three scales of stratal thickness (a proxy for time). First, we examine single sequences about 20 m (66 ft) thick (~100,000 yr), where peat accumulated in response to individual high-accommodation events. Second, we examine composite (grouped) sequences about 50–150 m (164–492 ft) thick (~500,000 yr), where peat accumulated preferentially during certain parts of longer base-level cycles. Third, we examine 1.5 km (0.9 mi) of basinal fill, where coal distribution (the "coal window") reflects long-term patterns of subsidence, climate, and basin hydrology.

GEOLOGICAL SETTING OF THE SYDNEY BASIN

The regional Maritimes Basin of Atlantic Canada (inset, Figure 1) is a late Paleozoic depocenter in the

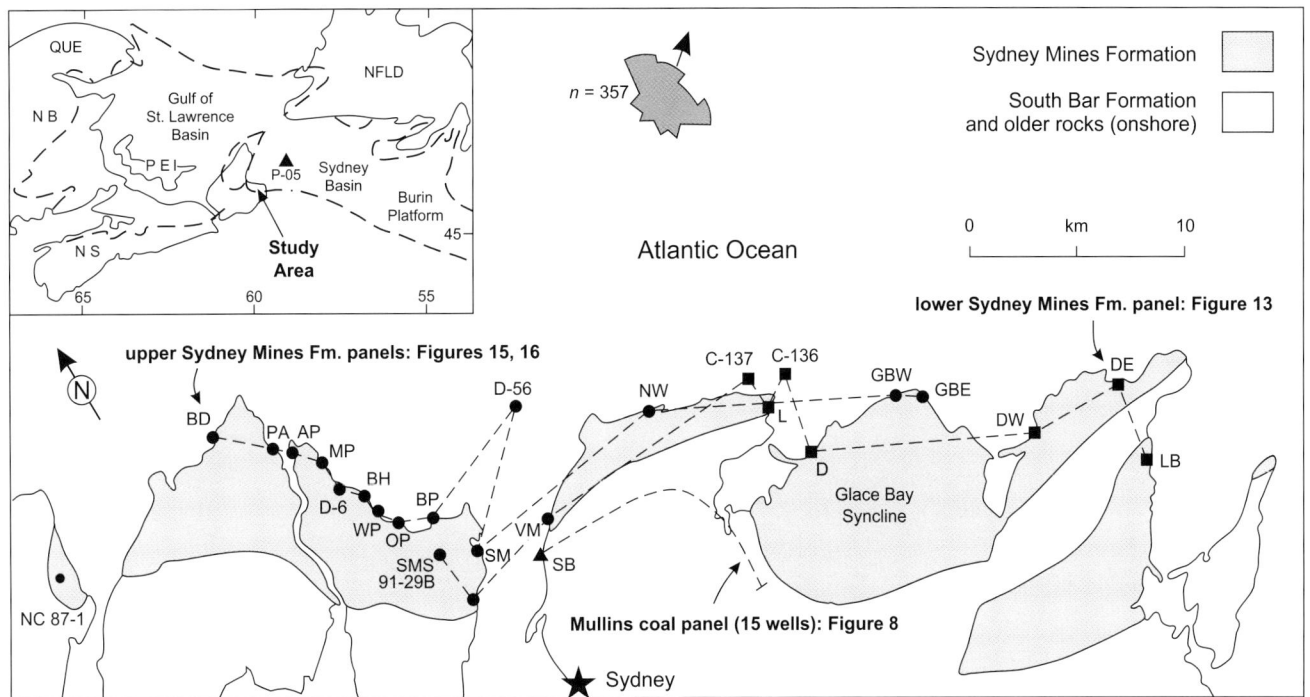

Figure 1. Location of measured sections and wells in the Sydney Mines and South Bar Formations, Sydney Basin, Nova Scotia. Rose diagram shows 357 trough cross-beds and vector mean for channel bodies of the two formations. BD = Bras d'Or; PA = Point Aconi; AP = Alder Point; MP = Merritt Point; BH = Bonar Head; WP = Wetneck Point; OP = Oxford Point; BP = Black Point; SM = Sydney Mines; VM = Victoria Mines; SB = South Bar; NW = New Waterford; L = Lingan; D = Dominion; GBW and GBE = Glace Bay West and East; DW and DE = Donkin West and East; LB = Long Beach. Inset map shows the area of the regional Maritimes Basin (enclosed by dashed lines), which includes the two major Upper Carboniferous to Permian depocenters of the Gulf of St. Lawrence and Sydney Basins. Triangle shows position of P-05 well offshore. NB = New Brunswick; NS = Nova Scotia; NFLD = Newfoundland; PEI = Prince Edward Island; QUE = Quebec.

northeastern Appalachian orogen. The basin is a structural remnant of a formerly more extensive region, including the Gulf of St. Lawrence and Sydney Basins, that lay near the equator during the Pennsylvanian.

The Pennsylvanian fill of the Sydney Basin comprises the Morien and Pictou Groups (Figure 2). The Morien Group (Duckmantian–Bolsovian to Cantabrian) is as much as 1800 m (5906 ft) thick and rests on Visean and Namurian strata, with an angular unconformity that correlates broadly with the Mississippian–Pennsylvanian unconformity of the Appalachian basin (Pascucci et al., 2000). The basal South Bar Formation is at least 860 m (2822 ft) thick onshore and consists of braided-fluvial sandstones with minor mudstone and coal (Rust and Gibling, 1990); uncommon, thick coals suggest marine incursion (Tibert and Gibling, 1999). The overlying Sydney Mines Formation is about 1000 m (3281 ft) thick and consists of stacked cyclothems of sandstone, gray and red mudstone, coal, dark limestone, and calcrete, deposited in alluvial to restricted marine conditions and attributed to glacioeustasy (Gibling and Bird, 1994). Eco-

nomic coals as much as 4.3 m (14 ft) thick are present throughout the formation (Figure 2). For the purposes of this study, each formation is divided into informal lower and upper units. Red mudstone and sandstone with minor coal of the Pictou Group [Stephanian to Permian(?)] is not exposed but may be 1000 m (3281 ft) thick under the nearshore area. Paleoflow for the Morien Group was broadly northeastward (Figure 1), and rivers that traversed the basin probably rose in the Appalachian orogen and drained eastward to fully marine basins in southern Europe (Gibling et al., 1992).

Marine influence during Morien Group deposition is indicated by the presence of high-sulfur coals (Hacquebard and Donaldson, 1969), glaucony (Batson and Gibling, 2002), and invertebrates (especially agglutinated foraminifera) with marine affinities. Faunal-concentrate beds contain abundant bivalves and ostracods, but open-marine faunal elements such as goniatites have not been identified.

Stratal continuity can be documented through well-exposed coastal sections along a 50-km (31-mi),

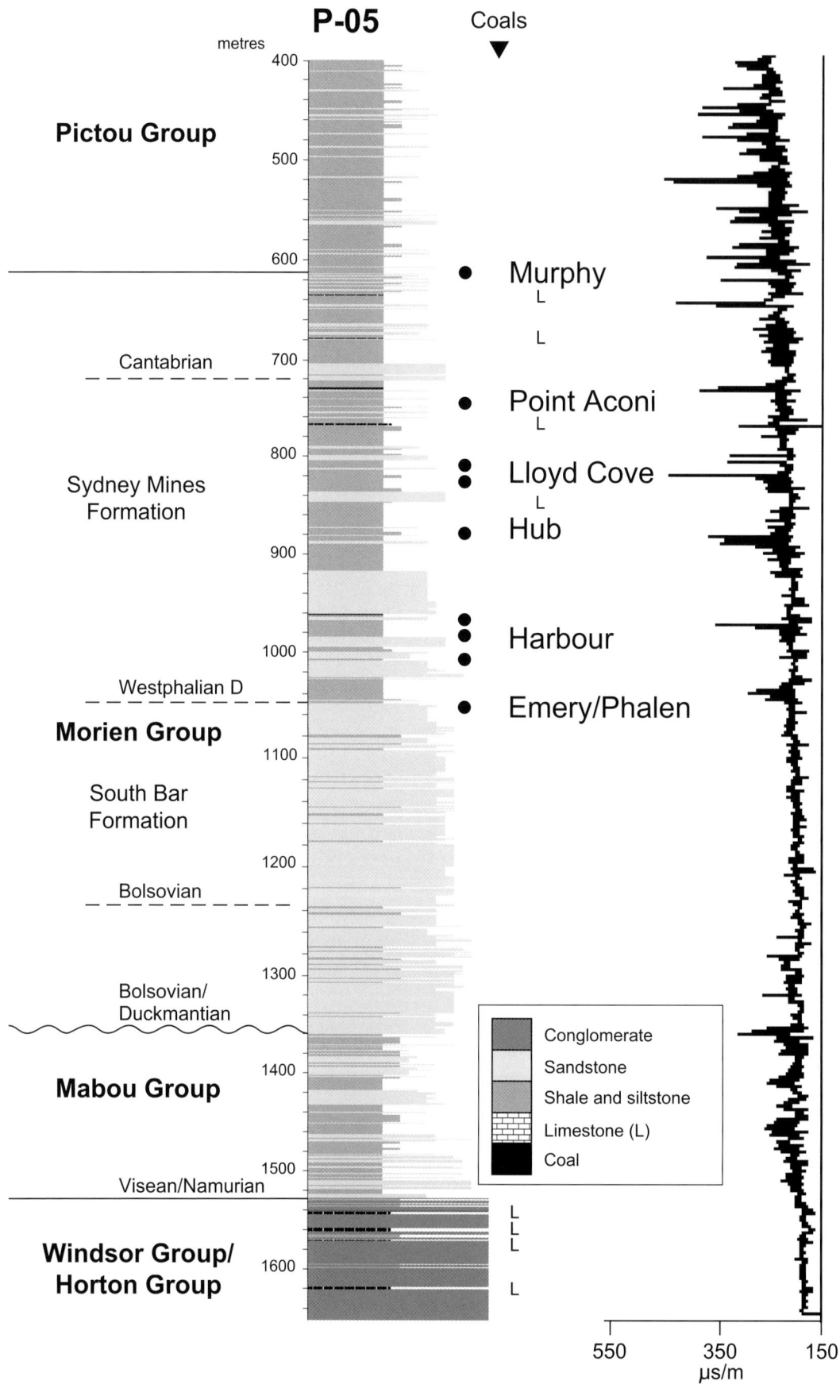

northwest-southeast traverse, which is nearly parallel to depositional strike as indicated from paleoflow (Figure 1). Identification of the major economic coals (Robb, 1876) allows stratigraphic intervals of less than 20 m (66 ft) to be reliably correlated over this distance. However, little information is available about proximal to distal trends, apart from limited information from the P-05 and adjacent F-24 wells about 50 km (31 mi) offshore (inset, Figure 1). Across the subsea part of the basin, the coal measures form a seismic reflector package that dips gently and onlaps basement rocks (Pascucci et al., 2000), a blanket stratal geometry that suggests minimal intrabasinal tectonism and a cratonic sedimentation style.

Palynological assessment for the base of the Westphalian D and Cantabrian suggests that about 520 m of strata represent the Westphalian D, which lasted for about 3 m.y. (Menning et al., 2000). Consequently, the long-term stratal accumulation rate (not corrected for compaction) was about 173 m/m.y. (568 ft/m.y.). This estimate is poorly constrained, assuming a constant subsidence rate and disregarding differential compaction and varied accumulation rates for individual facies. Menning et al. (2000) summarized Westphalian D stratal thickness for several European foreland and other basins, and the Sydney Basin estimate lies at the low end of this thickness range, in accord with a modest subsidence rate and cratonic style. Based on thickness and facies evidence, Morien strata accumulated in local paleotopographic lows that correspond to present-day synclines (e.g., Glace Bay syncline; Figure 1). This implies modest Pennsylvanian fault movement and/or compactional draping over underlying basement blocks and older extensional basins (Hacquebard, 1983; Gibling et al., 2002).

Formations that correspond closely in lithology, age, and coal distribution to those at Sydney underlie much of the Gulf of St. Lawrence, where they also rest on a profound middle Carboniferous unconformity (Rehill, 1996; Giles and Utting, 1999). Grant (1994) correlated a coal-reflector package across much of the Gulf area, and the two coal basins originally may have been united. Duckmantian–Bolsovian to Stephanian coal measures currently underlie greater than 100,000 km^2 (38,610 mi^2) in the Maritimes Basin, an area broadly comparable to that of the Illinois basin.

The Sydney seams are high-volatile C to medium-volatile bituminous coals, with vitrinite reflectance R$_o$ from 0.59 to 1.29% (Hacquebard and Donaldson, 1969; Hacquebard, 1983, 1998; Marchioni et al., 1996). They are banded humic coals with vitrinite predominant, low to moderate inertinite and liptinite, abundant pyrite, and average sulfur content from 0.8 to 6.2%. Common seam splits, high ash content (typically >10%), and extensive dull bands suggest that the mires were periodically inundated and rheotrophic (Marchioni et al., 1994; White et al., 1994; Calder et al., 1996).

SEQUENCE DESCRIPTION

The following sedimentological analysis emphasizes well-studied sections and places the coals in the context of sequences and sequence sets (van Wagoner et al., 1988; Mitchum and van Wagoner, 1991), extending our earlier analysis of certain intervals. We follow Muto and Steel (2000) in evaluating "accommodation" in terms of the thickness of preserved strata at a given place and time; in contrast, "potential accommodation" denotes the maximum of possible accommodation, which corresponds closely to the height of a water column at a given place and time.

Individual coals in the Morien Group fall into three main depositional settings (Figures 3, 4): (1) thin and reworked peat in alluvial settings on braided-river plains (lower and upper South Bar Formation); (2) coastal peat on braided-river plains (upper South Bar Formation); and (3) coastal peat on alluvial plains with meandering and distributary channels and valley fills (Sydney Mines Formation).

Lower South Bar Formation

The basal 430 m (1411 ft) of the South Bar type section (Figure 5) consists mainly of trough cross-bedded sandstone and pebbly sandstone with minor mudstone (Rust and Gibling, 1990). The strata comprise stacked channel bodies as much as 13 m (43 ft) thick (typically 1–5 m [3.3–16.4 ft]), with erosional lags of extrabasinal clasts, mudstone intraclasts, and plant and coal fragments. Laminated mudstones represent deposition of fines on bar and bank tops and in

Figure 2. Lithological column and sonic log for P-05 well in the offshore part of the Sydney Basin (Figure 1). Coals (identified from cuttings and sonic logs) were correlated with those onshore by Hacquebard (1976), mainly on petrographic grounds. Stage boundaries are from Barss et al. (1979) and differ slightly in position from assessments of equivalent strata onshore (Figures 5, 9). Note that the well was drilled on a basement high, and stratal thicknesses are reduced compared with those in onshore area.

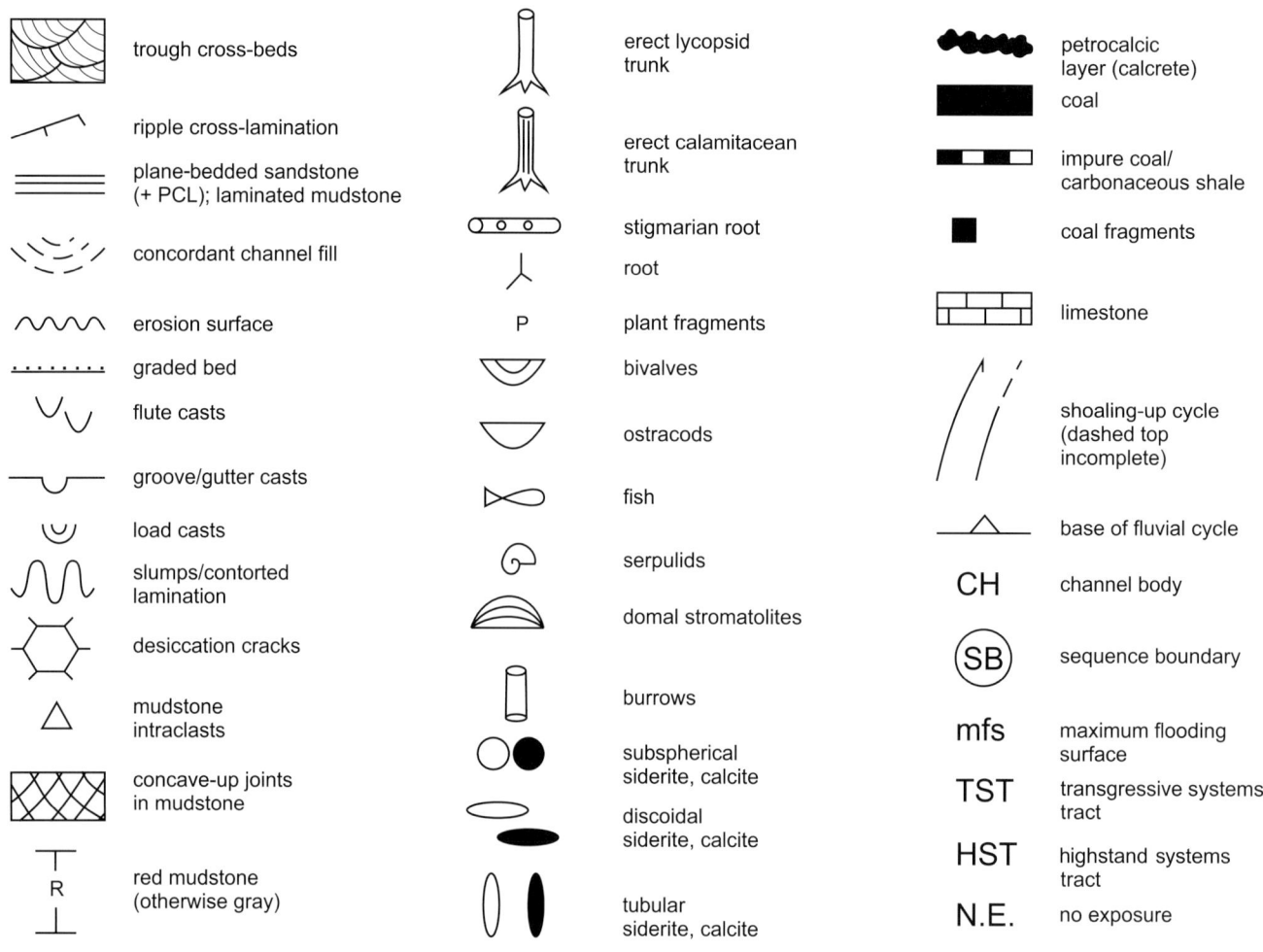

Figure 3. Legend for stratigraphic logs.

abandoned channels, whereas rotated and deformed mudstone lenses are channel-base slump blocks (Figure 5). The strata were laid down in proximal braided rivers that progressively filled bedrock valleys cut into underlying formations (Rust and Gibling, 1990). No clear stratal organization is apparent. Although some channel-base erosion surfaces could constitute sequence boundaries, they are uncommonly exposed for more than 100 m (328 ft) in coastal sections, and their regional significance cannot be judged.

Thin, discontinuous coals include the Clarke, Martin, Lorway, Le Cras, Round Island, Shoemaker, and McAulay coals, are typically decimeters thick, and are closely intercalated with sandstones (Robb, 1876; Hayes and Bell, 1923). None are apparently traceable for more than a few kilometers, and Rust and Gibling (1990) inferred that they formed in local backswamps in the braided-river belt. Peat was probably much more abundant than the present distribution of coal would suggest. Finely banded coal fragments (clearly distinguishable from plant fragments) were identi-

fied in channel-base lags at 38 levels in the South Bar type section (Figure 5) and are locally as much as 30 cm (12 in.) thick and 1.5 m (5 ft) long, with strongly splayed out terminations (Figure 6). The fragments are interpreted as remnants of peats that were undercut by channel migration.

Upper South Bar Formation

The 430–739-m (1411–2425-ft) interval of the upper South Bar Formation contains two thick mudstone-coal intervals (Figure 5). The Mullins coal is as much as 2.1 m (7 ft) thick and averages 5.9% sulfur. At South Bar (Figure 7), it rests on a thin-rooted mudstone that caps braided-fluvial sandstones and is overlain by a coarsening-upward unit capped by a thin, impure coal (Tibert and Gibling, 1999). Sparse agglutinated foraminifera were identified in mudstones associated with the coal. The Mullins extends for about 15 km (9 mi) along strike, shaling out eastward into the Glace Bay syncline (Figure 8) and probably eroded below channel sandstones west of South Bar.

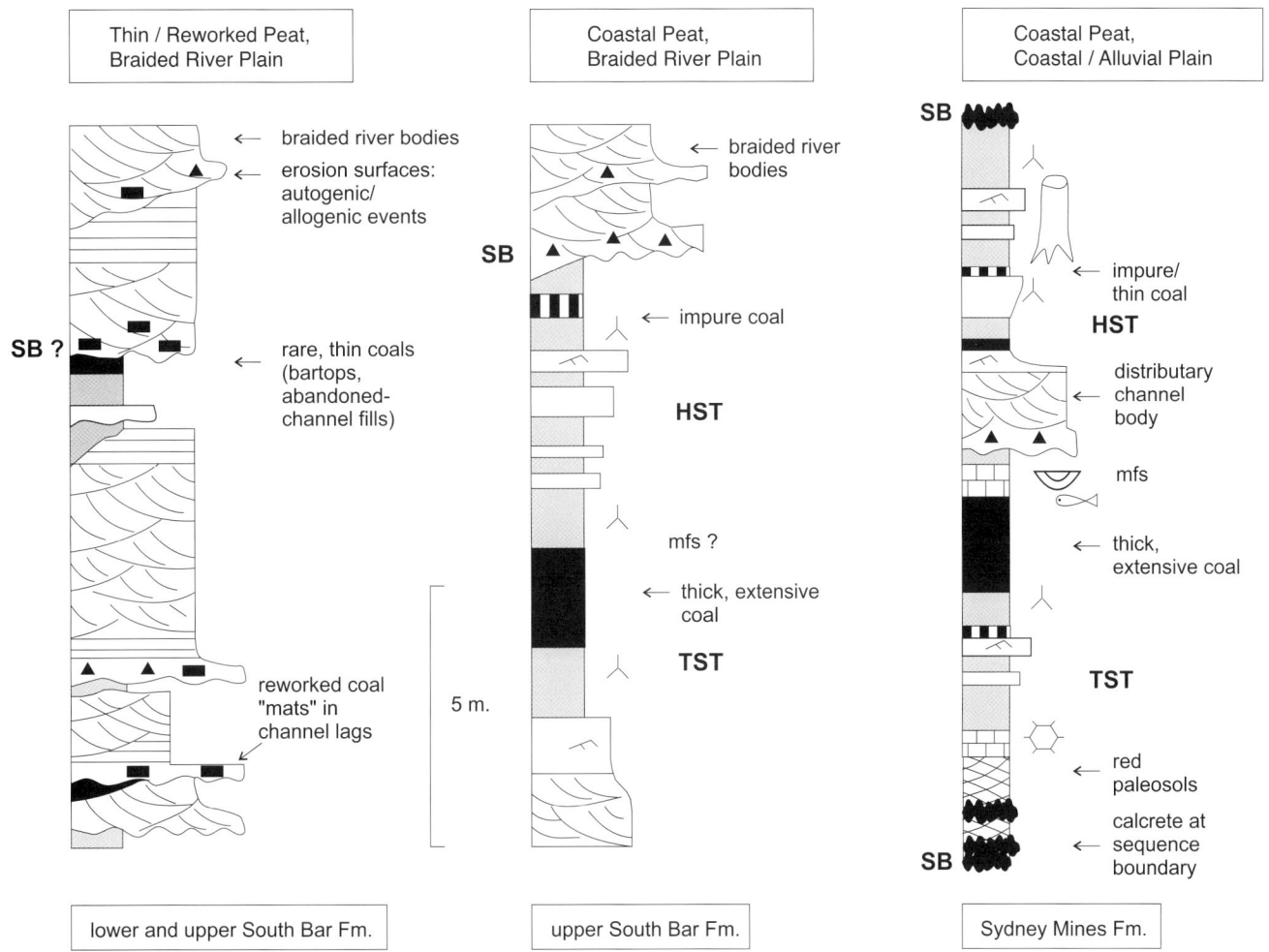

Figure 4. Summary of coal occurrence in the Morien Group, Sydney Basin. Legend in Figure 3. SB = sequence boundary; TST and HST = transgressive and highstand systems tracts, respectively; mfs = maximum flooding surface.

Tibert and Gibling (1999) interpreted the Mullins as a coastal peat blanket that accumulated late in transgression, close to a maximum flooding level. The incoming of braided-fluvial sandstones marks a strong basinward facies shift and sequence boundary (Figures 7, 8), although the South Bar braided rivers may have debouched directly into the sea. A stable, noncompactible sandstone platform was probably conducive to achieving a rate of potential accommodation creation suitable for peat accumulation (Nemec, 1992; Hampson et al., 1996). The overlying channel-sandstone interval shows no conspicuous stratal organization.

An uncommonly thick (15 m [49 ft]) mudstone lies at 650 m (2133 ft) in the South Bar section (Figure 5). The basal 7 m (23 ft) is dark gray with plant fragments, overlain by 3 m (10 ft) of red mudstone with a 30-cm (12-in.) calcic paleosol. Above this level, gray mudstone with thin sandstone beds is overlain

by channel sandstones with 20 cm (8 in.) coal fragments, suggesting the former presence of peat. At a comparable stratigraphic level, the 60-cm (24-in.) Ingraham coal is overlain by thick mudstone south of Sydney Mines (Robb, 1876; DeWolfe, 1906), and the 40-cm (16-in.), highly pyritic Henderson coal overlies 8 m (26 ft) of red mudstone in drill hole NC87-1 (Figure 1) (Robb, 1876; MacNeil, 1987). Although coal correlation is uncertain (Bell, 1923), a mudstone-coal interval may extend for greater than 20 km (12 mi) along strike, probably representing coastal peat accumulation.

Lower Sydney Mines Formation

Facies descriptions for the lower Sydney Mines Formation were described in detail in earlier papers (Gibling and Bird, 1994; Tandon and Gibling, 1997; Batson and Gibling, 2002). Coastal-plain facies are gray sandstones and shales with coals, hydromorphic

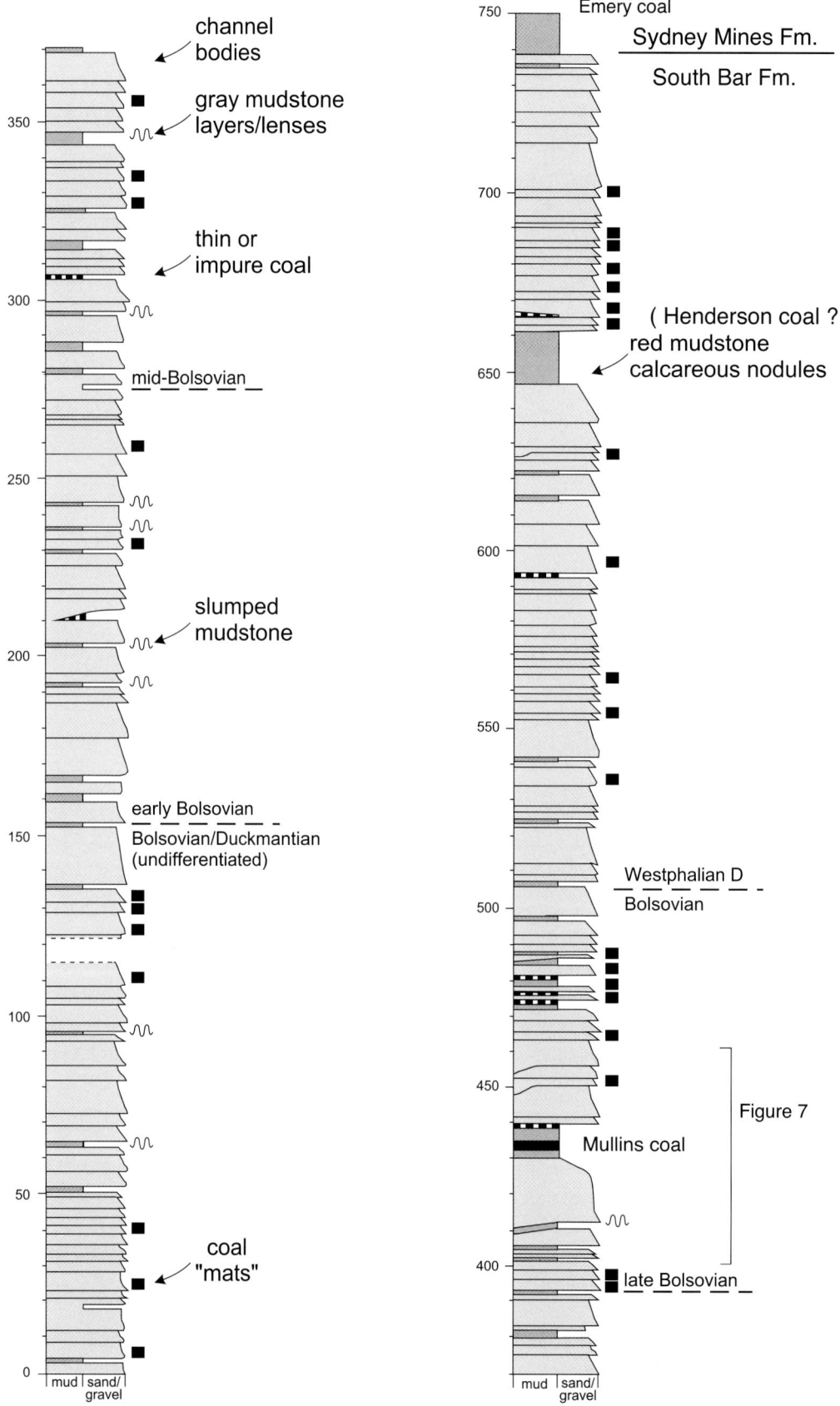

Figure 5. Stratal column for the upper 740 m (2428 ft) of the South Bar Formation at Victoria Mines (Figure 1). Stage boundaries from Dolby (1989). Mudstone unit at 650 m (2133 ft) may correspond with the Henderson coal near New Campbellton (MacNeil, 1987).

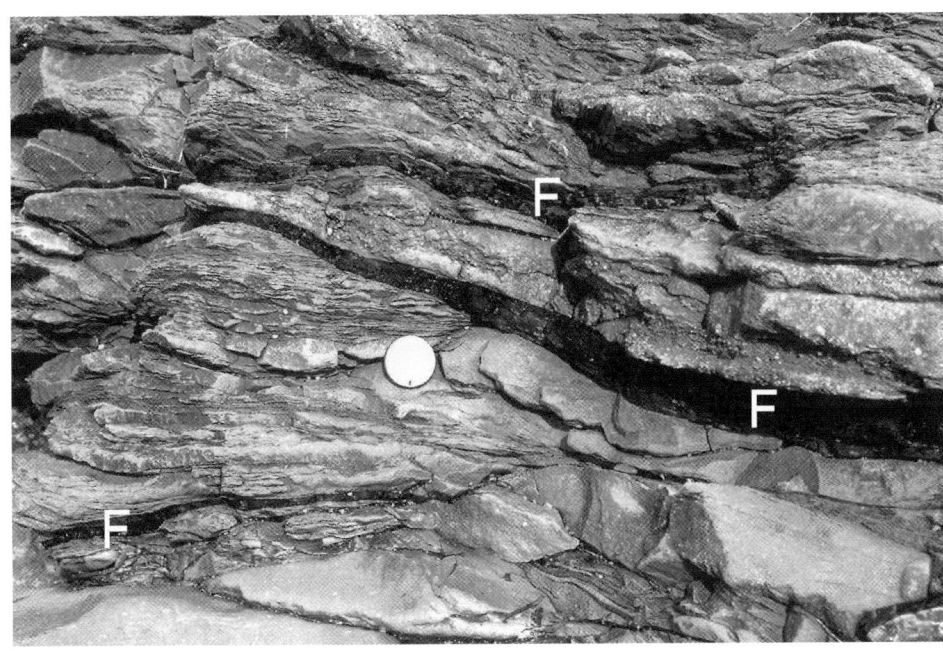

Figure 6. Coal fragments (F) in cross-bedded fluvial sandstones in the South Bar Formation at South Bar. Lens cap is 5 cm (2 in.) in diameter.

onshore were identified in wells (including P-05; Figure 2) as much as 50 km (31 mi) offshore (Hacquebard, 1976; Shimeld and Deptuck, 1998).

Sequence 1

The sequence includes the South Bar–Sydney Mines Formation contact, at which mudstone-dominated strata abruptly replace sandstone-dominated strata in a sustained base-level rise (Rust and Gibling, 1990), apparently without ravinement. The overlying 14 m (46 ft) of platy, unfossiliferous shales are probably lacustrine.

Sequence 2

The 22-m (72-ft) sequence commences with a pale, rooted paleosol overlain by a thin leaf of the Emery coal, which is highly split across the coalfield and averages 2.69% sulfur (Hacquebard and Donaldson, 1969). The maximum flooding surface is provisionally placed at the top of the thick, extensive topmost leaf of the Emery. The transgressive systems tract below this level is thin and coal dominated, with two shoaling-up units (parasequences) that commence with thin coals. At Lingan, the thick, upper Emery leaf rests on a gray mudstone with concave-up joints and desiccation cracks filled with black, coaly material, a candidate sequence boundary at a slightly higher level. The coal at Victoria Mines is overlain by a small distributary channel body (Batson and Gibling, 2002), and the highstand systems tract consists of a coarsening-up unit capped by dryland facies. The latter includes a 30-cm (12-in.) pale limestone with algal domes, ostracods, serpulids, and fish, a facies that contrasts strongly with the black fauna-rich limestones that overlie many younger coals.

Sequence 3

The boundary of this 14.5-m (48-ft) sequence is placed in a thick gray mudstone with roots, prominent concave-up joint sets, and (at Lingan) calcareous

paleosols, and dark limestone and shale with faunal-concentrate beds, attributed to bayfills, mires, and distributary channels (wetland settings). The facies closely resemble the bayfills and distributary channels of modern shelf-phase deltas such as the Mississippi, but are described here more generally as "coastal-plain deposits" because their three-dimensional form is unknown. Alluvial-plain facies are typically red mudstones and sandstones, with nodular calcrete, attributed to channels and flood plains with well-drained paleosols (dryland settings). Paleovalley fills of thick sandstone are incised into coastal-plain deposits. The fauna include the bivalves *Anthraconauta*, *Anthraconaia*, and *Naiadites*, the ostracods *Carbonita* and *Candona*, estheriids, branchiopods, the serpulid worm *Spirorbis*, agglutinated foraminifera and thecamoebians, fish fragments, algal stromatolites, and fragments of algae resembling *Girvanella*, *Garwoodia*, and *Ortonella* (see Copeland, 1959; Gibling and Bird, 1994; Calder, 1998). Some foraminifera, bivalves, and algae suggest marine influence (Calver, 1968; Mamet, 1992). Floral elements of the Morien Group are listed by Calder (1998).

The basal 145 m (476 ft) of the Bras d'Or section provides a near-complete transect through the formation onshore and comprises nine high-frequency sequences (Figure 9). The lower six sequences and the base of sequence 7 are especially well exposed at Victoria Mines (Figure 10), Lingan, and Dominion, where they average about 19 m (62 ft) in thickness (sequences 2–6). Although the extent of individual coals has not been established, most economic seams

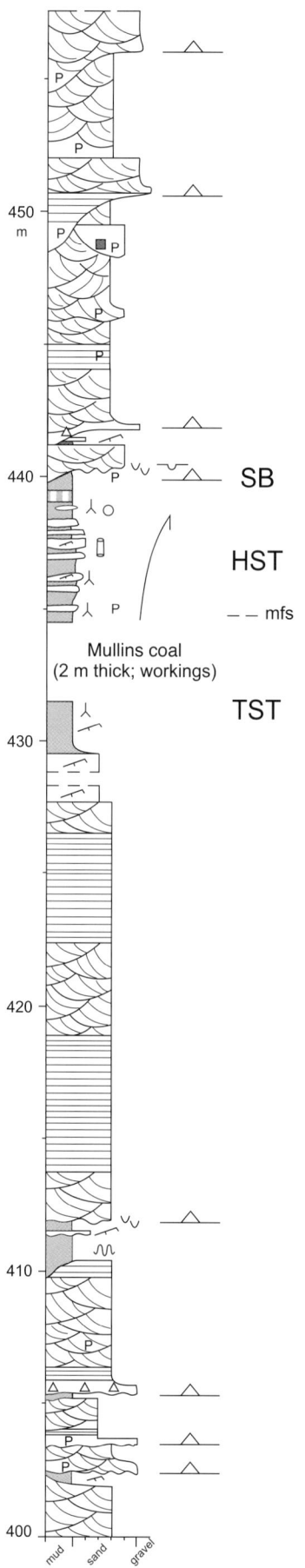

nodules. Similar beds were interpreted by Tandon and Gibling (1997) as calcic, vertisol-like paleosols, with the joint sets representing shrink-and-swell features formed under a seasonal climate. The mudstone is 4 m (13 ft) thick, and no single surface is easily identified as the sequence boundary. The paleosol is overlain by the Stony coal, which is capped by a black, bivalve-ostracod limestone with 8–10% organic matter, predominantly sporinite with minor telalginite (Gibling and Kalkreuth, 1991, who incorrectly identified the underlying coal as the Phalen). The limestone is taken to include the maximum flooding surface, so that the Stony coal forms most of the transgressive systems tract. The highstand systems tract comprises 12 m (39 ft) of wetland deposits, gray and carbonaceous shales, thin coals, rooted mudstones, and thin distributary channel bodies. At Lingan and Dominion, these strata include dark, faunal limestones, calamitaceans, and the stigmarian roots of lycopsid trees.

Sequence 4

The 23-m (75-ft) sequence commences with two 30-cm (12-in.) calcretes with a prominent vertic fabric, intepreted by Tandon and Gibling (1997) to represent calcrete development to stage II/III of Machette (1985). The upper calcrete separates mudstone strata with siderite nodules below from strata with predominantly calcareous nodules above and denotes a sustained ground-water fall. It is overlain by 10 m (33 ft) of red and gray mudstone with concave-up, slickensided joints and scattered calcareous nodules. Gray siderite-rich mudstone reappears just below the Phalen coal, the top of which is taken as the maximum flooding surface. Thus, the transgressive systems tract contains a thick succession of dryland soils. Equivalent paleosols at Lingan and Dominion include a pale limestone with bivalves, ostracods, fish, algal domes, stigmarian roots, and desiccation cracks, interpreted by Tandon and Gibling (1997) as a palustrine limestone with a low exposure index. The Phalen coal is as much as 2.5 m (8 ft) thick, averages 3.15% sulfur (Hacquebard and Donaldson, 1969; Hacquebard, 1998), and is overlain by flood-plain deposits with distributary-channel bodies, assigned to the highstand systems tract. At Dominion, a muddy abandoned-channel fill

Figure 7. Stratal column for the Mullins coal interval in the South Bar Formation (Figure 5) at Victoria Mines (Figure 1). Legend in Figure 3.

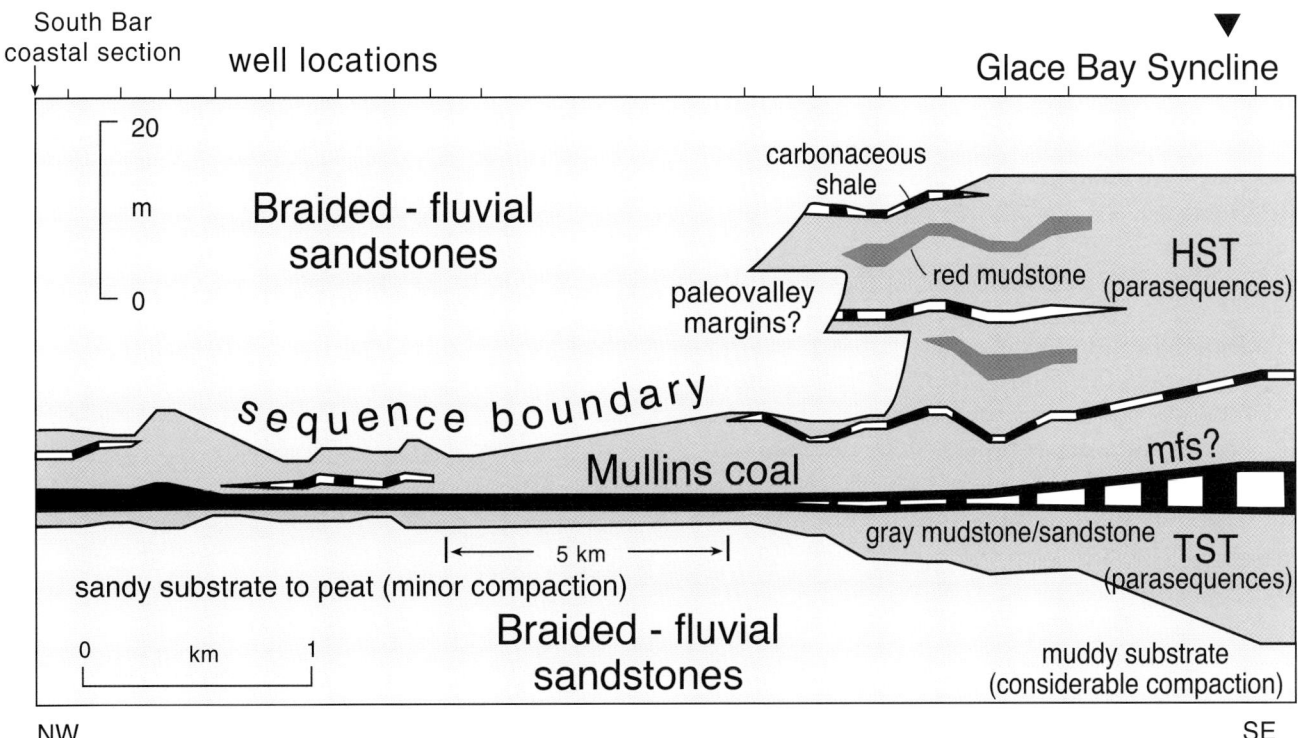

Figure 8. Correlation of Mullins coal interval in wells and coastal section (line of section in Figure 1). mfs = maximum flooding surface; TST and HST = transgressive and highstand systems tracts, respectively. Modified from Tibert and Gibling (1999).

4 m (13 ft) thick includes a shaly coal and a limestone with bivalves.

Sequence 5

The boundary of the 12-m (39-ft) sequence is taken at a rooted, slightly calcareous horizon at the top of a channel body, suggesting a slight break in sedimentation. The overlying carbonaceous shale and thin unnamed coal are capped by an extensive, 60-cm (24-in.) limestone with bivalves, ostracods, and desiccation cracks. The limestone is overlain by flood-plain deposits that include an uncommonly coarse, pebbly channel body at the 80-m (262-ft) level. Coal forms most of the thin transgressive systems tract.

Sequence 6

The 21.7-m (71-ft) sequence commences with one of the most prominent calcrete zones (2 m [7 ft] thick) in the formation (Figure 11a). Calcareous nodules are developed in the top of a channel body and in two overlying coarsening-up units. The nodules increase in abundance upward, and the topmost calcareous layer (taken as the sequence boundary) comprises coalesced nodular masses as much as 70 cm (28 in.) wide and 30 cm (12 in.) thick. Overlying strata in-

clude limestone and gray, calcareous mudstone with concave-up joints, and the transgressive systems tract thus includes a thick, dryland component.

The Backpit coal is a single, unsplit seam as much as 1.5 m (5 ft) thick across the entire onshore outcrop belt (Figure 12), averaging 5.2% sulfur, with two closely spaced leafs in wells 5–10 km (3–6 mi) offshore (White et al., 1994; Shimeld and Deptuck, 1998). Dulling-upward pulses in the upper part of the seam were interpreted by White et al. (1994) as inundation events that culminated in the drowning of the mire. The overlying dark limestone and shale (Backpit roof unit of White, 1992; Figure 12) is as much as 75 cm (30 in.) thick and extends for greater than 45 km (28 mi) along strike. The unit comprises (1) fine calcarenite with reworked shell fragments and a high sulfur content; (2) carbonate rich in single and articulated bivalve and ostracod shells; (3) massive, fine-grained dark carbonate; (4) dark gray to black shale; (5) siderite bands; and locally, (6) pale, calcareous mudstone with roots. At Victoria Mines, a basal calcarenite passes up into fossiliferous and massive carbonate, overlain by black shale. Fossils include bivalves, ostracods, branchiopods, serpulids, fish (including xenacanthids and lungfish), indeterminate bone fragments,

Figure 9. Stratal column for the lower 400 m (1312 ft) of the Sydney Mines Formation (to the highest exposed strata) at Bras d'Or (Figure 1). Coals below the Backpit coal are relatively thin and impure compared with their equivalents in the Victoria Mines to Glace Bay area. Stage boundary from Dolby (1988); see also Zodrow and Cleal (1988).

agglutinated foraminifera, and vascular plant fragments. Phosphatic nodules are present, at least some of which are coprolites. The unit contains as much as 11.3% by volume of pyrite as crystals and framboids, and total organic carbon averages 3.7% (as much as 20%), mainly liptinite (sporinite, cutinite, lamalginite, and liptodetrinite) with some vitrinite and inertinite. The unit is capped by coarsening-up organic-poor mudstone and siltstone with distributary channel bodies and (at Lingan and Dominion) thin coals, limestones, and lycopsid trunks.

The widespread basal layer of reworked shell fragments suggests that the Backpit roof unit rests on a transgressive ravinement surface on top of the coal, probably associated with a restricted-marine incursion. Carbonates and faunal-concentrate horizons accumulated under conditions of detrital sediment starvation, whereas black shale and overlying coarser beds suggest the resumption of detrital sedimentation. The maximum flooding surface is placed at the carbonate-shale boundary.

Sequence 7

The sequence boundary is placed at a thin mudstone with a prominent petrocalcic horizon. The overlying 20 m (66 ft) below the Bouthillier coal consists of mudstones that are mainly red below and gray above, with roots, concave-up joints, scattered calcareous nodules, and one pale fossiliferous limestone bed. At Sydney Mines, gray bayfills of sequence 6 are capped by a prominent nodular calcrete underlain by calcite-cemented bayfill sandstones and overlain by red vertic paleosols (Figure 11B). Similar prominent calcretes, locally overlain by redbeds, are also present at Bras d'Or and Dominion.

Sequences 5 and 6 and the basal part of sequence 7 are readily correlated across the area (Figure 13), with coals, capping limestones, bounding calcretes, and thin red mudstones traceable for greater than 30 km (19 mi) along strike. Stratal patterns are more complex at Donkin and Longbeach, where thick redbed units and more numerous calcretes appear. The unnamed coal and capping limestone

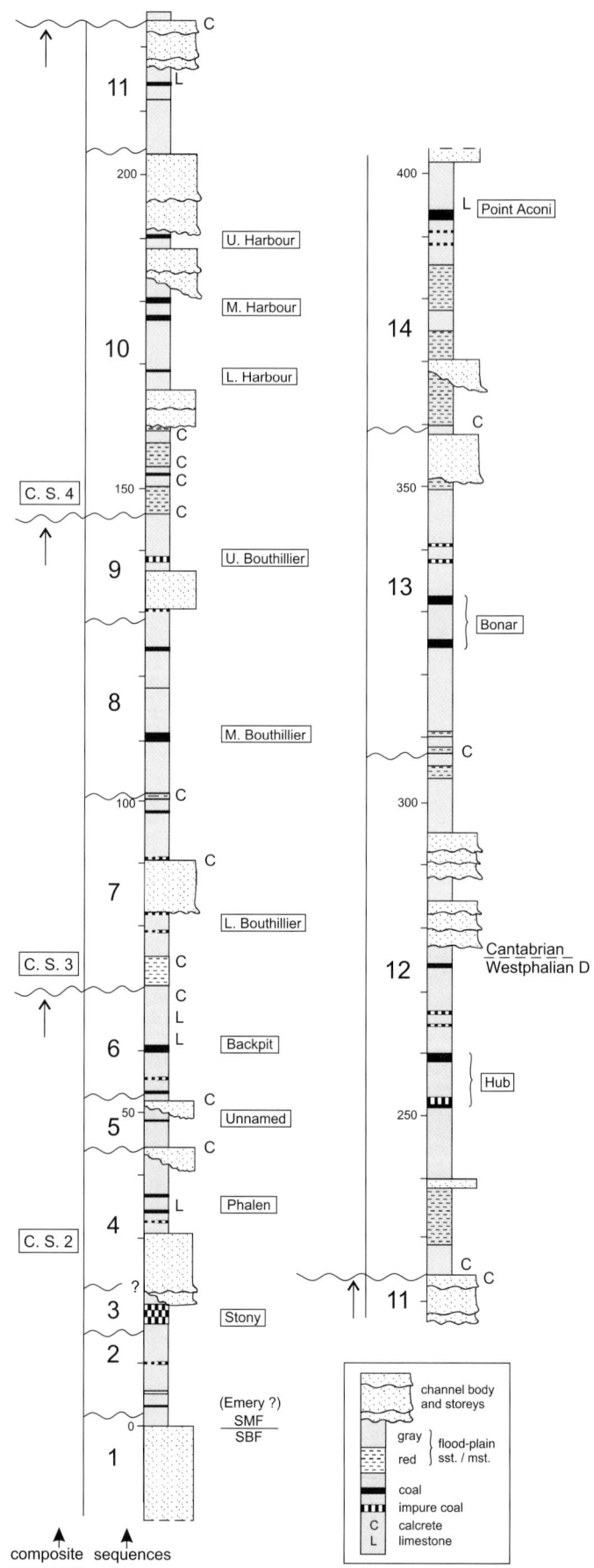

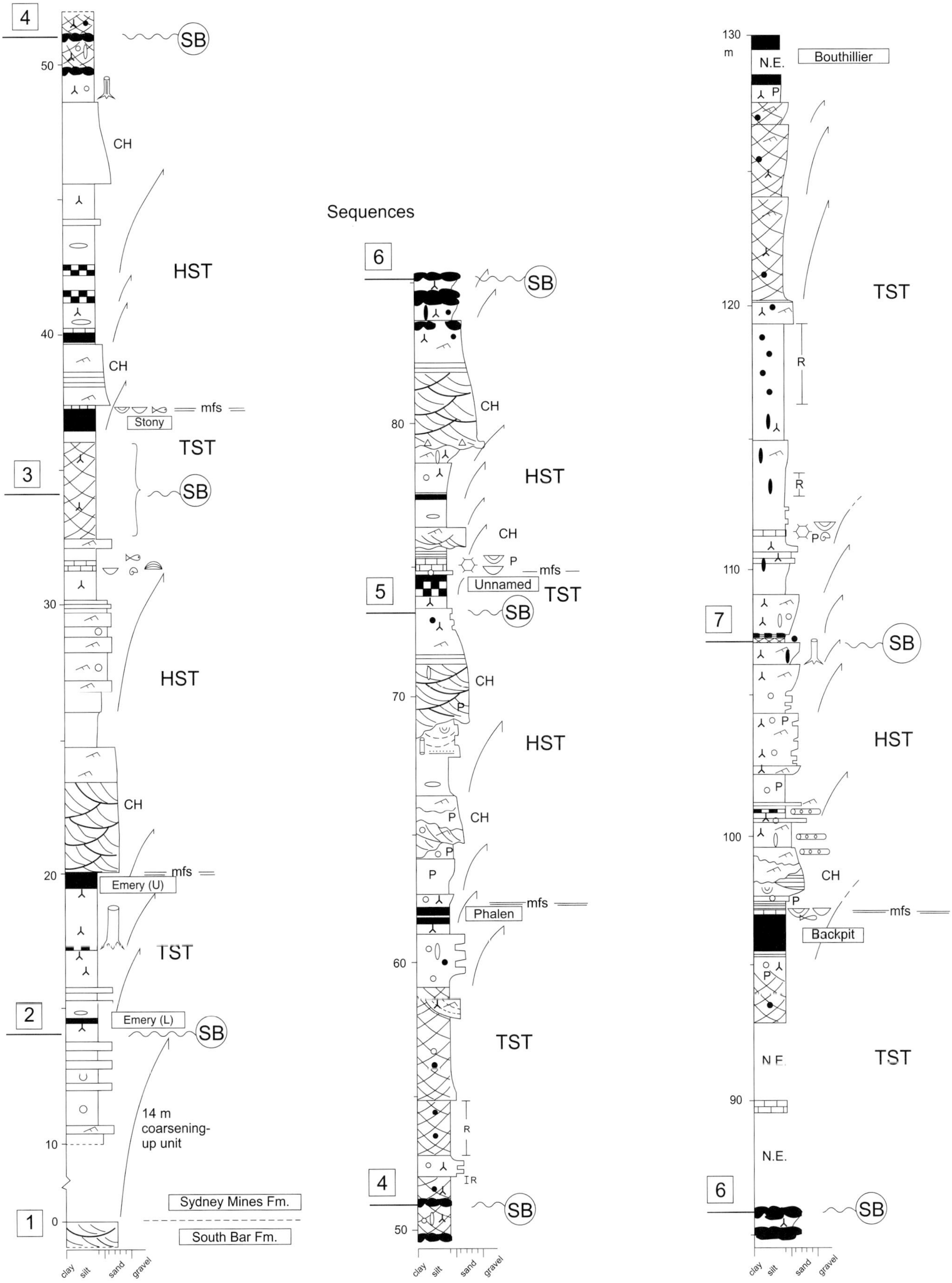

Figure 10. Stratal column for the lower Sydney Mines Formation at Victoria Mines (Figure 1). Legend in Figure 3.

Figure 11. (A) Nodular calcrete marking base of sequence 6, Victoria Mines. Sandstone with a slight degree of pedogenic alteration (S) passes upward into calcrete (C) with a strong vertic fabric and complete loss of stratification. The calcrete can be traced along strike for tens of kilometers. Scale is 10 cm (4 cm) long. (B) Base of sequence 7 at Sydney Mines (boundary of composite sequence 3). A 1-m (3.3-ft)-thick calcrete (C) marks the sequence boundary. It is underlain by gray bayfill deposits of sandstone and shale with siderite nodules (coastal plain deposits); sandstones below the calcrete (S) are calcite-cemented and resistant. The calcrete is overlain by red calcic vertisols (V) interbedded with splay sandstones (alluvial-plain deposits).

Figure 12. Backpit coal (B) in the Bras d'Or section. The resistant bed above the coal (L) is the Backpit roof unit, an organic-rich limestone and shale with faunal-concentrate horizons (see text). Scale is 1 m (3.3 ft) long.

die out toward this area, suggesting transition to a slightly elevated upland; however, thick coals such as the Backpit cross the area, indicating that relief was modest. Channel bodies are mostly relatively thin (as much as 5.5 m [18 ft]) and were interpreted by Batson and Gibling (2002) as distributary channel bodies associated with bayfills in the highstand systems tract. In contrast, a multistory valley fill 30 m (98 ft) thick near New Waterford cuts down 20 m (66 ft) from above the unnamed coal to the top of the Phalen coal and probably correlates with the uncommonly coarse channel body at 80 m (262 ft) in the Victoria Mines section (Batson and Gibling, 2002). At Lingan, the fill comprises stacked, erosionally based trough cross-sets, similar to the South Bar braided-fluvial facies.

Upper Sydney Mines Formation

The 145–400-m (476–1312-ft) interval in the Bras d'Or section (Figure 9) comprises four complete sequences (sequences 10–13) that are 20–85 m (66–279 ft) thick, averaging 55 m (180 ft), more than twice as thick as those in the lower part of the formation. Each sequence includes thick dryland (redbed) facies with numerous channel bodies.

The Hub-Bonar coal interval (sequences 12–13) can be divided into three assemblages (Figure 14). Assemblage 1 contains multistory channel-sandstone bodies as much as 15 m (49 ft) thick, with width/thickness ratios as much as 50:1 and lateral accretion sets. They are associated with sandstone sheets, gray mudstones, dark carbonaceous shale, and coal. The Hub coal is greater than 2 m (6.6 ft) thick, with as much as 4.5% sulfur; thinner coals are less common above. The assemblage represents coastal wetlands traversed by meandering rivers, probably part of a distributary channel network, and marine influence is indicated by high-sulfur coals and agglutinated foraminifera. Assemblage 2 consists mainly of red mudstone (paleosols of vertisol type), sandstone sheets, and narrow channel bodies, with a single, prominent calcrete as much as 1 m (3.3 ft) thick in most outcrops.

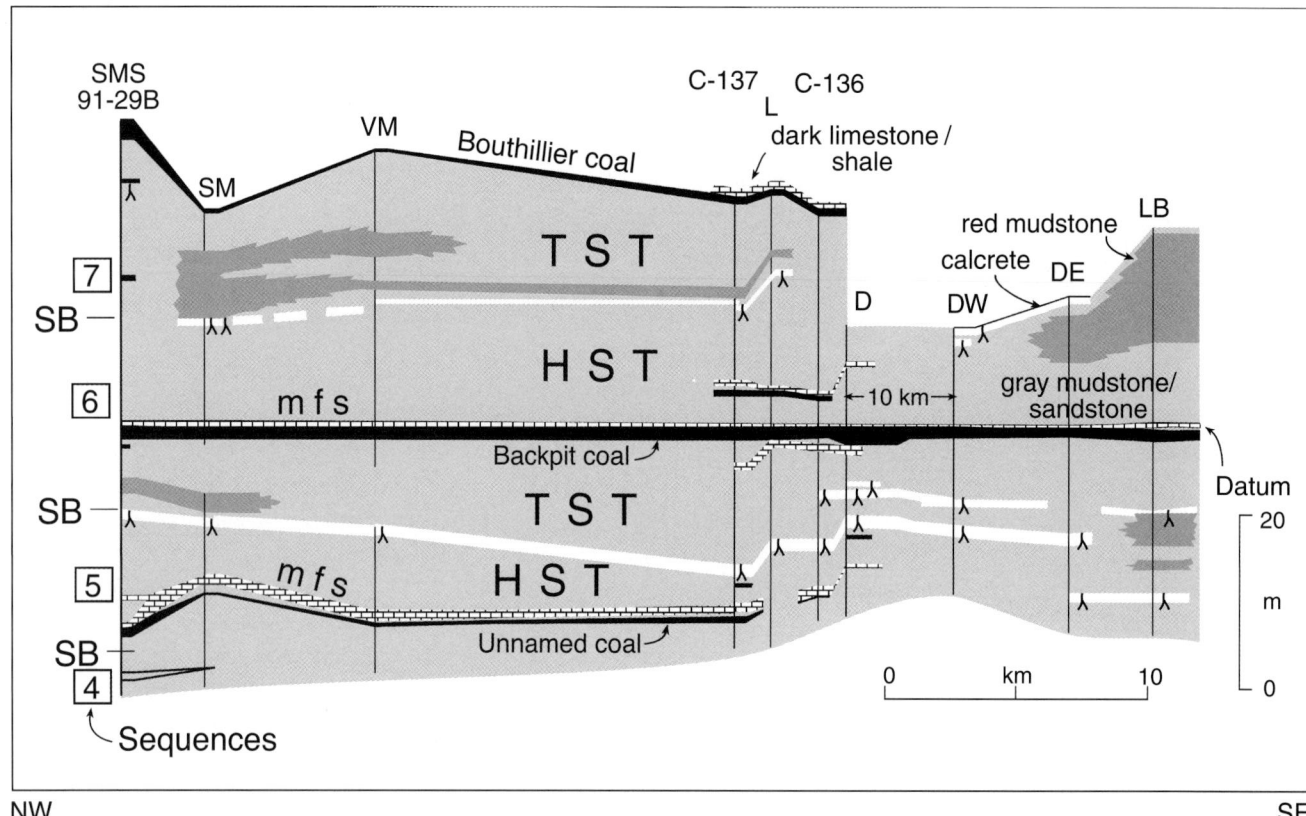

Figure 13. Correlation of unnamed to Bouthillier coal interval in coastal sections and wells (line of section in Figure 1). SB = sequence boundary; mfs = maximum flooding surface; TST and HST = transgressive and highstand systems tracts, respectively.

The assemblage formed on a stable land surface with incised channels and dryland paleosols. Assemblage 3 comprises gray and red siltstone and calcite-cemented channel bodies as much as 2.5 m (8 ft) thick, with width/thickness ratios of less than 30:1. The assemblage formed on a dryland, anastomosing river plain, probably under conditions of rising base level that culminated in peat accumulation (the Bonar coal, >2 m [>6.6 ft] thick).

The three assemblages can be traced for greater than 30 km (19 mi) across the basin, as far as outcrops extend (Figure 15). The Hub coal zone is highly split and is inferred to contain the maximum flooding surface, although its precise position is unclear. Assemblage 1 is attributed to the highstand systems tract of sequence 12, and the base of sequence 13 is marked by the calcrete and red vertic paleosols of assemblage 2. The overlying anastomosing river deposits of assemblage 3 are ascribed to the transgressive systems tract. In the Prince Colliery offshore, a 13-m (43-ft) multistory channel body that rests on the Hub coal (Gibling et al., 2000) is a candidate valley fill at the base of sequence 13.

In the broadly similar Bonar-Point Aconi coal interval (sequences 13 and 14; Figure 16), the Bonar coal is overlain by a dark, plant-rich shale (Zodrow, 1985), taken to approximate a maximum flooding level. A prominent calcrete marks the base of sequence 14, associated with a probable multistory valley fill and overlain by thick red vertic mudstones (Gibling and Wightman, 1994). A few thin, gray shales in the predominantly red interval yielded abundant foraminifera. A dark, plant-rich shale shortly above the Point Aconi coal (Zodrow and Cleal, 1988) is interpreted to approximate the maximum flooding surface of sequence 14.

Topmost Sydney Mines Formation and Pictou Group

Strata above the Point Aconi coal are known only from offshore wells. In P-05 (Figure 2), Hacquebard (1976) identified two additional seams, each about 1.5 m (5 ft) thick. The Murphy coal (his seam A) is about 105 m (344 ft) above the Point Aconi coal, and the intervening strata contain red and gray intervals, probably indicative of one or more thick sequences.

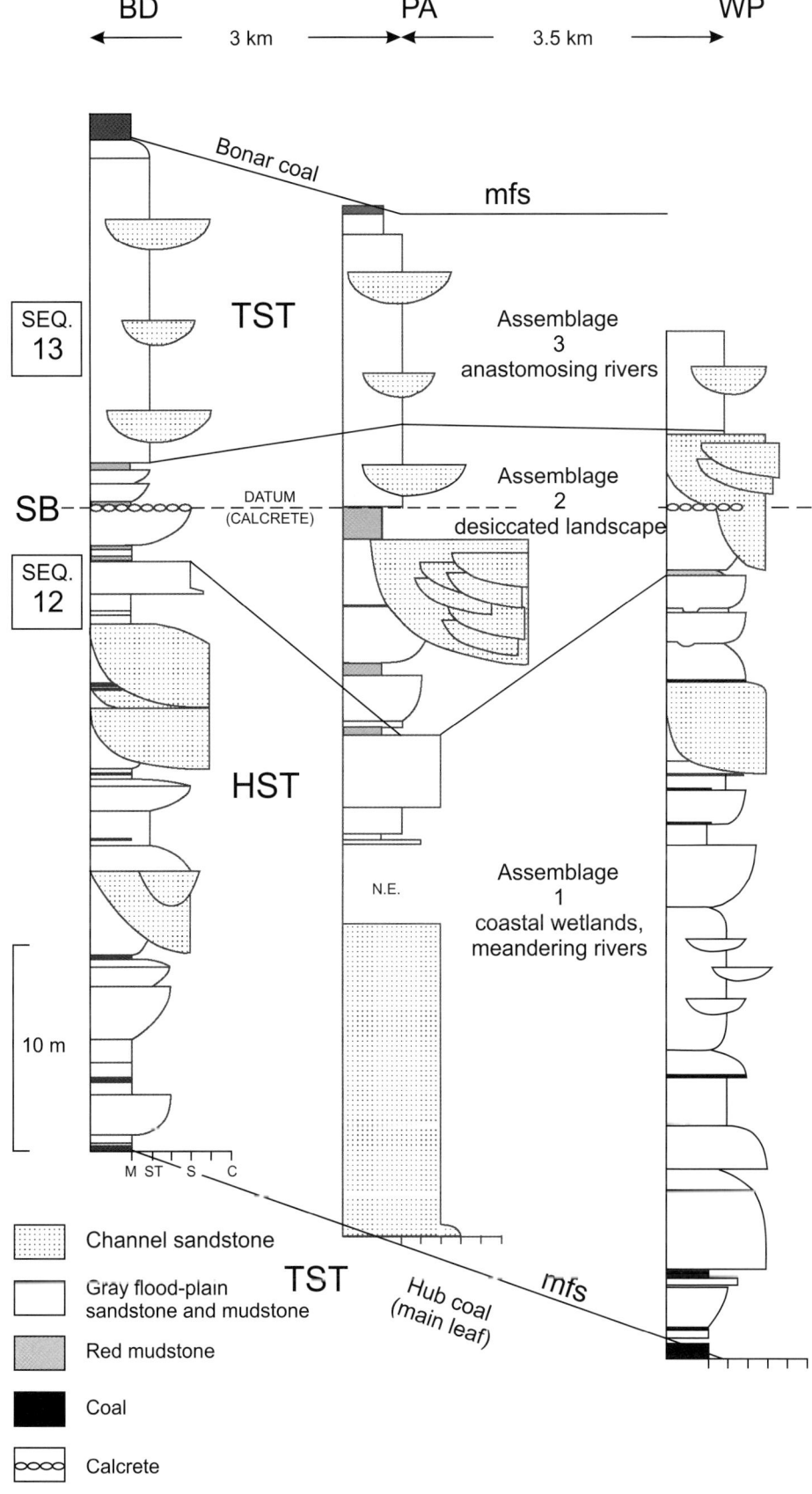

Figure 14. Stratal columns for sequences 12 and 13 in the Hub–Bonar coal interval of the upper Sydney Mines Formation (sections located in Figure 1). Modified from Saunders (1995); Gibling et al. (2000).

Above this level, the Pictou Group consists of more than 300 m (984 ft) of redbeds without coal.

MODEL FOR COAL IN SINGLE SEQUENCES

Sequence Boundaries and Alluvial Storage During Transgression and Falling Stage

Most sequence boundaries are marked by widespread, mature paleosols in the form of calcretes and/or thick calcareous paleosols of vertisol type. These paleosols indicate prolonged stabilization of the coastal plain during falling stage and lowstand of relative sea level, possibly for tens of thousands of years (Tandon and Gibling, 1997), and they also represent strongly seasonal climates, probably with considerably less precipitation than that which characterized the intervening coal-bearing intervals (see also Vosgerau et al., 2000). The common upward passage from calcrete to vertic mudstone suggests resumption of alluvial sedimentation on the lowstand surface, implying flood-plain storage during renewed transgression (Wright and Marriott, 1993). Precise boundary selection is problematic where thick vertic mudstones are present. Although their tops may represent stable land surfaces, sediment may also have accumulated during the lowstand. The proximal position of the outcrop belt and its along-strike orientation preclude the identification of lowstand shorelines. A few sequences (e.g., sequence 5) commence with paleosols that

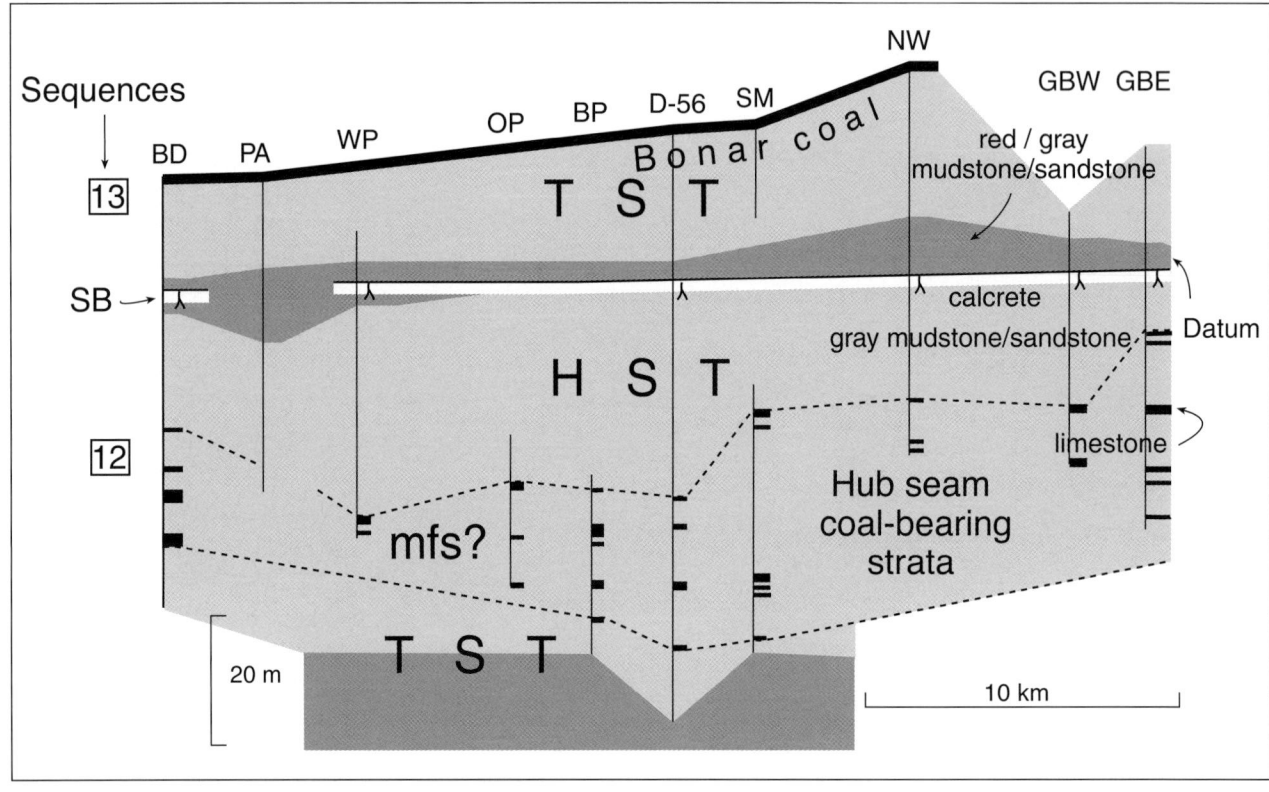

Figure 15. Correlation of strata in the Hub–Bonar coal interval in coastal sections and wells (line of section in Figure 1). SB = sequence boundary; mfs = maximum flooding surface; TST and HST = transgressive and highstand systems tracts, respectively.

are apparently only moderately mature, a problem in sequence identification (Aitken and Flint, 1995; Church and Gawthorpe, 1994; Hampson et al., 1996).

Sequence-bounding calcretes in the upper Sydney Mines Formation are overlain by redbeds tens of meters thick with channel bodies, implying prolonged periods of alluvial-plain sedimentation. These strata may reflect flood-plain storage in aggrading channel belts during rapid base-level rise (see Tornqvist, 1993 for a Quaternary analog). However, they may also represent low-accommodation periods in proximal settings, probably equivalent to multiple sequences in more distal settings (see fuller discussion later in this chapter).

Channel bodies commonly rest directly on coals, as noted by Robb (1876). Most are the fills of distributary channels that eroded down to tough fibrous peat. However, several deeply incised bodies show braided-fluvial or dryland characteristics and are paleovalley fills that mark a profound basinward facies shift.

The prevalence of sequence boundaries marked by valley fills and mature paleosols indicates that the rate of base-level fall repeatedly outpaced that of subsidence. Most sequences are explained reasonably by sea level fluctuation at equatorial latitudes linked to the coeval Gondwanan glaciation (Gibling and Bird, 1994). Although there is no evidence for major events, minor tectonic adjustments (see section on geological setting) modified facies expressions locally, as in the splitting and shaling out of the Mullins coal into the Glace Bay syncline (Figure 8) and the incoming of redbeds above the Backpit coal near Donkin (Figure 13). Coals such as the Unnamed coal that die out laterally reflect termination against local topography, instead of delta-lobe switching.

The prominence of sequence boundaries accords with the cratonic basinal style and modest sedimentation rate, under which high-frequency sea level fluctuations should be expressed as sequences instead of parasequences (Mitchum and van Wagoner, 1991). Their prominence also accords with the alluvial to coastal setting, where sea level fluctuations should have their maximum expression (MacNaughton et al., 1997). Parasequences are identified in all sequences, with flooding surfaces marked by carbonaceous shales

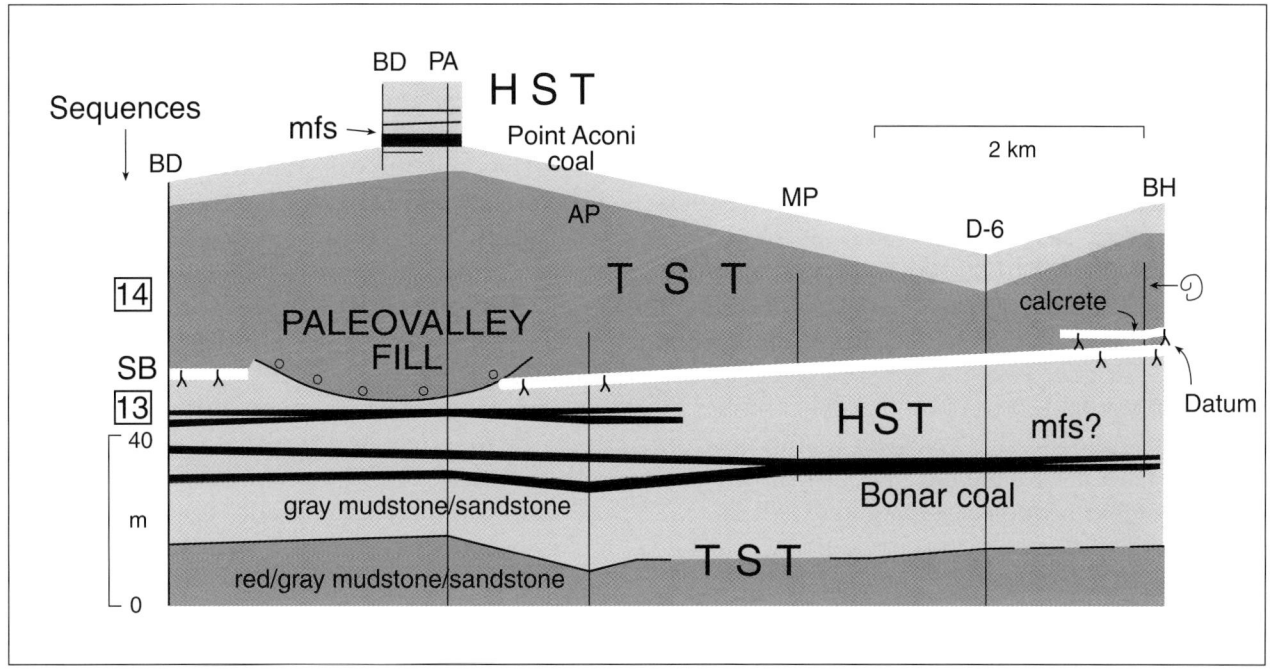

NW

SE

Figure 16. Correlation of strata in the Bonar–Point Aconi coal interval in coastal sections and wells (line of section in Figure 1). SB = sequence boundary; mfs = maximum flooding surface; TST and HST = transgressive and highstand systems tracts, respectively. Shell symbol shows location of foraminifera. Gibling and Wightman (1994) inferred that the paleovalley fill in the Point Aconi (PA) section was 20 m (66 ft) thick. However, if the pale paleosol at the right side of their figure 4B is equivalent to the basal calcrete of sequence 14, the paleovalley fill would be restricted to storey 3, 10 m (33 ft) thick. The valley maintained its position to generate superimposed storeys 4–5 as sediment built up on the interfluve.

and thin coals. The well-developed sequence architecture contrasts with that of the South Wales and Black Warrior foreland basins (Hartley, 1993; Pashin, 1998), where rapid subsidence promoted parasequence sets.

Dryland facies locally underlie sequence-bounding calcretes (e.g., top of sequences 3 and 5; Figure 9). Although boundary selection in these cases is difficult, base level apparently began to fall prior to landscape stabilization and sequence-boundary development, and the strata may be placed in the falling-stage systems tract (see Plint and Nummedal, 2000). Alternatively, these thin intervals of better-drained sediment may reflect slight landscape elevation, for example, along a channel margin. In the lower Sydney Mines Formation, calcretes at several localities are underlain by distinctive channel bodies with plane-bedded sandstones, desiccation features, and entombed standing trees. These features contrast with those of the majority of channel bodies and suggest a flood-dominated, seasonal fluvial regime that is compatible with falling to lowstand conditions prior to calcrete formation (Batson and Gibling, 2002).

Coals and Accommodation

As noted by Hartley (1993) and Hampson et al. (1999), the widespread extent of coal seams and groups and their occurrence at specific stratigraphic levels implies that peat accumulation was linked to allogenic controls, principally transgressive events (see Introduction). Most Sydney sequences contain several coals, but one coal (commonly the lowest) is typically especially thick. In several sequences, the thickest coal is overlain by a dark limestone-shale with concentrated fauna, the best example being the Backpit coal and roof unit. We infer that these organic-rich beds represent levels of maximum flooding and suppression of clastic supply to the basin (see Maynard et al., 1991; Church and Gawthorpe, 1994; Hampson et al., 1996, 1999; O'Mara and Turner, 1999). Their restricted faunal composition suggests that they are the updip, basin-margin equivalents of marine bands such as those of the European and United States basins (Calver, 1968; Wignall, 1987). Thus, we infer that most peat at this location on the Sydney Mines coastal plain accumulated while relative sea level was approaching the maximum extent of transgression

(Kosters and Suter, 1993; Greb and Chesnut, 1996; Bohacs and Suter, 1997).

Some thin coals lie below the inferred maximum flooding level, including some on sequence boundaries. They are probably the updip, coastal-plain equivalents of initial flooding surfaces on the shelf (Hampson et al., 1996). Other thin coals mark parasequence bases in the highstand systems tract. An upward change from thick to thin coals was taken to mark the point of maximum accommodation (Flint et al., 1995; Ulicny, 1999). A few thin coals occupy abandoned channels, but none have been noted in paleovalley fills. Peat did not form in the outcrop area during sea level lowstand, probably because of rapid drawdown of the coastal aquifer (Bohacs and Suter, 1997) and the move to a seasonal and possibly semiarid lowstand climate.

Multiple coals or coal zones with seam splits are common, especially in the upper Sydney Mines Formation. This suggests a relatively proximal site dominated by autogenic, channel effects (Hampson et al., 1999). Some thicker coals are overlain by plant-rich, carbonaceous shales that may mark maximum flooding levels, but in other cases, we can only infer that thick coals correspond broadly to maximal flooding (Flint et al., 1995). The Hub and Bonar coals each comprise two thick leaves in the study area (Figures 15, 16), which are possibly transgressive-regressive coal couplets with a basinward-facing split and the maximum flooding level between the leaves (Banerjee et al., 1996). However, split patterns in these seams appear complex, and this possibility remains untested.

Coal thickness correlates broadly with sequence thickness, although this is difficult to quantify. Thicker, more economic coals (Emery, Phalen, and Backpit) tend to lie in thicker sequences (sequences 2, 4, and 6), which also tend to have thicker strata of the transgressive systems tract (Figure 10). In the thin sequence 5 (Figure 13), a thin coal and capping limestone dies out laterally. These observations underscore the importance of potential accommodation in governing peat accumulation during transgression.

Available coal-petrographic information for Sydney coals is mainly unsuitable for investigating accommodation patterns and key surfaces (see Diessel et al., 2000). The general link between peat accumulation and transgression noted above implies a response to potential accommodation increase, and most seam bases should be paludification surfaces. However, the presence of agglutinated foraminifera in some seat earths may imply peat accumulation during a local stillstand or slight sea level lowering (Gibling and Bird,

1994), in which case coal bases should be terrestrialization surfaces. In the Backpit coal, the change at midseam level from brightening-up to dulling-up patterns (White et al., 1994) suggests an accommodation reversal surface, and the Backpit coal-roof unit contact represents a give-up transgressive surface. Similar dulling-up cycles were noted in Mesozoic coals by Petersen et al. (1998) and Diessel et al. (2000). For seams that lack a distinctive, overlying flooding unit, the coal top may mark a nonmarine flooding surface.

The scarcity of ravinement surfaces probably reflects the proximal location of the Sydney outcrops with their lack of shoreface deposits and concomitant coastal erosion (e.g., Liu and Gastaldo, 1992; Greb and Chesnut, 1996; Pashin, 1998). The best example is the erosive base of the Backpit roof unit, which may record the most proximal extent of transgression.

MODEL FOR COAL IN COMPOSITE SEQUENCES

Mitchum and van Wagoner (1991) defined composite sequences as successions of genetically related sequences (each erosionally based) that are bounded by especially prominent erosional surfaces with a pronounced basinward facies shift. The component sequences can be arranged into lowstand, transgressive, and highstand sequence sets with progradational, aggradational, and retrogradational stacking patterns, respectively. Sequence sets may represent a composite relative sea level curve governed by both long- and short-term cyclic patterns (Mitchum and van Wagoner, 1991; MacNaughton et al., 1997) or tectonic events (Xue, 1997; O'Mara and Turner, 1999).

Sydney Basin Composite Sequences

Four composite sequences are identified in the studied sections (Figure 17), as well as intervals that lack clear composite patterns. The lower South Bar Formation could comprise a single, thick sequence or a succession of stacked sequences, but exposures are inadequate to distinguish allogenic from autogenic erosional events. Because a proximal to distal traverse is not available in the basin, recognition of sequence sets must be judged by trends in vertical sections instead of from regional stacking patterns.

Composite sequence 1 commences at a prominent sequence boundary above the Mullins coal. The overlying 220-m (722-ft) interval mainly comprises stacked channel-sandstone bodies, but the presence at the top of the interval of redbeds and calcareous paleosols

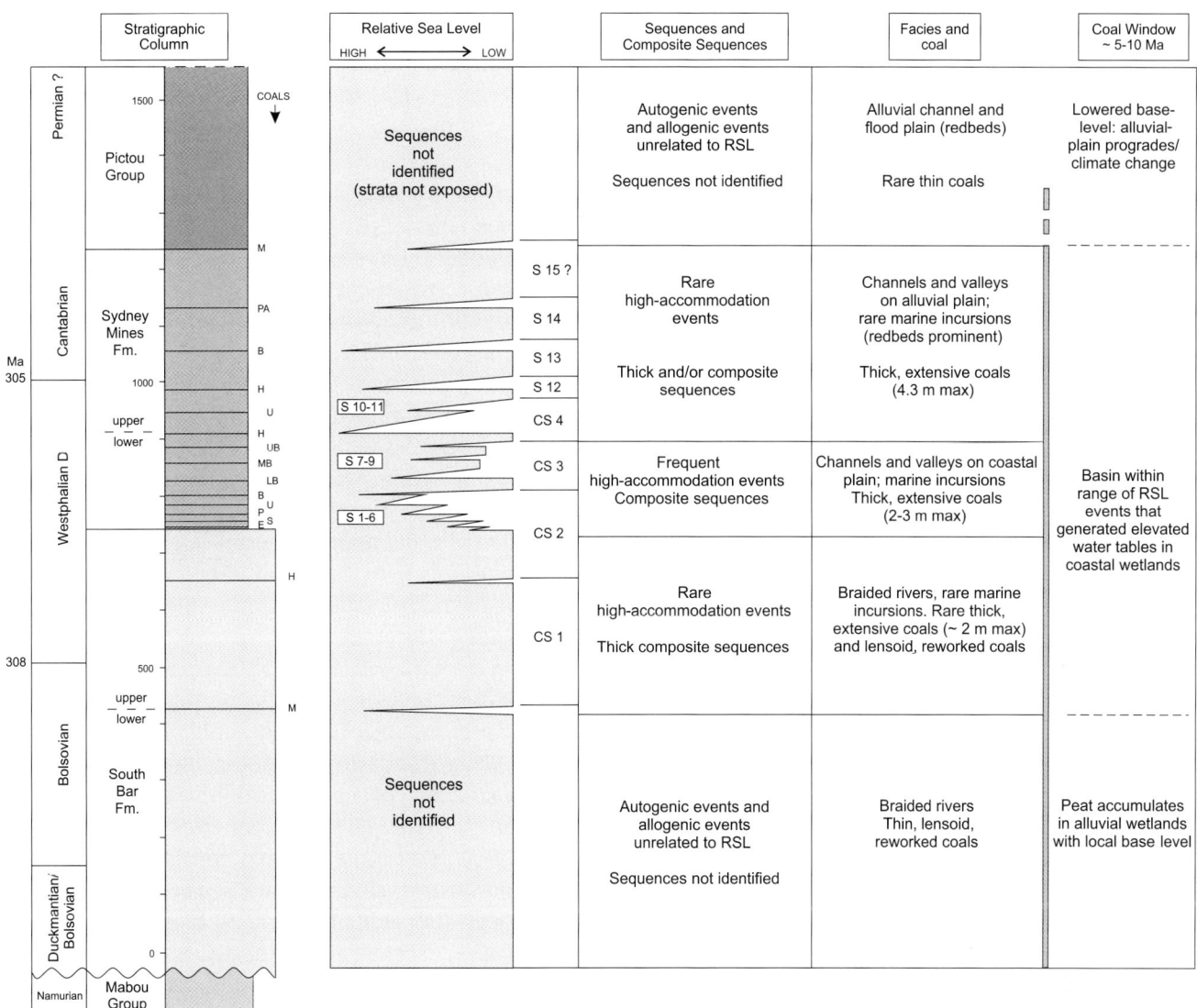

Figure 17. Summary of formations, coal distribution, and sequence-stratigraphic patterns for the Upper Carboniferous strata of the Sydney Basin. Coal names shown in Figures 2, 5, and 9; upper M = Murphy coal. Radiometric dates from Hess and Lippolt (1986). CS = composite sequence; S = sequence; RSL = relative sea level.

capped by reworked coal implies the presence of several sequences.

Composite sequence 2 comprises 150–190 m (492–623 ft) of the topmost South Bar Formation (above the Henderson coal level) and sequences 1–6 of the Sydney Mines Formation. The succession commences with thick channel deposits (Figure 5). Transgression, probably to a lacustrine setting, marks sequence 1, and sequences 2–6 show a marked thickening of the transgressive systems tract of successive sequences, culminating in the faunal-concentrate Backpit roof unit with its basal ravinement surface, the most landward transgressive level. The South Bar fluvial sandstones show no apparent stratal organization apart from multiple erosion surfaces and cannot readily be considered a lowstand sequence set. We infer that

sequences 1–6 form a transgressive sequence set, although aggradation appears nearly as significant as retrogradation. The maximal flooding level (Backpit roof unit) closely underlies the base of composite sequence 3, and a highstand sequence set is not apparent.

Composite sequences 3 and 4 have received only partial study (Figure 9; see figure 4 of Gibling and Bird, 1994). Composite sequence 3 is 75 m (246 ft) thick and commences with a prominent regional calcrete overlain by thick redbeds (Figures 11B, 13). The lower Bouthillier coal marks a return to a predominantly gray stratal interval with thin coals of the middle and upper Bouthillier, all highly split (Hacquebard and Donaldson, 1969) and mainly uneconomic. Composite sequence 4 is 80 m (262 ft) thick and commences with a regional calcrete-redbed interval. The overlying Harbour coal marks a major transgressive period, with thick economic coals and standing trees entombed in bayfills (Calder et al., 1996). Extensive, thick redbeds with calcretes reappear below the Hub coal. Composite sequences are not apparent in the upper Sydney Mines Formation and Pictou Group, where stacked, thick sequences lack strong indications of progradational or retrogradational patterns.

The composite sequence boundaries represent particularly strong, persistent basinward facies shifts, accompanied by base-level lowering (calcretes). However, these boundaries are not markedly different from other sequence boundaries, as noted by Posamentier and Allen (1999). Lowstand sequence sets are not readily identified, probably because much of the lowstand period is represented by mature paleosols at this proximal location. Composite sequence 2 has a well-marked transgressive sequence set. The abrupt change from a sandstone- to a mudstone-dominated succession at the South Bar–Sydney Mines Formation boundary has counterparts across the southern Gulf of St. Lawrence, and Rehill (1996) linked this regional change in sediment grade to Appalachian tectonic events (see Rogers, 1998, for a similar inference). Individual Sydney sequences probably reflect high-frequency glacioeustatic events, and some composite sequence boundaries could represent enhanced lowering of relative sea level during long-term eustatic cycles, as envisaged by Mitchum and van Wagoner (1991).

Composite sequences in coal-bearing successions have been discussed for Cretaceous strata of the western United States (Ryer, 1981, 1983; Cross, 1988; Shanley and McCabe, 1993; Bohacs and Suter, 1997), for Permian–Carboniferous strata worldwide (Church and Gawthorpe, 1994; Aitken and Flint, 1995; Hampson et al., 1996; Herbert, 1997; O'Mara and Turner, 1999), and for Tertiary lignite-bearing strata of Australia (Holdgate and Clarke, 2000). Our results for the lower Sydney Mines Formation broadly concur with these studies in that the thickest and most closely spaced coals lie in an aggradational to slightly retrogradational setting, suggesting a long-term balance between the rate of relative sea level change and sediment supply.

The Thick Sequence Problem

The prominent upward change in the Sydney Mines Formation from thin sequences (sequences 1–9) to thick sequences (sequences 10–14) could be explained by an increased subsidence rate approximately balanced by sediment supply or a change in

the Milankovitch frequency that governed relative sea level fluctuations, such as that which took place during the late Cenozoic (Raymo et al., 1989). Alternatively, following a strong, sustained progradation of the alluvial plain at midformation level (Figure 18), only major transgressive events would have inundated the alluvial plain (Collier et al., 1990). If so, a thick, proximal sequence could correlate distally with several thin sequences or a composite sequence. Subsequent, continued progradation would place the alluvial plain beyond reach of transgressions. Support for this hypothesis is provided by the identification of foraminifera in thin shales in the redbed intervals of thick sequences (Gibling and Wightman, 1994). Although undistinguished, these shales appear to represent the most proximal extent of transgressions across the alluvial plain. Their presence suggests that thick sequences include cryptic cyclic patterns and pass basinward into stacked thin sequences with well-developed architecture.

If the scenario of Figure 18 is correct, then thick and thin sequences appear superficially similar in their stratal organization but actually represent different accommodation patterns. Stacked thin sequences on the coastal plain (right side of Figure 19) represent a series of high-frequency transgressions which generated most of the potential accommodation. In contrast, stacked thick sequences on the alluvial plain (left side of Figure 19) represent a combination of (1) extreme accommodation events because of transgressions of short duration, capable of reaching far up the plain, and (2) low-accommodation periods of long duration, during which stratal accumulation was governed principally by processes other than sea level change (e.g., climatic, tectonic, and autogenic events). The fundamental architecture of the thick sequences would be governed by infrequent, high-accommodation events which allowed rapid stratal accumulation but represent only a small proportion of the time.

Based on an estimated accumulation rate for the Morien Group, thin sequences 2–9 average 105,000 yr and lie in the Milankovitch band. Thick sequences 12 and 13 are estimated at 0.3–0.5 m.y., and composite sequences 3 and 4 are estimated at 0.4–0.5 m.y. These similar estimated durations, which lie at the upper end of the Milankovitch band, support the model of Figures 18 and 19. Previous studies of Pennsylvanian cyclothems and sequences have found a similar general range of mean durations (Heckel, 1986; Maynard and Leeder, 1992). Composite sequences 1 and 2 include South Bar strata and are of much longer

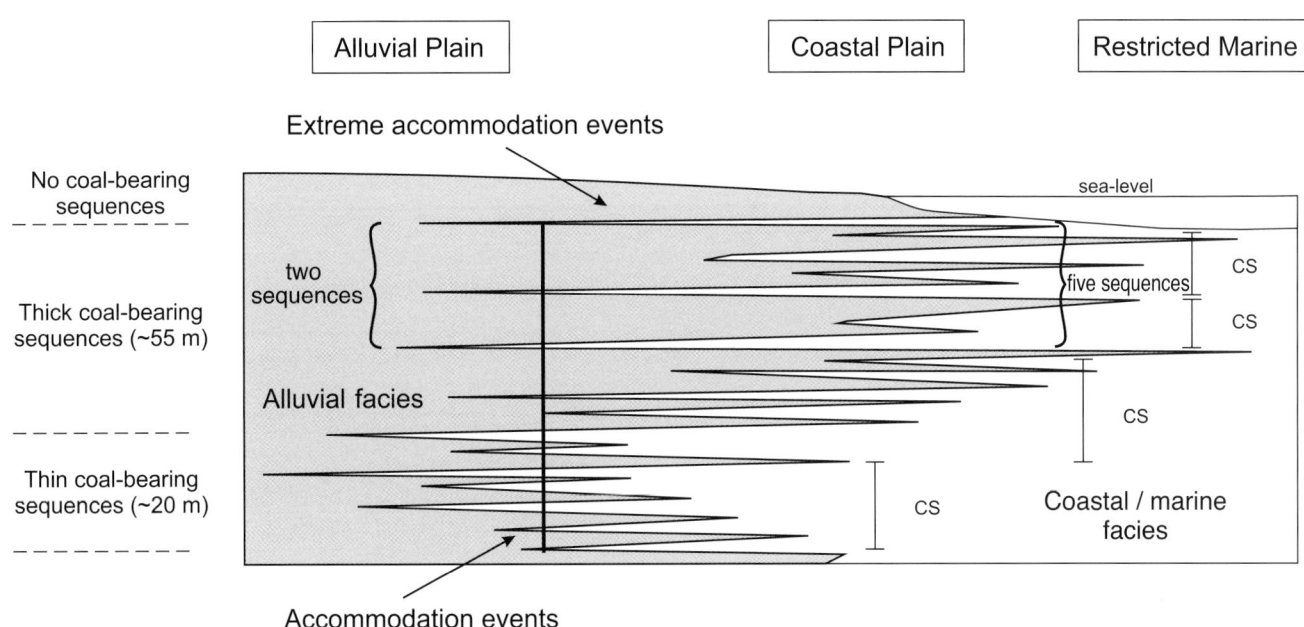

Figure 18. Simplified representation of sequence architecture for Sydney Mines Formation. Thick vertical line denotes the present outcrop position. Note the importance of extreme accommodation events in generating thick sequences in proximal areas of the basin. CS = composite sequence.

estimated duration (1.4 Ma and 0.9 m.y.), supporting the inference that they contain high-frequency sequences that have evaded identification to date.

MODEL FOR COAL AT BASINAL SCALE: THE COAL WINDOW

The near exclusion of detrital sediment from a depositional site allows a brief "window" favorable for peat accumulation, provided that suitable hydrological conditions prevail (Nemec, 1992). On a longer timescale, the stratigraphic distribution of coal measures in a basin fill represents the period during which basin hydrology was suitable for peat accumulation. Calder and Gibling (1994) and Calder (1994) defined the coal window to represent the stratigraphic distribution of coals in a basin.

The coal window in the Morien Group represents 5–10 Ma during which extensive peats formed on at least 16 occasions (Figure 17) and in three distinct settings (Figure 4). Uncommon, thin coals and abundant reworked coal fragments throughout the South Bar Formation imply that a suitable climate and ground-water regime existed for peat formation; however, peat formed only locally on the alluvial plain, where ground-water level was suitable and clastic sediment precluded, and the peats were mainly reworked by shifting channel belts. Thick, extensive coals in the upper South Bar and Sydney Mines Formations rep-

resent blanket coastal peats. In the latter formation, they form parts of well-defined sequences, and their presence implies a sufficiently humid climate and/or low seasonality. The systematic increase in the spacing of coal beds in the topmost Sydney Mines Formation and Pictou Group, with eventual loss of coal, represents a progressively more proximal location and/or a more seasonal and arid climate as Euramerica drifted northward (Schutter and Heckel, 1985). A coal window of similar duration is present across the Gulf of St. Lawrence area, with progressive loss of coals during the Stephanian (Rehill, 1996; Giles and Utting, 1999). Phillips et al. (1985) inferred that the Duckmantian to Westphalian D climate in the Appalachian basin was conducive for peat accumulation, whereas the Stephanian to Permian was progressively less suitable.

The Sydney Basin coal window corresponds largely to the period during which the Morien coastal and alluvial plain was within range of high-accommodation, relative sea level events that caused ground-water ponding and allowed widespread peat accumulation. However, many peats were formed but not preserved effectively in the South Bar Formation, and continent-scale climate change is strongly implicated in the termination of peat accumulation.

A key question is whether the petrographic composition of the Morien coals changed as the depositional setting evolved from braided rivers (South

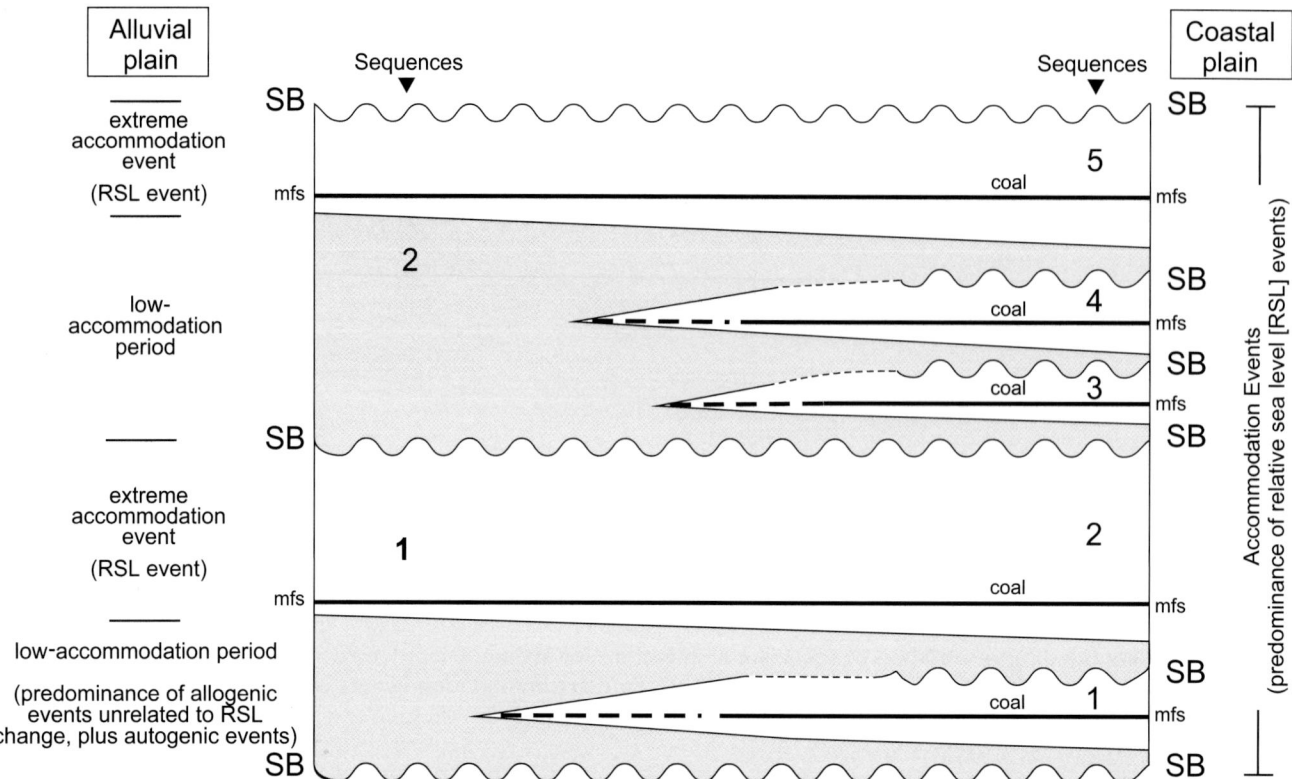

Figure 19. Model to illustrate the varied impact of accommodation events across coastal and alluvial plains, with reference to the Morien Group of the Sydney Basin. In the proximal, alluvial part of the basin, note the alternation of strata deposited during extreme accommodation events and low-accommodation periods. As a result, the succession can be divided into sequences, but only a small proportion of the strata reflect accommodation events originating in the coastal plain. SB = sequence boundary; mfs = maximum flooding surface; RSL = relative sea level.

Bar Formation) to coastal plain-alluvial plain (Sydney Mines Formation) to alluvial plain (Pictou redbeds). This was partially tested by Marchioni et al. (1996), who analyzed 32 coals at Bras d'Or (Stony to Point Aconi seams), as well as the Emery and Gardiner coals elsewhere; the Mullins and older South Bar coals were not studied. Seam petrography shows little change through this approximately 500-m (1640-ft) interval. Vitrinite increases slightly upsection, and an upward increase in the WD ratio (structured vitrinite + structured inertinite/inertodetrinite + discrete macrinite + sporinite) was interpreted to indicate more frequent flooding lower in the Sydney Mines Formation. Drier conditions in the Hub seam and above are suggested by an upward decrease in the structured inertinite ratio (semifusinite/fusinite) and by the high inertinite and semiinertinite content of the Murphy coal in the P-05 well (Hacquebard, 1976). Plots of maceral-derived parameters suggest a lower delta-plain setting for most coals, by comparison with data from Australian Permian coals (Diessel, 1992). In general, coal composition shows relative-

ly little change through the Sydney Mines Formation, implying that peat accumulated under broadly similar conditions regardless of stratigraphic position. This inference accords well with the sequence-stratigraphic analysis presented above, where coals in all well-developed sequences represent transgressive episodes on the Morien coastal plain. Drier conditions may have prevailed in the youngest mires.

CONCLUSIONS

Several key points emerge from this study of Pennsylvanian coal-bearing sequences of the Sydney Basin (Figure 20). Sequence architecture and peat accumulation was controlled largely by high-accommodation events of relative sea level rise with peat accumulation, followed by relative falls. This style is reasonably interpreted as the expression of glacioeustatic events in an equatorial, cratonic basin with a moderate subsidence rate. Although both thin and thick sequences show similar architecture, they differ principally in the presence of thick alluvial channel and redbed flood-plain deposits that overlie the basal calcretes of thick sequences. These alluvial deposits are inferred to represent prolonged, low-accommodation periods during which stratal architecture was controlled by channel switching, climate, and/or tectonism, whereas relative sea level change yielded only cryptic surfaces. Thick sequences represent the most landward transgressions and probably pass basinward into multiple thin sequences or composite sequences.

Most coals represent blanket coastal peats that accumulated just prior to maximum transgression. Maximum flooding levels are marked by thin limestone and shale with an abundant restricted-marine fauna, probably the updip equivalents of open-marine bands. A general correlation between coal and sequence thickness confirms a link between potential accommodation and peat accumulation. Some of the major economic coals lie in the transgressive sequence set of a composite sequence, the component sequences of which show a slightly retrogradational to aggradational style.

The stratigraphic distribution of coal in the basin fill is termed the coal window. Major economic coals correspond to the period during which the outcrop belt lay within range of high-accommodation, relative sea level events that promoted accumulation of thick coastal peat. The coals show little petrographic change upsection, suggesting that such events tended to generate peats with similar properties at any stratigraphic level. Thick braided-fluvial deposits lower in the basinal fill contain thin and reworked coals that have no obvious connection with relative sea level events. They probably formed where local base level was suitable for peat accumulation, on channel margins and in abandoned channels. Upward loss of coal reflects regional progradation of the alluvial plain, coupled with long-term climatic change as Euramerica drifted northward. Thus, the coal window reflects a combination of long-term accommodation driven by subsidence, high-frequency accommodation events driven by glacioeustasy, local accommodation sites on the alluvial plain, and broader climatic controls.

ACKNOWLEDGMENTS

We are grateful to John Calder, Don MacNeil, and Rob Naylor for discussion, Allen Archer and Tim Demko for their helpful reviews of the manuscript, and Jack Pashin and Bob Gastaldo for editorial assistance. Andy Henry and Dalhousie Graphics are thanked for their drafting skills. Financial assistance through a Natural Sciences and Engineering Research Council of Canada (NSERC) Research Grant to M. R. Gibling, an NSERC Scholarship to K. Saunders, and a Texaco Scholarship to J. White are gratefully acknowledged.

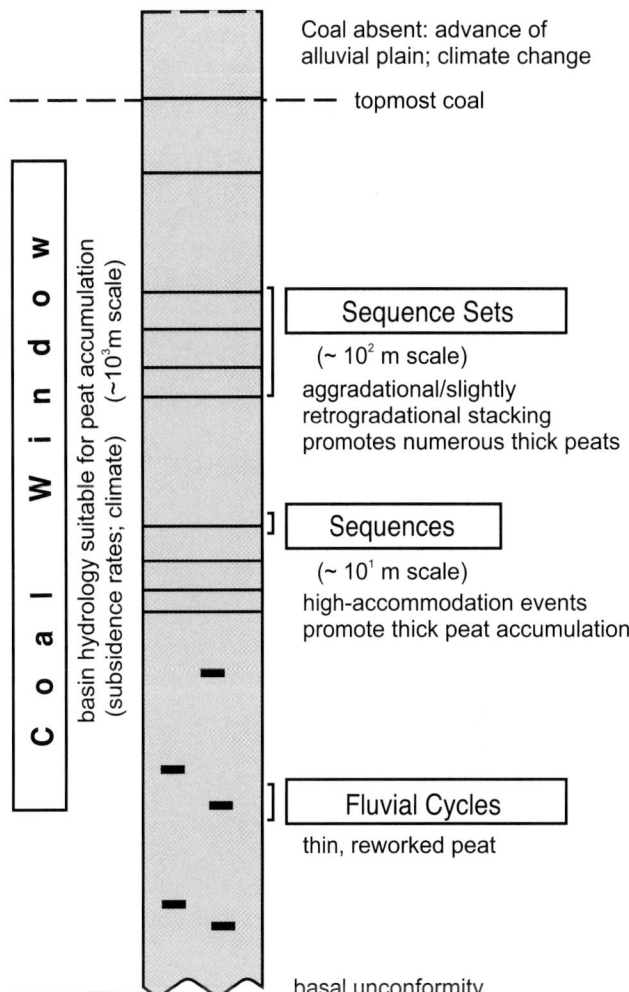

Figure 20. Summary of major factors that governed coal distribution in the Sydney coalfield.

REFERENCES CITED

Aitken, J. F., and S. S. Flint, 1995, The application of high-resolution sequence stratigraphy to fluvial systems: A case study from the Upper Carboniferous Breathitt Group, eastern Kentucky, U.S.A.: Sedimentology, v. 42, p. 3–30.

Arditto, P. A., 1991, A sequence stratigraphic analysis of the Late Permian succession in the southern coalfield, Sydney Basin, New South Wales: Australian Journal of Earth Science, v. 38, p. 125–137.

Banerjee, I., W. Kalkreuth, and E. H. Davies, 1996, Coal seam splits and transgressive-regressive coal couplets: A key to stratigraphy of high-frequency sequences: Geology, v. 24, p. 1001–1004.

Barss, M. S., J. P. Bujak, and G. L. Williams, 1979, Palynological zonation and correlation of sixty-seven wells, eastern Canada: Geological Survey of Canada Paper 78-24, 118 p.

Batson, P. A., and M. R. Gibling, 2002, Architecture of channel bodies and paleovalley fills in high-frequency Carboniferous sequences, Sydney Basin, Atlantic Canada: Bulletin of Canadian Petroleum Geology, v. 50, p. 138–157.

Bell, W. A., 1923, Stratigraphy of Great Bras d'Or coal district, Victoria County, Cape Breton, in A. O. Hayes and W. A. Bell, eds., The southern part of the Sydney coalfield, Nova Scotia (appendix): Geological Survey of Canada Memoir 133, p. 90–104.

Bohacs, K., and J. Suter, 1997, Sequence stratigraphic distribution of coaly rocks: Fundamental controls and paralic examples: AAPG Bulletin, v. 81, p. 1612–1639.

Calder, J. H., 1994, The impact of climate change, tectonism and hydrology on the formation of Carboniferous tropical intermontane mires: The Springhill coalfield, Cumberland Basin, Nova Scotia: Palaeogeography, Palaeoclimatology, Palaeoecology, v. 106, p. 323–351.

Calder, J. H., 1998, The Carboniferous evolution of Nova Scotia, in D. Blundell and A. C. Scott, eds., Lyell: The past is the key to the present: Geological Society (London) Special Publication 143, p. 296–331.

Calder, J. H., and M. R. Gibling, 1994, The Euramerican coal province: Controls on late Paleozoic peat accumulation: Palaeogeography, Palaeoclimatology, Palaeoecology, v. 104, p. 1–21.

Calder, J. H., M. R. Gibling, C. F. Eble, A. C. Scott, and D. J. MacNeil, 1996, The Westphalian D fossil lepidodendrid forest at Table Head, Sydney Basin, Nova Scotia: Sedimentary, paleoecology and floral response to changing edaphic conditions: International Journal of Coal Geology, v. 31, p. 277–313.

Calver, M. A., 1968, Distribution of Westphalian marine faunas in northern England and adjoining areas: Proceedings of the Yorkshire Geological Society, v. 37, p. 1–72.

Church, K. D., and R. L. Gawthorpe, 1994, High resolution sequence stratigraphy of the late Namurian in the Widmerpool Gulf (East Midlands, U.K.): Marine and Petroleum Geology, v. 11, p. 528–544.

Collier, R. E. L., M. R. Leeder, and J. R. Maynard, 1990, Transgressions and regressions: A model for the influence of tectonic subsidence, deposition and eustasy, with application to Quaternary and Carboniferous examples: Geological Magazine, v. 127, p. 117–128.

Copeland, M. J., 1959, Coalfields, west half Cumberland County, Nova Scotia: Geological Survey of Canada, Memoir 298, 89 p.

Courel, L., M. Donsimoni, and D. Mercier, 1986, La place du charbon dans la dynamique des sytemes sedimentaires des bassins houillers intramontaneux, in L. Courel and P. Vetter, eds., Les Bassins Houillers Limniques: La Societe Geologique De France, p. 37–50.

Cross, T. A., 1988, Controls on coal distribution in transgressive-regressive cycles, Upper Cretaceous, western interior, U.S.A., in C. K. Wilgus, B. S. Hastings, C. G. St. C. Kendall, H. W. Posamentier, C. A. Ross, and J. C. Van Wagoner, eds., Sea-level changes: An integrated approach: SEPM, p. 371–380.

DeWolfe, L. A., 1906, The structure and succession at North Sydney and Sydney Mines: Nova Scotia Institute of Science Proceedings and Transactions, v. 10, p. 289–323.

Diessel, C. F. K., 1992, Coal-bearing depositional systems: Berlin, Springer-Verlag, 721 p.

Diessel, C., R. Boyd, J. Wadsworth, D. Leckie, and G. Chalmers, 2000, On balanced and unbalanced accommodation/peat-accumulation ratios in the Cretaceous coals from Gates Formation, western Canada, and their sequence-stratigraphic significance: International Journal of Coal Geology, v. 43, p. 143–186.

Dolby, G., 1988, The palynology of the Morien Group, Sydney Basin, Cape Breton Island, Nova Scotia: Nova Scotia Department of Mines and Energy Report, 21 p.

Dolby, G., 1989, The palynology of the Morien Group, Sydney Basin, Cape Breton Island, Nova Scotia, Nova Scotia Department of Mines and Energy Report, 23 p.

Fielding, C. R., 1987, Coal depositional models for deltaic and alluvial plain sequences: Geology, v. 15, p. 661–664.

Flint, S., J. Aitken, and G. Hampson, 1995, Application of sequence stratigraphy to coal-bearing coastal plain successions: Implications for the U.K. coal measures, in M. K. G. Whateley and D. A. Spears, eds., European Coal Geology: Geological Society (London) Special Publication, p. 1–16.

Gastaldo, R. A., T. M. Demko, and Y. Liu, 1993, Application of sequence and genetic stratigraphic concepts to Carboniferous coal-bearing strata: An example from the Black Warrior Basin, U.S.A.: Geologische Rundschau, v. 82, p. 212–226.

Gibling, M. R., and W. D. Kalkreuth, 1991, Petrology of selected carbonaceous limestones and shales in Late Carboniferous coal basins of Atlantic Canada: International Journal of Coal Geology, v. 17, p. 239–271.

Gibling, M. R., and D. J. Bird, 1994, Late Carboniferous cyclothems and alluvial paleovalleys in the Sydney Basin, Nova Scotia: Geological Society of America Bulletin, v. 106, p. 105–117.

Gibling, M. R., and W. G. Wightman, 1994, Paleovalleys and protozoan assemblages in a Late Carboniferous cyclothem, Sydney Basin, Nova Scotia: Sedimentology, v. 41, p. 699–719.

Gibling, M. R., J. H. Calder, R. Ryan, H. W. van de Poll, and G. M. Yeo, 1992, Late Carboniferous and Early Permian drainage patterns in Atlantic Canada: Canadian Journal of Earth Sciences, v. 29, p. 338–352.

Gibling, M. R., A. T. Martel, M. H. Nguyen, A. M. Kennedy, F. Baechler, J. Shimeld, S. Forgeron, and B. Mackenzie, 2000, Fluid evolution and diagenesis of a Carboniferous channel sandstone in the Prince Colliery, Nova

Scotia, Canada: Bulletin of Canadian Petroleum Geology, v. 48, p. 95–115.

Gibling, M. R., W. Langenberg, W. D. Kalkreuth, J. W. F. Waldron, R. Courtney, J. Paul, and A. M. Grist, 2002, Deformation of Upper Carboniferous coal measures in the Sydney Basin: Evidence for late Alleghanian tectonism in Atlantic Canada: Canadian Journal of Earth Sciences, v. 39, p. 79–93.

Giles, P. S., and J. Utting, 1999, Maritimes Basin stratigraphy— Prince Edward Island and adjacent Gulf of St. Lawrence: Geological Survey of Canada Open-file Report 3732.

Grant, A. C., 1994, Aspects of seismic character and extent of Upper Carboniferous coal measures, Gulf of St. Lawrence and Sydney Basins: Palaeogeography, Palaeoclimatology, Palaeoecology, v. 106, p. 271–285.

Greb, F., and D. R. Chesnut, 1996, Lower and lower middle Pennsylvanian fluvial to estuarine deposition, central Appalachian basin: Effects of eustasy, tectonics, and climate: Geological Society of America Bulletin, v. 108, p. 303–317.

Hacquebard, P. A., 1976, Appraisal of coal intersected in Murphy off-shore well (North Sydney P-05) of Sydney coalfield, Nova Scotia: Geological Survey of Canada Report, 12 p.

Hacquebard, P. A., 1983, Geological development and economic evaluation of the Sydney Coal Basin, Nova Scotia: Geological Survey of Canada Paper 83-1A, p. 71–81.

Hacquebard, P. A., 1998, Petrographic, physico-chemical, and coal facies studies of ten major seams of the Sydney coalfield of Nova Scotia, Geological Survey of Canada Bulletin, v. 520, 46 p.

Hacquebard, P. A., and J. R. Donaldson, 1969, Carboniferous coal deposition associated with flood-plain and limnic environments in Nova Scotia, in E. C. Dapples and M. E. Hopkins, eds., Environments of coal deposition: Geological Society of America Special Paper 114, p. 143–191.

Hampson, G. J., T. Elliott, and S. S. Flint, 1996, Critical application of high resolution sequence stratigraphic concepts to the Rough Rock Group (Upper Carboniferous) of northern England, in J. A. Howell and J. F. Aitken, eds., High resolution sequence stratigraphy: Innovations and applications: Geological Society (London) Special Publication 104, p. 221–246.

Hampson, G., H. Stollhofen, and S. Flint, 1999, A sequence stratigraphic model for the lower coal measures (Upper Carboniferous) of the Ruhr district, north-west Germany: Sedimentology, v. 46, p. 1199–1231.

Hartley, A. J., 1993, A depositional model for the mid-Westphalian A to late Westphalian B coal measures of South Wales: Journal of the Geological Society (London), v. 150, p. 1121–1136.

Hayes, A. O., and W. A. Bell, 1923, The southern part of the Sydney coal field, Nova Scotia: Geological Survey Canada Memoir 133, 108 p.

Heckel, P. H., 1986, Sea-level curve for Pennsylvanian eustatic marine transgressive-regressive depositional cycles along midcontinent outcrop belt, North America: Geology, v. 14, p. 330–334.

Herbert, C., 1997, Relative sea level control of deposition in the Late Permian Newcastle coal measures of the Sydney Basin, Australia: Sedimentary Geology, v. 107, p. 167–187.

Hess, J. C., and H. J. Lippolt, 1986, $^{40}Ar/^{39}Ar$ ages of tonsteins and tuff sanidines: New calibration points for the improvement of the Upper Carboniferous time scale: Chemical Geology, v. 59, p. 143–154.

Holdgate, G. R., and J. D. A. Clarke, 2000, A review of Tertiary brown coal deposits in Australia— Their depositional factors and eustatic correlations: AAPG Bulletin, v. 84, p. 1129–1151.

Kosters, E. C., and J. R. Suter, 1993, Facies relationships and systems tracts in the late Holocene Mississippi delta plain: Journal of Sedimentary Petrology, v. 63, p. 727–733.

Liu, Y., and R. A. Gastaldo, 1992, Characteristics of a Pennsylvanian ravinement surface: Sedimentary Geology, v. 77, p. 197–213.

Lottes, A. L., and A. M. Ziegler, 1994, World peat occurrence and the seasonality of climate and vegetation: Palaeogeography, Palaeoclimatology, Palaeoecology, v. 106, p. 23–37.

Machette, M. N., 1985, Calcic soils of the southwestern United States: Geological Society of America Special Paper 203, p. 1–21.

MacNaughton, R. B., R. W. Dalyrmple, and G. M. Narbonne, 1997, Multiple orders of relative sea-level change in an earliest Cambrian passive-margin succession, Mackenzie Mountains, northwestern Canada: Journal of Sedimentary Research, v. 67, p. 622–637.

MacNeil, D. J., 1987, Log of drill hole NC-87-1, Sydney Basin, Nova Scotia: Department of Mines and Energy Report, 29 p.

Mamet, B. L., 1992, Paleogeographie des algues calcaires marines carboniferes: Canadian Journal of Earth Sciences, v. 29, p. 174–194.

Marchioni, D., W. Kalkreuth, J. Utting, and M. Fowler, 1994, Petrographical, palynological and geochemical analyses of the Hub and Harbour seams, Sydney coalfield, Nova Scotia, Canada— Implications for facies development: Palaeogeography, Palaeoclimatology, Palaeoecology, v. 106, p. 241–270.

Marchioni, D., M. R. Gibling, and W. D. Kalkreuth, 1996, Petrography and depositional environment of coal seams in the Carboniferous Morien Group, Sydney coalfield, Nova Scotia: Canadian Journal of Earth Sciences, v. 33, p. 863–874.

Maynard, J. R., and M. R. Leeder, 1992, On the periodicity and magnitude of Late Carboniferous glacio-eustatic sea-level changes: Journal of the Geological Society (London), v. 149, p. 303–311.

Maynard, J. R., P. B. Wignall, and W. J. Varker, 1991, A "hot" new shale facies from the Upper Carboniferous of northern England: Journal of the Geological Society (London), v. 148, p. 805–808.

Menning, M., D. Weyer, G. Drozdzewski, H. W. J. Van Amerom, and I. Wendt, 2000, A Carboniferous time scale 2000: Discussion and use of geological parameters as time indicators from central and western Europe: Geologische Jahrbuch, v. A156, p. 3–44.

Mitchum, R. M. Jr., and J. C. van Wagoner, 1991, High-frequency sequences and their stacking patterns: Sequence-stratigraphic evidence of high-frequency eustatic cycles: Sedimentary Geology, v. 70, p. 131–160.

Muto, T., and R. J. Steel, 2000, The accommodation concept in sequence stratigraphy: Some dimensional problems and possible redefinition: Sedimentary Geology, v. 130, p. 1–10.

Nemec, W., 1992, Depositional controls on plant growth and peat accumulation in a braidplain delta environment: Helvetiafjellet Formation (Barremian–Aptian), Svalbard, in P. J. McCabe and J. T. Parrish, eds., Controls on the distribution and quality of Cretaceous coals: Geological Society of America Special Paper 267, p. 209–226.

O'Mara, P. T., and B. R. Turner, 1999, Sequence stratigraphy of coastal alluvial plain Westphalian B Coal Measures in Northumberland and the southern North Sea: International Journal of Coal Geology, v. 42, p. 33–62.

Pascucci, V., M. R. Gibling, and M. A. Williamson, 2000, Late Paleozoic to Cenozoic history of the offshore Sydney Basin, Atlantic Canada: Canadian Journal of Earth Sciences, v. 37, p. 1143–1165.

Pashin, J. C., 1998, Stratigraphy and structure of coalbed methane reservoirs in the United States: An overview: International Journal of Coal Geology, v. 35, p. 209–240.

Petersen, H. I., J. A. Bojesen-Koefoed, H. P. Nytoft, F. Surlyk, J. Therkelsen, and H. Vosgerau, 1998, Relative sea-level changes recorded by paralic liptinite-enriched coal facies cycles, Middle Jurassic Muslingebjerg Formation, Hochstetter Forland, northeast Greenland: International Journal of Coal Geology, v. 36, p. 1–30.

Phillips, T. L., R. A. Peppers, and W. A. Dimichele, 1985, Stratigraphic and interregional changes in Pennsylvanian coal-swamp vegetation: Environmental inferences: International Journal of Coal Geology, v. 5, p. 43–109.

Plint, A. G., and D. Nummedal, 2000, The falling stage systems tract: Recognition and importance in sequence stratigraphic analysis, in D. R. Hunt, and R. L. Gawthorpe, eds., Sedimentary responses to forced regression: Geological Society (London) Special Publication 172, p. 1–17.

Posamentier, H. W., and G. P. Allen, 1999, Siliciclastic sequence stratigraphy— Concepts and applications: SEPM Concepts in Sedimentology and Paleontology, v. 7, 210 p.

Raymo, M. E., W. F. Ruddiman, J. Backman, B. M. Clement, and D. G. Martinson, 1989, Late Pliocene variation in northern hemisphere ice sheets and north Atlantic deep water circulation: Paleoceanography, v. 4, p. 413–446.

Rehill, T. A., 1996, Late Carboniferous nonmarine sequence stratigraphy and petroleum geology of the central Maritimes Basin, Eastern Canada: Ph.D. thesis, Dalhousie University, Halifax, Nova Scotia, 406 p.

Robb, C., 1876, Report on explorations and surveys in Cape Breton, Nova Scotia: Report of Progress for 1874-75, Geological Survey of Canada, p. 166–266.

Rogers, R. R., 1998, Sequence analysis of the Upper Cretaceous Two Medicine and Judith River Formations, Montana: Nonmarine response to the Claggett and Bearpaw marine cycles: Journal of Sedimentary Research, v. 68, p. 615–631.

Rust, B. R., and M. R. Gibling, 1990, Braidplain evolution in the Pennsylvanian South Bar Formation, Sydney Basin, Nova Scotia, Canada: Journal of Sedimentary Petrology, v. 60, p. 59–72.

Ryer, T. A., 1981, Deltaic coals of Ferron Sandstone member of Mancos shale: Predictive model for Cretaceous coal-bearing strata of western interior: AAPG Bulletin, v. 65, p. 2323–2340.

Ryer, T. A., 1983, Transgressive-regressive cycles and the occurrence of coal in some Upper Cretaceous strata of Utah: Geology, v. 11, p. 207–210.

Saunders, K. I., 1995, Sedimentology and depositional environments of the Pennsylvanian Hub cyclothem, Sydney Mines Formation, Cape Breton, Canada: M.S. thesis, Dalhousie University, Halifax, 171 p.

Schutter, S. R., and P. H. Heckel, 1985, Missourian (early late Pennsylvanian) climate in mid-continent North America: International Journal of Coal Geology, v. 5, p. 111–140.

Shanley, K. W., and P. J. McCabe, 1993, Alluvial architecture in a sequence stratigraphic framework: A case history from the Upper Cretaceous of southern Utah, U.S.A., in S. S. Flint and I. D. Bryant, eds., The geological modelling of hydrocarbon reservoirs and outcrop analogues: International Association of Sedimentologists Special Publication 15, p. 21–56.

Shimeld, J., and M. Deptuck, 1998, Lithostratigraphic correlation of the upper Sydney Mines Formation in the Sydney Basin (Donkin to Point Aconi), northeastern Nova Scotia: Geological Survey of Canada Openfile Report 3673.

Supardi, A. D. Subekty, and S. G. Neuzil, 1993, General geology and peat resources of the Siak Kanan and Bengkalis Island peat deposits, Sumatra, Indonesia, in J. C. Cobb and C. B. Cecil, eds., Modern and ancient coal-forming environments: Geological Society of America Special Paper 286, p. 45–61.

Tandon, S. K., and M. R. Gibling, 1997, Calcretes at sequence boundaries in Upper Carboniferous cyclothems of the Sydney Basin, Atlantic Canada: Sedimentary Geology, v. 112, p. 43–67.

Tibert, N. E., and M. R. Gibling, 1999, Peat accumulation on a drowned coastal braidplain: The Mullins coal

(Upper Carboniferous), Sydney Basin, Nova Scotia: Sedimentary Geology, v. 128, p. 23–38.

Titheridge, D. G., 1993, The influence of half-graben syndepositional tilting on thickness variation and seam splitting in the Brunner coal measures, New Zealand: Sedimentary Geology, v. 87, p. 195–213.

Tornqvist, T. E., 1993, Holocene alternation of meandering and anastomosing fluvial systems in the Rhine-Meuse Delta (central Netherlands) controlled by sea-level rise and subsoil erodibility: Journal of Sedimentary Petrology, v. 63, p. 683–693.

Udden, J. A., 1912, Geology and mineral resources of the Peoria quadrangle, Illinois: U.S. Geological Survey Bulletin, v. 506, p. 1–103.

Ulicny, D., 1999, Sequence stratigraphy of the Dakota Formation (Cenomanian), southern Utah: Interplay of eustasy and tectonics in a foreland basin: Sedimentology, v. 46, p. 807–836.

van Wagoner, J. C., H. W. Posamentier, R. M. Mitchum, P. R. Vail, J. F. Sarg, T. S. Loutit, and J. Hardenbol, 1988, An overview of the fundamentals of sequence stratigraphy and key definitions, *in* C. K. Wilgus, B. S. Hastings, C. A. Ross, H. W. Posamentier, J. van Wagoner, and C. G. St. C. Kendall, eds., Sea-level changes: An integrated approach, SEPM Special Publication 42, p. 39–45.

Vosgerau, H., J. A. Bojesen-Koefoed, H. I. Petersen, and F. Surlyk, 2000, Forest fires, climate, and sea-level changes in a coastal plain-shallow marine succession (early–middle Oxfordian Jakobsstigen Formation, north-east Greenland): Journal of Sedimentary Research, v. 70, p. 408–418.

Weisenfluh, G. A., and J. C. Ferm, 1984, Geologic controls on deposition of the Pratt seam, Black Warrior Basin, Alabama, U.S.A., *in* R. A. Rahmani and R. M. Flores, eds., Sedimentology of coal and coal-bearing se-

quences: International Association of Sedimentologists Special Publication 7, p. 317–330.

Weller, J. M., 1930, Cyclic sedimentation of the Pennsylvanian period and its significance: Journal of Geology, v. 38, p. 97–135.

White, J. C., 1992, Late Carboniferous cyclothems and organic facies in the Phalen-Backpit seam interval, Sydney coalfield, Nova Scotia: M.S. thesis, Dalhousie University, Halifax, 287 p.

White, J. C., M. R. Gibling, and W. D. Kalkreuth, 1994, The Backpit seam, Sydney Mines Formation, Nova Scotia: A record of peat accumulation and drowning in a Westphalian coastal mire: Palaeogeography, Palaeoclimatology, Palaeoecology, v. 106, p. 223–239.

Wignall, P. B., 1987, A biofacies analysis of the *Gastrioceras cumbriense* marine band (Namurian) of the central Pennines: Proceedings of the Yorkshire Geological Society, v. 46, p. 111–121.

Wise, D. U., E. S. Belt, and P. C. Lyons, 1991, Clastic diversion by fold salients and blind thrust ridges in coal-swamp development: Geology, v. 19, p. 514–517.

Wright, V. P., and S. B. Marriott, 1993, The sequence stratigraphy of fluvial depositional systems: The role of floodplain sediment storage: Sedimentary Geology, v. 86, p. 203–210.

Xue, L., 1997, Depositional cycles and evolution of the Paleogene Wilcox strata, Gulf of Mexico Basin, Texas: AAPG Bulletin, v. 81, p. 937–953.

Zodrow, E. L., 1985, *Odontopteris* Brongniart in the Upper Carboniferous of Canada: Palaeontographica Abteilung A, v. 196, p. 79–110.

Zodrow, E. L., and C. J. Cleal, 1988, The structure of the Carboniferous pteridosperm frond *Neuropteris ovata* Hoffmann: Paleontographica Abteilung B, v. 208, p. 105–124.

Pashin, J. C., 2004, Cyclothems of the Black Warrior Basin, Alabama,
U.S.A.: Eustatic snapshots of foreland basin tectonism, *in* J. C. Pashin
and R. A. Gastaldo, eds., Sequence stratigraphy, paleoclimate, and
tectonics of coal-bearing strata: AAPG Studies in Geology 51, p. 199–217.

9

Cyclothems of the Black Warrior Basin, Alabama, U.S.A.: Eustatic Snapshots of Foreland Basin Tectonism

Jack C. Pashin

Geological Survey of Alabama, Tuscaloosa, Alabama, U.S.A.

ABSTRACT

Flooding-surface-bounded depositional cycles were used to make three-dimensional models of total effective subsidence in the lower Pennsylvanian Pottsville Formation of the eastern Black Warrior Basin and demonstrate the utility of Pennsylvanian cyclothems for assessing controls on the development of accommodation space. The Black Warrior foreland basin underwent tectonic loading during Pottsville deposition, and the regional subsidence pattern reflects superposition of an Appalachian depocenter on an older Ouachita flexural moat. Modeling cycle duration within the constraints of biostratigraphy and foreland-basin geodynamics indicates a transition from dominant third-order cyclicity to dominant fourth- and fifth-order cyclicity that may be equivalent to the 0.4- and 0.1-m.y. Milankovitch orbital eccentricity periods. Changes in subsidence pattern from cycle to cycle are most pronounced in the distal part of the basin and reflect significant spatial and secular excursions of subsidence rate. These excursions indicate that high-frequency tectonic activity that generally has not been considered in stratigraphic and geophysical models of foreland basins can be detected by backstripping and mapping Pennsylvanian cyclothems.

INTRODUCTION

The importance of eustasy and tectonics in forming Pennsylvanian cyclothems has been debated vigorously since the landmark papers of Weller (1930, 1936) and Wanless and Shepard (1936). Advances in sedimentology, geodynamics, and paleoclimatology have resulted in increasingly sophisticated models of cyclothems that incorporate a delicate interplay among sedimentation, tectonics, and high-frequency glacial eustasy (e.g., Heckel, 1977, 1994; Cecil, 1990;

Klein, 1994). Regardless of tectonic setting, cyclothems typically represent intervals of geologic time compatible with Milankovitch orbital parameters (100–400 k.y.). For this reason, eustasy driven by Gondwanan glaciation is cited widely as the primary cause of cyclicity in the Pennsylvanian system (e.g., Heckel, 1986; Ross and Ross, 1988; Dickinson et al., 1994).

If this is true, then Pennsylvanian cyclothems provide a well-resolved stratigraphic record that facilitates four-dimensional modeling of the generation

Figure 1. Regional tectonic setting of the Black Warrior foreland basin (modified from Thomas, 1988).

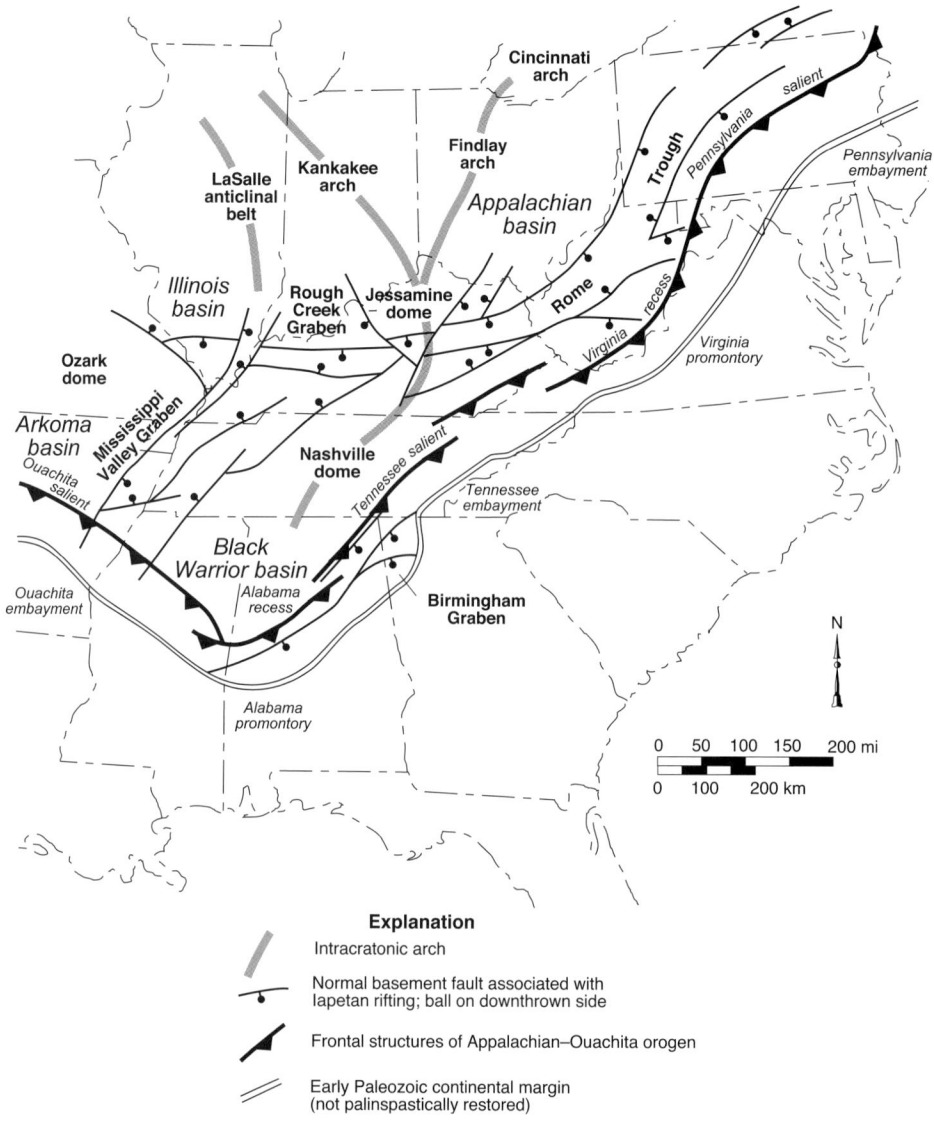

Explanation

~~~ Intracratonic arch

•— Normal basement fault associated with Iapetan rifting; ball on downthrown side

◢ Frontal structures of Appalachian–Ouachita orogen

// Early Paleozoic continental margin (not palinspastically restored)

of accommodation space in sedimentary basins. As a step toward realizing this potential, the objective of this chapter is to use stratigraphic interpretations of cyclothems based on geophysical well logs and cores to model the early Pennsylvanian subsidence history of the Black Warrior foreland basin in Alabama (Figures 1, 2). The Black Warrior Basin is a late Paleozoic foreland basin in Alabama and Mississippi that formed in response to converging thrust and sediment loads in the Appalachian–Ouachita orogen (e.g., Thomas, 1976, 1985, 1995). The basin has a triangular plan and occupies the structural recess between the Appalachian orogen on the southeast and the Ouachita orogen on the southwest. In a broad sense, the basin can be characterized as a faulted homocline that dips southwest toward the Ouachita orogen. The frontal folds and thrust faults of the Appalachian orogen are superimposed on the southeast part of the homocline (Figure 2).

The Appalachian–Ouachita orogen formed by deformation of the Alabama promontory, a major protuberance of the Laurentian continental platform that developed during Iapetan rifting (Thomas, 1991). The southwest margin of the promontory remained passive until the Mississippian, when the Black Warrior foreland basin was initiated by obduction of an Ouachita accretionary prism (Thomas, 1976; Viele and Thomas, 1989). Convergence along the southeastern, or Appalachian, margin of the promontory began during the Ordovician Taconic orogeny. However, the Black Warrior Basin was effectively sheltered from Appalachian tectonic activity and sediment sources by the Birmingham graben until the early Pennsylvanian, when a sediment source and subsidence center developed along the southeastern margin of the basin (Sestak, 1984; Pashin et al., 1991).

This study focuses on the lower Pennsylvanian Pottsville Formation, which is a siliciclastic succession containing virtually all the economic coal and coal-bed methane resources in the Black Warrior Basin (Figure 3). Understanding that coal beds are preserved in stratigraphic clusters, McCalley (1900) subdivided the Alabama Pottsville into a series of informal coal groups that have provided the basis for most later subdivisions (Butts, 1910, 1926; Culbertson, 1964; Metzger, 1965). Butts (1926) was first to recognize the evidence for repeated marine transgressions and regressions during Pottsville deposition. During the late 1960s and 1970s, the Alabama Pottsville played a central role in the development of facies models for Appalachian coal-bearing strata

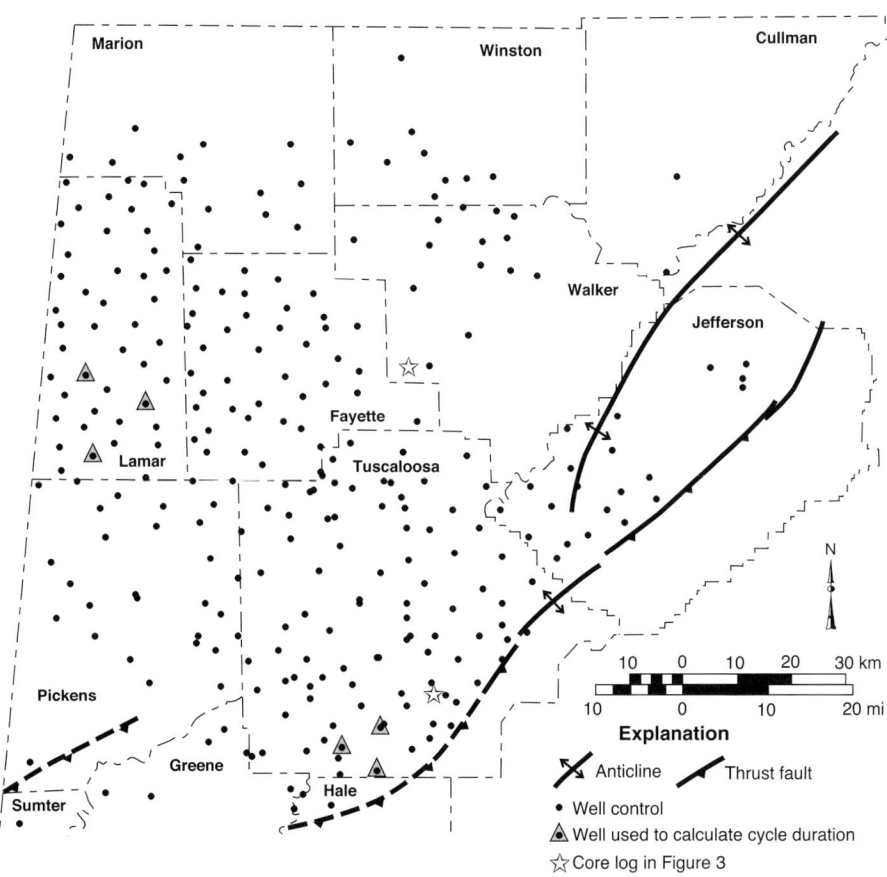

Figure 2. Index map of study area in the Black Warrior Basin of Alabama showing well control.

than 0.3 m (0.98 ft) and have been interpreted as condensed sections that mark maximum flooding. The ravinement surfaces can serve as cycle boundaries in outcrop and core, and the ravinements and condensed sections can be approximated in geophysical well logs by picking the bases of thick shale units just below the point of maximum gamma count (Figure 3). Thus, the cycle boundaries as picked in geophysical well logs are maximum flooding surfaces in the terminology of Van Wagoner et al. (1990), are genetic sequence boundaries in the terminology of Galloway (1989), and can accordingly serve as effective time lines.

Pottsville cycle boundaries have been used to develop balanced models of thin-skinned folds and faults (Wang, 1994; Pashin and Groshong, 1998), and these surfaces appear equally useful for characterizing foreland basin evolution. The principal methods used in this investigation are geophysical log analysis, backstripping, three-dimensional (3-D) computer gridding, and estimation of cycle duration and subsidence rate with respect to geochronologic and geodynamic constraints. The main reasons for this study were (1) to demonstrate the utility and limitations of regionally extensive flooding surfaces in sedimentary basin analysis and (2) to characterize spatial and temporal variation of subsidence in a foreland basin by taking advantage of the stratigraphic resolution afforded by Pennsylvanian cyclothems.

(Ferm et al., 1967; Horsey, 1981; Rheams and Benson, 1982; Ferm and Weisenfluh, 1989), and discussions focused mainly on autogenic variables. Wanless (1976) made passing mention of cyclicity in the Pottsville Formation, but it was not until the intensive exploration for conventional hydrocarbons and coal-bed methane in the 1980s that basinwide depositional cycles were confirmed (Sestak, 1984; Thomas, 1988; Pashin et al., 1991). The first cyclostratigraphic subdivision of the upper part of the Pottsville Formation was made by Pashin et al. (1991), who defined a series of basinwide coarsening- and coaling-upward cycles. Pashin (1994a) indicated that the average duration of Pottsville cycles was between 0.2 and 0.5 m.y., pointing toward glacial eustasy as the dominant causal mechanism.

Throughout the 1990s, work focused on applying sequence-stratigraphic principles to the Pottsville cycles in Alabama (Gastaldo et al., 1993; Pashin 1994b, 1998) (Figures 3, 4). A key advance was the recognition of ravinement surfaces, or transgressive surfaces of erosion, that formed during the maximum rate of rise of relative sea level (Liu and Gastaldo, 1992; Demko and Gastaldo, 1996). The ravinement surfaces are overlain by marine shell beds that are typically thinner

METHODS

Cores containing strata from the base of the Pottsville Formation through the Brookwood coal zone were described, logged graphically, and interpreted using standard methods (Figure 3). The core logs were

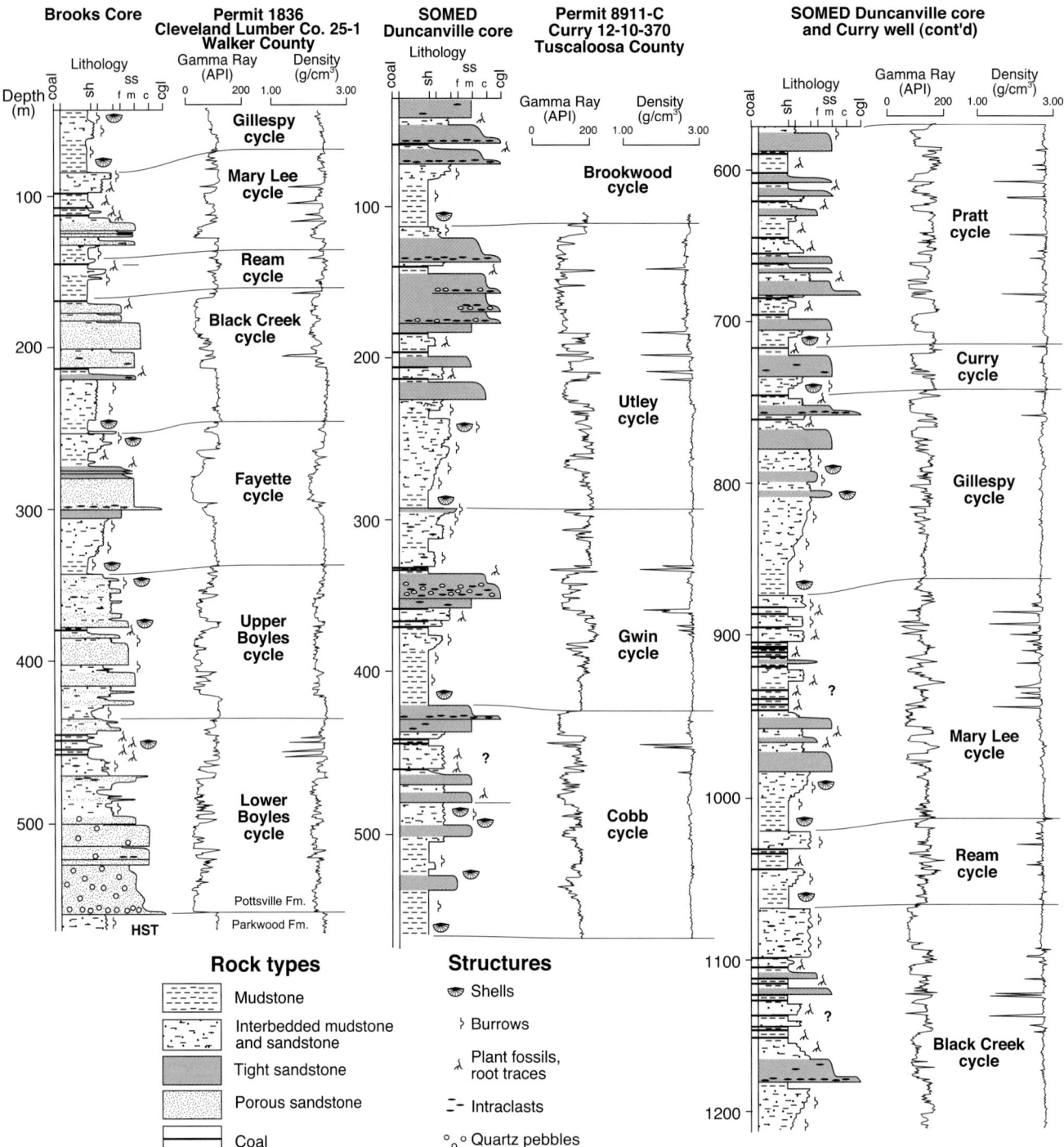

Figure 3. Graphic logs of long cores showing depositional cycles in the Pottsville Formation correlated with nearby geophysical well logs.

correlated with nearby gamma-ray and density logs to determine the geophysical response of each rock type and provide the most precise approximation of Pottsville cycle boundaries possible. The well logs were correlated with geophysical logs from 275 other coal-bed methane and conventional wells (Figure 2). These wells were chosen to provide the most uniform

spacing possible in the study area. Because of limited control outside the coal-bed methane and conventional hydrocarbon fairways, a well spacing of approximately 5 km (3 mi) was used where possible. Few wells have been drilled deeper than the Black Creek cycle in the coal-bed methane fairway of southern Tuscaloosa and western Jefferson Counties, and

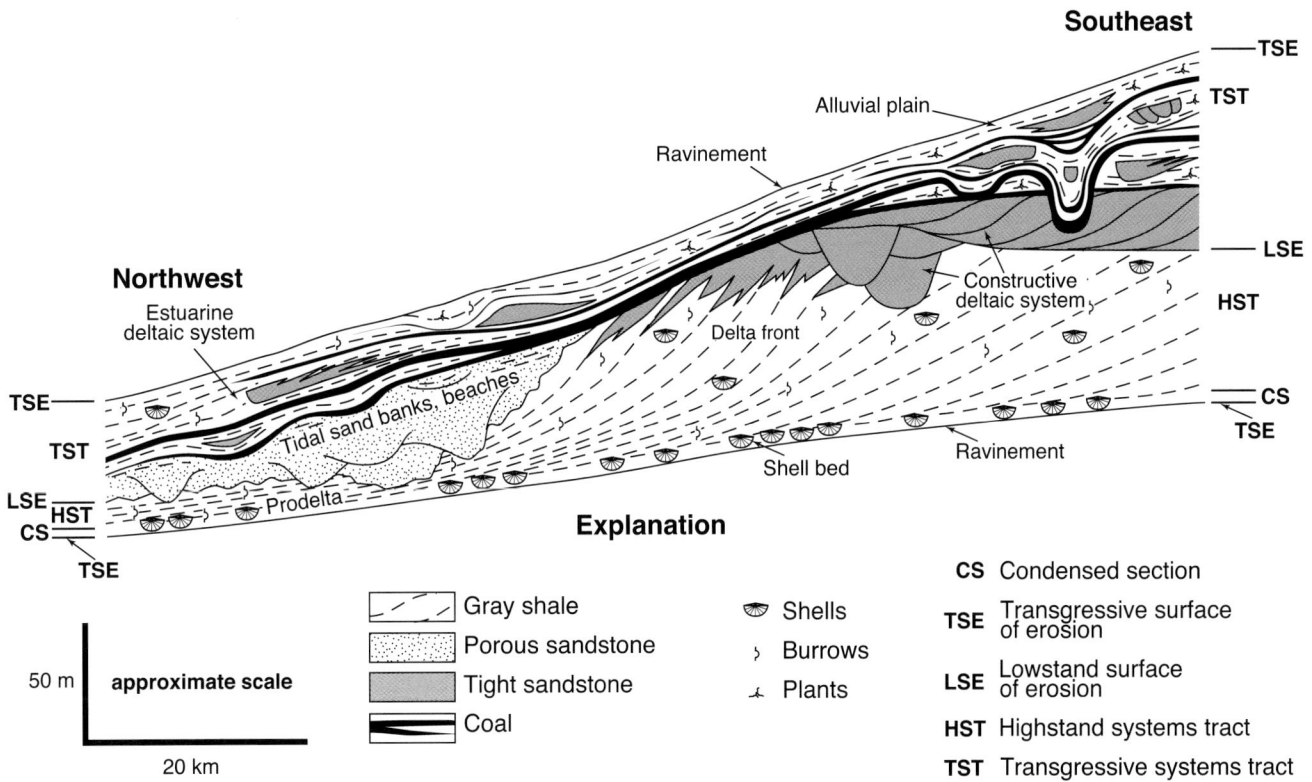

Figure 4. Idealized depositional cycle in the Pottsville Formation of the Black Warrior Basin in Alabama.

the principal control on the deep units in this area comes from deep salt-water disposal wells.

Well locations (latitude and longitude) were calculated from line calls in the database of the State Oil and Gas Board of Alabama using a digital land grid and the Geographix Exploration System (GES). Cycle boundaries were picked in the well logs, and cycle thickness was calculated. The only cycle boundary that was not picked near a ravinement or condensed section is the base of the lower Boyles cycle, which marks the base of the Pottsville Formation and is considered to be a prominent lowstand surface of erosion (Thomas, 1988; Pashin, 1993). In addition, to determine basin geometry prior to Pottsville deposition, the interval from the base of the Pottsville Formation to the top of the youngest, regionally extensive limestone marker in the underlying Parkwood Formation was mapped.

Calculation of subsidence comprised two major components: decompaction and computer gridding. Each cycle was decompacted using the approximate method of Van Hinte (1978). Because no systematic porosity-depth trends for shale and sandstone were identified in the geophysical well logs, original porosity values cannot be determined empirically. Therefore, the commonly used porosity-depth values and

compaction constants of Sclater and Christie (1980) for shale, sandstone, and limestone were used in this investigation to maintain uniformity of other investigations of burial history in the Black Warrior Basin (Hines, 1988; Thomas et al., 1991; Whiting and Thomas, 1994; Carroll et al., 1995). Decompaction of cycles to original thickness using other compaction constants (e.g., Bond et al., 1983) and other decompaction methods (e.g., Magara, 1980; Welte et al., 1996) yielded an uncertainty of ±20%. Coal was not backstripped because thickness cannot be determined accurately without high-resolution density logs (Pashin, 1994c), and because most compaction occurs during the initial stages of burial (Nadon, 1998). Most coal beds are thinner than 0.3 m (0.98 ft); hence, the effect of coal on estimates of total effective subsidence is minimal.

Estimates based on vitrinite-reflectance profiles indicate that 1.2–2.5 km (0.75–1.55 mi) of sediment has been eroded from the Black Warrior Basin since maximum burial during Appalachian–Ouachita orogeny (Telle et al., 1987; Hines, 1988; Carroll et al., 1995). Although these values have a large range of uncertainty, sediment is compacted sufficiently at these depths that the value used has minimal effect on decompaction to original thickness. Therefore, 2 km

(1.24 mi) was added to measured depth in all wells. Conversely, the thickness of poorly consolidated Mesozoic strata, which disconformably overlie the Pottsville Formation in the western part of the study area, was subtracted from measured depth. Where Mesozoic strata were not logged, thickness was estimated using a structural contour map of the Paleozoic–Mesozoic disconformity surface (Kidd, 1976). Original thickness was then calculated for each Pottsville cycle and the upper Parkwood Formation using the equation of Van Hinte (1978). Relatively few wells in the coal-bed methane fields penetrate the base of the synorogenic clastic wedge, and no wells reach crystalline basement; thus, calculation of the tectonic component of subsidence was not attempted.

At this point, georeferenced present and original thickness values for all 277 wells were loaded into the Isomap module of GES and gridded and contoured using a minimum-curvature algorithm. Each grid cell was approximately 2.3 km^2 (0.89 mi^2), making a field of 71 cells north by 70 cells east. Present cycle thickness was mapped to compare cycle isopach maps to maps of original cycle thickness. Isopach maps are not published here because contours effectively parallel those depicted in the subsidence maps. Original thickness was entered as a negative value so that Isomap would depict basin geometry correctly in three dimensions. Much of Pottsville deposition took place at or near sea level; therefore, no bathymetric correction is required to convert original cycle thickness to total effective subsidence. The final step was to plot total effective subsidence (i.e., original cycle thickness) as 3-D contoured grids for the upper Parkwood Formation and for each Pottsville cycle. All grids were constructed to be viewed from an azimuth of 220°, from 50° above horizontal, with a twist angle of 5°, and with a vertical exaggeration of 150.

Total effective subsidence can be converted to subsidence rate if the duration of each Pottsville cycle is known. The duration of lower Boyles–Brookwood deposition was estimated using the available biostratigraphic control and the timescale of Harland et al. (1990). Pottsville cycles maintain similar thickness proportions throughout the basin. Therefore, a rough estimate of the duration of each cycle can be made by combining information on cycle thickness with the accelerating subsidence of actively loaded foreland basins as modeled by Watts et al. (1982) and Stockmal et al. (1986). Six well logs recording the base of the Pottsville through the top of the Brookwood cycle without interruption were chosen (Figure 2). Three well logs are from southern Lamar County, which is

removed from the most active centers of late Paleozoic subsidence, and three are from southern Tuscaloosa County, which are in the most active area of subsidence. To minimize the effect of local variation (<10 m [<33 ft]), backstripped cycle thickness values from each set of three wells were averaged to construct a composite section. The results in each area then were compared to test the sensitivity of the calculations to location in the basin.

Thomas et al. (1991) and Whiting and Thomas (1994) modeled a 50–100% increase in tectonic subsidence rate from the start of Pottsville deposition through deposition of the Pratt cycle. As a baseline hypothesis of constant subsidence rate, cycle duration was first estimated for each area by multiplying the percentage each decompacted cycle constitutes of the lower Boyles–Brookwood interval by the estimated time span of the complete interval. To model accelerating subsidence and a doubling of subsidence rate, cycle duration was recalculated using a simple goal-seeking interpolation in which a unit of decompacted sediment thickness at the start of Pottsville deposition represents twice as much time as at the end of Brookwood deposition. The time estimates then were used to calculate total effective subsidence rates for each cycle, and additional sensitivity tests were conducted to evaluate the validity of those rates.

TOTAL EFFECTIVE SUBSIDENCE

Gridding and contouring total effective subsidence for the upper Parkwood Formation and each Pottsville cycle documents variation of the distribution of accommodation space (Figures 5–7). The upper Parkwood grid shows an increase of the magnitude of subsidence from negligible in the northeastern part of the study area to nearly 450 m (1476 ft) in the southwest part (Figure 5). Lower Boyles deposition marks a major change in the geometry of the eastern Black Warrior Basin. The lowest calculated subsidence value for this cycle is greater than 100 m (328 ft), and maximum subsidence values exceed 375 m (1230 ft). Irregular contour patterns in the northwest part of the study area reflect extreme internal facies heterogeneity, including channeling. The most striking changes relative to the upper Parkwood grid are that parts of many contours are oriented east-west with some northeast-southwest elements, and that the highest subsidence values are along the southern edge of the study area.

Subsidence values in the upper Boyles cycle range from slightly less than 100 to more than 300 m (328

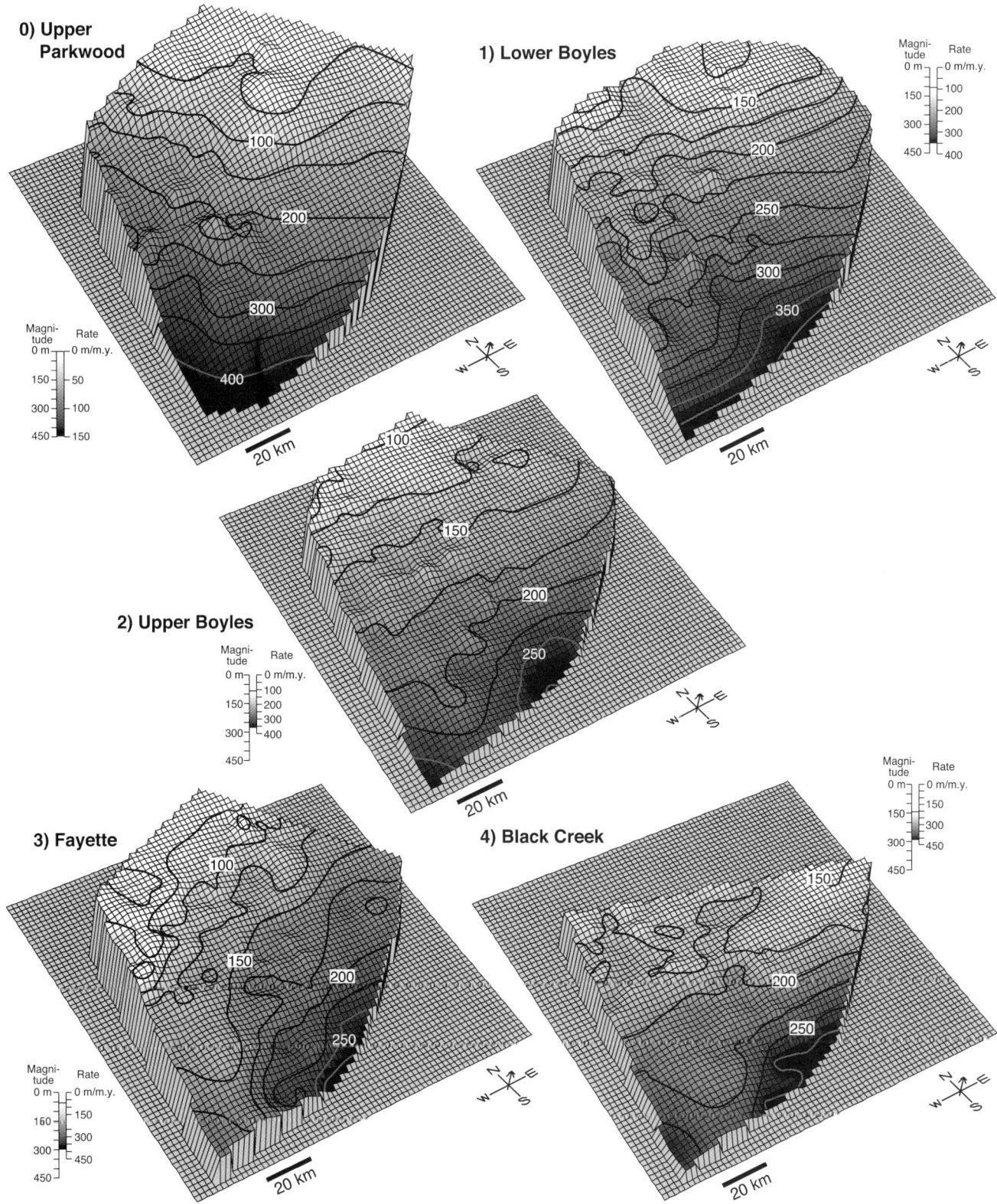

Figure 5. Grid plots of total effective subsidence in the upper Parkwood Formation through the Black Creek cycle of the Pottsville Formation. Contours are total effective subsidence (meters).

to >984 ft) (Figure 5). The most significant changes from lower Boyles deposition are that maximum subsidence values define a depocenter in the southeastern part of the study area (southern Tuscaloosa and northern Hale Counties) and that contours near the southwest corner again are oriented northwest. This

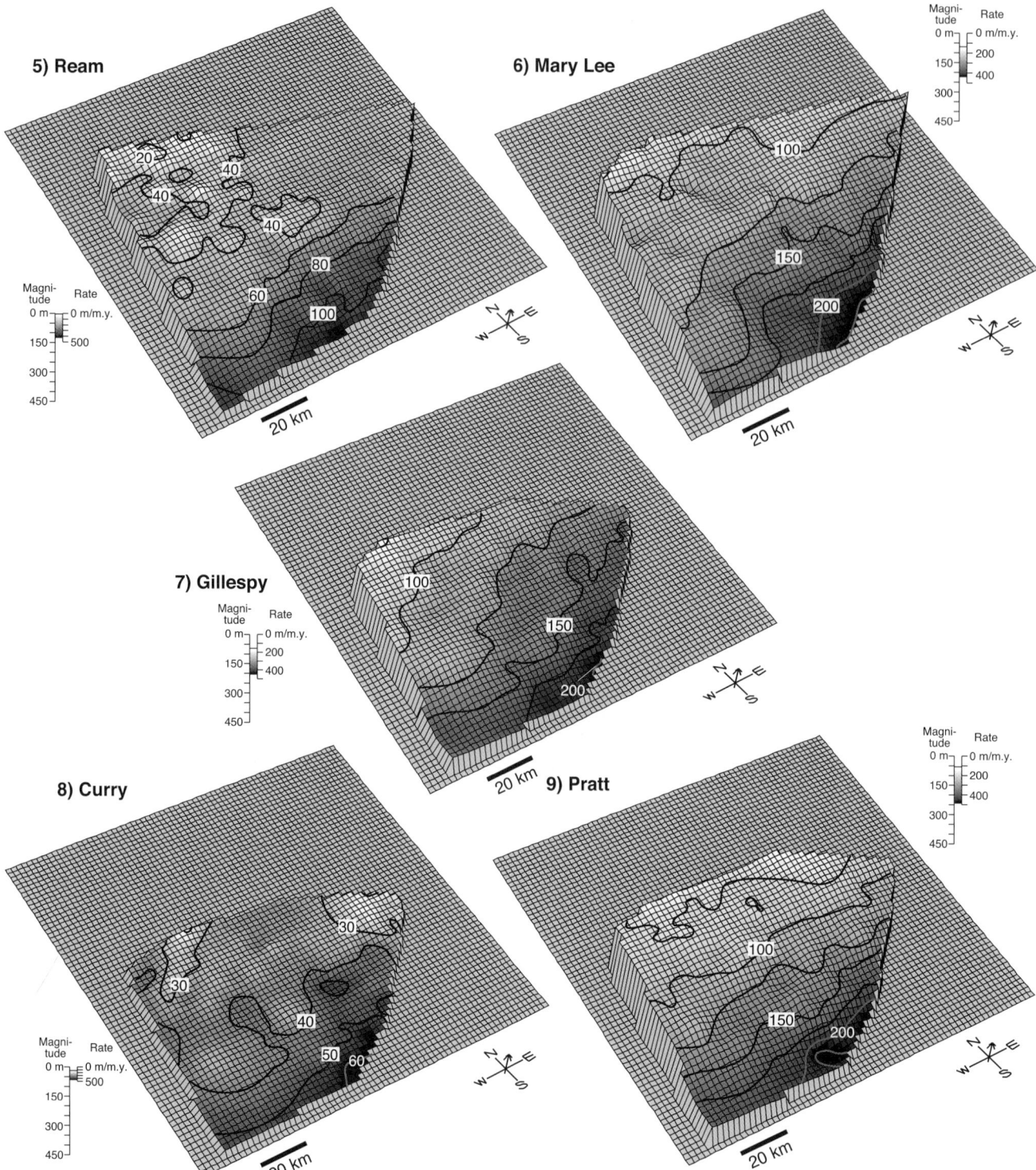

Figure 6. Grid plots of total effective subsidence in the Ream through Pratt cycles of the Pottsville Formation. Contours are total effective subsidence (meters).

general configuration persists in the remainder of the Pottsville cycles, although some differences between the grids are noteworthy (Figures 5–7). In the Fayette cycle, for example, many contours are nearly parallel to the southeast basin margin (Figure 5). The sub-

sidence pattern in the Black Creek through Mary Lee cycles resembles that in the upper Boyles (Figures 5, 6). The Ream cycle is notable, however, because subsidence values are low, ranging from 15 to 110 m (49 to 361 ft) (Figure 6). As in the lower Boyles cycle,

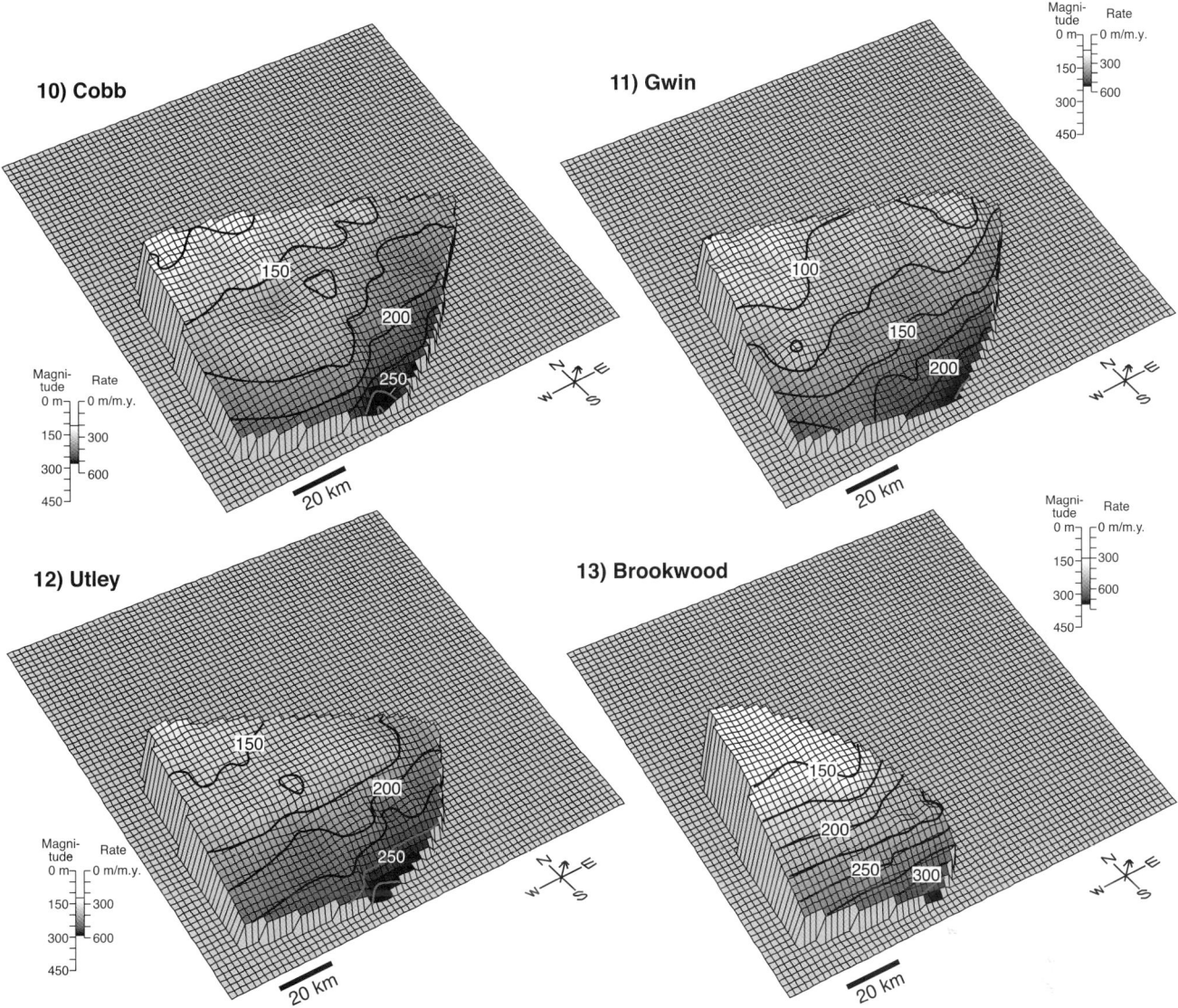

Figure 7. Grid plots of total effective subsidence in the Cobb through Brookwood cycles of the Pottsville Formation. Contours are total effective subsidence (meters).

irregular contour patterns in the northwest reflect facies heterogeneity in the Ream cycle.

Similar to the Fayette cycle, many contours in the Gillespy cycle are subparallel to the southeast margin of the basin (Figure 6). The Curry cycle tends to be the thinnest of all Pottsville cycles, and this is reflected in subsidence values ranging from 19 to only 63 m (62 to only 207 ft). Grid and contour patterns are similar in the Pratt through Gwin cycles (Figures 6, 7). A component of subsidence increasing toward the southwest in the Cobb and Gwin cycles persists despite a loss of stratigraphic control in the southwesternmost well, and contours in the distal part of the basin show a general clockwise rotation (Figure 7). In the Utley and Brookwood cycles, contours are oriented nearly east-west, and no

significant component of southwest thickening is apparent.

CYCLE DURATION

The magnitude of subsidence during deposition of Pottsville cycles is modeled readily in three dimensions (Figures 5–7), but quantifying the rate of subsidence is more uncertain because it requires an accurate determination of cycle duration. The upper Parkwood Formation is of latest Chesterian (Namurian A, ~321 Ma) through early Morrowan age (Butts, 1926; Jennings and Thomas, 1987). Miospore assemblages from the Appalachian thrust belt indicate that the uppermost Parkwood strata are mainly of Namurian C age and may contain some Langsettian strata in

Table 1. Sensitivity of calculated cycle duration (m.y.) based on decompacted sediment thickness in composite sections of Lamar and Tuscaloosa Counties.

Cycle	Lamar County*		Tuscaloosa County**		Difference	
	Constant Subsidence Rate	Doubling Subsidence Rate	Constant Subsidence Rate	Doubling Subsidence Rate	Constant Subsidence Rate	Doubling Subsidence Rate
Brookwood	0.64	0.45	0.58	0.40	0.06	0.05
Utley	0.66	0.50	0.64	0.48	0.02	0.02
Gwin	0.40	0.33	0.51	0.41	−0.11	−0.08
Cobb	0.59	0.51	0.59	0.51	0.00	0.00
Pratt	0.38	0.34	0.48	0.44	−0.10	−0.10
Curry	0.14	0.13	0.14	0.13	0.00	0.00
Gillespy	0.43	0.41	0.45	0.44	−0.02	−0.03
Mary Lee	0.50	0.50	0.52	0.53	−0.02	−0.03
Ream	0.18	0.19	0.28	0.29	−0.10	−0.10
Black Creek	0.85	0.92	0.63	0.69	0.22	0.23
Fayette	0.50	0.57	0.67	0.77	−0.17	−0.20
Upper Boyles	0.75	0.91	0.63	0.76	0.12	0.15
Lower Boyles	0.98	1.26	0.89	1.15	0.09	0.11
Minimum	0.14	0.13	0.14	0.13	−0.17	−0.20
Maximum	0.98	1.26	0.89	1.15	0.22	0.23
Mean	0.54	0.54	0.54	0.54	0.00	0.00
Median	0.50	0.50	0.58	0.48	0.00	0.00

*State Oil and Gas Board of Alabama permit numbers 2492, 3136, and 3305.
**State Oil and Gas Board of Alabama permit numbers 2617, 9635-C, and 7131-C.

the upper 10 m (33 ft) (Eble et al., 1991). For the purposes of this report, the Pottsville–Parkwood contact is placed at the Namurian C–Langsettian boundary (318 Ma).

The Pottsville Formation of Alabama has long been thought to be of lower Pennsylvanian age (Butts, 1926; Cropp, 1960; Upshaw, 1967; Eble and Gillespie, 1989). The Brookwood cycle contains palynomorph, plant megafossil, and marine invertebrate assemblages indicative of a late Morrowan (Langsettian) age (Metzger, 1965; Eble and Gillespie, 1989). As many as six coal zones above the Brookwood are preserved in the structurally deepest parts of the Black Warrior Basin (Telle et al., 1987; Henderson and Gazzier, 1989; Bodden, 1997), and the Morrowan–Atokan (Langsettian–Duckmantian) boundary may be in this section. However, samples from two coal zones younger than the Brookwood contain the same palynomorph assemblage as the Brookwood (R. E. Carroll, 1996, personal communication). According to Harland et al. (1990), the Morrowan–Atokan boundary has an age of 310 Ma. Because the top of the Brookwood cycle appears to be slightly older, it is assigned an age of 311 Ma. In summary, a working hypothesis using the timescale

of Harland et al. (1990) is that the Pottsville Formation represents a span of 7 m.y. from 311 to 318 Ma. Using different timescales (Hess and Lippolt, 1986; Menning et al., 2000), the Pottsville Formation would represent less than 3.5 m.y. (~313–316.5 Ma).

If the 13 Pottsville cycles are assumed to represent equal amounts of time, then each cycle was deposited in 0.5 m.y. This is almost certainly untrue, because variation of cycle thickness would necessitate unreasonable changes of subsidence rate for an actively loaded foreland basin. For example, the Curry cycle is typically one-fourth the thickness of the underlying Gillespy cycle; hence, if these cycles were deposited in the same amount of time, subsidence would have to slow by a factor of 4 in only 0.5 m.y. A second hypothesis is that subsidence rate remained constant throughout Pottsville deposition. If this was true, the percentage of Pottsville time represented by each cycle would be equal to the percentage of backstripped sediment thickness. Applying this principle to the wells in southern Lamar and southern Tuscaloosa Counties (Figure 2), individual cycles were deposited in 0.1–1.1 m.y. with a median duration of 0.5 m.y. (Table 1).

A more realistic estimation of cycle duration in an actively loaded foreland basin should take into account accelerating subsidence. Modeling a doubling of subsidence rate in the Black Warrior Basin increases the length of the oldest cycles, decreases the length of the youngest cycles, and has little effect on the cycles near the middle of the Pottsville Formation (Table 1). Accordingly, individual cycles were deposited in 0.1–1.3 m.y., again with a median duration of 0.5 m.y. Sensitivity of the calculations to location in the basin is approximately 0.2 m.y. in the upper Boyles through Black Creek cycles and is 0.1 m.y. or less in all other cycles. By this method, estimations of cycle duration in the relatively low-accommodation setting of Lamar County and the high-accommodation setting of southern Tuscaloosa County differ by 25% or less, the only exception being the estimates for the Ream cycle, which differ by 35%.

The results of this exercise indicate that the lower Boyles cycle represents more time than any other Pottsville cycle (1.2–1.3 m.y.), and that the upper Boyles through Black Creek cycles were deposited in 0.6–0.9 m.y. (Figure 8). The Curry cycle apparently was deposited the most rapidly of any Pottsville cycle, representing 0.1 m.y. All other cycles fall into a narrow range of 0.3–0.5 m.y. by this method.

SUBSIDENCE RATE

Having modeled cycle duration on the basis of stratigraphic and geodynamic constraints, the rate of total effective subsidence can be estimated by dividing the magnitude of said subsidence by cycle duration. Assuming a constant subsidence rate during deposition of the Lamar County composite section yields a local rate of 220 m/m.y. (722 ft/m.y.) and during deposition of the Tuscaloosa County section yields a local rate of 440 m/m.y. (1444 ft/m.y.). Modeling accelerating and doubling subsidence rate on the basis of backstripped cycle thickness yields an initial rate of approximately 155 m/m.y. (509 ft/m.y.) and a terminal rate of 310 m/m.y. (1017 ft/m.y.) for the Lamar County section, and an initial rate of 325 m/m.y. (1066 ft/m.y.) and a terminal rate of 650 m/m.y. (2133 ft/m.y.) for the Tuscaloosa County section (Figure 9).

The sensitivity of these calculations to the point of reference is assessed easily by recalculating the subsidence rate in the Lamar County composite section using the time estimate from the Tuscaloosa County composite section and vice versa (Figure 9).

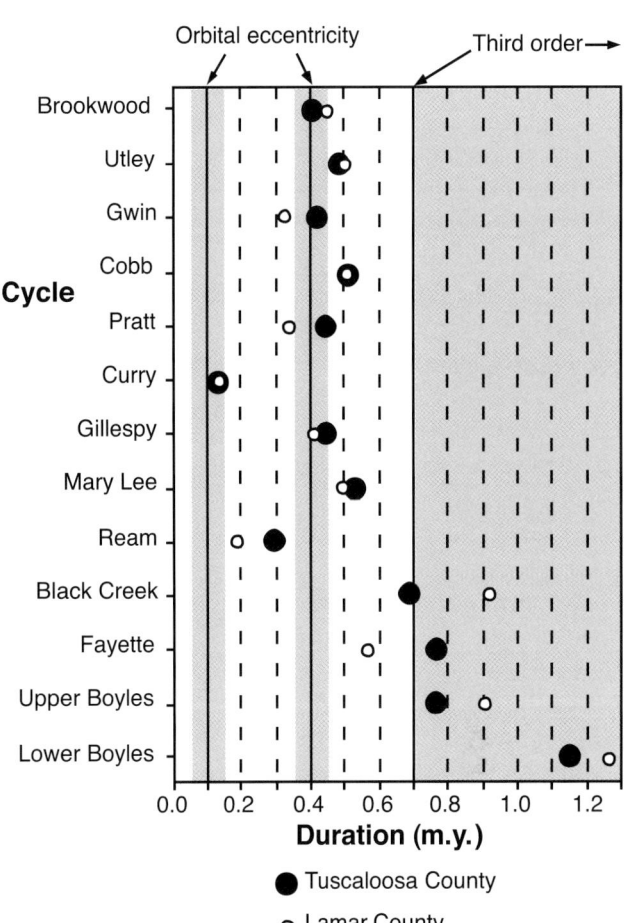

Figure 8. Estimated duration of Pottsville cycles modeled on the basis of decompacted cycle thickness and doubling subsidence rate. Results suggest a decrease of cycle duration related to a transition from third-order changes of relative sea level to fourth- and fifth-order changes during Pottsville deposition.

Variation of subsidence rate in Lamar County is at most moderately sensitive to a reference point in the depocenter of southern Tuscaloosa County, whereas subsidence rate in southern Tuscaloosa County is extremely sensitive to a reference point in Lamar County. Subsidence rate in the Lamar and Tuscaloosa sections deviates from the reference values by a maximum of 75 and 230 m/m.y. (246 and 755 ft/m.y.), respectively. This sensitivity also is reflected in percent variation. Subsidence rate in the Lamar section deviates from the reference rate by 36% or less, whereas subsidence rate in the Tuscaloosa section deviates from the reference rate by as much as 55%. This effect is amplified if one considers excursions of subsidence rate from cycle to cycle. From Black Creek to Ream deposition, for example, the calculated subsidence rate changes by 140 m/m.y. (459 ft/m.y.) in the Lamar section and by 325 m/m.y. (1066 ft/m.y.) in

Figure 9. Sensitivity of calculated subsidence rate to point of reference used to estimate cycle duration. Calculated subsidence rate is more sensitive to a reference point in the distal part of the basin (Lamar County composite section) than to one in the depocenter of southern Tuscaloosa and northern Hale Counties (Tuscaloosa County composite section).

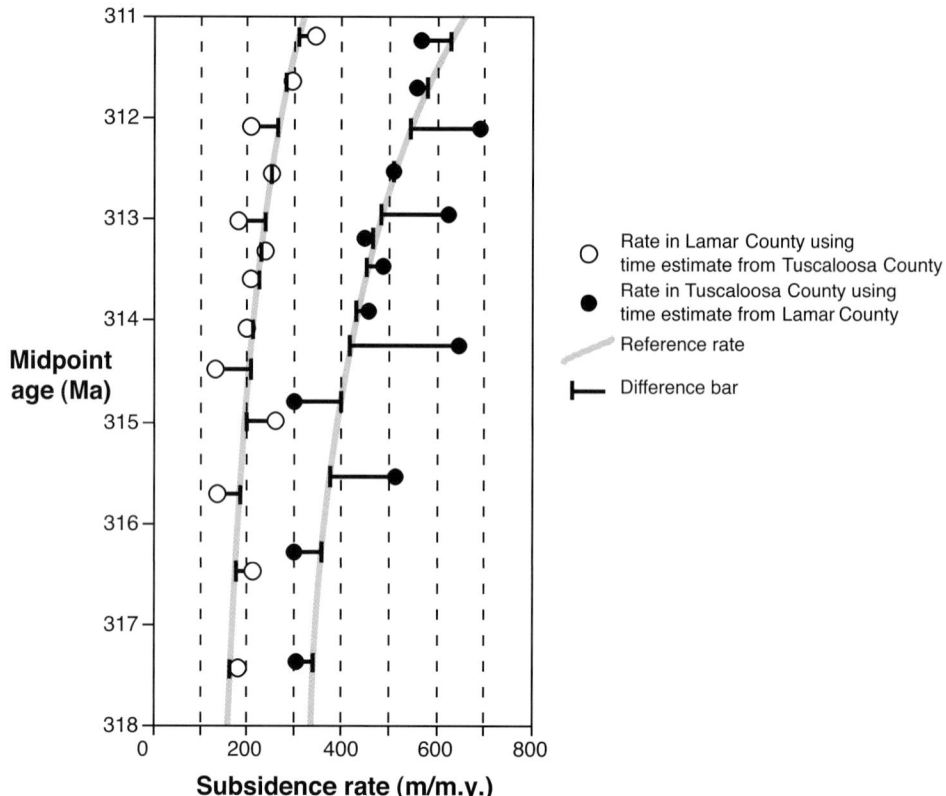

the Tuscaloosa section. Using the Tuscaloosa section as a reference, the only excursions greater than 50 m/m.y. (164 ft/m.y.) from the reference rates are negative excursions during Fayette, Ream, Pratt, and Gwin deposition, and a positive excursion during Black Creek deposition.

Clearly, point of reference is critical for modeling subsidence rate during Pottsville sedimentation. Using the Lamar County section as a geochronologic reference gives rise to unrealistic secular variation of subsidence rate in the depocenter of southern Tuscaloosa County. The relatively low sensitivity of the Lamar section to a point of reference in the depocenter, by comparison, suggests that the Tuscaloosa composite section is a superior reference for estimating cycle duration and subsidence rate (Figures 8, 9). Indeed, subsidence in proximal parts of a foreland basin is dominated ideally by thermal and isostatic compensation of the encroaching thrust and sediment loads, whereas distal parts can be more sensitive to far-field tectonic forces and autogenic sedimentary processes. In other words, using the Lamar composite to estimate cycle duration and subsidence rate is analogous to making a tail wag its dog.

Having identified the composite section in southern Tuscaloosa County as a working reference point, the range of subsidence rates was calculated for each cycle (Figures 5–7, 10). The lowest rate calculated is 25 m/m.y. (82 ft/m.y.) in the Ream cycle, and minimum rate tends to increase upward stratigraphically (Figure 10). However, this increase partly reflects preservation of younger cycles structurally deeper in the basin. Maximum rates for each cycle essentially follow the reference rates based on the Tuscaloosa

County composite section, and the highest rate in the eastern Black Warrior Basin was estimated to be 740 m/m.y. (2428 ft/m.y.) in the Brookwood cycle.

DISCUSSION

The grid plots provide evidence for considerable spatial and secular variation of the generation of accommodation space in the eastern Black Warrior Basin (Figures 5–7). The mechanisms that generated this space include tectonic subsidence, compaction, eustasy, regional topographic relief, and local depositional processes such as channel incision. Tectonic subsidence and compaction are factors that, in the absence of active sedimentation, favor marine transgression, and the effect of tectonic subsidence on the deposition of Pennsylvanian cyclothems is thought to be strongest near orogenic belts (Klein and Willard, 1989). The effect of these variables is to amplify eustatic rises of sea level and to reduce the effect of eustatic falls (Posamentier et al., 1988). Previous work in the Black Warrior Basin suggests that tectonic subsidence was related primarily to lithospheric flexure and accounted for approximately half of total effective subsidence (Hines, 1988; Thomas et al., 1991; Whiting and Thomas, 1994).

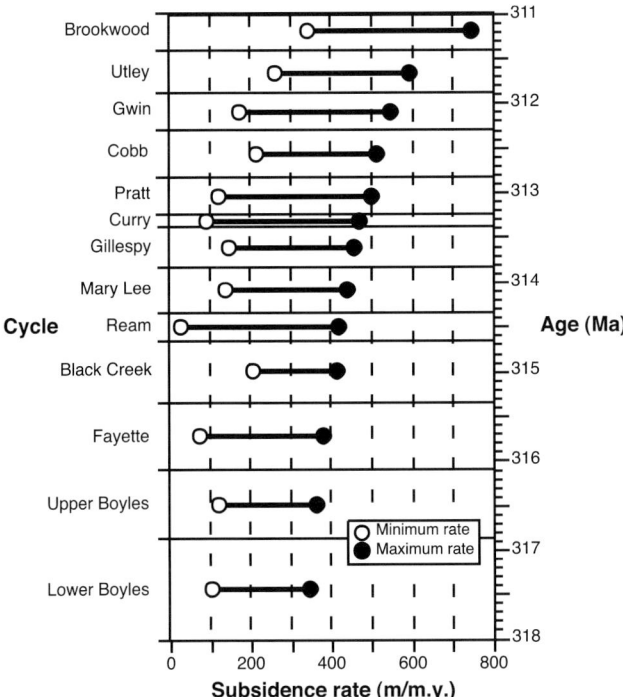

Figure 10. Range of estimated subsidence rate for each Pottsville cycle using the time estimate from the southern Tuscaloosa County composite section.

The ever-wet equatorial climate of the Euramerican coal belt facilitated high sediment flux during the early Pennsylvanian (Cecil, 1990), and subsidence, eustasy, and sediment flux functioned collectively to cause sediment starvation early in each Pottsville cycle and to fill the basin later in each cycle. Condensed shell beds (Figure 4) mark the time from the maximum rate of transgression to maximum flooding or initial highstand, and the thickness of the overlying deltaic deposits indicates significant deepening during this episode. Applying the method of Klein (1974) to backstripped sediment thickness in the Duncanville core of southern Tuscaloosa County (Figure 3) indicates maximum water depths ranging from 10 m (32.8 ft) during Curry deposition to 100 m (328 ft) during Gillespy and Cobb deposition. All Pottsville cycles include terrestrial or shore-zone strata that extend throughout the study area, suggesting that the available accommodation space in the eastern Black Warrior Basin was filled with sediment during or shortly after highstand. Evidence for valley incision in the proximal part of the basin, moreover, indicates overfilling during or after lowstand in some cycles. In the Mary Lee cycle, for example, Pashin (1994c) mapped a coal-filled dendritic paleovalley with a preserved depth of 20 m (66 ft). Allowing for compaction, the original depth of the valley was ap-

proximately 36 m (118 ft), which is a minimum estimate for overfilling.

The cycle-bounding ravinements are interpreted to have formed by shoreface erosion at or just below sea level, and the effect of this erosion was to redistribute sediment locally, thereby reducing paleotopographic relief (Liu and Gastaldo, 1992). Deepening following regional flooding was a temporary source of accommodation space that was filled completely during highstand. Similarly, valley incision was a temporary response to fallen sea level and overfilling of the basin with sediment, and no valleys are known to have remained unfilled prior to regional inundation. The only Pottsville cycles where underfilling and channeling may have affected total accommodation space beyond a local scale are (1) the Ream cycle in the northwest part of the study area, where variation of cycle thickness exceeds 20 m (66 ft) (Figure 6) and (2) the lower Boyles cycle, which typically contains sandstone at the base (Figure 3). This leaves tectonic and compactional subsidence as the primary processes governing the magnitude of accommodation space generated between basinwide flooding events (see Gastaldo et al., 2004). This confirms that total effective subsidence and accommodation space are equivalent for each cycle and that the grids of total effective subsidence provide a meaningful record of basin evolution.

Subsidence during deposition of the upper Parkwood Formation increased southwestward (Figure 5), confirming the existence of an Ouachita foreland basin with no significant effect related to Appalachian tectonism. The lower Boyles cycle has a subsidence pattern that is transitional between the upper Parkwood Formation and the younger cycles. Thus, the start of Pottsville deposition coincided with a major tectonic reorganization of the eastern Black Warrior Basin. By upper Boyles deposition, the depocenter of southern Tuscaloosa and northern Hale Counties was differentiated fully from the main flexural moat of the Black Warrior Basin, which is adjacent to the Ouachita orogen. The depocenter is adjacent to a bend in strike of the frontal thrust fault of the Appalachian orogen (Figure 2) and represents the principal effect of Appalachian thrust and sediment loading on subsidence of the Black Warrior Basin (Pashin et al., 1991; Pashin, 1994a). For most of the remainder of Pottsville deposition, subsidence in the study area was greatest in the depocenter, and a secondary locus of subsidence related to Ouachita orogenesis persisted in the southwest corner of the study area (Figures 5–7). The predominantly east-west contour patterns in the Utley and Brookwood cycles indicate that the depocenter may have merged with the larger Ouachita flexural moat late in Pottsville deposition, but this interpretation is provisional because of a lack of stratigraphic control in Sumter County (Figure 7).

Shifting grid and contour patterns indicate that the subsidence pattern continued to change significantly after the depocenter became separated from the Ouachita moat, especially in distal parts of the foreland basin (Figures 5–7). Northeast-oriented contours in the Fayette and Gillespy cycles, for example, provide evidence for the strongest influence of the Appalachian orogen on subsidence in the northern part of the study area. Irregular contour patterns in the Black Creek and Ream cycles, in contrast, suggest that distal parts of the basin at times had little sensitivity to orogenic activity, and contours oriented east-west in the Pratt through Brookwood cycles suggest that distal parts of the basin simply tilted southward in compromise to Ouachita and Appalachian flexure and regional epeirogenesis.

Regardless of the specific causes, tectonically driven changes in the regional subsidence pattern took place in the time frame of the Pottsville cycles. Estimation of cycle duration with respect to accelerating foreland-basin subsidence suggests that the mechanisms for cyclicity in the Pottsville Formation changed through time. Using the timescale of Harland et al. (1990) and the composite section of southern Tuscaloosa County as a reference point indicates a major decrease in cycle duration from lower Boyles through Brookwood deposition (Figure 8). The lower Boyles through Black Creek cycles have estimated lengths ranging from 0.7 to 1.2 m.y., suggesting that they formed in response to third-order changes of relative sea level as defined by Mitchum (1977) and Carter et al. (1991). The remaining cycles cluster around 0.4 m.y., although the Curry cycle has an estimated duration of only 0.1 m.y., suggesting that it was an exceptional high-frequency event. These estimates are compatible with fourth- and fifth-order changes of relative sea level and Milankovitch orbital parameters, specifically long- (0.4 m.y.) and short-periodicity (0.1 m.y.) orbital eccentricity. Pashin et al. (2003), moreover, recognized that the Mary Lee, Curry, Cobb, Gwin, and Utley cycles each contain three widespread internal flooding surfaces, suggesting that the short eccentricity signal is superimposed on the long eccentricity signal.

Third-order changes of relative sea level have been attributed to tectonoeustatic processes, including orogeny, changes in midoceanic ridge volume, and intraplate stress (Pitman, 1978; Cloetingh, 1986; Cathles and Hallam, 1991). Association of the Kaskaskia–Absaroka boundary with tectonic reorganization of the eastern Black Warrior Basin supports a component of tectonic forcing, although the precise causes of third-order cyclicity in the lower part of the Pottsville Formation are a matter for speculation. Increasing cycle frequency in the lower part of the Pottsville, however, marks a transition from the average 1.6-m.y. cyclicity identified in Chesterian strata of the Black Warrior Basin (Pashin, 1993; 1994b) to the 0.4-m.y. cyclicity that appears to have dominated later Pottsville deposition. Climatic variation associated with Milankovitch orbital parameters, including orbital eccentricity, is thought to have driven high-frequency glacial eustasy associated with major continental ice sheets during the Pleistocene and Permian–Carboniferous (Hays et al., 1976; Berger et al., 1984; Heckel, 1986). Accordingly, decreasing cycle duration may be attributed to expansion of the Gondwanan ice sheet at this time (Crowell 1978; Frakes et al., 1992). Recognizing the uncertainty in Pennsylvanian geochronology (Klein, 1990; Menning et al., 2000), correspondence to orbital eccentricity may be gratuitous. Even so, estimates of cycle duration using the timescale of Harland et al. (1990) tend to be longer than estimates using other timescales, and the 7-m.y. span assigned to the Pottsville Formation is a conservative estimate. Therefore, cycles in the upper part of the Pottsville Formation were deposited almost certainly at the high frequencies attributed to the classic Pennsylvanian cyclothems of the North American mid-continent.

Was eustasy the dominant cause of Pottsville cyclicity? The regional sediment budget was sufficient to fill the Black Warrior Basin during each cycle, with the possible exception of the Ream, yet total effective subsidence rates exceeding 300 m/m.y. (984 ft/m.y.) were sufficient to inundate the proximal part of the basin given any significant interruption of sedimentation. If tectonic subsidence is assumed to be 50% of total effective subsidence, then eustatic rises did not exceed 50 m (164 ft) (half of maximum water depth at highstand) during deposition of any cycle and may have been less than 5 m (16 ft) during deposition of the Curry cycle. This is similar to numbers calculated for the early Pennsylvanian by Maynard and Leeder (1992) and is well short of the 100-m (328-ft) eustatic rises of sea level postulated for middle and upper Pennsylvanian cyclothems (Veevers and Powell, 1987; Veevers, 1994; Soreghan and Giles, 1999). The high-accommodation setting of the Black

Warrior Basin relative to the Appalachian basin, which was dominated by valley incision during much of the Pennsylvanian (Aitken and Flint, 1994, 1995; Chesnut, 1994; Heckel et al., 1998; Nadon and Kelly, 2004; Martino, 2004; Greb et al., 2004), may have enhanced the effect of high-frequency changes of sea level. Perhaps the best way to consider the role of eustasy during Pottsville deposition is as a moderating force in which small to moderate eustatic rises were capable of interrupting sedimentation, thus facilitating regional flooding and condensation.

Pottsville cycles can be used to quantify the magnitude of subsidence in the eastern Black Warrior Basin (Figures 5–7), but estimating rates of subsidence hinges entirely on an assumption of geodynamic behavior at a geographic point of reference. Using a point of reference in the distal part of the basin gives a result for the depocenter that is incompatible with foreland-basin geodynamics, whereas a point of reference in the depocenter gives more consistent estimates for an actively loaded foreland basin (Figure 9). However, the assumptions employed in this study necessitate that the time estimates and subsidence rates be accepted as modeled and not determined. An interesting test for the time estimates and subsidence rates used in this study would be to apply the same methods to the thickest lower Boyles–Brookwood section adjacent to the Ouachita orogen in Mississippi.

Independent of these assumptions, an important outcome of this modeling is that, relative to any reference point in the eastern Black Warrior Basin, spatial and temporal variation of subsidence rate must have occurred in the time frame of Pennsylvanian cyclothems (Figures 5–7, 9, 10). During deposition of the lower and upper Boyles cycles, the changing subsidence pattern was dominated by diachronous deformation of the Alabama promontory and establishment of a depocenter in the southeastern part of the basin (Figure 5). Later changes were more subtle and, as noted previously, were expressed most prominently in distal parts of the basin as changes in strike and excursions of subsidence rate relative to the depocenter that locally exceeded 50 m/m.y. (164 ft/m.y.) (Figures 5–7, 9), or 20 m (66 ft) per long eccentricity period.

Geodynamic models of foreland basins account for isostatic and thermal compensation of tectonic and sediment loads in a time frame of 10–30 m.y. (e.g., Beaumont et al., 1987, 1988; Johnson and Beaumont, 1995; Coakley and Gurnis, 1995) and therefore are not intended to explain high-frequency changes

in the subsidence pattern like those observed in distal parts of the eastern Black Warrior Basin. Tectonic activity occurring within and below the Milankovitch band is beginning to be recognized in orogenic belts (Fortuin and de Smet, 1991; Ito et al., 1999), but the influence of this activity on subsidence of the foreland is only beginning to be acknowledged (Gastaldo et al., 2004). Grid and contour patterns suggest that the influence of the Appalachian orogen in distal parts of the basin was greatest during the Fayette and Gillespy cycles and was minimal during other cycles (Figures 5–7). One possibility is that adjustments in the Appalachian and Ouachita tectonic loads led to modest flexural interactions with crustal heterogeneity and far-field tectonic processes that influenced the regional subsidence pattern at time intervals analogous to third- to fifth-order eustatic cycles.

SUMMARY AND CONCLUSIONS

The Pottsville Formation of Alabama contains 13 flooding-surface-bounded depositional cycles that can be used to model total effective subsidence in an actively loaded foreland basin. Prior to Pottsville deposition, subsidence increased uniformly toward the southwest, confirming development of an Ouachita foreland basin with little or no effect of Appalachian tectonism. The basal contact of the Pottsville Formation coincides with initiation of a depocenter adjacent to the Appalachian thrust belt. This depocenter was fully formed by upper Boyles deposition and persisted northeast of the main Ouachita flexural moat for the remainder of Pottsville deposition. Following a general stabilization of the depocenter, 3-D grid and contour patterns define significant changes in the subsidence pattern from cycle to cycle, especially in distal parts of the basin.

Pottsville cycles maintain similar thickness proportions throughout the study area, and backstripped cycle thickness was used to estimate cycle duration with respect to biostratigraphic control and the geodynamic constraint of accelerating subsidence. This exercise suggests a transition from dominant third-order cyclicity during deposition of the lower Boyles through Black Creek cycles to dominant fourth- and fifth-order cyclicity in the Milankovitch eccentricity band for the rest of Pottsville deposition. Thus, cycles above the Black Creek formed at high frequencies associated with classic Pennsylvanian cyclothems. Eustatic sea level rises during Pottsville deposition were probably less than 50 m (164 ft), and eustasy is envisioned as a factor moderating rapid subsidence and high rates of sediment flux that gave rise to alternating episodes of condensation and basin filling during each cycle.

Modeling rates of total effective subsidence in the eastern Black Warrior Basin requires a geodynamic and geochronologic reference section. A reference section in the depocenter is essential for making a tectonically feasible interpretation of total effective subsidence rate, and maximum rates for each cycle are modeled as having increased from 345 to 740 m/m.y. (1132 to 2428 ft/m.y.) Regardless of point of reference, significant changes in the regional subsidence pattern occurred in the time frame of Pennsylvanian cyclothems, and excursions of more than 50 m/m.y. (164 ft/m.y.) from reference rates were modeled in some cycles. These excursions demonstrate that high-frequency tectonic processes that are not considered in geodynamic models may affect foreland basins. In the eastern Black Warrior Basin, changes in the regional subsidence pattern may reflect adjustments in tectonic loads and flexural interactions with crustal heterogeneity and far-field tectonic processes.

ACKNOWLEDGMENTS

Parts of this research were supported by the U.S. Department of Energy (contract DOE/BC/14448) and the U.S. Geological Survey (National Coal Resources Data System) under various cooperative agreements. Richard Carroll calculated well locations from line calls, and Diane Norris provided graphical assistance. Rupert Bodden contributed samples of coal from above the Brookwood coal zone for palynological analysis. Phillip H. Heckel and Timothy M. Demko provided insightful reviews that substantially improved the content of this paper.

REFERENCES CITED

Aitken, J. F., and S. S. Flint, 1994, High-frequency sequences and the nature of incised-valley fills of the Breathitt Group (Pennsylvanian), Appalachian foreland basin, eastern Kentucky: SEPM Special Publication, 51, p. 353–368.

Aitken, J. F., and S. S. Flint, 1995, The application of high-resolution sequence stratigraphy to fluvial systems: A case study from the Upper Carboniferous Breathitt Group, eastern Kentucky, U.S.A.: Sedimentology, v. 42, p. 3–30

Beaumont, C., G. M. Qunlan, and J. Hamilton, 1987, The Alleghanian orogeny and its relationship to the evolution of the eastern interior, North America: Canadian Society of Petroleum Geologists Memoir 12, p. 425–445.

Beaumont, C., G. M. Qunlan, and J. Hamilton, 1988. Orogeny and stratigraphy: Numerical models of the Paleozoic in the eastern interior of North America: Tectonics, v. 7, p. 389–416.

Berger, A., J. Imbrie, J. Hays, G. Kukla, and B. Saltzman, 1984, Milankovitch and climate: Understanding the response to orbital forcing: Boston, D. Reidel, 510 p.

Bodden, W. R., III, 1997, Coalbed methane production and desorption testing in the Black Warrior Basin of Alabama: Ph.D. dissertation, University of South Carolina, Columbia, South Carolina, 237 p.

Bond, G. C., M. A. Kominz, and W. J. Devlin, 1983, Thermal subsidence and eustasy in the lower Paleozoic miogeocline of western North America: Nature, v. 306, p. 775–779.

Butts, C., 1910, Description of the Birmingham quadrangle, Alabama: U.S. Geological Survey Atlas, Folio 175, 24 p.

Butts, C., 1926, The Paleozoic rocks: Alabama Geological Survey Special Report 14, p. 41–230.

Carroll, R. E., J. C. Pashin, and R. L. Kugler, 1995, Burial history and source-rock characteristics of Upper Devonian through Pennsylvanian strata, Black Warrior Basin, Alabama: Alabama Geological Survey Circular 187, 29 p.

Carter, R. M., S. T. Abbott, C. S. Fulthorpe, D. W. Haywick, and, R. A. Henderson, 1991, Application of global sea-level and sequence stratigraphic models in southern hemisphere Neogene strata from New Zealand: International Association of Sedimentologists Special Publication 12, p. 41–65.

Cathles, L. M., and A. Hallam, 1991, Stress-induced changes in plate density, Vail sequences, epeirogeny, and short-lived global sea level fluctuations: Tectonics, v. 10, p. 659–671.

Cecil, C. B., 1990. Paleoclimate controls on stratigraphic repetition of chemical and siliciclastic rocks: Geology, v. 18, p. 533–536.

Chesnut, D. R., 1994, Eustatic and tectonic control of deposition of the lower and middle Pennsylvanian strata of the central Appalachian basin: SEPM Concepts in Sedimentology and Paleontology 4, p. 51–64.

Cloetingh, S., 1986, Intraplate stresses: A new tectonic mechanism for fluctuations of relative sea level: Geology, v. 14, p. 617–621.

Coakley, B., and M. Gurnis, 1995, Far field tilting of Laurentia during the Ordovician and constraints on the evolution of a slab under an ancient continent: Journal of Geophysical Research, v. 100, p. 6313–6327.

Cropp, F. W., 1960, Pennsylvanian spore floras from the Warrior basin, Mississippi and Alabama: Journal of Paleontology, v. 34, p. 359–367.

Crowell, J. C., 1978, Gondwanan glaciation, cyclothems, continental positioning, and climate change: American Journal of Science, v. 278, p. 1345–1372.

Culbertson, W. C., 1964, Geology and coal resources of

the coal-bearing rocks of Alabama: U.S. Geological Survey Bulletin, v. 1182-B, 79 p.

Demko, T. M., and R. A. Gastaldo, 1996, Eustatic and allocyclic influences on deposition of the lower Pennsylvanian Mary Lee coal zone, Warrior Basin, Alabama: International Journal of Coal Geology, v. 31, p. 3–19.

Dickinson, W. R, G. S. Soreghan, and K. A. Giles, 1994, Glacio-eustatic origin of Permo-Carboniferous stratigraphic cycles: Evidence from the Cordilleran region: SEPM Concepts in Sedimentology and Paleontology, v. 4, p. 24–34.

Eble, C. F., and W. H. Gillespie, 1989, Palynology of selected coal beds from the central and southern Appalachian basin: Correlation and stratigraphic implications: Washington, D.C., American Geophysical Union, 28th International Geological Congress Guidebook T352, p. 61–66.

Eble, C. F., W. H. Gillespie, and T. W. Henry, 1991, Palynology, paleobotany, and invertebrate paleontology of Pennsylvanian coal beds and associated strata in the Warrior and Cahaba coal fields, *in* W. A. Thomas and W. E. Osborne, eds., Mississippian–Pennsylvanian tectonic history of the Cahaba synclinorium: Alabama Geological Society Guidebook, 28th Annual Field Trip, p. 119–132.

Ferm, J. C., R. Ehrlich, and T. L. Neathery, 1967, A field guide to Carboniferous detrital rocks in northern Alabama: 1967 Geological Society of America Coal Division Field Trip Guidebook, 101 p.

Ferm, J. C., and G. A. Weisenfluh, 1989, Evolution of some depositional models in Late Carboniferous rocks of the Appalachian coal fields: International Journal of Coal Geology, v. 12, p. 259–292.

Fortuin, A. R., and M. E. M. de Smet, 1991, Rates and magnitudes of late Cenozoic vertical movements in the Indonesian Banda Arc and the distinction of eustatic effects: International Association of Sedimentologists Special Publication 12, p. 79–89.

Frakes, L. A., J. E. Francis, and J. I. Syktus, 1992, Climate modes of the Phanerozoic: Glasgow, Cambridge University Press, 274 p.

Galloway, W. E., 1989, Genetic stratigraphic sequences in basin analysis: I. Architecture and genesis of flooding-surface bounded depositional units: AAPG Bulletin, v. 73 p. 125–142.

Gastaldo, R. A., T. M. Demko, and Y. Liu, 1993, Application of sequence and genetic stratigraphic concepts to Carboniferous coal-bearing strata: An example from the Black Warrior Basin, U.S.A.: Geologische Rundschau, v. 82, p. 212–226.

Gastaldo, R. A., I. Stevanovic-Walls, and W. N. Ware, 2004, Erect forests are evidence for coseismic base-level changes in Pennsylvanian cyclothems of the Black Warrior Basin, U.S.A., *in* J. C. Pashin and R. A. Gastaldo, eds., Sequence stratigraphy, paleoclimate, and tectonics of coal-bearing strata: AAPG Studies in Geology 51, p. 219–238.

Greb, S. F., D. R. Chesnut, and C. F. Eble, 2004, Temporal changes in coal-bearing depositional sequences (Lower and Middle Pennsylvanian) of the central Appalachian Basin, U.S.A., *in* J. C. Pashin and R. A. Gastaldo, eds., Sequence stratigraphy, paleoclimate, and tectonics of coal-bearing strata: AAPG Studies in Geology 51, p. 89–120.

Harland, W. B., R. L. Armstrong, A. V. Cox, L. E. Craig, A. G. Smith, and D. G. Smith, 1990, A geologic time scale 1989: London, Cambridge University Press, 131 p.

Hays, J. D., J. Imbrie, and N. J. Shackleton, 1976, Variations in the Earth's orbit: Pacemaker of the ice ages: Science, v. 194, p. 1121–1132.

Heckel, P. H., 1977, Origin of phosphatic black shale facies in Pennsylvanian cyclothems of midcontinent North America: AAPG Bulletin, v. 61, p. 1045–1068.

Heckel, P. H., 1986, Sea-level curve for Pennsylvanian eustatic transgressive-regressive depositional cycles along midcontinent outcrop belt, North America: Geology, v. 14, p. 330–334.

Heckel, P. H., 1994, Evaluation of evidence for glacio-eustatic control over marine Pennsylvanian cyclothems in North America and consideration of possible tectonic effects: SEPM Concepts in Sedimentology and Paleontology, v. 4, p. 65–87.

Heckel, P. H., M. R. Gibling, and N. R. King, 1998, Stratigraphic model for glacial-eustatic Pennsylvanian cyclothems in highstand nearshore detrital regimes: Journal of Geology, v. 106, p. 373–383.

Henderson, K. S., and C. A. Gazzier, 1989, Preliminary evaluation of coal and coalbed gas resource potential of western Clay County, Mississippi: Mississippi Bureau of Geology Report of Investigations 1, 31 p.

Hess, J. C., and H. J. Lippolt, 1986, $^{40}Ar/^{39}Ar$ ages of tonstein and tuff sanidines: New calibration points for the improvement of the Upper Carboniferous time scale: Isotope Geoscience, v. 59, p. 143–154.

Hines, R. A., Jr., 1988, Carboniferous evolution of the Black Warrior foreland basin, Alabama and Mississippi: Ph.D. dissertation, University of Alabama, Tuscaloosa, Alabama, 231 p.

Horsey, C. A., 1981, Depositional environments of the Pennsylvanian Pottsville Formation in the Black Warrior Basin of Alabama: Journal of Sedimentary Petrology, v. 51, p. 799–806.

Jennings, J. R., and W. A. Thomas, 1987, Fossil plants from Mississippian–Pennsylvanian transition strata in the southern Appalachians: Southeastern Geology, v. 27, p. 207–217.

Johnson, D. D., and C. Beaumont, 1995, Preliminary results from a planform kinematic model of clastic foreland basin stratigraphy: SEPM Special Publication 52, p. 3–24.

Ito, M., T. Nishikawa, and H. Sugimoto, 1999, Tectonic control of high-frequency depositional sequences with durations shorter than Milankovitch cyclicity: An example from the Pleistocene paleo-Tokyo Bay, Japan: Geology, v. 27, p. 763–766.

Kidd, J. T., 1976, Configuration of the top of the Pottsville Formation in west-central Alabama: Alabama State Oil and Gas Board Oil and Gas Map 1, 1 sheet, scale 1:250,000.

Klein, G. D., 1974, Estimating water depths from analysis of barrier island and deltaic sedimentary sequences: Geology, v. 2, p. 409–412.

Klein, G. D., 1990, Pennsylvanian time scales and cycle periods: Geology, v. 18, p. 455–457.

Klein, G. D., 1994, Depth determination and quantitative distinction of the influence of tectonic subsidence and climate on changing sea level during deposition of midcontinent Pennsylvanian cyclothems: SEPM Concepts in Sedimentology and Paleontology, v. 4, p. 35–50.

Klein, G. D., and D. A. Willard, 1989, Origin of the Pennsylvanian coal-bearing cyclothems of North America: Geology, v, 17, p. 152–155.

Liu, Y., and R. A. Gastaldo, 1992, Characteristics of a Pennsylvanian ravinement surface: Sedimentary Geology, v. 77, p. 197–213.

Magara, K., 1980, Comparison of porosity-depth relationships of shale and sandstone: Journal of Petroleum Geology, v. 3, p. 286–307.

Martino, R. L., 2004, Sequence stratigraphy of the Glenshaw Formation (middle–late Pennsylvanian) in the central Appalachian Basin, in J. C. Pashin and R. A. Gastaldo, eds., Sequence stratigraphy, paleoclimate, and tectonics of coal-bearing strata: AAPG Studies in Geology 51, p. 1–28.

Maynard, J. R., and M. R. Leeder, 1992, On the periodicity and magnitude of Late Carboniferous glacio-eustatic sea-level changes: Journal of the Geological Society (London), v. 149, p. 303–311.

McCalley, H., 1900, Report on the Warrior coal basin: Alabama Geological Survey Special Report 10, 327 p.

Menning, M., D. Weyer, G. Drodzewski, and H. W. J. van Amerom, 2000, A Carboniferous time scale 2000: Discussion and use of geological parameters as time indicators from central and western Europe: Geologische Jährbüch Hanover A, v. 156, p. 3–44.

Metzger, W. J., 1965, Pennsylvanian stratigraphy of the Warrior Basin, Alabama: Alabama Geological Survey Circular 30, 80 p.

Mitchum, R. M., Jr., 1977, Part eleven: Glossary of terms used in seismic stratigraphy: AAPG Memoir 26, p. 205–212.

Nadon, G. C., 1998, Magnitude and timing of peat-to-coal compaction: Geology, v. 26, p. 727–730.

Nadon, G. C., and R. R. Kelly, 2004, The constraints of glacial eustasy and low accommodation on sequence-stratigraphic interpretations of Pennsylvanian strata, Conemaugh Group, Appalachian Basin, U.S.A., in J. C. Pashin and R. A. Gastaldo, eds., Sequence stratigraphy, paleoclimate, and tectonics of coal-bearing strata: AAPG Studies in Geology 51, p. 29–44.

Pashin, J. C., 1993, Tectonics, paleoceanography, and paleoclimate of the Kaskaskia sequence in the Black Warrior Basin of Alabama, in J. C. Pashin, ed., New perspectives on the Mississippian system of Alabama: Alabama Geological Society 30th Annual Field Trip Guidebook, p. 1–28.

Pashin, J. C., 1994a, Flexurally influenced eustatic cycles in the Pottsville Formation (lower Pennsylvanian), Black Warrior Basin, Alabama: SEPM Concepts in Sedimentology and Paleontology, v. 4, p. 89–105.

Pashin, J. C., 1994b, Cycles and stacking patterns in Carboniferous rocks of the Black Warrior foreland basin: Gulf Coast Association of Geological Societies Transactions, v. 44, p. 555–563.

Pashin, J. C., 1994c, Coal-body geometry and synsedimentary detachment folding in Oak Grove coalbed-methane field, Black Warrior Basin, Alabama: AAPG Bulletin, v. 78, p. 960–980.

Pashin, J. C., 1998, Stratigraphy and structure of coalbed methane reservoirs in the United States: An overview: International Journal of Coal Geology, v. 35, p. 207–238.

Pashin, J. C., and R. H. Groshong Jr., 1998, Structural control of coalbed methane production in Alabama: International Journal of Coal Geology, v. 38, p. 89–113.

Pashin, J. C., R. E. Carroll, R. H. Groshong Jr., D. E. Raymond, M. R. McIntyre, and W. J. Payton, 2003, Geologic screening criteria for sequestration of CO_2 in coal: Quantifying potential of the Black Warrior coalbed methane fairway, Alabama: Annual Technical Progress Report, U.S. Department of Energy, National Technology Laboratory, contract DE-FC-00NT40927, 190 p.

Pashin, J. C., W. E. Ward II, R. B. Winston, R. V. Chandler, D. E. Bolin, K. E. Richter, W. E. Osborne, and J. C. Sarnecki, 1991, Regional analysis of the Black Creek–Cobb coalbed-methane target interval, Black Warrior Basin, Alabama: Alabama Geological Survey Bulletin, v. 145, 127 p.

Pitman, W. C., 1978, Relationship between eustacy and stratigraphic sequences on passive margins: Geological Society of America Bulletin, v. 89, p. 1389–1403.

Posamentier, H. W., M. T. Jervey, and P. R. Vail, 1988, Eustatic controls on clastic deposition: I. Conceptual framework: SEPM Special Publication 42, p. 109–124.

Rheams, L. G., and D. J. Benson, 1982, Depositional setting of the Pottsville Formation in the Black Warrior Basin: Alabama Geological Society 19th Annual Field Trip Guidebook, 94 p.

Ross, C. A., and J. R. P. Ross, 1988, Late Paleozoic transgressive-regressive deposition: SEPM Special Publication 42, p. 227–247.

Sclater, J. G., and P. A. F. Christie, 1980, Continental stretching: An explanation of the post-mid-Cretaceous subsidence of the central North Sea basin: Journal of Geophysical Research, v. 85, p. 3711–3739.

Sestak, H. M., 1984, Stratigraphy and depositional environments of the Pennsylvanian Pottsville Formation in the Black Warrior Basin, Alabama and Mississippi:

M.S. thesis, University of Alabama, Tuscaloosa, Alabama, 184 p.

Soreghan, G. S., and K. A. Giles, 1999: Amplitudes of late Pennsylvanian glacioeustasy: Geology, v. 27, p. 255–258.

Stockmal, G. S., C. Beaumont, and R. Boutillier, 1986, Geodynamic models of convergent tectonics: The transition from rifted margin to overthrust belt and consequences for foreland basin development: AAPG Bulletin, v. 70, p. 181–190.

Telle, W. R., D. A. Thompson, L. K. Lottman, and P. G. Malone, 1987, Preliminary burial-thermal history investigations of the Black Warrior Basin: Implications for coalbed methane and conventional hydrocarbon development: Tuscaloosa, Alabama, University of Alabama, 1987 Coalbed Methane Symposium Proceedings, p. 37–50.

Thomas, W. A., 1976, Evolution of the Ouachita–Appalachian margin: Journal of Geology, v. 84, p. 323–342.

Thomas, W. A., 1985, The Appalachian–Ouachita connection: Paleozoic orogenic belt at the southern margin of North America: Annual Review of Earth and Planetary Sciences, v. 13, p. 175–199.

Thomas, W. A., 1988, The Black Warrior Basin, in L. L. Sloss, ed., Sedimentary Cover— North American Craton: Geological Society of America, The geology of North America, v. D-2, p. 471–492.

Thomas, W. A., 1991, The Appalachian–Ouachita rifted margin of southeastern North America: Geological Society of America Bulletin, v. 103, p. 415–431.

Thomas, W. A., 1995, Diachronous thrust loading and fault partitioning of the Black Warrior foreland basin within the Alabama recess of the late Paleozoic Appalachian–Ouachita thrust belt: SEPM Special Publication 52, p. 111–126.

Thomas, W. A., B. A. Ferrill, J. L. Allen, W. E. Osborne, and D. E. Leverett, 1991, Synorogenic clastic-wedge stratigraphy and subsidence history of the Cahaba synclinorium and the Black Warrior foreland basin, in W. A. Thomas and W. E. Osborne, eds., Mississippian–Pennsylvanian tectonic history of the Cahaba synclinorium: Alabama Geological Society 28th Annual Field Trip Guidebook, p. 37–39.

Upshaw, C. F., 1967, Pennsylvanian palynology and age relationships, in J. C. Ferm, R. Ehrlich, and T. L. Neathery, eds., A field guide to Carboniferous detrital rocks in northern Alabama: Alabama Geological Society, 1967 Geological Society of America Coal Division Field Trip Guidebook, p. 16–20.

Van Hinte, J. E., 1978, Geohistory analysis— Application of micropaleontology in exploration geology: AAPG Bulletin, v. 62, p. 201–222.

Van Wagoner, J. C., R. M. Mitchum, K. M. Campion, and V. D. Rahmanian, 1990, Siliciclastic sequence stratigraphy in well logs, cores, and outcrops: Concepts for high-resolution correlation of time and facies: AAPG Methods in Exploration, v. 7, 55 p.

Veevers, J. J., 1994, Evolution of a supercontinent and its consequences for Earth's paleoclimate and sedimentary environments: Geological Society of America Special Paper 288, p. 13–23.

Veevers, J. J., and Powell, C. M., 1987, Late Paleozoic glacial episodes in Gondwanaland reflected in transgressive-regressive depositional sequences in Euramerica: Geological Society of America Bulletin, v. 98, p. 475–487.

Viele, G. W., and W. A. Thomas, 1989, Tectonic synthesis of the Ouachita orogenic belt, in R. D. Hatcher, W. A. Thomas, and G. W. Viele, eds., The Appalachian–Ouachita orogen in the United States: Geological Society of America, The geology of North America, v. F-2, p. 695–728.

Wang, S., 1994, Three-dimensional geometry of normal faults in southeastern Deerlick Creek coalbed methane field, Black Warrior Basin, Alabama: M.S. Thesis, University of Alabama, Tuscaloosa, Alabama, 77 p.

Wanless, H. R., 1976, Appalachian region: U.S. Geological Survey Professional Paper 853-C, p. 17–62.

Wanless, H. R., and F. P. Shepard, 1936, Sea level and climatic changes related to late Paleozoic cycles: Geological Society of America Bulletin, v. 47, p. 1177–1206.

Watts, A. B., G. D. Karner, and M. S. Steckler, 1982, Lithospheric flexure and the evolution of sedimentary basins: Philosophical Transactions of the Royal Society of London, v. A-305, p. 249–281.

Weller, S., 1930, Cyclic sedimentation of the Pennsylvanian period and its significance. Journal of Geology, v. 38, p. 97–135.

Weller, S., 1936, Argument for diastrophic control of late Paleozoic cyclothems: AAPG Bulletin, v. 41, p. 195–207.

Welte, D. H., B. Horsfield, and D. R. Baker, eds., 1996, Petroleum and basin evolution: New York, Springer-Verlag, 535 p.

Whiting, B. M., and Thomas, W. A., 1994, Three-dimensional controls on subsidence of a foreland basin associated with a thrust-belt recess, Black Warrior Basin, Alabama and Mississippi: Geology, v. 22, p. 727–730.

10

Gastaldo, R. A., I. Stevanoviç-Walls, and W. N. Ware, 2004, Erect forests are evidence for coseismic base-level changes in Pennsylvanian cyclothems of the Black Warrior Basin, U.S.A., *in* J. C. Pashin and R. A. Gastaldo, eds., Sequence stratigraphy, paleoclimate, and tectonics of coal-bearing strata: AAPG Studies in Geology 51, p. 219–238.

Erect Forests Are Evidence for Coseismic Base-level Changes in Pennsylvanian Cyclothems of the Black Warrior Basin, U.S.A.

Robert A. Gastaldo

Department of Geology, Colby College, Waterville, Maine, U.S.A.

Ivana Stevanoviç-Walls

Department of Earth and Environmental Science, University of Pennsylvania, Philadelphia, Pennsylvania, U.S.A.

William N. Ware

1109 Wynterhall Land, Dunwoody, Georgia, U.S.A.

ABSTRACT

Examination of the plant taphonomic character and sedimentological processes responsible for preservation of an in situ, erect forest above the Pennsylvanian Blue Creek coal of the Mary Lee coal cycle, Alabama, provides evidence for rapid generation of accommodation space by coseismic subsidence. Standing vegetation is preserved at least to 4.5 m (15 ft) in height above the coal and includes lycopsids, regenerative calamites, tree ferns, and seed ferns (pteridospermous gymnosperms); the forest-floor litter is preserved as an adpression assemblage directly above the coal. Sediments entombing the standing trees, burying both the peat mire and forest-floor litter, and casting the erect vegetation consist of rhythmically bedded tidalites. Neap-spring-neap tidalite patterns indicate that entombment occurred on the order of a few decades, whereas burial of the mire and forest-floor litter happened on the order of weeks, if not days. Comparison with documented Holocene rates of eustatic and tectonic base-level changes indicates that eustatic processes alone cannot account for the generation of the accommodation required to provide a basis for the sedimentologic and taphonomic characteristics of the assemblage. Instead, coseismic subsidence of very high magnitude is determined to be the mechanism responsible for preservation. Hence, erect forests buried by estuarine tidal deposits provide evidence for rapid coseismic basinal subsidence. These criteria can be used to identify similar coseismic subsidence events beginning in the middle Paleozoic and provide constraints on the magnitude of event-driven base-level change in various basinal regimes.

INTRODUCTION

Various models and mechanisms have been proposed to account for cyclicity in the sedimentary character of marine and continental rocks in the Upper Carboniferous of Euramerica, and the rubric of these controversies is reviewed elsewhere (see Pashin, 2004; Cecil, 1990; Klein and Kupperman, 1992; Klein, 1994). There is a general agreement that the generation of accommodation space to allow for cyclothem deposition is eustatic-driven (Dewey and Pitman, 1998), and timescales under which such fluctuations in sea level are compatible with those that are orbitally mediated (e.g., Dickinson et al., 1994). However, it has been argued that both allocyclic and autocyclic processes can operate in cyclothems (Klein and Willard, 1989; Klein and Kupperman, 1992), and the sedimentary signatures encapsulated therein can provide clues to decipher which process is responsible for a particular part of that stratigraphic interval (Demko and Gastaldo, 1996). New data from the Blue Creek coal in the Mary Lee coal zone, Black Warrior Basin, Alabama, are now available that provide a basis on which tectonic versus eustatic changes in base level can be evaluated and identified in a single cyclothem. Hence, these data can be used not only for constraining the magnitude of tectonic base-level change, but also for providing criteria that can be used to identify similar coseismic coastal subsidence in other parts of the stratigraphic record.

Because of the discontinuous nature of the stratigraphic record, basinal subsidence rates have been calculated on several gross time intervals. These rates generally are presented as modeled subsidence curves plotting change-in-depth vs. stratal age (e.g., Stephenson et al., 1992) or as calculations to reflect meters of displacement per million years or centimeters per millennium (e.g., Fortuin and de Smet, 1991). Although several authors (e.g., Thommeret et al., 1981; Bloom and Yonekura, 1985) have suggested that localized basin subsidence and uplift occur in a spasmodic manner, and published data support this assertion (e.g., Plafker and Rubin, 1967; Plafker and Savage, 1970), little evidence has been provided, to date, allowing for identification and demonstration of these processes in the rock record. The purpose of this contribution is to provide such data and the rationale for identifying earthquake-induced base-level changes of at least 4.5 m (15 ft) in a foreland basin setting.

MARY LEE COAL ZONE

The Mary Lee coal zone is one of nine basinwide economic coal-bearing cycles identified in the Pottsville Formation (Langsettian = Westphalian A; early Pennsylvanian) of the Black Warrior Basin, a triangular foreland basin located between the Appalachian and Ouachita orogenic belts (Thomas, 1988) (Figure 1). Pashin (2004) has characterized the basin as a faulted homocline dipping southwestward toward the

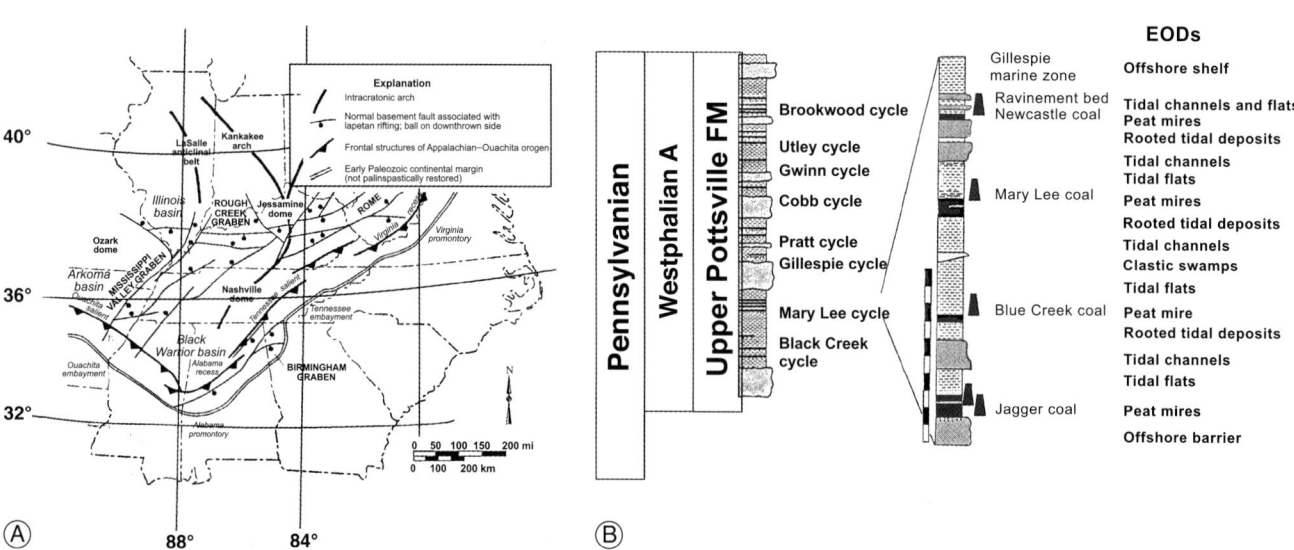

Figure 1. (A) Regional tectonic setting of the Black Warrior Basin (after Thomas, 1988) and (B) generalized stratigraphic column of the upper part of the Carboniferous section in the Black Warrior Basin. The Mary Lee coal cycle is expanded, illustrating the stacking pattern of siliciclastic deposits and coal, with interpreted environments of deposition (EOD) indicated (after Gastaldo et al., 1990).

Ouachita orogen, with the frontal folds and faults of the Appalachian system superimposed on the southeastern margin. Thirteen cycles in this sequence (Pashin, 1994a, b) are bounded by ravinement surfaces formed during the maximum rate of sea level rise (Liu and Gastaldo, 1992a), which can be identified not only in outcrop and core (Demko and Gastaldo, 1996), but also in geophysical well logs (Pashin, 1994a). Hence, these have been considered by Gastaldo et al. (1993) and Pashin (1994a, 2004) coincident with the genetic sequence boundaries of Galloway (1989) and the transgressive maximum flooding surfaces of Van Wagoner et al. (1990).

Gastaldo et al. (1991, 1993) defined the Mary Lee coal zone as consisting of the Jagger, Blue Creek, Mary Lee, and Newcastle coal seams (the latter a coal-seam split restricted to northwest Walker County, Alabama) underlain by a thick fine- to medium-grained sublitharenite informally called the "Jagger bedrock" sandstone. The coal zone crops out more than at least a 1000 km^2 (386 mi^2) area as surface mine highwalls (now mostly reclaimed), road cuts, and natural exposures. Mine highwalls extended laterally for several kilometers, having allowed for the development of a three-dimensional perspective of facies architecture (Demko, 1990a). Demko and Gastaldo (1996) described the vertical succession of facies in the following sequence of environments (Figure 1):

1) Tide-influenced shelf to lower shoreface as recorded in the Jagger bedrock sublitharenite. This sandstone is characterized by large-scale trough cross-stratification organized in large dune and sand-wave megaforms organized as shore-parallel bars, the tops of which have undergone incipient pedogenesis.

2) Discontinuous coastal peat mires, clastic swamps (Gastaldo, 1987), and tidal mud flats represented by the Jagger coal and overlying clastic facies. The Jagger coal occurs as isolated, lenticular bodies that may be as thick as 2.3 m (8 ft) (in troughs of the underlying Jagger bedrock sandstone) or less than 0.1 m (0.3 ft) (overlying megaform crests) across a distance less than 1 km (0.6 mi). A persistent carbonaceous shale parting splits the Jagger coal, which is low in sulfur (0.9–1.0%) and moderate in ash (10–15%; Barnett, 1986), and erect lycopsid trees are preserved in situ. The overlying interval is characterized by a pinstripe laminated, fine-grained sandstone-and-siltstone sequence as much as 9 m (30 ft) thick in which neap-spring tidal cyclicity has been recognized (Demko et al., 1991).

3) The Blue Creek coal is the most extensive coal bed in the basin, known in the subsurface close to the Appalachian structural front and westward into Mississippi and in outcrop exposure in northwest Walker County. This coal marks the position of lowstand in the coal zone and is overlain by aggrading clastic swamps (Demko and Gastaldo, 1992). The overlying sedimentary sequence is variable, including heterolithic pinstripe laminated siltstones; entisols with in situ stigmarian axes and rootlets, overlain by erect lycopsids, calamites, pteridosperms, and pteridophytes; and fine-grained sandstones in channel-form geometries, the tops of which also are rooted and on which the Mary Lee mire formed.

4) The peat mire deposits of the Mary Lee coal are overlain by tidal-influenced fluvial and deltaic deposits. The Mary Lee coal is split in the northwestern part of Walker County into a rider seam, the Newcastle coal, with an intervening sequence of channel-form sandstone bodies and mudstone lithologies that show sedimentary structures characteristic of tidal-influenced regimes (Gastaldo et al., 1990). Overlying the coals are shallow, widesheet, and channel-form bodies of fine sandstone and siltstone displaying rhythmic bedding cycles. These occur in accretionary bars wherein pinstripe and flaser-bedding structures have been identified (Liu and Gastaldo, 1992a, b). This interval was interpreted by Liu (1990) to represent primarily fluvial environments that experienced some tidal influence.

5) The Newcastle coal and/or the channel-form fine sandstones and siltstones are truncated by a planar erosional surface that can be traced in outcrop and the subsurface for greater than 1800 km^2 (695 mi^2). Liu and Gastaldo (1992a) described the sedimentological and paleontological variability of the thin overlying lithologies, interpreting these rocks as a condensed section, the result of ravinement processes. The overlying marine sediments are part of the Gillespy cycle.

Hence, this sequence reflects terrigenous sedimentation along a low-gradient coastal plain in one fourth- or fifth-order cycle (Pashin, 2004), at frequencies associated with classical Pennsylvanian cyclothems.

Blue Creek Mire

The thin (0.3–0.5 m [1–1.6 ft]), continuous Blue Creek coal is a very low-sulfur (0.6–0.7%) and ash (13–14%; Barnett, 1986) bright-banded coal with a

Figure 2. Index map of Alabama on which Walker County is indicated (black), and location of the Drummond Brothers, Cedrum mine in which the present study was conducted (SW 1/4, T14S, R8W, Sec. 18, U.S. Geological Survey, Townley, Alabama, 7.5′ quadrangle).

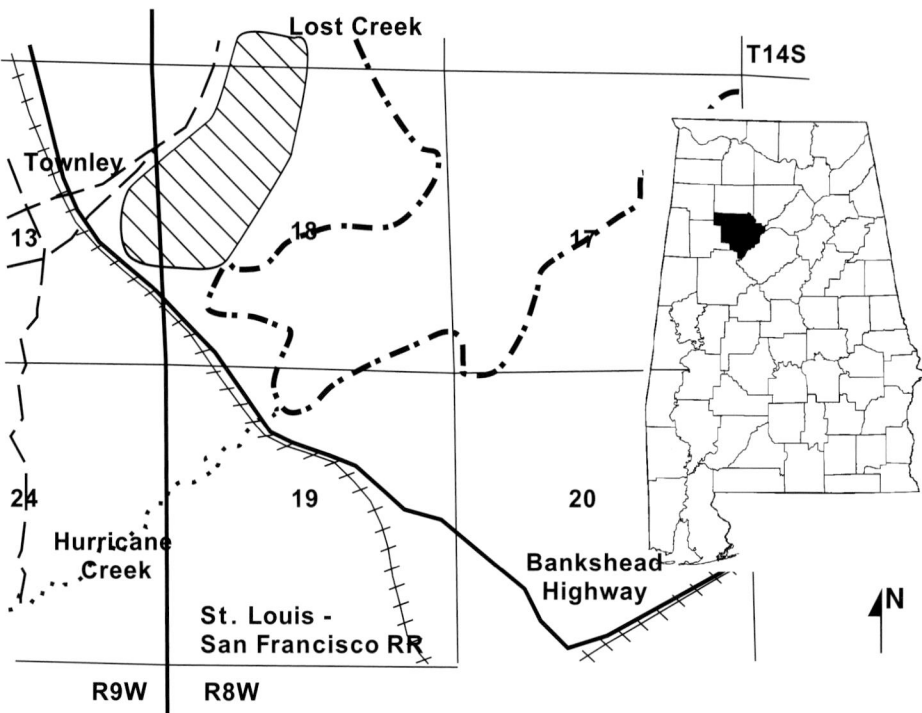

small fusain component considered to be degradofusinite (C. Eble, 1991, personal communication). Winston (1990a) determined that the coal was composed of 49% lycophyte trunks, 36% rootlets, 9% pteridosperm, 5% fern, and 1% degraded biomass in samples taken from the Cedrum mine in Townley, Alabama. Preserved at the coal-clastic contact is an autochthonous assemblage consisting of (1) erect lycophyte (Gastaldo, 1990), calamite (Gastaldo, 1992), pteridosperm, and pteridophyte axes; and (2) an exceptionally well-preserved adpression fossil-plant assemblage, with elements lying parallel or slightly inclined to bedding. Fluvial overbank processes were implicated to account for the burial of this autochthonous assemblage (Gastaldo, 1990). The plant fossils are surrounded and cast by the heterolithic siltstone-sandstone lithology in which neap-spring tidalite deposition has been recognized (Gastaldo, 1992).

Materials and Methods

The data presented herein are based on concurrent studies in the Drummond Brothers Company's Cedrum mine in Townley, Alabama (Townley, 7.5-ft quadrangle, Sec. 7, T14S, R8W, and Sec. 18, T14S, R8W; Stevanoviç-Walls, 2001; Ware, 2001) (Figure 2). Observations were made, and samples were collected primarily during 1999–2000 in a 1.36-km (0.8-mi)-long surface mine (western margin, N 33E49′ 868″, W 87E25′384″; eastern margin, N 33E49′380″, W 87E24′942″), although reported specimens and observations also originate from other mining operations in the area (e.g., Gastaldo et al., 1990; Gastaldo, 1992). A photomontage of the highwall was compiled, allowing for recognition of large-scale features above the Blue Creek coal. In addition, thin sections were made of

the 4–5-cm-thick layer of the forest-floor litter in what was field identified as mudstone, 0–10 cm (0–4 in.) directly above the coal seam. Oriented thin sections are perpendicular to bedding and were examined using standard petrographic techniques to evaluate the sediment character between litter-created bedding planes, the presence of primary sedimentary structures, and other features that may reflect depositional processes responsible for the preservation of the forest.

Siderite concretions identified in ground thin sections were analyzed for carbon and oxygen isotopes to characterize the sedimentological regime under which they formed. Billets were subsampled with a fine drill at approximately 5-mm (0.2-in.) intervals along six siderite nodules, with each microsample label indicating coal, collection site, and sample number, and subsample in the transect (e.g., sample BC VI 1Aa indicates the Blue Creek coal [BC], sample site VI, the first sample from the site [1], first part of sample 1 [A], and the consecutive microsamples [a–h]). Powdered samples were reacted with five drops of 100% phosphoric acid at 80°C in a Finnigan Kiel II automated reaction system for approximately 30 min. The CO_2 produced was analyzed on a Finnigan MAT 251 isotope-ratio mass spectrometer in the Department of Geology and Geophysics at Texas A&M University. For calibration to the Peedee belemnite (PDB) standard, the carbonate standard NBS-19 ($\delta^{13}C$ = 1.95‰, $\delta^{18}O$ = −2.20‰) was used (i.e., Vienna PDB).

Table 1. Taxonomic composition of canopy, subcanopy, and ground cover/liana plants preserved in the Blue Creek mire.

Canopy	Subcanopy	Ground Cover/Liana
Lepidodendron aculeatum	Pecopteris arborescens	Alloiopteris sp.
Lepidodendron obovatum	Cardiopteridium sp.	Diplothmema sp.
Lepidophloios laricinus	Eremopteris Rhodea type	Lyginopteris hoeninghausii
Sigillaria elegans	Eremopteris sp.	Palmatopteris furcata
Sigillaria ichthyolepis	Eusphenopteris lobata	Sphenophyllum emarginatum
Sigillaria scutellata	Sphenopteris brongniarti	Sphenophyllum cuneifolium
Calamites cisti	Alethopteris cf. valida	Sphenopteris cf. schatzlarensis
Calamites suckowi	Alethopteris lonchitica	Sphenopteris herbacea
Artisia	Neuralethopteris elrodi	Sphenopteris pseudocristata
	Neuralethopteris pocahontas	
	Neuralethopteris schlehani	
	Neuralethopteris smithsii	
	Neuropteridium sp.	

Plant Taphonomy of the Blue Creek Forest

Adpression and Prostrate Cast Assemblage

A concentrated accumulation (sensu Krasilov, 1975) of exclusively aerial plant parts is preserved directly above the Blue Creek coal in the first 5–10 cm (2–4 in.) of siliciclastics (this bed may be as thin as 3 cm (1.2 in.) and as thick as 15 cm (6 in.), depending on the locality). The thickness of the plant-fossil assemblage at any particular site is controlled spatially by the underlying topography of the Jagger sandstone (Demko, 1990a, b). The plant debris consists of randomly oriented trunks, stems, and branches of canopy and subcanopy elements (lycophytes and calamiteans); juvenile and mature foliage of canopy elements and their reproductive cones either terminally attached to branches or disseminated (lycophytes and calamiteans); mature foliage of subcanopy taxa and occasional bare rachial elements (pteridophytes and medullosan pteridosperms), as well as reproductive structures (pollen organs, fruits, and seeds); and ground cover/liana forms consisting of small-diameter axes with attached leaves [sphenophyllaleans, lyginopterid pteridosperms, medullosan and callistophytalean(?) pteridosperms, and pteridophytes; Table 1]. The plant parts are preserved primarily as adpressions, which Shute and Cleal (1987) have defined as any plant fossil showing a mixture of compression states (plant parts compressed by sediment where some original or chemically altered tissue is still preserved) and impression states (an imprint of the fossil plant on sediment or rock surface).

Fossil-plant detritus collected from the coal-siliciclastic interface is not preserved well and commonly appears as an organic coalified film without morphological or cellular detail. When viewed in thin section (Figure 3A), this material is similar in petrographic character to the underlying and contiguous coal. Leaves and small-diameter branches appear as individual, isolated vitrinite laminae, whereas some evidence exists for charcoalification (fusain) and cellular preservation of larger-diameter (>1 cm [>0.4 in]) axial debris near the uppermost interface of the coal (Figure 3A). Commonly, it is impossible to determine the systematic affinity of the plant below the major plant group.

Well-preserved plant fossils occur less than 1 cm (0.4 in.) above the upper coal contact. Thick, robust aerial stems are preserved mainly by coalification and commonly exhibit some three-dimensional topography in the rock. Prostrate lycophyte and calamitean trunks as much as 1 m (3 ft) in compressed width (equivalent to the original tree diameter; Rex and Chaloner, 1983; Thomas, 1986) generally are preserved slightly convex, unless they occur as casts, in which case they may exhibit a variety of cross-sectional lenticular shapes dependent on the quantity of siliciclastic fill (Gastaldo et al., 1989). Juvenile lycopsid trunks (sensu Kosanke, 1979) have been found lying flat with spirally attached leaves (Lepidophylloides sp.), each of which may be as much as 1 m (3 ft) in length. Smaller-diameter axes, such as medullosan trunks that are as much as 15 cm (6 in.) in compressed width, also may be preserved prostrate in three dimensions or may crosscut the entombing lithology and, in a few instances, extend into the overlying, plant-barren beds.

Leaves are found in a variety of stages of completeness and preservation. Entire fronds (central bifurcate

Figure 3. Thin sections across the upper contact of the Blue Creek coal and overlying sediments. (A) Thin section in which the Blue Creek coal (BCC) and overlying forest-floor litter can be seen to be buried by thin laminations of very fine sandstone/coarse siltstone and siltstone/mudstone. A fusainized axis (FA) is near the top of the coal. (B) Thin section of the forest-floor litter wherein aerial plant parts, including rachial axes (R) and pinnules (P), define bedding planes. (C) Thin section of tidalite sequence overlying the forest-floor litter horizon. Scale for all photomicrographs is in millimeters.

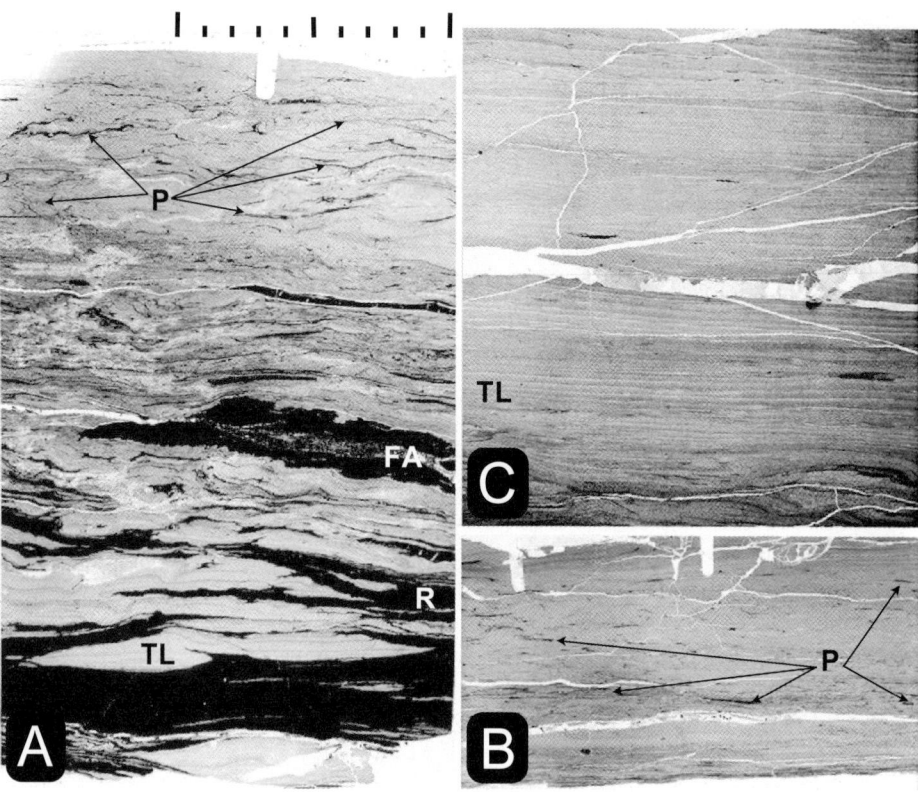

or nonbifurcate rachis with attached lateral pinnae including pinnules), as well as frond fragments (entire pinnae or pinnae apices) and isolated individual pinnules (i.e., medullosan pteridosperms), coexist at all stratigraphic levels in the bed, preserving the assemblage. Irregardless of the completeness of the leaf (frond), pinnules from thin and delicate taxa (e.g., *Sphenopteris*), large, robust, and thick taxa (e.g., *Neuralethopteris*, *Alethopteris*), or small, robust, and thick taxa (i.e., *Lyginopteris*), all exhibit nondessicated features that include well-defined outlines that may be in three dimensions, distinct venation patterns, and in some cases, cuticles and cellular details. Additionally, pinnae of larger fronds commonly were found across successive stratigraphic levels, defining the bedding planes in those instances. In thin section, pinnules are encountered as coalified laminae that may be parallel or at some angle to bedding (Figure 3B).

The taphonomic character of plant detritus changes upsection into the more siliciclastic-rich interval overlying the coal, where there is a marked decrease in plant-part concentration. Here, the fossil assemblage is dispersed; there is a complete loss of all prostrate canopy and subcanopy axes, as well as any evidence of complete and partial leaves attributable to subcanopy taxa. Only occasionally are medullosan pinnules found preserved in rocks more than 10 cm (4 in.) above the concentrated assemblage, wherein they appear as isolated coalified laminae in thin section (Figure 3C). A concentrated, coalified plant assemblage

may be encountered at the stratigraphic level that defines the uppermost limit of the erect trees (Demko and Gastaldo, 1996) (Figure 4).

Erect, In situ Trees

The standing forest is cast above the Blue Creek coal and consists of erect lycopsids (Gastaldo, 1990), calamites (Gastaldo, 1992), pteridophytes, and pteridosperms; no erect cordaitean axes have been encountered. Basal trunks of lycopsids, ranging from 0.20 to 0.95 m (0.7 to 3 ft) in diameter, are congruent with the top of the coal, and with stigmarian axes extending into the underlying mire. Subterranean axes and "rootlets" commonly are not cast by siliciclastics; instead, these are coalified and can be traced in the uppermost part of the coal bed when exposed prior to exploitation. Lycopsid trunks are cast by the same lithology responsible for entombing the trees (see below), and the bark tissues (periderm) have been coalified as a surrounding vitrain band. Individual trees are preserved for various heights above the Blue Creek coal, and the total height of any tree may be dependent on the underlying topographic relief of the Jagger bedrock sandstone (Demko and Gastaldo, 1996) or the exposure in the highwall. Individual trees range from less than 0.5 to more than 4.5 m (1.6 to >15 ft) in height (Figure 4), but the maximum

Figure 4. Erect, unidentifiable lycopsid at the Drummond Brothers mine, Townley, Alabama. (Sec. 24, T14S, R9W, Townley 7.5-ft quadrangle) extending 4.5 m (15 ft) above the Blue Creek coal. A siltstone-cast lycopsid axis (at arrow) can be seen at the stratigraphic level where the standing tree terminates, and lateral accretion beds onlap the tree (at arrow).

height recorded in any particular part of the mine represents the height to which all living vegetation was entombed. In most instances, lycopsids are not preserved perpendicular to the coal, but instead are at some slight angle from vertical. Most are typified by an exterior vertical fissuring, indicative of bark sloughing during diameter increase, making assignment to a specific taxon difficult. Average tree diameter in the Cedrum mine was reported to be 42 cm (17 in.) (Gastaldo, 1990).

Calamites are encountered as either isolated, erect pith casts or in small clusters. Most specimens are oriented at a slight angle from perpendicular to nearly 45° from vertical, and pith casts are surrounded by coalified aerial tissues (wood and bark). Several examples exist where individual plants have undergone regeneration following burial (Gastaldo, 1992), and in these cases, helically arranged roots that originated at buried nodes crosscut entombing primary bedding structures (Figure 5A). Piths are cast by the same lithology surrounding the aerial axes, and axes gen-

erally remain constant in diameter from the base to the top of the specimen, with a maximum observed diameter of 0.28 m.

It is unusual that both isolated, sparse ferns (pteridophytes) and seed ferns (pteridosperms) are encountered erect and cast above the coal. Several stems of *Psaronius* cf. *simplicaulis* (DiMichele and Phillips, 1977), a distichously branched tree fern, have been recovered that exhibit both external morphology and internal anatomy (Figure 6A) (in the absence of internal anatomy, specimens with these features would be assigned to *Megaphyton*; Pfefferkorn, 1976). Tree fern stems are as much as 9 cm (4 in.) in diameter and were at least greater than 2 m (6.6 ft) in height when excavated from the highwall, although it was not possible to recover an entire tree. Each specimen has U-shaped leaf scars arranged in two vertical rows on opposite sides of the stem along with adventituous roots directed geopedally. In addition, each stem also exhibits internal vascular trace architecture, with coalified conducting cells preserved in a siliciclastic matrix (Figure 6B, C). Pteridosperm axes are much thinner in diameter than all other erect vegetation, with maximum diameters approaching less than 10 cm (4 in.). They generally are found inclined at angles as much as 45° originating from the upper surface of the Blue Creek coal and may have adventituous roots directed geopedally. The axes commonly are compressed, with a minimum of sediment infill.

Sedimentology of the Blue Creek Forest

Based on field characterization, the forest litter is preserved in a mudrock, ranging from 3 to 15 cm (1 to 6 in.) in thickness, that is distinguished by numerous bedding planes created by the preferential orientation of the plant fossils. The lowermost 3–5 mm (0.1–0.2 in.) of the interval encompasses the top of the coal, is dark gray (N2) to black (N1), and consists of concentrated degraded organic matter. The remainder of the mudstone is light to medium gray (N3), appears to show no primary sedimentary structures in hand sample, and contains the bulk of identifiable adpressions. Depending on the site sampled, sideritic nodules with authigenically cemented plant material may or may not be present. Where present, siderite nodules are more concentrated directly above the coal and decrease in size and frequency upsection.

Isotope analyses of siderite nodules indicate that carbon ratios range from 6.04 to 9.96‰ and oxygen ratios range from −4.16 to −3.02‰ (Figure 7). High carbon isotopic values are diagnostic of formation in the zone of methanogenesis (Gautier, 1982; Moore

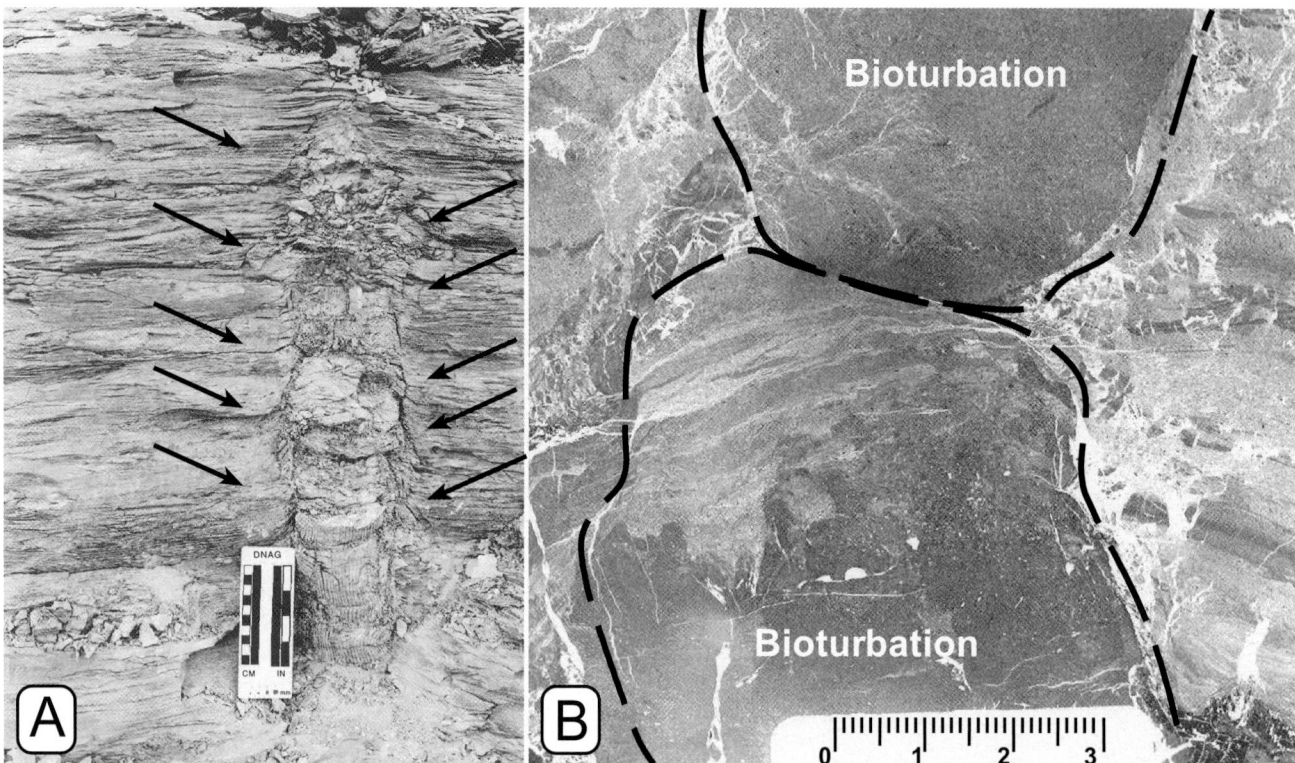

Figure 5. Erect, standing regenerated calamitean axis with secondary roots crosscutting tidalite sediments. (A) Standing axis at Coal Systems, Inc., Lost Creek mine (T13S, R9W, Sec. 33/38, U.S. Geological Survey, Nauvoo, Alabama 7.5-ft quadrangle; Demko and Gastaldo, 1992) with secondary rooting structures indicated at arrows, originating from the nodes of the aerial axis. Scale is 10 cm (3.9 in.). (B) Longitudinal section of pith cast in which tidalite deposits are displayed above and below zones of bioturbation. Dashed lines outline the longitudinally sectioned calamite axis (see Gastaldo, 1992). Rhythmites can be seen adjacent to axis in entombing sediments. Scale in millimeters.

et al., 1992), which in freshwater systems is accomplished primarily by acetate fermentation (Whiticar et al., 1986). The kinetic carbon-isotope fractionation associated with methanogenesis can produce $\delta^{13}C$ values as high as 10‰, which are within the range of values obtained. The high values found in the siderite nodules above the coal probably are caused by the abundant plant debris in the roof shale and high rates of methanogenesis associated with their preservation. However, because C3 plant carbon is isotopically light (~26‰), the $\delta^{13}C$ data reflect fractionation overprinting by bacterial methanogenesis instead of the original isotopic signature of the organic matter. Similar values of sideritic $\delta^{13}C$ have been reported in freshwater Holocene marshes of the Mississippi delta (Moore et al., 1992) and in a shale parting of the Foord coal seam, Nova Scotia (Zodrow and Cleal, 1999). The $\delta^{18}O$ data reflect isotope values consistent with nonmarine pore waters (Mozley and Wersin, 1992), and the small range in values is indicative of early diagenetic siderite precipitation. Additionally, depleted $\delta^{18}O$ values in sites II and V (Figure 7) correlate with what is interpreted as a "wetter" setting based on the

ratio of *Sigillaria* to *Lepidophloios* debris. Gautier (1982) and Moore et al. (1992) interpreted similar data as representative of fresh pore-water environments. Hence, the sediment-laden waters responsible for burial and preservation of the forest were freshwater in origin.

Overlying the forest litter is a tidalite facies (Demko and Gastaldo, 1996) distinguished on the presence of pinstripe interlaminated, dark-gray to medium-gray (N3–N4) mudstone and very fine-grained sandstone (N7–N8). Laminations range from 0.1 to 3 mm (0.004–0.12 in.) in thickness, and silt-sized muscovite and sideritic mudstone clasts are minor constituents. Sandstone laminae thicker than 1 mm (0.04 in.) commonly are draped with silt- to sand-sized, coalified, comminuted plant detritus, which also may be found concentrated in the troughs of current ripple structures. Mudstone and sandstone laminations range in thickness throughout the interval in a pronounced cyclicity (Demko, 1990b). Primary sedimentary structures include (1) horizontal, parallel, bedding; (2) microscale cross-lamination; (3) tool marks; (4) rill marks; (5) rippled surfaces; (6) raindrop imprints; and (7) soft-sediment deformational features. Paleocurrent

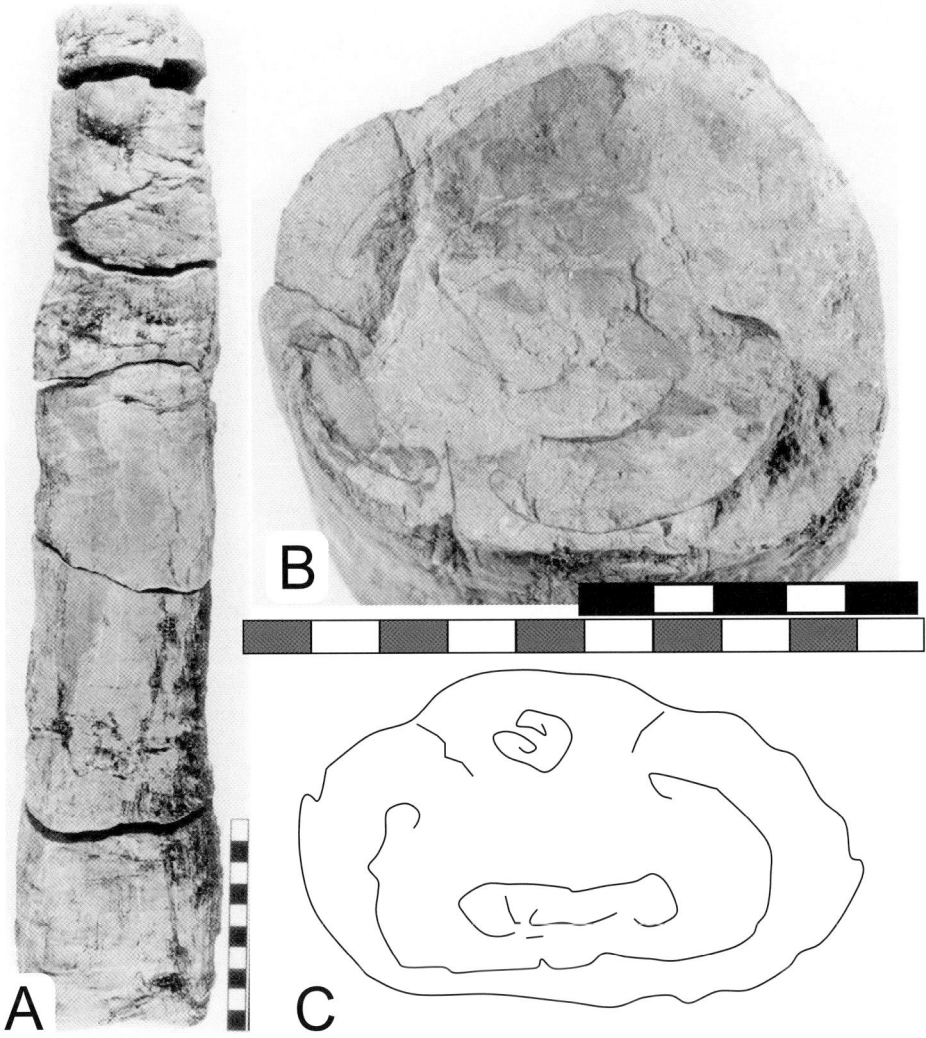

Figure 6. Erect marattialean fern assigned to *Psaronius simplicaulis*. (A) Partial segment of fern axis exhibiting leaf scars and adventitious roots. (B) Cross section of stem in which the vascular architecture (stele) has been preserved in vertical orientation. (C) Line illustration of stem cross section illustrating the vascular architecture of the tree. All scales in centimeters.

vertical dwelling (*Rosselia*) and resting (*Lingulichnus*, *Lockiea*) traces, feeding burrows (*Parahaentzschelinia*, *Helminthopsis*), surface trails (e.g., *Kouphichnium*, *Cincosaurus*), and grazing traces (*Haplotichnus*) (Rindsberg, 1990).

In thin section (Figure 3A), the fossiliferous mudstone is recognizable because of the occurrence of bedded, three-dimensionally distributed plant detritus. Plant fossils vary in their disposition relative to bedding and may be found anywhere from horizontal to greater than 45° from horizontal. The majority of preserved organic matter, representative of leaves, is no greater than 0.2 mm (0.008 in.) thick, whereas thicker organic layers, which represent compressed plant axes, attain thicknesses of greater than 3 mm (0.12 in.) and are orientated horizontal to 20° from horizontal. A point count of the sediment clasts indicates that the interval is dominated by quartz and clay, with clasts ranging in size from 0.01 to 0.005 mm

indicators (current ripples, oriented macrodetritus, tool marks) record orientations to the southeast to southwest. Time-series analysis of data from the Jagger to Blue Creek interval using the maximum entropy method of power-spectrum analysis reveals strong periodic components that occur at 18 and 200 laminations per cycle (Demko et al., 1991). Trace fossils include horizontal burrows (*Paleophycus*, *Treptichnus*),

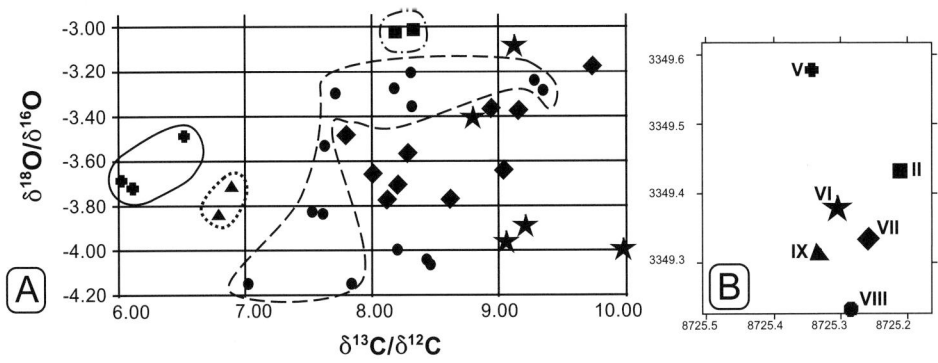

Figure 7. Isotope analyses of siderite carbonate nodules recovered from within and above the preserved Blue Creek forest floor. (A) Plot of $\delta^{13}C/^{12}C$ vs. $\delta^{18}O/^{16}O$ isotope data. (B) Distribution of samples' sites in the Cedrum mine.

(0.0004 to 0.0002 in.) (fine- to very fine-grained silt). The only primary sedimentary structures observable consist of thin, discontinuous parallel laminae, which are identical to the tidalite facies. Laminations are found in and overlying the Blue Creek coal, as well as in the forest-litter hollow voids in degraded plant axes (Figure 3A).

In thin section (Figure 3C), the tidalite facies is recognizable by the presence of uninterrupted lamination of alternating very fine sandstone and siltstone and/or siltstone and mudstone. A near-total absence of organic matter occurs in these rhythmites but where present, consists of either isolated pinnules, axes, or degraded and dispersed organic clasts ranging in size from 0.001 to 0.005 mm (0.00004 to 0.0002 in.). Notably, however, the composition and texture of the sediment is identical to that in the forest-floor litter horizon below, the contact of which is gradational (Figure 3).

Preservation of the Blue Creek Mire

The preserved forest-floor litter and standing forest indicate very rapid sedimentation, with burial of the leaf litter on the order of days to weeks and burial of erect trees probably on the decadenal scale. Based on Holocene studies of plant decay in tropical clastic swamps and mires (Burnham, 1993; Gastaldo and Staub, 1999 and references therein), the pristine preservational state of aerial plant parts in the upper part of the litter horizon indicates that these elements were not exposed to decay processes for any extended period of time. This is in contrast with the debris preserved at the contact between the coal and the roof-shale flora (sensu Gastaldo et al., 1995b).

In Holocene tropical forest litters, decay rates and leaf half-lives are highly variable and dependent on the composition of the leaf (both cellular configuration and biochemistry; table 2 of Gastaldo and Staub, 1999). In humid tropical forests, k-values (decay constants) have been reported to range between 1 ($k = 1$ indicates that a leaf will be degraded completely in 1 yr) and 4 (degraded completely in 1.5 months, e.g., Cuevas and Medina, 1988). Although k-values of 7.5 have been reported for particular taxa (Bernhard-Reversat, 1972), other reports indicate very low decomposition rates on the order of $k = 0.4–0.5$ (e.g., Irmler and Furch, 1980). These latter rates probably are the result of low nutritional value in the decomposing litter (Edwards, 1977; Klinge, 1978). In alluvial and dipterocarp forests of equatorial southeast Asia, half-lives have been reported to be as little as a few

weeks (e.g,. Lavelle et al., 1993) to more than half a year (Anderson et al., 1983).

Leaf-litter fall in the tropics occurs throughout the year (e.g., Burghouts et al., 1992), as leaves become nonfunctional and are lost physiologically. Equatorial forests have been reported to produce total litter fall (including branches) between 9.3 and 10.9 t ha^{-1} yr^{-1}, whereas leaf-litter fall (sensu stricto) has been reported to average 6.7 t ha^{-1} yr^{-1} (Bray and Gorham, 1964). Total fine-litter and leaf-litter productivity in lowland dipterocarp forests of peninsular Malaysia range from 7.5 and 5.4 (Gong and Ong, 1983) to 10.6 and 6.3 t ha^{-1} yr^{-1} (Ogawa, 1978), whereas Proctor et al. (1983) reported values of 8.8 and 5.4 t ha^{-1} yr^{-1} for lowland dipterocarp forests in Sarawak. A pronounced peak in litter fall may occur when either dry (e.g., Gong and Ong, 1983), wet (Ogawa, 1978), or very windy conditions prevail (Proctor et al., 1983). It can be assumed that Carboniferous peat-mire forests produced similar quantities of biomass per year depending on the vegetational mosaic, and that both physiological and traumatic (Gastaldo and Staub, 1999) litter production was responsible for peat accumulation.

The residency time of leaves on tropical forest floors is generally short (on the order of weeks) because of bacterial, fungal, and saprophagous interactions, as well as the adherence of, and degradation by, fine roots of the underlying superficial root mat (Cuevas and Medina, 1988). This combination results in rapid cellular deterioration and destruction of internal anatomy, the interlocking of leaves via fungal growth, and leaf skeletonization (the loss of soft tissues between the "veins;" Gastaldo, 1994). The superficial root mat captures or extracts (via mycorrhizal associations) nutrients from decomposing litter (Stark and Jordan, 1978). The organic residuum is incorporated into the peat matrix.

Tropical forests experience varied leaf-fall patterns and generally do not go through one large pulse of leaves entering the system (Stout, 1980), although leaf flushing may correlate with either the wet or dry season (e.g., Leigh and Windsor, 1996). Because there is no change in tropical temperatures throughout the year that may alter the rate of microbial decay (Stout, 1980), the near-constant addition of canopy parts replenishes the leaf-litter mat, maintaining a relatively constant thickness over time. In addition, as decay rates are also constant because of prevailing climate, degraded organic matter continually is contributed to the peat, accounting for the high accumulation rates as observed in the literature (Anderson, 1983). Therefore, the mechanism(s) responsible for the preservation of unaltered, nondecayed leaves (those without signs of cellular catabolysis, skeletonization, or fungal penetration) in the forest-floor litter of a peat mire must provide for a bypass of the day-to-day degradation processes operating therein. Hence, the "fresh" state of the Blue Creek roof "shale" flora (Gastaldo et al., 1995b), comprising both "delicate" and "robust" taxa (e.g., *Sphenopteris* cf. *schatzlarensis* and *Neuralethopteris schlehani*, respectively), is indicative of an extremely short residence time at the forest-floor surface, probably on the order of several weeks prior to burial and preservation. Based on a physical count of the tidal couplets found in the leaf litter (Figure 3A), a minimum of 1 month must have transpired before the leaf-litter horizon was buried completely.

Continued rapid burial for some extended time is further supported by the presence of the in situ trunks of all major systematic groups, except *Cordaites* (*Artisia* pith casts have been identified in the forest litter, but aerial stems have not been encountered; Stevanoviç-Walls, 2001; Ware, 2001). Erect lycopsid trunks are cast by the same pinstripe lithology found in the entombing tidalite sequence (Gastaldo, 1992). These trees, composed of more bark (periderm) than either wood or soft (parenchyma) tissues, were buried to heights of at least 4.5 m (15 ft) prior to death, resulting in the presence of a second canopy litter horizon at this level (Demko and Gastaldo, 1996) and in decay. Once the structural integrity of the trunk was compromised by rot, the stem and canopy above the level of burial would have fallen over, exposing the hollowed entombed axis. Sediment transported into the system then filled each void (Gastaldo, 1986a, b), resulting in the tree cast (Figure 4). Because of the differential rates of decay for the variety of plant tissues found in a lycopsid trunk (Gastaldo, 1986b), it is not possible to place any constraint on the amount of time from burial to casting.

Similarly, *Calamites* stems are cast by tidalite sedimentation (Figure 5B) (Gastaldo, 1992). However, unlike the lycopsids that were determinate in growth strategy, calamiteans had regenerative capabilities (Gastaldo, 1992). Buds, which would have been capable of developing into aerial leaves and/or branches with leaves at nodes, redifferentiated into rooting structures that grew downward following burial across primary sedimentary structures (Figure 5A). Additionally, new aerial stems grew from buried nodes upward out of the sediment (Gastaldo, 1992) (Figure 2), resulting in another stand of calamiteans at a higher stratigraphic level. These, too, ultimately were pre-

served by burial and casting in tidalite sediments (Gastaldo, 1992, Figure 6). The central piths of these plants are hollowed by cellular breakdown during growth, and this central void is not the result of tissue decay following plant death. Therefore, it is not possible to place a constraint on timing between burial and casting of these forest constituents.

The most instructive erect elements are the Marattiallean tree ferns. Living marattialeans possess very few thick-walled structural cells (sclerenchymatous tissue) in their stems; instead, the principal structural cell is the main-stem, water-conducting system (cauline vasculature of primary xylem) that "floats" in soft spongy tissue (cortical tissue of undifferentiated parenchyma). These ferns also produce a sheath of secondary (adventituous) roots surrounding the trunk, adding support for these small trees. Such features also are found in Carboniferous representatives (Stewart and Rothwell, 1993). The life habit of these tree ferns is such that the trunk can be essentially dead as much as 1 m (3.3 ft) or so from the apex, and the plant continues to produce photosynthetic leaves. The stem is commonly dead and rotted away near the base of large plants; hence, the inner adventitious roots are contributing to the support of the plant, but probably do not play any additional role. In essence, the entire tree is really a short plant on top of a platform made up of formally living stem and roots (G. Rothwell, 2002, personal communication).

Erect Blue Creek tree ferns retain their vascular architecture (Figure 6B) in the siliciclastic internal cast. This feature indicates that following burial, sufficient time passed for the soft, fleshy parenchyma tissues of the upper "living" tree to have decayed, but the more resistant conducting cells remained essentially erect prior to sediment infill. Parenchymal tissue decay correlates generally with a decay half-life (Gastaldo and Staub, 1999), which can be as short as a few weeks in the wet tropics. Hence, these standing tree ferns allow for a constraint on how quickly internal decay of erect vegetation occurred following (or coincident with) burial and the timing of subsequent infill, which would have been on the order of a few months.

Demko (1990b) evaluated a continuous 5-m (16-ft) drill core of tidalite sediments recovered from between the Jagger and Blue Creek coals at the Hope Galloway mine, Walker County, Alabama. Ninety neap-spring-neap tidal cycles were identified in the first 1340 sandstone laminae (1.75 m [6 ft] in thickness), from which a semidiurnal system in some neap-spring-neap cycles and diurnal tides recorded during other cycles was interpreted (Demko et al., 1991). Present neap-spring-neap tidal cycles are 14.77 days, with Carboniferous neap-spring-neap cycles estimated to have been 15 days (A. Archer, 2002, personal communication). Using this estimate, it would have taken approximately 20 yr to bury the tallest lycopsid encountered (not considering the decompacted mud portion of each cycle; if the mudstone is decompacted, burial may have been on the order of a decade). Therefore, the sedimentation event responsible for burial of the forest-floor litter, removing it from decay processes operating at the surface of the peat mire, and the entombment of the standing forest, which thereafter underwent decay and casting, can be constrained to the order of a decade or less.

DISCUSSION

The burial of the peat forest by tidal sedimentation was the result of a rapid transgression that generated sufficient accommodation space to preserve standing trees to at least a height of 4.5 m (15 ft). In addition, accommodation was generated to overcome the thickness of the Blue Creek peat body, interpreted as a planar, immature, raised mire (Demko and Gastaldo, 1992). Decompaction of the Blue Creek coal based on the relative plant-part composition of the peat (Winston, 1990a) results in an original peat-body thickness ranging between 3 and 6 m (10 and 20 ft) over an outcrop area of at least 1100 km² (425 mi²) (Demko and Gastaldo, 1992). The paleogeographic position of the Cedrum mine has been placed some 60–100 km (37–62 mi) from the shoreline at the time of Blue Creek mire accumulation (J. Pashin, 2002, personal communication). Hence, when the thickness of the peat body in Walker County is added to the documented maximum lycopsid-trunk height, a minimum of 11 m (36 ft) of accommodation is required for preservation of this assemblage.

Demko and Gastaldo (1996) concluded that eustasy was the dominant allocyclic mechanism responsible for the emplacement of fully marine sediments above ravinement indicators (Liu and Gastaldo, 1992b). Conversely, the emplacement of tidal facies between peat mires was ascribed to autocyclic mechanisms as the result of peat compaction following catastrophic flooding of the mire (Gastaldo, 1990) and burial of the forest (Demko and Gastaldo, 1996). Based on the present analyses, this model requires reinterpretation, and allocyclic processes are implicated.

Accommodation space can be generated by either eustatic rise or tectonic subsidence, the magnitude of which still is debated (e.g., Klein, 1994) and probably cannot be generalized in cyclothems. This is because the mechanism is dependent on basinal history (Klein and Willard, 1989) and regional variation in tectonically active areas (Fortuin and de Smet, 1991). What must be taken into consideration in the Black Warrior Basin is not only the rapidity at which the Blue Creek peat mire was placed below base level but also the depth below base level that allowed for preservation of the forest under an estuarine meso- to macrotidal system, the regime in which tidal rhythmite deposits are generated (Tessier et al., 1995). Given the constraints of tropical plant decay, the final series of events responsible for base-level change occurred on the order of decades and possibly less.

Recent papers (e.g., Clark et al., 2002; Blanchon et al., 2002) indicate that eustatic rise during the present interstadial has been pulsed, with rapid deepening occurring globally. The global meltwater pulse IA (MWP-IA) at the beginning of the last deglaciation about 14 ka is recorded in corals in Barbados (Bard et al., 1990) and elsewhere, resulting in an average rise of 20 m (66 ft) in less than 1000 yr (Clark et al., 2002). However, because of the proximity of any geographical site to melting of the ice cap, actual sea level rise might be as little as about 75% or as much as about 142% of the eustatic mean. With meltwater contribution from Antarctica and data from Barbados and the Sunda Shelf, Clark et al. (2002) demonstrated an approximate 25-m (82-ft) rise in eustatic sea level in the tropics over a period of only 700 yr. If this rise was continuous, the rate would exceed 40 mm/yr (1.6 in./yr) (Clark et al., 2002). Other meltwater pulses have been identified in the Holocene record and interpreted on similar timescales.

Blanchon et al. (2002) have interpreted relic reefs on Grand Cayman, Barbados, St. Croix, St. Thomas, and northern Florida to have stopped accreting within 160 yr of each other, whereas modern reefs became established at depths 4–9 m (13–30 ft) higher upslope within 100 yr of the former's demise (if the older radiocarbon dates, uncorrected for metabolic fractionation, are accurate). The change in elevation of reef growth is explained by a rapid rise in sea level of at least 6 m (20 ft); hence, evidence exists for a circum-Caribbean backstepping response on a century scale, with the rate of sea level rise on the order of 60 mm/yr (2.4 in./yr). The modal age of Holocene deltas around the globe also record initiation over a 400-yr-interval, backstepping upslope (Stanley and Warne, 1994) that coincides with the reef demise.

Figure 8. Features of Holocene coastal-plain and deltaic transgression under eustatic control in Kalimantan, Indonesia, and Sarawak, east Malaysia. (A) Transgression of coastal-plain deposits south of the Tandjung Tambangotngot, Mahakam River delta. Wave action has eroded organic-rich sediments from above a rooted inceptisol and deposited shell- and organic-rich beach sediments. Photo taken October 1988. (B) Oblique aerial photograph of coastal mangrove transgression in the Rajang River delta, showing undercutting of rooting systems, felling of vegetation, and transport of woody debris into the South China Sea. Photo taken by K. Bartram, July–August 1993.

An eustatic rise in sea level, similar to documented backstepping, cannot account for the rapid generation of accommodation necessary for Blue Creek forest preservation even if the peat-mire surface, some 60–100 km (37–62 mi) inland, was within 1 m (3.3 ft) or so of base level (hence, reducing required accommodation to ~5 m (~16 ft); see below). Water level would have had to rise at a rate between 500 and 1000 mm/yr (0.5–1.0 m/yr [1.6–3.3 ft/yr]) to generate accommodation, allowing for initiation of sedimentation, burial of standing trees by diurnal tidal deposits, and subsequent decay and infill of the axial voids. Holocene rates are an order of magnitude lower than that required to bury and preserve the forest, and where coastal transgression presently is occurring in the tropics, trees either are being felled and transported oceanward or are covered with a thin veneer of beach sediment (Figure 8). If eustatic rise occurred at documented Holocene rates, the lycopsid aerial canopy parts shed after death would be preserved in close proximity to the forest-floor litter. Instead, these assemblages are preserved along with mud-cast prostrate trees at the stratigraphic level, where the standing trees terminate (Figure 4). Hence, eustatic sea level rise alone would have been too slow to generate water depths to account for both the observed sedimentological sequence and the plant taphonomic character of the assemblage.

The alternative allocyclic mechanism to account for the requisite accommodation is tectonism. Phillips et al. (1994) and Phillips and Bustin (1996) reported the transgression of a Holocene coastal mangrove peat as the result of earthquake-induced subsidence in a Panamanian microtidal system. The back-barrier Changuinola peat deposit (Cohen et al., 1989) covers approximately 60 km^2 (23 mi^2) onshore and another 20 km^2 (8 mi^2) offshore beneath shallow-marine sediments of Almirante Bay. The ^{14}C AMS age of the peat, −8.10 m (−26.5 ft) beneath the present mire surface, is reported as 3.04 ± 0.08 ka, with nearly 40% of the peat body presently beneath sea level (Phillips and Bustin, 1996). Cohen et al. (1985) report an approximate age of the basal peat body as between 4 and 5 ka, and Phillips and Bustin (1996) concur; their basalmost age determination is approximately 2 m (6.6 ft) above the contact with underlying sand. An earthquake of M_s = 7.5 on April 22, 1991, led to a minimum of 30 cm (12 in.) and a maximum of 50–70 cm (20–28 in.) of coseismic subsidence and flooding or drowning of the southeastern margin of the mire. Phillips and Bustin (1996) concluded that the rate of tectonically driven, punctuated subsidence in the area is between 2.2 and 2.6 mm/yr (2.2–2.6 m/k.y.), and the disposition of marginal peat records this style of subsidence for at least the past 2 k.y. and probably longer.

Earthquake-induced subsidence of 1 m (3.3 ft) or more per event is documented along the Chilean (Plakker and Savage, 1970), northwestern United States (Atwater and Yamaguchi, 1991), and Alaskan (Plafker and Rubin, 1967; Plafker, 1969) coasts. Central Chile was devastated by an M_s = 7.5 (May 21, 1960) earthquake followed by two closely spaced M_s = 8.5 (May 22, 1960) earthquakes during which time both coseismic uplift and subsidence relative to sea level occurred along the coast over an area of 200,000 km^2 (77,220 mi^2). Vertical displacements were determined by differences in the height of pre- and postquake positions of the lower vegetational growth limit. Plafker and Savage (1970) reported that subsidence occurred in an area that was 75–100 km (47–62 mi) wide and at least 800 km (497 mi) in length. Local maximum coseismic subsidence was as much as −2.7 m (−8.8 ft) on unconsolidated sediments near Valdivia, with this area experiencing an average subsidence of −1.8 m (6 ft). Adjacent coastal farmland was submerged permanently by at least 2 m (7 ft), whereas submergence at bedrock sites on Isla Chiloé was as much as 2.4 m (8 ft). Similar coseismic displacements (1 m [3.3 ft] average, 2.2 m [7.2 ft] maximum subsidence) followed the 1964 Alaskan Earthquake (Plafker, 1965; Plafker and Rubin, 1967), and submerged forests along the coast of Washington state are indicative of at least 1-m (3.3-ft) subsidence (Atwater and Yamaguchi, 1991). In some Chilean coastal areas, post-earthquake subsidence of a few decimeters continued for several years.

Historical records provide data on earthquake frequency along the Chilean coast (Lomnitz in Plafker and Savage, 1970). Earthquakes in the Valdivia region with estimated $M_s > 7.5$ occurred in 1575, 1737, and 1837; major earthquakes centered on Concepción occurred in 1570, 1575, 1751, and 1835; and Arauco Indian folklore notes at least two large earthquakes were accompanied by submergence in the Lake Budi area, north of Valdivia. Hence, the recurrence interval for high-magnitude earthquakes along this part of the Chilean coast is somewhat less than 100 yr since the early part of the 16th century. If the average coseismic subsidence associated with each of these high-magnitude events is a conservative 1.5 m (5 ft), then 6–7.5 m (20–25 ft) of cumulative subsidence could have occurred in less than 450 yr. Hence, if the cumulative effect of such short-term subsidence events can lower the base of a peat body while the mire is accumulating organic material, without concurrent siliciclastic contamination (Phillips and Bustin, 1996), then siliciclastic sediments will be deposited only when the peat-mire forest floor finally is submerged below sea level.

Therefore, coseismic, tectonically induced subsidence associated with high magnitude activity is the most parsimonious explanation to account for the preservation of the Blue Creek mire. This model involves transgression of the coastal zone by backstepping generated by century- to millennial-scale coseismic events, placing the top of the mire close to base level. The actual coastline would have been kilometers distant of the Cedrum mine, where coeval tidal sediments would have been deposited. Although marine or estuarine waters would have been in close proximity to the study site, peat continued to accumulate under freshwater conditions because of the hydraulic head inherent in planar and raised mires (Winston, 1994). Subsequently, an anomalous high-magnitude earthquake (or series of closely spaced earthquakes) resulted in the displacement of the forest floor to at least 5 m (16 ft) below sea level (accommo-

dation required to account for the maximum lycopsid height). Such vertical displacement would have affected fluvial distribution patterns in the area and, possibly, increased sediment loading from slope failure in the hinterland. Such sediment loading may account for the rapid deposition of very fine sand and coarse silt in tidal rhythmites found within and above the mire, with redistribution of sediment-water interface clastics into the vertical voids produced when entombed trees decayed. Sediment transported and deposited occurred in a freshwater regime, as indicated by isotopic data from siderite concretions. The presence of tidal deposits in the upper reaches of estuaries and in fluvial-dominated systems is common where meso- to macrotidal ranges occur (e.g., Mahakam delta, Gastaldo et al., 1995a; Rajang delta, Staub et al., 2000; Fly River Delta, Baker et al., 1995).

Confirmation of Tectonic Influence on Black Warrior Sedimentation

Tectonic influence on sediment accumulation in the greater Black Warrior Basin first was suggested by Thomas (1968), and Weisenfluh and Ferm (1984) recognized this phenomenon as affecting peat distribution during accumulation of the Pratt coal seam. They recognized basement faulting contemporaneous with peat accumulation as evidenced by the presence of increasing sediment thicknesses on fault blocks from the edge of the basin successively southward. Such contemporaneous faulting influenced the peat distribution across the basin, with thick merged coals occurring on the upthrown sides of fault blocks and thin coal splits on the downthrown sides (Weisenfluh and Ferm, 1984, Figure 4). They tested this hypothesis using data recovered from 200 core logs and underground seam measurements in a small geographic area. Although never stated directly as to whether the observed displacement was restricted only to contemporaneous faulting, lithologies that intervene between coal splits on the downthrown side of the fault block may be 45–60 m (148–197 ft) in thickness. This would imply that localized subsidence, contemporaneous with peat accumulation, could account for the rapid change in base level, drowning, and preservation of forests in this basin. Penecontemporaneous faulting associated with the development of the Mary Lee cycle has been documented by Pashin (1994c).

Implications for Subsidence Rates

In an analysis of Cenozoic tectonism in the Banda Sea, Indonesia, Fortuin and de Smet (1991, Figure 5) calculated the average subsidence and uplift rates since the Miocene, but cautioned that the actual subsidence rates may be higher because neither compaction nor isostasy caused by sediment loading were considered for the sake of simplicity. It is evident from their data that the timing, rates, and magnitudes of vertical crustal movement vary from place to place across the Banda Sea. For example, average subsidence rates at Buru (early to middle Miocene) are 30 cm/ka (12 in./ka), Buton (middle–late Miocene) experienced rates of greater than 75 cm/ka (30 in./ka), whereas Seram (late Miocene–late Pliocene) experienced a rate slightly less than 75 cm/ka (30 in./ka). They argued, however, that local subsidence is episodic near convergent margins over very short temporal scales and concluded that processes operating over several orders of magnitude of time result in an intricate pattern of both uplift and subsidence. Pashin (2004) has modeled subsidence rates for the Mary Lee cycle ranging from a minimum of 12 cm/ka (5 in./ka) to a maximum of 45 cm/ka (18 in./ka), with increasing rates higher in the Pottsville Formation. These figures fall in the lower range of values calculated by Fortuin and de Smet (1991) and are slightly more than half of the fastest subsidence rate. Composite sections used in Pashin's (2004) analysis originated from either Lamar or Tuscaloosa Counties, approximately 65 and 100 km (40 and 62 mi) southwest and south of the Cedrum mine, respectively. At the Cedrum mine, the Jagger, Blue Creek, and Mary Lee coals are in closer stratigraphic proximity to each other.

Pleistocene sediments in the Rajang River delta consist of beach/terrace deposits, some 5–7 m (16–23 ft) higher than mean sea level (MSL), adjacent to upland areas which Staub and Gastaldo (2003) interpret as the remains of the VIIa highstand surface of 125 ka, some 6 m (20 ft) above present MSL (Chappell and Shackleton, 1986). In the lower delta plain, near Daro, subsurface peat recovered from a depth of 78 m (256 ft) below MSL has a ^{14}C age date of 40,370 (+2944/–2150; Staub and Gastaldo, 2003). This peat is part of the IIIb highstand of 40 ka, which occurred at 41 ± 4 m (135 ± 13 ft) below present MSL (Bloom et al., 1974; Chappell and Shackleton, 1986). These relationships imply that approximately 40 m (131 ft) of subsidence has occurred since then, averaging nearly 1 m/ka (3.3 ft/ka). Hence, it is not impossible to envision a major tectonic event, such as the Alaskan (Plafker, 1969) or Chilean (Plafker and Savage, 1970) earthquake, wherein the extent of basinal subsidence would have been equal to the cumulative average displacement for 10 ka or more.

Calculated average subsidence rates impart an assumption that base-level changes are gradual and extended over temporal durations of millennial scale. Such paradigms are not substantiated fully by processes operating in tectonically active areas, and there is reason to believe that coseismic changes in base level have had their effect on the stratigraphic record of various basinal configurations. Although average subsidence rates provide a means by which to evaluate overall long-term basinal history, the recognition and identification of in situ erect forests preserved in estuarine deposits are a key in understanding not only the pulsed nature of basinal response, but also provide a means to evaluate the magnitude of coseismic subsidence in the deep past. The height to which these forests are preserved serves as a proxy for localized, tectonically rapid subsidence.

CONCLUSIONS

The low-ash, low-sulfur (Winston, 1990b) Blue Creek mire was deposited at the inflection point of the sea level curve recorded in the Mary Lee cycle (Gastaldo et al., 1993). Eustasy alone cannot explain tidalite deposition within and above the Blue Creek mire. Documented Holocene rates of sea level rise (Bard et al., 1990; Blanchon et al., 2002; Clark et al., 2002) when applied to the Carboniferous are too slow to account for the burial and preservation of the in situ lycopsids, calamiteans, and particularly, the tree ferns. Such a mechanism could account for buried forests in which tree trunks are less than 1 m (3.3 ft) in height (e.g., Gastaldo, 1986a), but not for trees preserved to heights of several meters. Given the fact that progressive burial of the margins of a peat body can occur during backstepping (either eustatic or tectonic) without affecting the mire's sulfur and clastic content (the overlying clastics are either a roof shale or coal parting depending on subsequent events; Gastaldo et al., 1995b; Phillips and Bustin, 1996), it is possible that some part of the westward and northwestward extension of the Blue Creek mire may have been submarine, effectively bringing the top of the peat closer to base level, prior to the time when the standing forest was preserved in Walker County. Hence, it is not necessary to invoke a single, short-term event to account for 11 m (36 ft) of subsidence to bury the peat body and the forest; instead, total subsidence would be less than this requisite base-level change, but would still be on the order of 5 m (16 ft) to account for burial of the tallest in situ trees.

Rapid, coseismic subsidence, however, can significantly change base level on local (e.g., Weisenfluh and Ferm, 1984; Staub and Gastaldo, 2003) or regional scales (Plafker, 1965; Plafker and Savage, 1970; Fortuin and de Smet, 1991). Although average long-term subsidence rates are relied on to characterize basinal history, neotectonic investigations have demonstrated that long-term uplift and subsidence rates tend to be considerably lower than those calculated for shorter durations (e.g., Tjia, 1981; Bloom and Yonekura, 1985). Depending on the magnitude of any single tectonic event, subsidence may be less than 1 m (3.3 ft) (e.g., Phillips and Bustin, 1996) or as much as 4 m (13 ft) (Prince William Sound, Alaska; Plafker, 1969) of vertical displacement. Evidence exists in the Mary Lee cycle for tremor-induced liquifaction of sand bodies (Demko, 1990a, b), indicating that effects of tectonic loading associated with the Appalachians and Ouachitas (Thomas, 1988, 1995) are recorded in this sedimentological record. Therefore, it is most parsimonious that earthquake-induced subsidence was responsible for the documented dramatic change in the elevation of the Blue Creek mire, reducing it to several meters below base level, and allowing for tidal processes operating in a freshwater regime to bury and preserve this forest. The model proposed by Gastaldo (1990) and discussed by Demko and Gastaldo (1996) for the burial and preservation of the Blue Creek forest via catastrophic high-magnitude fluvial processes for leaf-litter burial and preservation is untenable. Although autocyclic compaction (Demko and Gastaldo, 1996) may have played some role in positional change of base level, allocyclic coseismic processes are implicated as the primary mechanism for overriding and preserving the Blue Creek peat mire.

The recognition that sedimentation following coseismic subsidence is the mechanism responsible for entombment and burial of in situ erect forests above coastal-plain mires allows for the use of such terrestrial plant-fossil assemblages as a proxy for understanding the magnitude of episodic base-level change in the stratigraphic record. Hence, short-term tectonic subsidence events can be identified in terrestrial/transitional regimes by close examination of the sedimentological features associated with in situ forests, irregardless of geologic age, and using such evidence, a better understanding of the short-term spasmodic changes in base level that are time averaged into long-term subsidence rates can be ascertained.

ACKNOWLEDGMENTS

The authors would like to thank the following funding agencies for support of research that led to this publication: National Science Foundation-EAR 8618815 (R.A.G.), Geological Society of America Grant-in-Aid of Research (W.N.W. and I.M.S.-W.), Paleobiological Fund (W.N.W. and I.M.S.-W.), the Gulf Coast Association of Geological Societies (W.N.W. and I.M.S.-W.), and the Paleontological Society (I.M.S.-W.). The authors are indebted to Drummond Brothers Coal Company, Jasper, Alabama, for permission to access their operations, and particularly to Mike Hendon and Perry Hubbard, and Ethan Grossman, Department of Geology and Geophysics, Texas A&M University, for access to his laboratory instrumentation. The authors also thank George D. Klein and Richard Carroll for insightful reviews that improved the quality of this paper.

REFERENCES CITED

Anderson, J. A. R., 1983, The tropical swamps of western Malaysia, in A. P. J. Gore, ed., Mires, swamp bog, fen, and moor. Ecosystems of the world, regional studies 4B: Amsterdam, Elsevier, p. 181–200.

Anderson, J. M., J. Proctor, and H. W. Vallack, 1983, Ecological studies in four contrasting lowland rainforest in Gunung Mulu National Park, Sarawak: 3. Decomposition processes and nutrient losses from leaf litter: Journal of Ecology, v. 71, p. 503–527.

Atwater, B. F., and D. K. Yamaguchi, 1991, Sudden, probably coseismic submergence of Holocene trees and grass in coastal Washington state: Geology, v. 19, p. 706–709.

Baker, E. K., P. T. Harris, J. B. Keene, and S. A. Short, 1995, Patterns of sedimentation in the macrotidal Fly River delta, Papua New Guinea, in B. W. Flemming and A. Bartholomä, eds., Tidal signatures in modern and ancient sediments: International Association of Sedimentologists Special Publication 24, p. 193–211.

Bard, E., B. Hamelin, R. G. Fairbanks, and A. Zindler, 1990, Calibration of the ^{14}C timescale over the past 30,000 years using mass spectrometric U-Th ages from Barbados corals: Nature, v. 345, p. 405–409.

Barnett, R. L., 1986, Coal occurrence and general geology in the Townley 7.5-minute quadrangle, Walker County, Alabama: Alabama Geological Survey Special Map 166, 79 p.

Bernhard-Reversat, F., 1972, Décomposition de la litiPre de feuilles en forLt ombrophile de basse Côte-d'Ivoire: Oecologia Plantarum, v. 7, p. 279–300.

Blanchon, P., B. Jones, and D. C. Ford, 2002, Discovery of a submerged relic reef and shoreline off Grand Cayman further support for an early Holocene jump in sea level: Sedimentary Geology, v. 147, p. 253–270.

Bloom, A. L., and N. Yonekura, 1985, Coastal terraces generated by sea-level change and tectonic uplift, *in* R. J. N. Devoy, ed., Models in geomorphology: Winchester, Massachusetts, Allen & Unwin, p. 139–154.

Bloom, A. L., S. W. Broeker, J. M. A. Chappell, R. K. Matthews, and K. J. Mesolella, 1974, Quaternary sea level fluctuations in a tectonic coast, New ^{230}Th/^{234}U from the Huon Peninsula, New Guinea: Quaternary Research, v. 4, p. 185–205.

Bray, J. R., and E. Gorham, 1964, Litter production in forests of the world, *in* J. B. Cragg, ed., Advances in ecological research: New York, Academic Press, v. 2, p. 101–157.

Burghouts, T., G. Ernsting, G. Korthals, and T. Devries, 1992, Litterfall, leaf litter decomposition and litter invertebrates in primary and selectively logged dipterocarp forest in Sabah, Malaysia: Philosophical Transactions of the Royal Society of London, B, v. 335, p. 407–416.

Burnham, R. J., 1993, Reconstructing richness in the plant fossil record: Palaios, v. 8, p. 376–384.

Cecil, C. B., 1990, Paleoclimate controls on stratigraphic repetition of chemical and siliciclastic rocks: Geology, v. 18, p. 533–536.

Chappell, J., and N. J. Shackleton, 1986, Oxygen isotopes and sea level: Nature, v. 324, p. 137–140.

Clark, P. U., J. X. Mitovica, G. A. Milne, and M. E. Tamislea, 2002, Sea-level fingerprinting as a direct test for the source of global meltwater pulse IA: Science, v. 295, p. 2438–2441.

Cohen, A. D., R. Raymond Jr., S. Mora, A. Alverado, and L. Malavassi, 1985, Economic characterization of the peat deposits of Costa Rica, preliminary study, *in* B. Wade, ed., Tropical peat resources—Prospects and potential: International Peat Society, p. 146–169.

Cohen, A. D., R. Raymond Jr., A. Ramirez, Z. Morales, and F. Ponce, 1989, The Changuinola peat deposit of northwest Panama, a tropical, back-barrier peat (coal) forming environment: International Journal of Coal Geology, v. 12, p. 157–192.

Cuevas, E., and E. Medina, 1988, Nutrient dynamics within Amazonian forests: II. Fine root growth, nutrient availability and leaf litter decomposition: Oecologia, v. 76, p. 263–273.

Demko, T. M., 1990a, Depositional environments of the lower Mary Lee coal zone, lower Pennsylvanian "Pottsville" Formation, northwestern Alabama, *in* R. A. Gastaldo, T. M. Demko, and Y. Liu, eds., Carboniferous coastal environments and paleocommunities of the Mary Lee coal zone, Marion and Walker Counties, Alabama: Geological Society of America, Southeastern Section, Field Trip Guidebook, p. 5–20.

Demko, T. M., 1990b, Paleogeography and depositional environments of the lower Mary Lee coal zone, Pottsville Formation, Warrior Basin, northwestern Alabama: M.S. thesis, Auburn University, Auburn, Alabama, 195 p.

Demko, T. M., and R. A. Gastaldo, 1992, Paludal environ-ments of the Mary Lee coal zone, Pottsville Formation, Alabama, stacked clastic swamps and peat mires: International Journal of Coal Geology, v. 20, p. 23–47.

Demko, T. M., and R. A. Gastaldo, 1996, Eustatic and autocyclic influences on deposition of the lower Pennsylvanian Mary Lee coal zone, Warrior Basin, Alabama: International Journal of Coal Geology, v. 31, p. 3–19.

Demko, T. M., J. Jirikowic, and R. A. Gastaldo, 1991, Tidal cyclicity in the Pottsville Formation, Warrior Basin, Alabama: Sedimentology and time-series analysis of an interlaminated sandstone-mudstone interval: Geological Society of America Abstracts with Program, v. 23, p. 287.

Dewey, J. F., and W. C. Pitman III, 1998, Sea-level changes, mechanisms, magnitudes and rates, *in* J. L. Pindell and C. L. Drake, eds., Paleogeographic evolution and non-glacial eustacy, northern South America: SEPM Special Publication 58, p. 1–16.

Dickinson, W. R., G. S. Soreghan, and K. A. Giles, 1994, Glacio-eustatic origin of Permo-Carboniferous stratigraphic cycles; evidence from the southern Cordilleran foreland region: SEPM Concepts in Sedimentology and Paleontology, v. 4, p. 25–34.

DiMichele, W. A., and T. L. Phillips, 1977, Monocyclic *Psaronius* from the lower Pennsylvanian of the Illinois Basin: Canadian Journal of Botany, v. 55, p. 2514–2524.

Edwards, P. J., 1977, Studies of mineral cycling in a montane rain forest in New Guinea: Journal of Ecology, v. 65, p. 971–992.

Fortuin, A. R., and M. E. M. de Smet, 1991, Rates and magnitudes of late Cenozoic vertical movement in the Indonesian Banda Arc and the distinction of eustatic effects: International Association of Sedimentologists Special Publication 12, p. 79–89.

Galloway, W. E., 1989, Genetic stratigraphic sequences in basin analysis: I. Architecture and genesis of flooding-surface bounded depositional units: AAPG Bulletin, v. 73, p. 125–142.

Gastaldo, R. A., 1986a, Implications on the paleoecology of autochthonous lycopods in clastic sedimentary environments of the early Pennsylvanian of Alabama: Palaeogeography, Palaeoclimatology, Palaeoecology, v. 53, p. 191–212.

Gastaldo, R. A., 1986b, An explanation for lycopod configuration, "Fossil Grove" Victoria Park, Glasgow: Scottish Journal of Geology, v. 22, p. 77–83.

Gastaldo, R. A., 1987, Confirmation of Carboniferous clastic swamp communities: Nature, v. 326, p. 869–871.

Gastaldo, R. A., 1990, Early Pennsylvanian swamp forests in the Mary Lee coal zone, Warrior Basin, Alabama, *in* R. A. Gastaldo, T. M. Demko, and Y. Liu, eds., Carboniferous coastal environments and paleocommunities of the Mary Lee coal zone, Marion and Walker Counties, Alabama: Geological Society of America, Southeastern Section, Field Trip Guidebook, p. 41–54.

Gastaldo, R. A., 1992, Regenerative growth in fossil horsetails following burial by alluvium: Historical Biology, v. 6, p. 203–220.

Gastaldo, R. A., 1994, The genesis and sedimentation of phytoclasts with examples from coastal environments, *in* A. Traverse, ed., Sedimentation of organic particles: Cambridge, United Kingdom, Cambridge University Press, p. 103–127.

Gastaldo, R. A., and J. R. Staub, 1999, A mechanism to explain the preservation of leaf litters lenses in coals derived from raised mires: Palaeogeography, Palaeoclimatology, Palaeoecology, v. 149, p. 1–14.

Gastaldo, R. A., T. M. Demko, Y. Liu, W. D. Keefer, and S. L. Abston, 1989, Biostratinomic processes for the development of mud-cast logs in Carboniferous and Holocene swamps: Palaios, v. 4, p. 356–365.

Gastaldo, R. A., T. M. Demko, and Y. Liu, 1990, Carboniferous coastal environments and paleocommunities of the Mary Lee coal zone, Marion and Walker Counties, Alabama: A Guidebook for Field Trip VI, 39th Annual Meeting Southeastern Section of the Geological Society of America, Tuscaloosa, Alabama. Geological Survey of Alabama, 139 p.

Gastaldo, R. A., T. M. Demko, and Y. Liu, 1991, A mechanism to explain persistent alternation of clastic and peat-accumulating swamps in carboniferous sequences: Bulletin Societé Geologié France, v. 8, p. 155–161.

Gastaldo, R. A., T. M. Demko, and Y. Liu, 1993, Application of sequence and genetic stratigraphy concepts to Carboniferous coal-bearing strata, an example from the Black Warrior Basin, U.S.A.: Geologische Rundschau, v. 82, p. 212–226.

Gastaldo, R. A., G. P. Allen, and A. Y. Huc, 1995a, The tidal character of fluvial sediments of the Recent Mahakam River delta, Kalimantan, Indonesia, *in* B. W. Flemming, and A. Bartholomä, eds., Tidal signatures in modern and ancient sediments: International Association of Sedimentologists Special Publication 24, p. 171–181.

Gastaldo, R. A., H. W. Pfefferkorn, and W. A. DiMichele, 1995b, Taphonomic and sedimentologic characterization of roof-shale floras, *in* P. C. Lyons, E. D. Morey, and R. H. Wagner, eds., Historical perspective of early twentieth century Carboniferous paleobotany in North America: Geological Society of America Memoir 185, p. 341–351.

Gautier, D. L., 1982, Siderite concretions: Indicators of early diagenesis in ancient sediments by analogy with processes in modern diagenetic environments, *in* D. A. McDonald and R. C. Surdam, eds., Clastic diagenesis: AAPG Memoir 37, p. 111–123.

Gong, W. K., and J. E. Ong, 1983, Litter production and decomposition in a coastal hill dipterocarp forest, *in* S. L. Sutton, T. C. Whitmore, and A. C. Chadwick, eds., Tropical rain forest, ecology and management: Oxford, United Kingdom, Blackwell Scientific Publications, p. 275–285.

Irmler, U., and K. Furch, 1980, Weight, energy and nutrient changes during decomposition of leaves in the emersion phase of Central Amazonian inundation forests: Pedobiologia, v. 20, p. 118–130.

Klein, G. D., 1994, Depth determination and quantitative distinction of the influence of tectonic subsidence and climate on changing sea level during deposition of mid-continent Pennsylvanian cyclothems: SEPM Concepts in Sedimentology and Paleontology, v. 4, p. 35–50.

Klein, G. D., and D. A. Willard, 1989, Origin of Pennsylvanian coal-bearing cyclothems of North America: Geology, v. 17, p. 152–155.

Klein, G. D., and J. B. Kupperman, 1992, Pennsylvanian cyclothems: Methods of distinguishing tectonically induced changes in sea level from climatically induced changes: Geological Society of America Bulletin, v. 106, p. 166–175.

Klinge, H., 1978, Litter production in tropical ecosystems: Malayan Nature Journal, v. 30, p. 415–422.

Kosanke, R. M., 1979, A long-leaved specimen of Lepidodendron: Geological Society of America Bulletin, v. 90, p. 431–434.

Krasilov, V. A., 1975, Paleoecology of terrestrial plants, Basic principles and techniques: New York, Wiley Publishing, 283 p.

Lavelle P., E. Blanchart, A. Martin, and S. Martin, 1993, A hierarchical model for decomposition in terrestrial ecosystems, applications to soils of humid tropics: Biotropica, v. 25, p. 130–150.

Leigh, E. G. Jr., and D. M. Windsor, 1996, Forest production and regulation of primary consumers in Barro Colorado Island, *in* E. G. Leigh Jr., A. S. Rand, and D. M. Windsor, eds., The ecology of a tropical forest, seasonal rhythms and long-term changes, 2d ed.: Washington, D.C., Smithsonian Institution, p. 111–122.

Liu, Y., 1990, Depositional environments of the Mary Lee coal zone, lower Pennsylvanian Pottsville Formation, northwestern Alabama, *in* R. A. Gastaldo, T. M. Demko, and Y. Liu, eds., Carboniferous coastal environments and paleocommunities of the Mary Lee coal zone, Marion and Walker Counties, Alabama: Geological Society of America, Southeastern Section, Field Trip Guidebook, p. 21–40.

Liu, Y., and R. A. Gastaldo, 1992a, Characteristics of a Pennsylvanian ravinement surface: Sedimentary Geology, v. 77, p. 197–213.

Liu, Y., and R. A. Gastaldo, 1992b, Characteristics and provenance of log-transported gravels in a Carboniferous channel deposit: Journal of Sedimentary Petrology, v. 62, p. 1072–1083.

Moore, S. E., R. E. Ferrell, and P. Aharon, 1992, Diagenetic siderite and other ferroan carbonates in a modern subsiding marsh sequence: Journal of Sedimentary Petrology, v. 62, p. 357–366.

Mozley, P. S., and P. Wersin, 1992, Isotopic composition of siderite as an indicator of depositional environment: Geology, v. 20, p. 817–820.

Ogawa, H., 1978, Litter production and carbon cycling in Pasoh Forest, Malaysia: Malayan Nature Journal, v. 30, p. 367–373.

Pashin, J. C., 1994a, Flexurally influenced eustatic cycles in the Pottsville Formation (lower Pennsylvanian), Black

Warrior Basin, Alabama, tectonic and eustatic controls on sedimentary cycles, *in* J. M. Dennison and F. R. Ettensohn, eds., SEPM Concepts in Sedimentary and Paleontology, v. 4, p. 89–105.

Pashin, J. C., 1994b, Cycles and stacking patterns in Carboniferous rocks of the Black Warrior foreland basin: Gulf Coast Association of Geological Societies Transactions, v. 44, p. 555–563.

Pashin, J. C., 1994c, Coal-body geometry and synsedimentary detachment folding in Oak Grove coalbed-methane field, Black Warrior Basin, Alabama: AAPG Bulletin, v. 78, p. 960–980.

Pashin, J. C., 2004, Cyclothems of the Black Warrior Basin, Alabama, U.S.A.: Eustatic snapshots of foreland basin tectonism, *in* J. C. Pashin and R. A. Gastaldo, eds., Sequence stratigraphy, paleoclimate, and tectonics of coal-bearing strata: AAPG Studies in Geology 51, p. 199–218.

Pfefferkorn, W. H., 1976, Pennsylvanian tree fern compressions *Caulopteris*, *Megaphyton*, and *Artistophyton* gen. nov. in Illinois: Illinois State Geological Survey Circular 492, p. 1–31.

Phillips, S., and R. M. Bustin, 1996, Sedimentology of the Changuinola peat deposit; organic and clastic sedimentary response to punctuated coastal subsidence: Geological Society of America Bulletin, v. 108, p. 794–814.

Phillips, S., R. M. Bustin, and L. E. Lowe, 1994, Earthquake-induced flooding of a tropical coastal peat swamp: A modern analogue for high-sulfur coals?: Geology, v. 22, p. 929–932.

Plafker, G., 1965, Tectonic deformation associated with the 1964 Alaska earthquake: Science, v. 148, p. 1675–1678.

Plafker, G., 1969, Tectonics of the March 27, 1964, Alaska Earthquake: U.S. Geological Survey Professional Paper 543-I, p. I1–I74.

Plafker, G., and M. Rubin, 1967, Vertical tectonic displacements in south-central Alaska during and prior to the great 1964 earthquake: Journal of Geoscience, Osaka City University, v. 10, p. 53–66.

Plafker, G., and J. C. Savage, 1970, Mechanism of the Chilean earthquake of May 21 and 22, 1960: Geological Society of America Bulletin, v. 81, p. 1001–1030.

Proctor, J., J. M. Anderson, S. C. L. Fogden, and H. W. Vallack, 1983, Ecological studies in four contrasting lowland rain forests in Gunung Mulu National Park, Sarawak: II. Litter fall, litter standing crop and preliminary observations on herbivory: Journal of Ecology, v. 71, p. 261–283.

Rindsberg, A. K., 1990, Freshwater to marine trace fossils of the Mary Lee coal zone and overlying strata (Westphalian A) Pottsville Formation of northern Alabama, *in* R. A. Gastaldo, T. M. Demko, and Y. Liu, eds., Carboniferous coastal environments and paleocommunities of the Mary Lee coal zone, Marion and Walker Counties, Alabama: Geological Society of America, Southeastern Section, Field Trip Guidebook, p. 82–96.

Rex, G. M., and W. G. Chaloner, 1983, The experimental formation of plant compression fossils: Palaeontology, v. 26, p. 231–252.

Shute, C. H., and C. J. Cleal, 1987, Palaeobotany in museums: Geological Curator, v. 4, p. 553–559.

Stanley, D. J., and A. G. Warne, 1994, Worldwide initiation of Holocene marine delta by deceleration of sea-level rise: Science, v. 265, p. 228–231.

Stark, N., and C. F. Jordan, 1978, Nutrient retention by the root mat of an Amazonian rain forest: Ecology, v. 59, p. 434–437.

Staub, J. R., and R. A. Gastaldo, 2003, Late Quaternary incised-valley fill and deltaic sediments in the Rajang River Delta, *in* H. F. Sidi, D. Nummedal, P. Imbert, H. Darman, and H. W. Posamentier, eds., Tropical deltas of Southeast Asia—Sedimentology, stratigraphy, and petroleum geology: SEPM Special Publication 76, p. 71–87.

Staub, J. R., H. L. Among, and R. A. Gastaldo, 2000, Seasonal sediment transport and deposition in the Rajang River delta, Sarawak, east Malaysia: Sedimentary Geology, v. 133, p. 249–264.

Stephenson, R. A., J. Boerstoel, A. F. Embry, and B. D. Ricketts, 1992, Subsidence analysis and tectonic modeling of the Sverdrup Basin, *in* D. K. Thurston and K. Fujita, eds., Proceedings International Conference on Arctic Margins: Washington, D.C., Minerals Management Service, U.S. Department of the Interior, p. 149–154.

Stevanoviç-Walls, I. M., 2001, A paleontological study of an early Pennsylvanian forest preserved above the Blue Creek coal seam, Pottsville, Formation, northwestern Alabama: M.S. thesis, Auburn University, Auburn, Alabama, 98 p.

Stewart, W. N., and G. W. Rothwell, 1993, Paleobotany and the evolution of plants: Cambridge, United Kingdom, Cambridge University Press, 521 p.

Stout, J., 1980, Leaf decomposition rates in Costa Rican lowland tropical rainforest streams: Biotropica, v. 12, p. 264–272.

Tessier, B., A. W. Archer, W. P. Lanier, and H. R. Feldman, 1995, Comparison of ancient tidal rhythmites (Carboniferous of Kansas and Indiana, U.S.A.) with modern analogues (the Bay of Mont-Saint-Michel, France), *in* B. W. Flemming and A. Bartholomä, eds., Tidal signatures in modern and ancient sediments: International Association of Sedimentologists Special Publication 24, p. 259–274.

Thomas, W. A., 1968, Contemporaneous normal faults on flanks of Birmingham anticlinorium, central Alabama: AAPG Bulletin, v. 52, p. 2123–2136.

Thomas, B. A., 1986, The formation of large diameter plant fossil moulds and the Walton theory of compaction: Geological Journal, v. 21, p. 381–385.

Thomas, W. A., 1988, The Black Warrior Basin, *in* L. L. Sloss, ed., The geology of North America: Geological Society of America, v. D-2, p. 471–492.

Thomas, W. A., 1995, Diachronous thrust loading and

fault partitioning of the Black Warrior foreland basin within the Alabama recess of the late Paleozoic Appalachian—Ouachita thrust belt: SEPM Special Publication 52, p. 111–126.

Thommeret, Y., J. Laborel, L. F. Montaggioni, and P. A. Pirazzoli, 1981, Late Holocene shoreline changes and seismotectonic displacements in western Crete (Greece): Zeitschrift Geomorphologie N.F. Supplement 40, p. 127–149.

Tjia, H. D., 1981, Examples of young tectonism in eastern Indonesia, *in* A. J. Barber and S. Wiryosujono, eds., The geology and tectonics of eastern Indonesia: Geological Resources and Development Centre Special Publication 2, p. 89–104.

Van Wagoner, J. C., R. M. Mitchum, K. M. Campion, and V. D. Rahmanian, 1990, Siliciclastic sequence stratigraphy in well logs, cores, and outcrops: AAPG Methods in Exploration Series 7, 55 p.

Ware, W. N., 2001, A paleoecological study of an in situ peat-forming Carboniferous forest in the Black Warrior Basin, Alabama: M.Sc. thesis, Auburn University, Auburn, Alabama, 81 p.

Weisenfluh, G. A., and J. C. Ferm, 1984, Geologic controls on deposition of the Pratt Seam, Black Warrior Basin, Alabama, U.S.A., *in* R. A. Rahmani and R. M. Flores, eds., Sedimentology of coal and coal-bearing sequences: International Association of Sedimentologists Special Publication 7, p. 317–330.

Whiticar, M. J., E. Faber, and M. Schoell, 1986, Biogenic methane formation in marsh and freshwater environments: CO_2 reduction vs. acetate fermentation-isotope evidence: Geochemica et Cosmochimica Acta, v. 50, p. 693–709.

Winston, R. B., 1990a, Coal paleobotany of the Mary Lee coal bed of the Pennsylvanian "Pottsville" Formation near Carbon Hill, Walker County, Alabama, *in* R. A. Gastaldo, T. M. Demko, and Y. Liu, eds., Carboniferous coastal environments and paleocommunities of the Mary Lee coal zone, Marion and Walker Counties, Alabama: Geological Society of America, Southeastern Section, Field Trip Guidebook, p. 55–64.

Winston, R. B., 1990b, Preliminary report on coal quality trends in upper Pottsville Formation coal groups and their relationships to coal resource development, coal-bed methane occurrence, and geologic history in the Warrior coal basin, Alabama: Alabama Geological Survey Circular 152, 53 p.

Winston, R. B., 1994, Models of the geomorphology, hydrology, and development of domed peat bodies: Geological Society of America Bulletin, v. 106, p. 1594–1604.

Zodrow, E. L., and C. J. Cleal, 1999, Anatomically preserved plants in siderite concretions in the shale split of the Foord Seam: Mineralogy, geochemistry, genesis (Upper Carboniferous, Canada): International Journal of Coal Geology, v. 41, p. 371–393.